Progress in Botany/Fortschritte der Botanik 41

Progress in Botany

Morphology · Physiology · Genetics
Taxonomy · Geobotany

Fortschritte der Botanik

Morphologie · Physiologie · Genetik
Systematik · Geobotanik

Editors/Herausgeber

Heinz Ellenberg, Göttingen
Karl Esser, Bochum
Klaus Kubitzki, Hamburg
Eberhard Schnepf, Heidelberg
Hubert Ziegler, München

Springer-Verlag
Berlin Heidelberg New York 1979

With 23 Figures

ISBN-13: 978-3-642-48635-7   e-ISBN-13: 978-3-642-48633-3
DOI: 10.1007/978-3-642-48633-3

Softcover reprint of the hardcover 1st edition 1979

2131/3130-543210.

Contents

List of Editors

Section A: Professor Dr. E. SCHNEPF, Lehrstuhl für Zellenlehre der Universität Heidelberg, Berliner Str. 15, D 6900 Heidelberg

Section B: Professor Dr. H. ZIEGLER, Institut für Botanik, Technische Universität München, Arcisstr. 21, D 8000 München 2

Section C: Professor Dr. K. ESSER, Lehrstuhl für Allgemeine Botanik Ruhr-Universität Bochum, Postfach 10 2148, D 4630 Bochum 1

Section D: Professor Dr. K. KUBITZKI, Institut für Allgemeine Botanik und Botanischer Garten, Universität Hamburg, Postfach 302 722, D 2000 Hamburg 36

Section E: Professor Dr. Dr. h.c. Dr. h.c. H. ELLENBERG, Lehrstuhl für Geobotanik, Systematisch-Geobotanisches Institut, Untere Karspüle 2, D 3400 Göttingen

A. Morphology

I. Special Cytology

Cytology and Morphogenesis of the Fungal Cell

By Manfred Girbardt

1. Books, Monographs etc.

The second edition of the *Fundamentals* has appeared (BURNETT, 1976). Symbiotic fungi are described by COOKE (1977) and sexual interactions by VAN DEN ENDE (1976). Electron micrographs of outstanding quality are collected in atlases of fungal ultrastructure (BECKETT et al., 1974) and of biodegradation (OLAH et al., 1978). Behavioral genetics of *Phycomyces* is reviewed by CERDA-OLMEDO (1977) and the papers of a symposium on nuclear division in fungi are edited by HEATH (1978a).

2. Cell Nucleus

a) Interphase

α) Staining and Quantitative Determinations. Exact cytophotometric analysis of Feulgen-stained nuclei has been performed with myxomycetes (HASKINS and THERRIEN, 1978). Yeast chromosomes are assumed to be countable (GALEOTTI and WILLIAMS, 1978). Fluorescent staining is possible without hydrolysis with a benzimidazol derivative binding selectively to DNA (LEMKE et al., 1978). A specific "sandwich technique" is used to detect β-D-galactan in nuclei (HORIS-BERGER et al., 1978a).

β) Chromatin. LAUER and KLOTZ (1975) have suggested the whole yeast genome to be composed of *one* continuous DNA molecule. The mitotic and meiotic behaviour of chromosomes seems, however, to contradict this assumption, and more recently *pieces* of DNA in the range of 10^9 daltons are interpreted as representing one chromosome (LAUER et al., 1977).

It is generally agreed that the "beads on a string" structure (nucleosomes) is also realized in fungal chromatin. The repeat lengths of the nuclease products are mostly shorter than in higher eukaryotes and vary (154 to 187 base pairs) according to differing linkers [*Aspergillus*: MORRIS (1976); *Neurospora*: NOLL (1976); *Physarum*: COMPTON et al. (1976); *Dictyostelium*: PARISH et al. (1977); BAKKE et al. (1978)]. The latter contains a 137 base pairs core and a 50 base pairs linker DNA. The diameter of the nucleosome is 98.6 A. This is surprising, as this organism contains only *four* histones. Neurospora with similar nucleosomes contains a normal complement of histones (HAUTALA et al., 1977).

All these data underline the supposition that fungi with small genomes [the genome size of yeast is about 3.3 times that of *E. coli*, the DNA of the smallest chromosome of *Schizophyllum* is 0.5 mm long (HAAPALA and NIENSTEDT (1976)] are well suited for investigations of fundamental questions related, for example, to variations of chromatin during the cell cycle. PINON and SALTS (1977) succeeded in isolating chromatin complexes (4600 and 3000 S) from yeast which seem to represent folded genomes in the G2 and G1 stages, respectively. These results have been derived without synchronization.

Supercoiled DNA is frequently reported. It is not yet clear, however, whether these supertwists are generated by disruption of the nucleosomal DNA or are normal components of the genome (McCREADY et al., 1977) responsible for chromosomal condensation which is again denied for yeast chromosomes (GORDON, 1977). Strand separation of chromosomal DNA is observed during meiosis in yeast (KLEIN and BYERS, 1978) and seems to be correlated with processes of replication and recombination.

γ) *Nucleolus.* It would be very significant to investigate "nucleolus-less" species. For a long time rusts seemed to be qualified as they showed either complete or partial absence of the nucleolus. Careful investigations (serial sections), however, showed that in cowpea rust a nucleolus is present in the interphase nucleus at all stages of development (HEATH and HEATH, 1978a). Another aspirant is *Pyronema.* Nuclei in paraphyses are said to lack nucleoli (HUNG, 1977). This specimen should be reinvestigated.

The nucleolus of *Physarum* is quite similar in fine structure to that of higher organisms. During mitosis the nucleolus persists after being fragmented into globular bodies. They contain ribosomal DNA which is composed of linear molecules (2.5 μm long) and a small number of circular ones (BOHNERT et al., 1975). Nucleolar DNA-synthesis is not limited to the S-period but persists during the G2-phase (LORD et al., 1977). In yeast a particular region of chromatin is associated with the nucleolus (STEVENS, 1976).

δ) *Pores.* One of the questions which puzzles cell biologists is how nucleocytoplasmic exchange of macromolecules is regulated. The idea has long been favoured that regulation occurred by expansion and contraction of the pores (WILLISON and RAJARAMAN, 1977), though many papers revealed that methodological pitfalls had not been considered, and the pore diameter in fact shows little variations (FRANKE, 1974). SEVERS and JORDAN (1978) confirm this conception with synchronized yeasts. Pore diameters of S-phase cells exhibited the same frequency distribution as those of resting cells. WILLISON and JOHNSTON (1978) compared the pore diameters of nuclei from a stationary phase population (G1 arrested by nitrogen starvation) with those of an exponentially growing population. They found, contrary to the foregoing results, that in growing populations the pore diameter was much smaller (75-115 nm) than in a nitrogen-starved one (120-160 nm). Further investigations are needed. Yeast cells should be used as they can be freeze-fractured without prefixation or cryoprotection.

ε) *Movement.* Fast nuclear movements, generally termed migrations, occur in filamentous fungi after mating of compatible strains. Nuclei reach a velocity 10-20-fold their normal. An extreme case has been reported by ROSS (1976). In *Coprinus congregatus* the nuclei migrate with more than 100-fold velocity (600 μm/min). It has still to be explained whether the infection of this strain with a mycoplasma-like organism (inhibiting meiosis) bears some causal relationship (dissolution of septa?) with this phenomenon, but nevertheless any model to explain the mechanism of migration will have to accommodate these high speeds.

In heterobasidiomycetous yeast (*Rhodosporidium*), sexual differentiation can be artificially induced by a secreted sex hormone (A factor) of the compatible strain. The nucleus moves into the mating tube and

behaves in the same manner as during normal sexual reproduction. This
indicates that the mechanism for nuclear movement is identical both in
factor-induced differentiation and the normal mating process (ABE et
al., 1977).

> Dedikaryotization is frequently reported in basidiomycetes (GINTEROVA, 1973).
> It can be induced by mechanical agitation of the culture, toxic ions and griseo-
> fulvin (RAUDASKOSKI and HUTTUNEN, 1977). It should be considered whether moving
> mechanisms of the two nuclei are different and nuclei therefore react different-
> ly to culture conditions (SEMERDZIEVA and MUSILEK, 1977).

Significance of microtubules for nuclear movement has been stressed
many times, but the exact meaning of the close spatial relationship
between the two components (including the nucleus-associated organelle
NAO) is as yet obscure. It seems to be obvious that cytoplasmic microtubules (MT) during movement of nuclei from the basidium through the
sterigmata into basidiospores are engaged, though primarily there is
no connection with the NAO. This is realized, however, during later
stages [NAKAI and USHIYAMA (1978), *Lentinus edodes*]. In infection structures of *Uromyces phaseoli* (HEATH and HEATH, 1978b) no MT radiate from
the NAO, but many are lying closely adjacent to the nuclear envelope.
For movement the interaction of a filamentous (3-6 nm, actin?) structure and these MT is considered. Similar results are reported during
pseudo-mycelium outgrowth in *Candida* (SOLL et al., 1978).

b) Division

There is now far-reaching agreement that many variants of a true eukaryotic mitosis are realized in fungi. As generally mechanisms for
moving the chromosomes are poorly understood (LITTLE et al., 1977),
studies of fungal mitotic and meiotic divisions might be expected to
contribute to fundamental questions. Comprehensive reviews have been
published on this subject (FULLER, 1976; WELLS, 1977).

α) *Evolution.* The many attempts to find evidence for phylogenetical development of the kinetic apparatus are mainly based on the following
assumptions: (i) pro- and eukaryotes have a common ancestor (genetic
code), the mechanism for distribution of the genetic material in higher organisms therefore should be derivable from that of prokaryotes;
(ii) interim stages, also in recent species should be found in phylogenetic niches.

Provided these assumptions are correct, one should find characteristic
mitotic substances and structures in prokaryotes (CAVALIER-SMITH, 1978).
Tubular structures are found in *Anabaena* (JENSEN and AYALA, 1976). In
spirochetes 15-25 nm microtubules, stainable with fluorescent antibodies, and probably composed of tubulin, have been shown (MARGULIS
et al., 1978). Another component of growing interest is actin, which
seems to work, beside its cytoskeletal function (POLLARD, 1976), also
in the spindle. It has been found in bacteria (MINKOFF and DAMADIAN,
1976; OHTOMO et al., 1976). Its presence in slime moulds has been well
established (ADELMAN, 1977; ZECHEL and WEBER, 1978). The involvement
of the nuclear membrane in the distribution of chromosomes as a bacterial relict seems to be realized at least partly in "closed" divisions of fungi [Symposium on the evolution of mitosis in eukaryotic
microorganisms, MARGULIS and TAYLOR (1975)].

(ii) Niches with aberrant mitosis (partly fungi-like) became evident
also in other groups [dinoflagellates: KUBAI (1975); diatoms: PICKETT-
HEAPS and TIPPIT (1978); red algae: DEMOULIN (1974); BRONCHART and

DEMOULIN (1977), raising again the question of red algal ancestors of fungi]. Sporozoa (*Plasmodium*) shows striking similarities during nuclear division with ascomyceteous yeast (SCHREVEL et al., 1977) and Plasmodiophoromycetes show fenestrated poles (DYLEWSKI et al., 1978).

β) *Nucleus-Associated-Organelle, NAO (= SPB)*. The reasons for election of the term NAO are discussed in GIRBARDT (1978). The microtubule organizing centre (MTOC)-activity of this organelle is either strictly cell-cycle-dependent (holobasidiomycetes) or almost completely independent of it, e.g., in ascomycetous yeast. It seems to contain RNA (UNGER, 1977): Ribonuclease A appears to *stimulate* the formation of microtubules (BYERS et al., 1978). A proteinaceous trypsin-sensitive component is indispensable for acting as initiating site.

One of the open questions is: How do the NAO's of compatible nuclei behave at karyogamy? The answer seems to be given by LU (1978) for *Coprinus*: the NAO of later-fusing nuclei remain monoglobular. After nuclear fusion the two globular entities form one complete diglobular NAO (including middle-plate). This time-course seems plausible, but should await further confirmation before being generally accepted. The same holds for the persistency of part of the middle-plate during prometaphase I and II in basidia of *Pholiota* as reported by WELLS (1978). So far nobody has clearly demonstrated a "division" of the NAO, and one should be careful in using this term (WRIGHT et al., 1978; SERNA and STADLER, 1978). In *Trametes* the "separation" of the preexisting two globular entities seems to be a passive process after dissolution of the connecting middle-plate (GIRBARDT and HÄDRICH, 1975).

> Counting microtubules in mitosis of *Saccharomyces cerevisiae* indicates that at least *one* microtubule is available for one linkage group (PETERSON and RIS, 1976). The number of microtubules in haploid cells is about half that in diploid cells.

NAO's are isolated and their MTOC-activity in vitro has been proven (BYERS et al., 1978; HYAMS and BORISY, 1978). The number of microtubules per NAO is limited, probably due to limited initiation sites. The microtubules are bipolar, "closed" at the initiation site and "open" at the distal end. Dependence on the cell cycle is evident as NAO's isolated from exponential cells induce less microtubules than those isolated from stationary cells.

γ) *Experimental Interfering with Mitosis*. One of the systemic fungicides most effective against phytopathogenic fungi is the benzimidazole derivative benomyl (= benlate). It works mainly via microtubular assemblage (DAVIDSE, 1975), but seems also to affect membrane integrity (BOURGOIS et al., 1977). It might be used to distinguish between haploid and diploid strains of *Aspergillus* (UPSHALL et al., 1977). The test is limited, however, as it cannot be applied for selection of single ascospores.

> A detoxification of the fungicide has been shown by DAVIDSE (1976) and YASUDA et al. (1973). Both sensitive and resistant strains metabolize the compound, indicating that differences in metabolic activity cannot account for the different sensitivity.

Probable interference with cytoplasmic microtubules is reported for *Fusarium* (HOWARD and AIST, 1977). D_2O counteracts all the observed effects (displacement of mitochondria, disappearance of the Spitzenkörper), and causes fragmentation of nuclei. Very interesting is the enlargement of the NAO under the influence of the fungicide (KÜNKEL and HÄDRICH, 1977). The discoidal entities are favoured in the accumulation of substances. Besides this enlargement a multiplication of NAO's is possible. Nuclei with up to four complete bipolar NAO's are formed.

> The influence of temperature is demonstrated in the meiosis of *Coprinus cinereus*. High temperatures (35°C) shorten diplotene from 4.5 h to less than 1 h. Lowering (5°C) extends pachytene and causes an increase of the recombination frequency, which therefore is a function of time (LU and CHIU, 1978). *Mitotic* crossing over and gene conversion can be induced in *Saccharomyces* by the fungicide Bavistin (SINGH, 1978) and mitotic delay in *Physarum* by cycloheximid (SCHEFFEY and WILLE, 1978).

Weak electromagnetic fields (EMF) cause mitotic delay in *Physarum* (MARRON et al., 1978). The delay in onset of mitosis persists after mixing with unexposed microplasmodia. It is supposed that EMF interferes with energy-generating processes or transport of metabolites. Regulation of mitotic processes by calcium ions is reported by HOLMES and STEWART (1977). In *Physarum* cyclic uptake and release of $^{45}Ca^{2+}$ occurring during interphase is altered during mitosis.

Some antifungal agents available for clinical use (KOBAYASHI and MEDOFF, 1977) are of interest for their mitotic efficiency. Griseofulvin attacks tubulin polymerization and its antimitotic properties seem in some cases specific for fungal spindles (GULL and TRINCI, 1974; HEATH, 1978b). The reason may be that it does not interact with tubulin, like colchicine and other "antitubulins", but with microtubule-associated protein (MAP) which promotes assembly of purified tubulin (ROOBOL et al., 1977).

3. Cell Cycle

a) General

> The ingenious inductor of the genetic approach to questions of the cell cycle has again reviewed the investigations mostly of his own group (HARTWELL, 1978). The same has been done from a somewhat different, critical point of view by PRINGLE (1978). He emphasizes that conclusions about translation of events into gene products should be regarded as tentative until the exact molecular mechanism is known. Five fungal specimens are now in use (*Saccharomyces, Schizosaccharomyces, Aspergillus, Ustilago, and Physarum*).

b) Volume-Regulation of Cell Division

Relationship between cell size and cell division is not only realized in *Schizosaccharomyces* (NURSE, 1975), but also in *Saccharomyces* (CARTER and JAGADISH, 1978). JOHNSTON et al. (1977) interpret their results in that growth is more important for division than completion of single stages (e.g., DNA synthesis) and specific early events in G1 can be completed only after reaching a critical size. The size (average cell volume) is remarkably independent of environmental conditions. Only glucose limitation yields cells reduced in size (ADAMS, 1977) and exhibits *no* size difference between haploids and diploids. Probably a constant critical size must be reached before the bud becomes initiated. Coordination of growth and division can be described by a model which is similar to that of prokaryotes.

The lengths of the cell cycle phases in yeast behave differently during growth on different carbon sources. The duration of S- and M-phase remains constant whereas G1 and G2 varies considerably (BARFORD and HALL, 1976). Volume-related regulation of cell division is dependent in *Schizosaccharomyces* on intracellular Mg^{2+} concentration. Chelating agents (EDTA) cause a continuous fall of MG^{2+} and stop nucleic acid synthesis. Protein synthesis continues, however. About 50% of the cells reach the point where they divide immediately after transfer to fresh medium. Synchronization is possible in this way (AHLUWALIA et al., 1978).

c) Cytokinesis (Septum Formation)

Disturbance of normal cytokinesis in the ascus can lead to an unequal number of ascospores in *Saccharomycopsis lipolytica* (ESSER and STAHL, 1976). It is assumed that meiosis and cleavage are independent processes. This assumption is contradicted, however, by many mono- and dikaryotic basidio- and ascomycetes. In these species both locus and time of septum formation are induced by the mitotic dividing nuclei (PATTON and MARCHANT, 1978). In *Trametes* (= *Polystictus*) a remarkable regularity is realized: the dividing nucleus triggers during meta-anaphase the incipient septum region (probably by alteration of a ring-like area of the plasmalemma). It is only when nuclear division is completed that a peripheral belt of microfilaments (4-7 nm) is laid down and cross-wall formation begins. The microfilamentous belt entraps microvesicles and mitochondria. Correspondance of these events with those at the "contractile ring" in animal cells is striking (GIRBARDT, 1977).

In yeast a highly ordered filamentous ring appears in advance of nuclear division and disappears during cytokinesis (BYERS and GOETSCH, 1976). It is questionable whether these structures correspond to those in *Trametes*. At any rate both may be composed of actin. This contractile protein has been demonstrated in *Neurospora* (ALLEN and SUSSMAN, 1978).

During cytoplasmic cleavage in zoosporangia of *Thraustochytrium* (KAZAMA, 1975) there is no indication of microfilaments but subplasmalemma microtubules cross the furrows. The difference between longitudinal and cross-wall is evident in that the latter does not contain α-galactomannan (HORISBERGER et al., 1978b).

Acknowledgements. The invaluable help of Mrs. I. LANGE, Mrs. I. BÄHRING, and Mrs. H. MÜNKEL in processing the literature and in careful typing is gratefully acknowledged.

References

ABE, K., TSUCHIYA, E., KUSAKA, I., FUKUI, S., KUSANAGI, A.: J. Gen. Appl. Microbiol. 23, 175-181 (1977). - ADAMS, J.: Exp. Cell Res. 106, 267-275 (1977). - ADELMANN, M.R.: Biochemistry 16, 4862-4871 (1977). - AHLUWALIA, B., DUFFUS, J.H., PATERSON, L.J., WALKER, G.M.: J. Gen. Microbiol. 106, P. 2, 261-264 (1978). - ALLEN, E.D., SUSSMAN, A.S.: J. Bacteriol. 135, 713-716 (1978).

BAKKE, A.C., WU, J.-R., BONNER, J.: Proc. Natl. Acad. Sci. USA 75, 705-709 (1978). - BARFORD, J.P., HALL, R.J.: Exp. Cell Res. 102, 276-284 (1976). - BECKETT, A., HEATH, I.B., McLAUGHLIN, D.J.: An Atlas of Fungal Ultrastructure. London: Longman 1974. - BOHNERT, H.J., SCHILLER, B., BÖHME, R., SAUER, H.W.: Eur. J. Biochem. 57, 361-369 (1975). - BOURGOIS, J.-J., BRONCHART, R., DELTOUR, R., de BARSY, Th.:

Pestic. Biochem. Physiol. 7, 97-106 (1977). - BRONCHART, R., DEMOULIN, V.: Nature (London) 268, 80-81 (1977). - BURNETT, J.H.: Fundamentals of Mycology, 2. ed. London: Arnold 1976. - BYERS, B., GOETSCH, L.: J. Cell Biol. 69, 717-721 (1976). - BYERS, B., SHRIVER, K., GOETSCH, L.: J. Cell Sci. 30, 331-352 (1978).

CARTER, B.L.A., JAGADISH, M.N.: Exp. Cell Res. 112, 15-24 (1978). - CAVALIER-SMITH, T.: Biosystems 10, 93-114 (1978). - CERDA-OLMEDO, E.: Annu. Rev. Microbiol. 31, 535-548 (1977). - COMPTON, J.L., BELLARD, M., CHAMBON, P.: Proc. Natl. Acad. Sci. USA 73, 4382-4386 (1976). - COOKE, R.: The Biology of Symbiotic Fungi. London, New York: Wiley 1977.

DAVIDSE, L.C., in: Microtubules and Microtubuli Inhibitors, eds. M. BORGERS, K. DeBRABANDER, pp. 483-495. Amsterdam: North-Holland 1975; - Pestic. Bioch. Physiol. 6, 538-546 (1976). - DEMOULIN, V.: Bot. Rev. 40, 315-345 (1974). - DYLEWSKI, D.P., BRASELTON, J.P., MILLER, C.E.: Am. J. Bot. 65, 258-267 (1978).

ESSER, K., STAHL, U.: Mol. Gen. Genet. 146, 101-106 (1976).

FRANKE, W.W.: Int. Rev. Cytol. Suppl. 71-236 (1974). - FULLER, M.S.: Int. Rev. Cytol. 45, 113-153 (1976).

GALEOTTI, C.L., WILLIAMS, K.L.: J. Gen. Microbiol. 104, 337-341 (1978). - GINTEROVA, A.: Folia Microbiol. 18, 277-285 (1973). - GIRBARDT, M.: Proc. II. Int. Mycol. Congr. Tampa, 227, 1977; - In: Nuclear Division in the Fungi, ed. I.B. HEATH, 1-20. New York: Academic Press 1978. - GIRBARDT, M., HÄDRICH, H.: Z. Allg. Mikrobiol. 15, 157-173 (1975). - GORDON, C.N.: J. Cell Sci. 24, 81-93 (1977). - GULL, K., TRINCI, A.P.: Protoplasma 81, 37-48 (1974).

HAAPALA, O.K., NIENSTEDT, I.: Hereditas 84, 49-60 (1976). - HARTWELL, L.H.: J. Cell Biol. 77, 627-637 (1978). - HASKINS, E.F., THERRIEN, C.D.: Exp. Mycol. 2, 32-40 (1978). - HAUTALA, J.A., CONNER, B.H., JACOBSON, J.W., PATEL, G.L., GILES, N.H.: J. Bact. 130, 704-713 (1977). - HEATH, I.B.: (ed.) Nuclear Division in the Fungi. New York: Academic Press 1978a; - In: Nuclear Division in the Fungi, 89-176. New York: Academic Press 1978b. - HEATH, I.B., HEATH, M.C.: Cytobiologie 16, 393-411 (1978b). - HEATH, M.C., HEATH, I.B.: Can. J. Bot. 56, 648-661 (1978a). - HOLMES, R.P., STEWART, R.P.: Nature (London) 269, 592-597 (1977). - HORISBERGER, M., FARR, D.R., VONLANTHEN, M.: Biochim. Biophys. Acta 542, 308-314 (1978a). - HORISBERGER, M., VONLANTHEN, M., ROSSET, J.: Arch. Microbiol. 119, 107-111 (1978b). - HOWARD, R.J., AIST, J.R.: Protoplasma 92, 195-210 (1977). - HUNG, C.-Y.: Mycologia 69, 321-327 (1977). - HYAMS, J.S., BORISY, G.G.: J. Cell Biol. 78, 401-414 (1978).

JENSEN, T.E., AYALA, R.P.: J. Ultrastruct. Res. 57, 185-193 (1976). - JOHNSTON, G.C., PRINGLE, J.R., HARTWELL, L.H.: Exp. Cell Res. 105, 79-98 (1977).

KAZAMA, F.Y.: J. Cell Sci. 17, 155-170 (1975). - KLEIN, H.L., BYERS, B.: J. Bacteriol. 134, 629-635 (1978). - KOBAYASHI, G.S., MEDOFF, G.: Annu. Rev. Microbiol. 31, 291-308 (1977). - KUBAI, D.F.: Int. Rev. Cytol. 43, 167-227 (1975). - KÜNKEL, W., HÄDRICH, H.: Protoplasma 92, 311-323 (1977).

LAUER, G.D., KLOTZ, L.C.: J. Mol. Biol. 95, 309-326 (1975). - LAUER, G.D., ROBERTS, T.M., KLOTZ, L.C.: J. Mol. Biol. 114, 507-526 (1977). - LEMKE, P.A., KUGELMAN, B., MORIMOTO, H., JACOBS, E.C., ELLISON, J.R.:

J. Cell Sci. 29, 77-84 (1978). - LITTLE, M., PAWELETZ, N., PETZELT, C., PONSTINGL, H., SCHROETER, D., ZIMMERMANN, H.-P. (eds.): Mitosis: Facts and Questions. Berlin-Heidelberg-New York: Springer 1977. - LORD, A., NICOLE, L., LAFONTAINE, J.G.: J. Cell Sci. 23, 25-42 (1977). - LU, B.C.: J. Cell Biol. 76, 761-766 (1978). - LU, B.C., CHIU, S.M.: J. Cell Sci. 32, 21-30 (1978).

MARGULIS, L., TAYLOR, F.J.R.: Biosystems 7, 295-297 (1975). - MARGULIS, L., TO, L., CHASE, D.: Science 200, 1118-1124 (1978). - MARRON, M.T., GOODMAN, E.M., GREENEBAUM, B.: Experientia 34, 589-591 (1978). - McCREADY, S.J., COX, B.S., Mc LAUGHLIN, C.S.: Exp. Cell Res. 108, 473-478 (1977). - MINKOFF, L., DAMADIAN, R.: J. Bacteriol. 125, 353-365 (1976). - MORRIS, N.R.: Cell 8, 357-363 (1976).

NAKAI, Y., USHIYAMA, R.: Can. J. Bot. 56, 1206-1211 (1978). - NOLL, M.: Cell 8, 349-355 (1976). - NURSE, P.: Nature (London) 256, 547-551 (1975).

OHTOMO, T., CHUN-HSIUNG, H., IIZUKA, H.: J. Gen. appl. Microbiol. 22, 113-117 (1976). - OLAH, G.M., REISSINGER, O., KILBERTUS, G.: Biodégradation et Humification. Atlas ultrastructural. Quebec, Paris: Presses de l'Univ. Laval Vuibert 1978.

PARISH, R.W., STALDER, J., SCHMIDLIN, S.: FEBS Lett. 84, 63-66 (1977). - PATTON, A.M., MARCHANT, R.: Arch. Microbiol. 118, 271-277 (1978). - PETERSON, J.B., RIS, H.: J. Cell Sci. 22, 219-242 (1976). - PICKETT-HEAPS, J.D., TIPPIT, D.H.: Cell 14, 455-467 (1978). - PINON, R., SALTS, Y.: Proc. Natl. Acad. Sci. USA 74, 2850-2854 (1977). - POLLARD, T.D.: J. Supramol. Struct. 5, 317-334 (1976). - PRINGLE, J.R.: J. Cell Physiol. 95, 393-406 (1978).

RAUDASKOSKI, M., HUTTUNEN, E.: Mykosen 20, 339-348 (1977). - ROOBOL, A., GULL, K., POGSON, C.I.: FEBS Lett. 75, 149-156 (1977). - ROSS, I.K.: Mycologia 68, 418-422 (1976).

SCHEFFEY, C., WILLE, J.J.: Exp. Cell Res. 113, 259-262 (1978). - SCHREVEL, J., ASFAUX-FOUCHER, G., BAFORT, J.M.: J. Ultrastruct. Res. 59, 332-350 (1977). - SEMERDZIEVA, M., MUSILEK, V.: Folia Microbiol. 22, 303-307 (1977). - SERNA, L., STADLER, D.: J. Bacteriol. 136, 341-351 (1978). - SEVERS, N.J., JORDAN, E.G.: Experientia 34, 1007-1011 (1978). - SINGH, I.: Indian J. Exp. Biol. 16, 393-399 (1978). - SOLL, D.R., STASI, M., BEDELL, G.: Exp. Cell Res. 116, 207-215 (1978). - STEVENS, B.J.: J. Cell Biol. 70, 365a (1976).

UNGER, E.: Z. Allg. Mikrobiol. 17, 487-490 (1977). - UPSHALL, A., GIDDINGS, B., MORTIMORE, I.D.: J. Gen. Microbiol. 100, 413-418 (1977).

VAN DEN ENDE, H.: Sexual Interactions in Plants. The role of specific substances in sexual reproduction. New York: Academic Press 1976.

WELLS, K.: Meiotic and mitotic divisions in the basidiomycotina. In: Mechanisms and Control of Cell division, eds. T.L. ROST, E.M. GIFFORD. Stroundsburg: Dowden, Hutchison and Ross 1977; - Protoplasma 94, 83-108 (1978). - WILLISON, J.H.M., JOHNSTON, G.C.: J. Bacteriol. 136, 318-323 (1978). - WILLISON, J.H.M., RAJARAMAN, R.: J. Microsc. 109, 183-192 (1977). - WRIGHT, R.G., LENNARD, J.H., DENHAM, D.: Trans. Br. Mycol. Soc. 70 , 229-237 (1978).

YASUDA, Y., HASIMOTO, S., SOEDA, Y., NOGUCHI, T.: Ann. Phytopathol. Soc. Jpn. 39, 49-62 (1973).

ZECHEL, K., WEBER, K.: Eur. J. Biochem. <u>89</u>, 105-112 (1978).

Professor Dr. MANFRED GIRBARDT
Zentralinstitut für Mikrobiologie
und experimentelle Therapie der
Akademie der Wissenschaften der DDR
Beuthenbergstraße 11
DDR 69 Jena

B. Physiology

I. Plant Water Relations

By Otto L. Lange and Rainer Lösch

An immense number of publications appeared during the review period
1975-1978 in the broad area of plant water relations. It is impossible
to cover here in detail the different aspects of water biology, which
range from biochemical and biophysical foundations to ecological and
phytogeographical implications and which also touch on problems of
applied botany. We attempt rather to report only general trends and
major findings of recent research and to list representative papers,
in order to gain an overview of a subject that is steadily increasing
in its scope and importance.

An extremely useful and comprehensive reference source is the *Water-in-Plants
Bibliography* which lists more than 1000 relevant titles per year beginning with
the literature of 1975 (POSPIŠILOVÁ and SOLÁROVÁ, 1977, 1978, 1979). Recent
books cover fluid behavior in biological systems (LEYTON, 1975), water deficit
and plant growth (KOZLOWSKI, 1976), and plant responses of saline environments
(POLJAKOFF-MAYBER and GALE, 1975). Various aspects of the role of water in
plant life are presented by LANGE et al. (1976). The booklet *Water and Plants*
by MEIDNER and SHERIFF (1976) has proven very useful for student courses. Mete-
orological implications of water exchange in vegetation are considered by
MONTEITH (1975, 1976; see also CAMPBELL, 1977). An overall picture of the importance of water is outlined in a symposium volume called *Water: Planets,
Plants and People* (MCINTYRE, 1978).

1. Water Relations of Cells and Tissues

Review articles and books treat general aspects of water relations of plant
cells (DAINTY, 1976; ZIMMERMANN and STEUDLE, 1978), physics and biochemistry
of turgor- and osmoregulations (ZIMMERMANN, 1978; KAUSS, 1977; CRAM, 1976).
In a monograph HAMMEL and SCHOLANDER (1976) discuss problems of osmosis and
tensile solvent.

For description of water relations in plant cells and tissues, RICHTER
(1978a) suggests a new diagram that might also be suitable for use in
textbooks. It adapts the classical "Höfler-diagram" to the water-po-
tential concept, a concept which seems to be widely accepted at present - though one has to be aware of possible difficulties when the
total water potential (ψ_w) is split into its components of osmotic
(ψ_s), matric (ψ_m), and hydrostatic pressure potential (ψ_p), see DAINTY
(1976). Terminology for describing water status based on the concept
of water potential in comparison with the suction-tension concept is
analyzed by RICHTER (1977). An equilibrium model of leaf water potentials, which separates intra- and extracellular potentials, is developed by ACOCK (1975).

During the period under review, interest in theoretical considerations
of cell and tissue water status declined. Most work done concerned
physiological and ecological problems. New methodological approaches

were introduced, and established techniques were improved and extended (for review see KREEB, 1977; CRAFTS, 1976).

It has become obvious that psychrometric methods for determination of water potential requiring excised tissue contain significant error (BAUGHN and TANNER, 1976a,b). Internal equilibration within the leaf disc (OERTLI et al., 1975), cut edge to volume ratio of the sample (NELSEN et al., 1978; TALBOT et al., 1975), and equilibrium time between air inside the psychrometer and disc (MILLAR, 1974) may influence the results (see also BROWN, 1976). An improvement over methods requiring excision is the in situ measurement of leaf water potential by silver-foil psychrometers (ZANSTRA and HAGENZIEKER, 1977). However, due to problems caused by thermal gradients, application still seems to be restricted to controlled conditions (HOFFMAN and HALL, 1976). NEUMANN et al. (1974) find hygrometrically determined water potentials in intact leaves linearly related to transpiration rates, i.e., to actual water contents. Extrapolated values of leaf ψ_s at zero transpiration are equal to the water potentials of the solutions applied to the roots. This clearly demonstrates the value of in situ water status measurements. Further progress may possibly be expected from other nondestructive techniques as well, such as electrical resistance measurements (DIXON et al., 1978) or stem diameter investigations (HUCK and KLEPPER, 1977; see Sect. 3c).

Pressure-chamber techniques are carefully examined, and shortcomings such as xylem cavitations (WEST and GAFF, 1976) and exclusion errors (MILLAR and HANSEN, 1975) are considered (for comparison of different methods see SEPASKHAH, 1977; IKE et al., 1978). Pressure-volume curves (Prog. Bot. 37, 79) generated from pressure-chamber measurements are now widely used for determination of the original bulk osmotic potential of the symplast, osmotic potentials at incipient plasmolysis, and cell wall elasticity of leaves and branches. These parameters are closely associated with the extent to which a tissue can osmoregulate or conserve water within a certain range of water potential change (CHEUNG et al., 1975; POWELL and BLANCHARD, 1976; for possible sources of errors of the method see CHEUNG et al., 1976). Investigations of seasonal and ontogenetical changes in water relations in different tree species demonstrate the versatility of the method (TYREE et al., 1978; ROBERTS and KNOERR, 1977). It is even possible to detect membrane damage in leaves caused by freezing or by phytotoxins (TURNER, 1976; G.N. BROWN et al., 1977). TALBOT et al. (1975; also RICHTER, 1978b) suggest an interesting extension of the method. They determine total water potential in leaves during the course of drying by thermocouple psychrometry. By plotting leaf water saturation deficit (WSD) against $1/\psi_w$, diagrams of the pressure-volume curve type result. By extrapolation to the intercept with the axis (WSD = 0) the bulk osmotic potential can be determined. This seems to yield more reliable results for determination of plant tissue osmotic potential than the standard cryoscopic methods because the problem of cell sap dilution is avoided.

Discussion of the possible existence of negative turgor pressure in leaf cells under extreme water stress has again been stimulated through improvement of experimental design and equipment. Several authors recently purport to have demonstrated negative turgor in stressed plants (TUNSTALL and CONNOR, 1975, in *Acacia harpophylla*; KAPPEN et al., 1975, in *Hammada scoparia*; JOHNSON and BROWN, 1977, and MAXWELL and REDMANN, 1978, in *Agropyron* species; MARGARIS, 1977, in mediterranean xerophytes). However, TYREE (1976) concludes from the shape of pressure-volume curves with a linear regression at high WSD values that negative turgor does not exist. This is confirmed by KYRIAKOPOULOS and RICHTER (1977): they found no indication of negative turgor in drying leaves of *Quercus ilex*, when evaluating pressure-volume curves, whereas ψ_w of the same plant material has 20 bar lower values than ψ_s when measured with the usual methods (KYRIAKOPOULOS and LARCHER, 1976). It appears that this difference is due to error involved when osmotic potential is determined with excised leaf samples (matric potential

as possible source of error is discussed by SHEPHERD, 1975). Even if
the fact is disappointing, it will have to realized that arid-zone
plants most probably do not have the ability to increase their suction
tension substantially by negative turgor.

Water status conditions during ontogenetic development of plants and
within individual organs have attracted special attention. During
expansion growth significant water potential gradients persist in
seedlings, even though the external water potential and transpiration
are zero. This growth-induced water potential is required to move
growth-associated water over macroscopic distances. It indicates that
cells have different wall and/or osmotic properties at different positions in the tissue in order for organized growth to occur (MOLZ and
BOYER, 1978; also MCNEIL, 1976). In growing leaves there are marked
differences with regard to the relationship between water content and
water potential: component potentials change for immature and mature
leaves (WENKERT et al., 1978; also KASSAM and ELSTON, 1976). Individual leaves show gradients of ψ_w and its components from base to tip
in the order of several bars, not only during transpiration (WIEBE
and PROSSER, 1977; MEIRI et al., 1975), but also when water status is
near saturation (MANOHAR, 1977).

Plant cells that lose water to the air or cells that are subjected to
changes in external solute concentration respond with osmoregulation
which is brought about by metabolic processes. Opinions differ as to
the nature and sites of primary influences through which changes in
cell water relations affect plant metabolism (OERTLI, 1976a). Recently, evidence has been increasing which suggests that hydraulic pressure might be the decisive component of total cell water potential.
This has led to intensive discussions of the possible mechanism responsible for perception of turgor pressure changes on the order of
1 bar. Since it is obvious that chemical reactions cannot be involved
in the initial steps of turgor-sensing, ZIMMERMANN (1977) suggests an
electromechanical model. Mechanical force, created either by pressure
gradients across membranes or by absolute pressure, results in com-
pression of membranes perpendicular to their planes, or in stretching.
Such pressure-induced changes in the geometrical dimensions of membranes are then transformed into changes in the intrinsic electric
field distribution within the membranes. This in turn might alter activity of membrane-bound enzymes and membrane transport properties.
There is evidence that turgor pressure can affect ion flux directly.
Most work on this problem so far is done with giant algal cells (Prog.
Bot. 40, 94). However, there are indications that similar mechanisms
may also operate in higher plants. LÜTTGE et al. (1975) suggest that
diurnal oscillations due to Crassulacean acid metabolism in leaf
cells of Bryophyllum daigremontianum are controlled by turgor pressure
(see also LÜTTGE et al., 1977). Several improvements in the techniques
for direct measurement and manipulation of hydraulic pressure within
individual cells should stimulate further investigations of turgor-
sensing mechanisms and turgor-triggered processes, as well as of cell
properties involved, such as cell wall elasticity and cell hydraulic
conductivity.

ZIMMERMANN (unpublished data, quoted in HÜSKEN et al., 1978) introduced gas
bubbles directly into the cell interior and calculated pressure relations from
changes in the dimensions of the bubbles. FERRIER and DAINTY (1977, 1978) applied external forces to a group of cells by means of a displacement transducer,
and drew conclusions about turgor-related parameters from cell deformation and
recovery. By means of a pressure-probe (ZIMMERMANN, 1977) it is now possible to
measure turgor directly in single tissue cells larger than 20 μm, such as cells
in the fruit of Capsicum annuum (HÜSKEN et al., 1978) or in large epidermal

bladder cells (STEUDLE et al., 1977). With a similar device, MEIDNER and EDWARDS (1975) determined turgor pressure in stomatal cells. Certainly, it can be expected that further important results in this field of research will lead to the elucidation of key processes in plant water relations at the cellular level.

2. Water Relations of Germination

The metabolic processes of germination begin with seed imbibition. In air-dry seeds of *Sinapis arvensis* which have only a 4%-6% water content and seed surface water potentials lower than -762 bar, metabolism is very low. However, if seed water content is increased, mitochondrial activity accelerates and ^{14}C-leucine incorporation into protein increases drastically (EDWARDS, 1976). Imbibition evokes many biochemical and cytomorphological changes. The physicochemical processes occurring during seed swelling can be described by mathematical models developed on the basis of water absorption properties of colloids (COLLIS-GEORGE and MELVILLE, 1975). But various tissues within the seeds undergo distinct size changes (PONOMARENKO, 1975) as structural reorganization of membranes and organelles proceeds (WEBSTER and LEOPOLD, 1977). Within 10-20 min after wetting, these different structures become discernible and appear better arranged than before. In dry seeds, SATO and ASAHI (1975) were able to distinguish three types of mitochondria on the basis of chemical differences in their respective membranes. During water uptake the mitochondria become active and stable as protein is assembled in the membranes. The membrane restoration taking place during imbibition may be based on manifold interactions between the surface of the cell structures and water molecules (NMR-measurements: SAMUILOV et al., 1976). The rearrangement of membranes occurs within the initial 5 min of imbibition during which time the membranes change from a relatively porous to a less permeable condition (PARRISH and LEOPOLD, 1977). This phase is followed by a slower, linear rate of water uptake. As imbibition proceeds, a sharp moving front separates wet and dry regions of the seed (WAGGONER and PARLANGE, 1976). Slight differences in water potential existing between the hull and other parts of the seed affect the water transfer. The pregerminative production of solutes brings about a decrease in seed- (in the special case: caryopse-) water potential: MCDONOUGH (1974). Impermeability to water in some legume seeds is based on biochemical (MARBACH and MAYER, 1975) and structural peculiarities of the testa (BALLARD, 1976). Strophiolar conductivity is also associated with these seed coat properties. Temperature treatments by radiofrequency electric fields soften such seeds. They then become permeable at the strophiole (BALLARD et al., 1976).

The consequence of water stress on postgermination depends on the stage of the seedling development (CREVECOEUR et al., 1976; HEGARTY and ROSS, 1978). In order to avoid damage under water stress, desiccation-protecting soluble sugars are transferred from the endosperm or the Fabacean cotyledons to the seedling (LIVSHIN and SAMYGIN, 1974). There are many investigations about the influence of different water potential values on seed viability. The germination of wild species was often more affected by lowered medium water potentials than that of seeds of cultivated plants, whereas the emerged wild seedlings were more resistant to low availability of water (BASSIRI and SIONIT, 1975; BASSIRI et al., 1977; HOGON and CHAN, 1977).

Some metabolic processes initiated by seed imbibition are discussed by DELL'AQUILA et al. (1978). Protein synthesis and growth in embryos

subjected to hydration-dehydration treatment are not impaired heavily
if dehydration commences earlier than 9 h after soaking and is followed by rewetting (SEN and OSBORNE, 1974). This presoaking can even improve the viability of the seeds (SAVINO et al., 1976). Enzymatic activities have been found depressed, enhanced, or unaffected by the
moisture conditions during germination, depending on tissues and species (VORA et al., 1975; MALI and MEHTA, 1977a).

From the standpoint of practical application, an understanding of the moisture
and temperature conditions optimal for germination is most essential. Forest
tree species have been investigated with respect to these aspects of germination by KAUFMANN and ECKARD (1977: *Picea engelmannii, Pinus contorta*) and BARON
(1978: *Pinus lambertiana*). Concerning temperature and moisture effects on germinating crop seeds see, e.g., MANOHAR and MATHUR (1975: *Lycopersicon esculentum,
Dolichos lablab*) and SRIVASTAVA and SINGH (1975: *Triticum aestivum*). Seedling
emergence in seeds of *Brassica oleracea* and *Daucus carota* sown in soil under
conditions of either low or moderate moisture stress, was faster when temperature changed in a smooth and continuous daily cycle than when it remained constant. Seeds sown in highly moisture-stressed soil showed no germination at
constant temperatures. In contrast to *Brassica*, the germination of carrot seeds
was unaffected by temperature treatments in osmotic solutions (HEGARTY, 1975).
LINDSTROM et al. (1976) propose a mathematical model predicting winter wheat
emergence as affected by soil temperature, water potential, and depth of plant-
ing. The effect of the interaction of water potential and temperature on germination in three semi-arid grassland species has been studied by SHARMA (1976a).
Different germination responses to temperature and water stress conditions influence species distribution within a short-grass prairie (BOKHARI et al.,
1975). The occurrence of several species of meadow plants at different microsites was attributed to differences in germination vigor in response to micro-
site water content (OOMES and ELBERSE, 1976): *Chrysanthemum leucanthemum* and
Achillea millefolium which germinate in drier microsites than *Prunella vulgaris,
Rumex acetosa, Plantago lanceolata,* and *Hypochoeris radicata*, also have a higher
water use efficiency than the other species. The germination of the three last-
named species was optimal at more protected sites.

3. Water Movement Through Plants

a) Water Uptake

Compared with other fields of study which fall within the general area
of plant water relations, the topic "water uptake from the soil" is
characterized by a prevalence of theoretical studies and a relative
lack of experimental data. Water potential gradients in the bulk soil
are smaller than those in the roots in the upper 90% of the water
availability range (MOLZ, 1976; REICOSKY and RITCHIE, 1976). Major
water potential gradients within the root-free water supplying zone
(pararhizal) are unlikely to occur, but soil water depletion around
individual roots (perirhizal) may result in considerable water potential depression (FAIZ and WEATHERLEY, 1977). Depending on the particular case, stress consequences may or may not arise for the plant. On
the average, large water potential gradients between soil and roots
seem to be restricted to situations involving very low root densities
and high extraction rates in relatively dry soils (WILLIAMS, 1974,
1976). TINKER (1976), reviewing the principles of water transport
from soil to plants, emphasizes the existence of locally variable
water potential and transport gradients difficult to incorporate in
models simulating average conditions. Indeed, HERKELRATH et al.
(1977a,b) were able to reconcile discrepancies between experimental

results and theoretical calculations of root water uptake by assuming
a reduction in root/soil contact resulting in increased root membrane
resistance to water uptake. However, generalizing equations may be
very useful in predicting water extraction patterns of crops (FEDDES
et al., 1976; VAN BAVEL and AHMED, 1976). Soil water exploitation occurs dynamically through the growth and proliferation of roots (HILLEL
and TALPAZ, 1976). The depletion effectiveness increases as younger
roots reach deeper, wetter, and less crowded soils (STONE et al.,
1976). As soil dries, the zone of maximum root water uptake shifts
downward and outward (NNYAMAH and BLACK, 1977: *Pseudotsuga menziesii*;
STURGES, 1977: *Artemisia tridentata*). Concerning soil water extraction
patterns under alfalfa see BRUN and WORCESTER (1975) and KOHL and
KOLAR (1976). CALDWELL (1976) discusses thoroughly the dynamics of
root extension and water absorption patterns. Young roots provide a
considerable amount of the water used by the whole plant. Consequently compact soil layers (e.g., from plow-soles) which impair root growth
can influence negatively the water supply (PRIHAR et al., 1975). Water
stress, on the other hand, modifies root morphology (DA et al., 1977).
Increases in root density and permeability prove to be as effective
in coping with the water demand of the plant as is an increase in
soil water content or a decrease in evaporation power of the atmosphere (HILLEL et al., 1975). If the ratio, leaf surface : root surface,
increases, the root system is able to increase and adapt its water
uptake capacity to the changed conditions (RICHARDS, 1977).

It is assumed that root water uptake is proportional: (1) to root
density (2) to soil hydraulic conductivity and (3) to bulk soil/root
surface water potential difference. The validity of the first was
proven experimentally by TAYLOR and KLEPPER (1975). Determinations
of the other two are necessary in order to estimate root resistance.
After the water has reached the root surface, it has to overcome the
transport resistance of the "radial-flow pathway" after which it enters
the xylem. In an article summarizing what is known about this transport NEWMAN (1976a) states that the main pathway for water flow has
yet to be established with certainty. As possible alternatives he
enumerates: (1) the vacuolar pathway (2) the free-space/endodermis
pathway and (3) the symplast pathway. From calculations based on
maize root properties he assumes the symplast pathway to be the most
likely (see, however, Prog. Bot. $\underline{37}$, 81). The electrical equipotenti-
ality of all cytoplasmic phases of roots (GINSBURG and GINZBURG, 1974)
indicates a continuity within and between the cortical and stelar
symplasms. Root resistance to water flow is not constant. PARSONS and
KRAMER (1974) observed diurnal cyclings presumably controlled by signals from the shoot. Phytohormones could be involved in such regula-
tions (COLLINS and KERRIGAN, 1974; HONG and SUCOFF, 1976; GLINKA,
1977), but environmental influences, like temperature (SHIRAZI et al.,
1975) must also be taken into consideration. The impairment of host
water uptake by pathogens has been studied by PRIEHRADNY (1976) and
YOUNG and GARNSEY (1977). The effects of various osmotica on permeability of root cells have been studied by GREENWAY (1974). ALEKSEEVA
et al. (1977) assume that lipid peroxidation plays a role in regulating wheat root permeability. SHALHEVET et al. (1976) found that root
hydraulic conductance is not affected by salinity, and stated that
experimental data agreed well with theoretical analyses of coupled
water and ion uptake. According to such a theoretical treatment by
DALTON et al. (1975), the ratio of the osmotic pressure of the nutrient and xylem solutions and of the solute flux depends upon the rate
of water uptake. A general model for coupled solute and water flow
through plant roots is proposed by FISCUS (1975, 1977); critical discussion: NEWMAN (1976b). An increase in water flow rate into the root
also enhances the ^{32}P movement into the xylem (MATTEUCCI, 1975). This

uptake rate diminishes with plant age corresponding to rooting depth
(BOLE, 1977). Suberized roots show less water and ^{32}P uptake than un-
suberized ones (CHUNG and KRAMER, 1975).

The passage of water into the root xylem is energy-consuming (MOZHAEVA
and PIL'SHCHIKOVA, 1976). Thereafter water flow has to overcome only
small root xylem resistances (BUSSCHER and FRITTON, 1978 - theoretical
calculation; STONE and STONE, 1975 - measurements in lateral roots of
Pinus resinosa). Root xylem cavitation increases this resistance, however (BYRNE et al., 1977). Energy-consuming root pressure is required
to move the water along the xylem, if transpirational water transport
is hampered. Only in this way is an adequate Ca^{2+} supply to head leaves
of cabbage assured (PALZKILL and TIBBITTS, 1977). SHONE and WOOD (1977)
detected nutrient and pesticide losses during the course of longitudinal xylem sap movement in barley roots.

Water uptake (CAPESIUS and BARTHLOTT, 1975) and transport (HAAS, 1975) in aerial
roots of orchids occurs by means of submicroscopic structures located in the
velamen radicum and in passage cells of the exodermis. In *Vanilla planifolia*,
the aerial roots lack the ability to absorb water, and are also structurally
different from those of other genera.

That water uptake through plant leaves and shoots could occur in a saturated
atmosphere, provided the SPAC water potential gradient is reversed, was demonstrated by SHEIN (1974). Some South-African succulent xerophytes are able to
absorb water through the epidermis, through dead paperlike stipules, or through
trichomes on the tip of the leaf (BARTHLOTT and CAPESIUS, 1974). A physical
process depending on the hygroscopicity of dead shield cells is the basis of
the well-known ability of *Tillandsia* species to resorb atmospheric moisture
(DE SANTO et al., 1976). Though water uptake from ambient humidity alone is
not sufficient for growth, under adverse conditions it may be important for
the survival of these species already adapted to xeric habitats by CAM metabolism (DE SANTO et al., 1977).

b) Stomatal Behavior

During the last years four topics in stomatal physiology have received
particular emphasis: (1) the biochemical processes within guard cells
producing and distributing the solutes needed for turgor regulation
(2) the effects on stomata of abscisic acid accumulated in leaves suf-
fering from water stress (3) the water relations within the epidermis
(4) the influences on stomatal behavior of plant internal and external
factors investigated under field and laboratory conditions. In a handbook article RASCHKE (1979) presents a comprehensive review and interpretation of what is known to date about guard cell metabolism. Within
the same series HSIAO (1976) considers the ion transport belonging to
stomatal movements. For more detailed information on this material the
reader is referred to these reviews. Only some principles of stomatal
metabolism will be emphasized here, together with supplements of some
papers published most recently.

It has now been fairly well established that potassium (DAYANANDAN and KAUFMAN,
1975; ZLOTNIKOVA et al., 1977a) and either malate (DITTRICH and RASCHKE, 1977a,
b; TRAVIS and MANSFIELD, 1977; OUTLAW and KENNEDY, 1978) or chloride (PENNY et
al., 1976; SCHNABL and ZIEGLER, 1977; SCHNABL, 1978) are the dominant solutes
responsible for turgor changes leading to stomatal movements. Normally the sub-
sidiary cells serve as potassium reservoirs (measurement of K^+-gradients along
several neighboring cells: PENNY and BOWLING, 1974; discussion of some yet weak
points of the K^+-shuttle hypothesis: MAIER-MAERCKER, 1979d), but in some cases
one has to reckon with peculiarities - for instance, nearly K^+-free guard cells
in *Paphiopedilum leeanum*: NELSON and MAYO, 1977; distinct ion-absorbing struc-

tures positioned on the polar parts of guard cell inner walls (STEVENS and
MARTIN, 1977a,b) in ferns with diacytic stomata (LÖSCH and BRESSEL, 1979) and
in *Tradescantia pallidus* (STEVENS and MARTIN, 1977c). Under steady-state conditions one always finds a linear correlation between guard cell potassium content and stomatal aperture, no matter which controlling factor is responsible
for a certain aperture degree (LÖSCH, 1978). During the movement process in
stomatal responses to humidity, however, an adjustment in the guard cell potassium content following changes in aperture occurs only after a delay (LÖSCH
and SCHENK, 1978).

In general, the positive charge of potassium is balanced by malate ions produced in the guard cells (PEARSON, 1975; THORPE et al., 1978; VAN KIRK and
RASCHKE, 1978a). The organic ion can be replaced, however, by chloride (PENNY
et al., 1976; VAN KIRK and RASCHKE, 1978b; SCHNABL, 1978), if this is offered
sufficiently (RASCHKE and SCHNABL, 1978). Similarly in exceptional cases K^+
may be replaced by Na^+: RAGHAVENDRA et al. (1976). DITTRICH et al. (1979) propose a mechanism to explain the Cl^-/malate balance accompanied by proton fluxes
occurring in guard cell solute turnover. OGAWA et al. (1978) have investigated
the action spectrum of malate formation in guard cells. SKAAR and JOHNSSON
(1978) assume that the blue light receptor most effective in influencing stomatal behavior is localized in the guard cell membranes. Not as clear is the
fate of malate when stomata close. DITTRICH and RASCHKE (1977b) enumerate
three possibilities for malate disappearance: release into the surroundings,
turnover within the tricarboxylic acid cycle and gluconeogenesis. WILLMER and
RUTTER (1977) supply some evidence that gluconeogenesis from malate can occur.
The production of malate and its distribution - as well as the electrochemical-
ly linked potassium transport - are possibly mediated by differences and changes in pH during stomatal movements (microelectrode measurements: PENNY and
BOWLING, 1975; ZLOTNIKOVA et al., 1977b).

In his review RASCHKE (1979) supplies evidence that the CO_2 fixation
within the guard cells occurs along the CAM pathway. RASCHKE and
DITTRICH (1977) stated that guard cell chloroplasts are not capable
of photosynthetic reduction of CO_2 in *Commelina communis* and *Tulipa ges-
neriana* (similarly in *Commelina cyanea*: THORPE et al., 1978; contrarily,
however, in *Vicia faba*: LURIE, 1977, 1978).

It is now evident from many investigations (e.g., BEARDSELL and COHEN,
1975; WALTON et al., 1977) that abscisic acid, naturally produced
under water stress (probably within the mesophyll chloroplasts:
LOVEYS, 1977), effects stomatal closure (review: MANSFIELD et al.,
1978). The closing effect also occurs when ABA is applied externally
(DAVIES and KOZLOWSKI, 1975; SEN and CHAWAN, 1975; DAVIES, 1978),
promising perhaps a useful application as a natural antitranspirant
(TALHA and LARSEN, 1975). Aftereffects of stress may stem from the
inhibition of photosynthesis by phaseic acid, a relative of ABA
(KRIEDEMANN et al., 1975). FENTON et al. (1976) - similarly: STUART
and COKE (1975) - investigate farnesol, which like ABA accumulates
within water-stressed plants and other chemically allied substances
in an effort to find antitranspirant effects. ITAI and MEIDNER (1978)
show that living epidermal cells are needed for the ABA-transport to
and the effect on stomata (WEYERS and HILLMAN, 1979; ITAI et al.,
1978). These authors advance a hypothesis that explains also the occasionally observed failure of the phytohormone to affect stomatal
closure (LANCASTER et al., 1977) by supposing that ABA affects the
stomatal mechanism not only by triggering solute efflux from guard
cells, but essentially also by enhancing the solute uptake by epidermal cells. RASCHKE (1975) - see also RASCHKE et al. (1976); DUBBE et
al. (1978) - reports that higher CO_2 concentrations alone are not effective in closing stomates, but that ABA must also be present. Concerning age-dependent ABA content and stomatal conductances see
RASCHKE and ZEEVAART (1976).

A fair amount of progress has been made in the analysis of the water potential gradients and water transport pathways within the leaf and their implications for stomatal behavior. MEIDNER (1975, 1976b) put forward the hypothesis that in some plants the epidermal inner walls are the primary locations of water evaporation, not the mesophyll walls, as the classical view postulates. (Concerning the water relations of the mesophyll walls: MEDERSKI et al., 1975; FARQUHAR and RASCHKE, 1978). MEIDNER (1976a) was able to substantiate his supposition experimentally (further information: SHERIFF and MEIDNER, 1975) by measuring the water turnover rate of *Tradescantia virginiana* leaf epidermis and mesophyll separately. SHERIFF (1977a) demonstrated the possibility that distillation processes within the substomatal airspace depend on minimal transversal temperature gradients within the leaf (further investigations: SHERIFF, 1979). That this might be the case has also been postulated theoretically by COWAN (1977). The water turnover of the cell walls surrounding the substomatal cavity affects the water relations of the guard cells. MEIDNER and EDWARDS (1975), EDWARDS et al. (1976), and EDWARDS and MEIDNER (1977) measured directly the turgor pressure within guard cells. They concluded that the low osmotic potentials of guard cells may be necessary to balance the very low matric potentials of their walls. Based on water potential measurements and on EM-observations EDWARDS and MEIDNER (1978) presented evidence that the water relations of the guard cells themselves are affected (SHERIFF, 1977b) when stomata respond to humidity changes. LÖSCH and SCHENK (1978) demonstrated that such alterations in the water relations of guard cells are followed by secondary adjustment of their solute content. MAIER-MAERCKER (1979b) emphasizes that hydroactive processes must be involved in stomatal responses to humidity, otherwise they would be only transient movements.

Studies of such stomatal responses to changed humidity conditions, independent of plant water status, have received much attention in the last years in investigations of stomatal behavior as affected by environmental factors (LAWLOR and MILFORD, 1975; ASTON, 1976; LÖSCH, 1977, 1979; SHERIFF and KAYE, 1977a,b; LANGE and MEDINA, 1979 - theoretical foundation: "feedforward regulation": COWAN, 1977; COWAN and FARQUHAR, 1977; FARQUHAR, 1978; see Sect. 5 - guard cell/subsidiary cell relations in this context: MAIER-MAERCKER, 1979a,c).

Ecophysiologists analyzed extensively the interactions of the variables influencing the plant stomata in its natural habitat (SCHULZE et al., 1974, 1975a,b; MARCHAND, 1975; JARVIS, 1976) and, for comparison, under controlled conditions (e.g., BEADLE et al., 1978; DAVIES, 1977). Many of these findings are presented and discussed by HALL et al. (1976), and for further information the reader is referred to this comprehensive treatise.

The effect of light on stomata, in combination with water stress and/or microclimatic factors, is studied in many investigations of gas-exchange by determinations of stomatal resistances or conductances (e.g., DÜRING, 1976: light and water stress/grape; ALLEN et al., 1976: light and water stress/peanuts; STERNE et al., 1977: light, water stress, and air humidity/avocado; WONG et al., 1978: light and CO_2 partial pressure/*Eucalyptus pauciflora*; PEREIRA and KOZLOWSKI, 1977b: light and temperature/broadleaved trees; IMBAMBA and TIESZEN, 1977: light and temperature/C_3 compared with C_4 plants). Normally, under low light conditions a linear relationship exists between light intensity and stomatal aperture, if plant water potential is sufficient to allow an opening of stomata. BROGÅRDH (1975) - see also BROGÅRDH and JOHNSSON (1975) - found both a slow and a fast stomatal response to stepwise changes in light intensity in *Avena*, the slow one mediated by the leaf internal CO_2 concentration. The rapid stomatal response to light changes seems to reflect a blue light-sensitive process acting directly on the ion transport properties of guard cell membranes (ZEIGER and HEPLER, 1977). In *Zea mays* stomatal conductance remained nearly unaffected by the enormous differences in the intensity of the incident light coming either directly or through the mesophyll. This indicates that in this species light effects are mediated mainly by lowered CO_2 concentrations (RASCHKE et al., 1978).

Temperature effects on stomata are investigated mostly in combination with other factors. Some of these investigations are cited above. Normally, higher temperatures bring about wider stomatal apertures (NEILSON and JARVIS, 1975). But the effect of temperature depends on the preexperimental growing conditions (e.g., HALL et al., 1975; PARSONS, 1978; LÖSCH, 1977). There is also evidence that the occurrence of wider pore opening with increasing temperature in well-watered plants is reversed under water stress (SCHULZE et al., 1973; TENHUNEN and GATES, 1975; JENSEN, 1976; LÖSCH, 1979). PEMADASA (1977) investigated stomatal opening resulting from high temperatures in darkness, and found an accompanying increase in starch hydrolysis and K^+ accumulation in guard cells (see also LÖSCH, 1978).

Within the context of stomatal responses to environmental factors a book by GRACE (1977) which deals with the influence of wind on plants and their microclimate should also be mentioned. The study by DAVIES et al. (1978) in which different humidity sensitivities were found in stomata of *Cytisus scoparius* subspecies, depending on whether or not they were exposed to wind, reveals delicate interrelationships between stomatal behavior and plant distribution: the prostrate subspecies which drastically increases stomatal resistances when wind steepens the humidity gradient between leaf and air, colonizes windswept areas. The erect subspecies which is unable to regulate its stomata under such conditions is confined to sheltered places.

Interest in the changes of stomatal behavior during ontogenesis is growing. Studies relating to this topic are reported by DAVIS et al. (1977, VÁCLAVÍK (1975), JORDAN et al. (1975), TURNER and HEICHEL (1977) and EL AOUNI (1976). Differences in the behavior of ad- and abaxial stomata have been measured by VÁCLAVÍK (1974), SIONIT and KRAMER (1976), NAGARAJAH (1978) and ASTON (1978).

Studies investigating applied aspects of stomatal biology deal with the impact of air pollutants on the leaf (SO_2: KONDO and SUGAHARA, 1978; O_3: RUNECKLES and ROSEN, 1977; NO_2: SRIVASTAVA et al., 1975; principally: TAYLOR, 1978) and with attempts to improve the water use of plants by antitranspirantia. The artificial reduction of plant water expenditure is possible by two means: by spraying the leaves with films which have high resistances to water permeation (DAVENPORT et al., 1974 - films; FROMMHOLD, 1975 - silicone oils; ANDERSON and KREITH, 1978 - films and silicone oils) or with chemicals which affect the guard cell physiology (various commercial antitranspirants: OLOFINBOBA et al., 1974; PMA: PATIL and RAJAT DE, 1976; MILLER and ASHBY, 1978; inhibitors of cyclic photophosphorylation: RAGHAVENDRA and DAS, 1977; acetyl-salicylic acid: LARQUÉ-SAAVEDRA, 1978).

c) Water Flow and Resistances Within the Soil-Plant-Atmosphere Continuum (SPAC)

The expression, SPAC, is widely used to indicate the thermodynamic continuum of water potential from the soil across the plant into the atmosphere. However, as pointed out by OERTLI (1976b), the term "continuum" cannot be applied to pressure and osmolality: plant membranes are locations of discontinuities. Since these are important factors of plant water relations, he prefers the term "soil-plant-atmosphere system".

The efficiency with which plants transport water is related to the water potential difference required to drive fluxes from the soil to the leaves (see CAMACHO-B et al., 1974). The SPAC is viewed as a series of flow resistances. The major resistance lies in the gaseous phase where the strategically placed stomata control the rate of flow through the system (WEATHERLEY, 1976; see Sect. 3b). Water loss results in a decrease in leaf water potential which provides the driving force for the movement of water along a pathway in which frictional resistances and gravitational potentials have to be overcome. As stressed in a review by JARVIS (1975) leaf water deficits occur not only when water loss from leaves through evaporation exceeds the supply from roots, as is so often stated. Rather, a water potential gradient is maintained normally under steady-state transpiration conditions even when water uptake and water loss are equal. Recent research has focused on quantitatively measuring and simulating actual flow rates through individual plants, as well as through plant stands, and on linking these to soil water properties, plant parameters and atmospheric conditions. Many publications discuss the flow resistances and the physiological and ecological implications of resulting changes in water potential and turgor pressure (see review by OERTLI, 1976b).

HINCKLEY et al. (1978) summarize what is known about temporal and spacial variations in the water status of forest trees. Plant-atmosphere interactions of conifers and coniferous forests are treated by JARVIS et al. (1976; for other ecosystems cf. MONTEITH, 1976). Methodological and physiological aspects of sap transfer in tree trunks are reviewed by GRANIER (1977; see also ZIEGLER, 1977). The frequently used pulse technique for measuring sap velocity in stems has been improved (ČERMÁC et al., 1976a; KUČERA et al., 1977). STONE and SHIRAZI (1975) present a theoretical analysis of the dependency between velocity of temperature signal and sap flow in the xylem. Heat pulse velocity is in many cases directly correlated with actual transpiration rate. Measurements at various sectors around the stem of a tree indicate differential rates of water loss from various parts of the crown (LASSOIE et al., 1977; also SHAW and GIFFORD, 1975). However, water extraction from storage tissues should be taken into account (see below). Stem diameter recording together with only occasional destructive sampling permits continuous monitoring of plant water within the conduction system (HUCK and KLEPPER, 1977). SHERIFF (1976) combines this method with an improved magnetohydrodynamic sap flux meter (see also ROA and PICKARD, 1975). A close relationship is found between soil moisture, stem circumference, and xylem pressure potential in oak trees in the field (HINCKLEY and BRUCKER-HOFF, 1975). Even *Opuntia occidentalis* shows diurnal fluctuations in stem diameter, but decreases occur in the dark in accordance with the stomatal response of this CAM plant (SCHRÖDER, 1975). Investigations of xylem water status by means of acoustic detection of cavitations have been continued (MILBURN and MCLAUGHLIN, 1974). In isolated and freely suspended vascular bundles of *Plantago major* cavitations marked by jerking take place already at sap tensions of 5-12 bar. Cavitations seem to occur normally even at moderate transpiration load in the field *(Plantago m., Tussilago farfara)*. Root pressure might be necessary to refill the vessels during the night.

The hydraulic system of intact plants proves to act as a true continuum. NULSEN et al. (1977) subject roots of corn plants to changes in hydrostatic pressure. Leaf water potential responses are consistently in agreement with increased or decreased pressure in the root zone. Within this hydraulic system liquid-phase flow resistances are located at the soil-root interface and in the roots (see Sect. 3a), in the xylem of stems, petioles and leaves, and in the mesophyll of the leaves themselves. Several authors confirm that in different plant types the root system accounts for the major part of the total resistance, even under unlimited soil water conditions (DUBE et al., 1975; ROBERTS, 1977; TESHA and KUMAR, 1977). However, there is still much discussion and uncertainty as to the magnitude and location of the dominant resistance (ROSE et al., 1976; STONE and STONE, 1975). These may differ from species to species (JOHNSON and CALDWELL, 1976). To be considered, on the other hand, is that the absolute magnitude of partial resistances and therefore their partitioning in the total resistance is a function of the flow itself. ROBERTS and KNOERR (1978) confirm that for a tree *(Ilex opaca)* apparent resistance drops, by a factor of more than 7, from low to high water flow rates. In apple trees the ratio of root to stem flow resistance alters from 2:1 at high transpiration rates to 1:1 at lower rates (LANDSBERG et al., 1976). BOYER (1974) showed that resistance in sunflower plants is about 30 times greater at low fluxes than at high fluxes, mostly due to changes in leaf mesophyll resistance. He attributes this difference to the fact that leaf protoplast pathway with high resistance determines water movement at low water uptake. There might also be other flow-dependent changes in resistances. Thus, in disagreement with Poiseuille's law, GIORDANO et al. (1978) found no linear relationship between flux and applied pressure in single vessels. Further, JOHNSON (1977) showed by EM that cell walls are drawn in against the helical thickenings of xylem vessels in leaf petioles under water stress. Such pressure-dependent deformations might explain variable resistances to the flow of water.

As an example of a synthesizing model, a paper by WARING and RUNNING (1976) should be mentioned, in which a general concept for movement of water through individual trees *(Pseudotsuga menziesii)* is presented, taking into account resistances and capacitances. The importance of internal water storage is stressed: there is an extensible tissue reserve (fine roots, cambium cells, leaves, etc.) as well as the mature wood reserve. It is estimated that for a mature Douglas-fir tree the water available from the sapwood alone is equivalent to that required for transpiration over at least a 10-day period in midsummer. Internal storage accounts for the time lag between transpiration and water uptake that can be observed in the field and that is quantitatively simulated by the model (see also ČERMÁC et al., 1976b).

d) Transpiration Patterns

Transpirational water loss has been determined in many investigations under various conditions. Four to five types of daily courses of transpiration can be distinguished (BATANOUNY, 1974): (1) dome shaped curves (2) single-peaked curves (3) single-peaked curves with a shoulder (4) two-peaked curves and (5) daily courses with more than two peaks. Types 1-4, in that order, are associated with increasingly adverse climatic and edaphic conditions. However, transpiration changes in well-watered plants cannot always be explained by environmental influences; physiological processes are also involved (ALEKSEENKO, 1975). In addition to diurnal fluctuations in transpiration resulting from endogenous rhythms in stomatal aperture (PALLAS et al., 1974), oscillations of much shorter duration are often observed (BRAVDO, 1977; GULYAEV, 1977) and mathematically described (SHIRAZI et al., 1976a,b). Varietal differences in transpiration rate occur (POCHARD and SERIEYS, 1974; PEGELOW et al., 1977) and can sometimes be ascribed to morphological peculiarities (e.g., bloom layers: CHATTERTON et al., 1975). Awns contribute moderately to photosynthesis, respiration, and transpiration of barley spikes (JOHNSON et al., 1975). The cooling effect of transpiration is important for fruit temperature regulation (PROKOV'EV et al., 1974).

The particular transpiration patterns associated with different plant species (ELIAS, 1975) and life-forms (RAKHMANINA, 1976) have phytogeographical implications. There is a declining north-south gradient in the transpiration rate of many trees (SOLNTSEVA and SUTULOVA, 1975).

The winter transpiration capacity of the shoots of woody species may in part determine their northern distribution boundaries (BYLINSKA, 1975). Transpiration rates of tropical forest species show significant differences (WEAVER, 1975) and it is possible to correlate them with the sapwood area (JORDAN and KLINE, 1977).

4. Implications of Water Shortage

The most important aspect of plant water relations concerns plant performance under conditions of limited water supply (reviews: HSIAO et al., 1976a,b). Much research has focused on metabolic responses to momentary stress situations at the level of organelles and cells. However, adaptation to drought involves more than short-term effects. SCHULZE and HALL (1979) have demonstrated that stomatal response is affected by integrated drought experience. Undoubtedly long-term drought influences other momentary physiological events, including enzyme activities and, as well, morphogenetic development, e.g., root growth or leaf structure. Our present understanding of these regulatory, time-integrating processes in particular, and of overall plant response to drought in general, remains limited, despite many advances in recent years.

Water stress causes dramatic changes in cell nitrogen metabolism, which seem to be controlled by several phytohormones (DHINDSA and CLELAND, 1975; ARAD and RICHMOND, 1976). Accumulation of different types of free amino acids such as γ-aminobutyric acid, asparagine, glutamic acid, and especially proline, is reported for a large variety of species (HANOWER and BRZOZOWSKA, 1975; RAI and BAPAT, 1977; BASKIN and BASKIN, 1974). In higher plants, proline usually does not accumulate until the water deficits are quite severe. In cotton the threshold is about -15 to -17 bar water potential (MCMICHAEL and ELMORE, 1977). However, proline concentration may even exhibit diurnal fluctuations due to changes of the water content of intact leaves (RAJAGOPAL et al., 1977). In some cases, a specific ion effect on proline accumulation is observed when stress is induced by external osmotic potential (barley: CHU et al., 1976; diatoms: SCHOBERT, 1974). In general, a capacity for free proline accumulation seems to indicate drought tolerance in many plant types. For example, in *Sorghum bicolor* it is correlated with the ability of cultivars to recover following the removal of stress (BLUM and EBERCON, 1976). Therefore, proline content is used in screening for drought-tolerant varieties of crop plants (MALI and MEHTA, 1977b; MEHKRI et al., 1977).

> Explanations at the metabolic pathway level of organization for proline accumulation remain contradictory, possibly due to differences between species. Accumulation has been discussed in connection with protein synthesis, inhibited proline oxidation, stimulated proline dehydrogenase activity, photosynthetic assimilation products, and respiratory metabolism (STEWART and BOGGESS, 1977; STEWART et al., 1977; BOGGESS et al., 1976; WRENCH et al., 1977; JAEGER and MEYER, 1977; LAWLOR and FOCK, 1977b). In addition to amino acids ("proline regulatory type" of plants: PALFI et al., 1974a) other nitrogen dipols, organic acids, or polyols may accumulate under water deficits. In certain halophytes (Chenopodiaceae, Graminaceae) the level of betaine increases markedly with salt-dependent water stress (STOREY and WYN JONES, 1975; STOREY et al., 1977; also HANSON and NELSEN, 1978). WYN JONES et al. (1977a,b) suggest that organic compounds are involved in cytoplasmic osmoregulation. Although osmotic potential of vacuole solution can be adjusted with inorganic ions this might be toxic to the cytoplasm where ionic strength does not normally exceed

200-250 mM when K^+ is the dominant cation. It has therefore been suggested that when a lower cytoplasm osmotic potential is required in order to meet stress conditions, molecules such as betaine or proline accumulate in the cytoplasm, resulting in minimal perturbation of protein-water interactions. On the other hand, SCHOBERT (1977a,b) doubts the role of proline and polyols as osmotic regulators. She assumes that proline associates via its hydrophobic part with hydrophobic side chains of cytoplasmic constituents. Water can then be bound to the carboxylic and imino groups by hydrogen bonding forces. A smaller number of water molecules is thereby sufficient for hydration of the biopolymers. Proline considerably increases the amount of strongly bound water in the cell (PALFI et al., 1974b). Possibly, these different mechanisms act jointly to adapt cells to situations of water-deficiency.

Many other metabolic responses to water stress which have been studied cannot be discussed here in detail. Cotyledon tissue of pumpkin seedlings shows a reduction in percentage of polyribosome within 5 min following desiccation-induced stress (RHODES and MATSUDA, 1976). RNase activity is increased by water stress (GENKEL et al., 1974; BRANDLE et al., 1977; BLEKHMAN, 1977; CHEN and LI, 1977), as is the content of free nucleotides and nucleic acids (ALI-ZADE and ABDULLAEV, 1974). In the drought-tolerant moss *Tortula ruralis*, messenger RNA is able to withstand desiccation (DHINDSA and BEWLEY, 1978; see Sect. 7). Many enzyme systems are affected soon after the onset of water stress in homoiohydric plants (e.g., PLAUT, 1974; review by VIEIRA DA SILVA, 1976).

The effects of water stress on photosynthesis have been of special interest during the past years (review see BOYER, 1976). It is evident that plant water status can affect virtually all aspects of carbon metabolism. Quantitative assessment of relative effects of water shortage on the various processes affecting overall net photosynthesis is difficult. This is partly due to methodological shortcomings and also due to the lack of understanding of the interactions of the single processes. There is evidence indicating that the initial inhibition in net photosynthesis which occurs as water stress is imposed, results from stomatal closure (see Sect. 3b). However, BUNCE (1977b) finds that for species from a variety of habitats, the onset of drought-induced increase in stomatal resistance is accompanied by an increase in mesophyll resistance as well. The latter may be effected through influences of water conditions on the CO_2-transport mechanisms operating between the intercellular air spaces and the site of carboxylation, and between the organelles involved in carbon metabolism within the cells. Metabolite transport may also be stress-dependent. In addition, total mesophyll resistance depends on the activity of the photosynthetic and respiratory apparatus itself. True photosynthesis, photorespiration, and mitochondrial respiration can be directly and differently influenced by water stress (LAWLOR and FOCK, 1975). On the other hand, in vitro investigations of the biochemical photosynthetic partial processes may not quantitatively reflect their in vivo responses in stressed leaves, as a result of rehydration prior to assay (BEADLE and JARVIS, 1977) or due to peculiarities in their metabolite exchange situation under drought conditions, which cannot be reproduced in vitro.

It is difficult to generalize from particular stress responses because of enormous species-specific differences in behavior. For example, under ambient CO_2- and O_2-conditions, net photosynthesis in sunflower leaves reaches zero at -18 bar water potential with evolution of CO_2 occurring at higher stress (LAWLOR and FOCK, 1975). In contrast, the moisture compensation point of the desert lichen *Ramalina maciformis* is only reached at a water potential of -290 bar (LANGE et al., 1975b). Mesophyll resistance in response to leaf desiccation differs in subspecies of the same species (*Kochia prostrata*; RAKHIMOV, 1975), and even young and old leaves of the same individual are differently affected

(ACKERSON et al., 1977; VOLODARSKII and BYSTRYKH, 1976). Drought hard-
ening may (wheat seedlings: GENEROZOVA, 1976) or may not (*Eucalyptus
socialis*: COLLATZ et al., 1976) change the nonstomatal photosynthetic
characteristic. Different kinds of regulating mechanisms within the
whole plant (HSIAO et al., 1976b), which might be mediated by hormonal
control (VAADIA, 1976; ITAI and BENZIONI, 1976) complicate still further the question of how water deficit affects photosynthetic mesophyll resistance. Even the method by which stress is imposed on a
plant, whether by withholding irrigation or osmotically, by using
salt solutions or polyethyleneglycol, results in different effects on
mesophyll resistance at the same leaf water potential. Further, the
base for calculation of responses is a source of error, because water
stress also changes the parameters to which photosynthetic activities
can be related (KENNEDY, 1977). All of these explain why in recent
literature interpretations of nonstomatal affects of lowered water
potential on photosynthesis are often controversial and why it is
not yet possible to fit the known components into a comprehensive
whole.

Water stress affects chloroplast development in greening leaves (FREEMAN and
DUYSEN, 1975), and reduces the formation rate of light-harvesting chlorophyll
a/b-protein (ALBERTE et al., 1975). The ultrastructure of chloroplasts in mature leaves shows characteristic alterations after even slight dehydration
(FELLOWS and BOYER, 1976; KURKOVA and MOTORINA, 1974). In *Zea mays*, a reduction in the lamellar content of chlorophyll a/b-protein takes place, a rather
specific target for water stress (ALBERTE et al., 1977). Leaf water potential
affects the photochemical portion of photosynthesis as well as the enzyme activities of the dark reaction. This may account for the reduction of photosynthesis which occurs already with small water deficits (BOYER, 1976; KECK
and BOYER, 1974; O'TOOLE et al., 1977). As demonstrated in sunflower (MOHANTY
and BOYER, 1976) and bean (POSPIŠILOVÁ et al., 1976), quantum yield is affected by water stress, confirming that changes occur close to the primary photochemical events of photosynthesis. However, such responses depend decisively
on plant history. In the field (Death Valley), *Larrea divaricata* showed no inhibition of quantum yield under severe drought conditions (-50 bar) whereas
laboratory grown plants were significantly affected when water potentials fell
below -20 bar (MOONEY et al., 1977). According to LAWLOR (1976), in wheat leaves
CO_2 compensation concentration is increased at water potentials lower than -5
bar. Mitochondrial respiration remains constant but increases relative to decreasing true photosynthesis. In this case photorespiration decreased in proportion to photosynthesis. However, LAWLOR and FOCK (1977a,b) show that in
sunflower water stress results in an increase in photorespiration in proportion
to photosynthesis (see also BUNCE and MILLER, 1976). They investigate changes
in the photosynthetic products caused by water deficiency. Stress conditions
seem to shift the proportion of primary products of the C_4 and C_3 pathways in
C_4 plants (KENNEDY, 1977) and stimulate CAM-behavior in succulents (see Sect.
5). Stress-induced changes in photosynthetic properties also influence plant
responses to other environmental factors: NOBEL et al. (1978) report effects
on the temperature optimum of net CO_2 exchange in desert plants.

One of the most important features of drought resistance in plants is
the ability to maintain or to adjust turgor under conditions of water
deficiency (JOHNSON and BROWN, 1977). As a result of osmotic regulation and changes in elastic tissue characteristics, stress-adapted
crop varieties show higher tissue water content and turgor pressure
at a given water potential than control plants (JONES and TURNER,
1978; see also FERERES et al., 1978). Morphogenetic changes may directly contribute to this adaptation. CUTLER et al. (1977) show that
under water deficit, *Gossypium hirsutum* develops leaves with smaller
cells and thicker walls. The authors conclude from simulations with
a model that smaller cells maintain turgor at lower values of water

potential than larger cells. Such developmental responses apparently
take place under hormonal control. In wheat, water stress as well as
an experimental increase in the tissue ABA-level decreases cell size
(QUARRIE and JONES, 1977).

> Many other morphological changes in plants grown under water shortage have been
> studied. Drought conditions decrease leaf elongation (BUNCE, 1977a; SANDS and
> CORRELL, 1976; LUDLOW and NG, 1977) and cause an increase in leaf succulence
> (tomato: DUMBROFF and BREWER, 1977; pea: MANNING et al., 1977). Leaf initiation
> is reduced (CLOUGH and MILTHORPE, 1975; MARC and PALMER, 1976). In *Pinus resi-*
> *nosa*, shoot extension is governed by the moisture conditions which exist in the
> middle period of the previous year and in the early period of the current year
> (GARRETT and ZAHNER, 1973). Cob formation in *Zea mays* is controlled by water
> relations during certain sensitive stages of development (DAMPTEY and ASPINALL,
> 1976), and the phasic development of annual *Medicago* species is influenced by
> periods of water stress (CLARKSON and RUSSELL, 1976). Root formation and root
> growth are decisively determined by water conditions. As shown with maize
> (SORIANO and GINZO, 1975) and soybean (SILVIUS et al., 1977), soon after onset
> of drought plants already exhibit alterations in the distribution of carbon
> assimilates among plant parts which favor the root system. Thus drought-adapted
> plants exhibit a special root morphology (MCINTYRE, 1976; DA et al., 1977),
> faster root growth (TORSSELL, 1976), and increased root/shoot ratio (ASHENDEN
> et al., 1975). This is true for individuals of the same genotype grown under
> different moisture regimes, as well as for members of different populations
> adapted to habitats of different water availability.

Morphogenetic and metabolic adaptations in response to long-term and
changing water stress affect plant performance, including biomass production and yield, in a very complex manner (KOZLOWSKI, 1976). The extremely dynamic nature of plant water status, the dependence of stress
effects on duration, severity, and timing during ontogeny, and the
interplay with other environmental variables make analyses and prediction difficult (review: HSIAO et al., 1976b). Different aspects of
these processes are studied. SMITH and NOBEL (1978) discuss the consequences of drought-induced, seasonal changes in leaf morphology for
productivity of the desert broadleaf *Encelia farinosa*. Seed yield as influenced by water stress occurring during different periods of development - vegetative growth, flowering, fruit formation, fruit filling -
is investigated for many crop plants (JOHNSON and MOSS, 1976; ANDREWS
et al., 1977; SIONIT, 1977; SIONIT and KRAMER, 1977). Attempts are
made to quantify the effect of environmental drought on production
using plant parameters. SCHULZE and HALL (1979) successfully correlate CO_2 assimilation and yield in different types of plants with a
cumulative pre-dawn plant water potential ("bar days"). More sophisticated parameters may be developed which will help to characterize and
to integrate plant stress experience.

5. Water Use and Productivity in Higher Plants

The relationship between water use and photosynthetic productivity
has gained interest from physiological, ecological, as well as agronomic points of view. "Water-use efficiency" (P/T; P for photosynthesis, T for transpiration), i.e., the ratio between mass of CO_2
fixed (or dry matter produced or yield) and mass of water loss (transpiration or evapotranspiration), and its reciprocal, the "transpiration ratio" (T/P) are expressions for water use (P and T both expressed in the same units; BIERHUIZEN, 1976). They can be related to momen-

tary gas exchange values, to integrated diurnal courses, or to total
growing periods (FISCHER and TURNER, 1978).

Actual water-use efficiency depends on specific plant properties and/or environmental conditions. For example, it differs in tree species originally adapted to different habitats (BRAUN, 1976, 1977). In cabbage and millet plants the
rate of water use changes at different stages of growth (NELSON and HWANG, 1976;
KASSAM and KOWAL, 1975). Water-use efficiency in desert broad leaf shrubs differs between sun and shade leaves and varies with seasonal changes in leaf morphology (SMITH and NOBEL, 1977a,b). Water-use efficiency of crops is of decisive
importance for plant production especially in arid and semi-arid zones. There is
a large bulk of information accumulating in this area (review by FISCHER and
TURNER, 1978).

Stomatal Action and Water-Use Efficiency. Increasingly, stomatal functioning is understood as a mechanism for "resolving the dilemma of oppos-
ing priorities" (RASCHKE, 1976): preventing excessive water loss while
ensuring sufficient CO_2-uptake. Two major, variable feedback control
loops, dependent on internal CO_2-concentration and on bulk leaf water
status, tend to achieve maximal photosynthesis and to avoid desiccation (RASCHKE, 1979). However, negative feedback alone is not sufficient to optimize water-use efficiency. This is supported by COWAN
and FARQUHAR (1977) who have developed the theoretical optimization
requirements for diurnal courses of gas exchange. With respect to
time, the partial differential quotient, $\partial T/\partial P = \gamma$ must be constant
for optimal stomatal behavior. The term, γ, which determines the integrated amount of water loss per day, then may vary according to
growth and ontogenetic development. Clearly, for mechanisms to provide optimal control it is necessary that plants directly and sufficiently sense the external environment. A feedforward control (as defined by COWAN, 1977), based on the stomatal response to the evapora-
tive demand of the ambient air, would conserve water (COWAN, 1978).
In fact, stomatal response to humidity improves water-use efficiency
in apricot trees cultivated under arid conditions (SCHULZE et al.,
1975b), and diurnal courses of desert plants with midday-depression
of gas exchange as a consequence of stomatal closure (LANGE et al.,
1975a) resemble the theoretically computed curves for optimization
(COWAN and FARQUHAR, 1977). Such a control system requires that the
stomatal aperture responds to a considerable extent independently
of the hydraulic conditions in the rest of the leaf (cf. LÖSCH, 1979).
This was shown to be the case in hazel where short-term fluctuations
in bulk leaf water potential did not affect leaf conductance and only
long-term water stress mediated the stomatal responses to humidity
(SCHULZE and KÜPPERS, 1979). Optimization of water use through humi-
dity-controlled stomatal responses might be widespread among plants
(in *Opuntia inermis*: OSMOND et al., 1979; in *Tillandsia recurvata*: LANGE
and MEDINA, 1979; in *Populus tremuloides*: BROWN and MC DONOUGH, 1977;
and in whole plants of barley in contrast to single leaves: RAWSON
et al., 1977).

There are interesting aspects concerning the role of photorespiration in connection with the requirement of plant water relations. When stomata are closed
in the light, the intercellular CO_2 partial pressure stabilizes at the CO_2 compensation point. Lack of photorespiration would result in a strong reduction
in internal CO_2 concentration under these conditions. POWLES and OSMOND (1978)
showed that leaves in CO_2-free N_2 with 1% O_2 and thus suppressed photorespiration suffer from irreversible photoinhibition of their CO_2 assimilation. They
put forward the hypothesis that photorespiration serves to protect against such
damage when leaves are deprived of CO_2 in the light, that is, when stomata are
closed.

Table 1. Summary of gas exchange parameters for CAM, C_4, and C_3 plants (SZAREK and TING, 1975)

		T/P $g \cdot g^{-1}$	R $s \cdot cm^{-1}$	P $mg\ CO_2 \cdot dm^{-2}\ h^{-1}$
CAM,	dark	25–150	2–10	10–15
	light	150–600	6	3–20
C_4		250–350	1	30–60
C_3		450–600	1	20–40

Water Relations and CO₂ Fixation Types. Crassulacean acid metabolism (CAM) and C_4 pathway of photosynthetic carbon fixation are primarily bio-chemical adaptations to arid environments (KLUGE and TING, 1978; OSMOND, 1978; DOLINER and JOLLIFFE, 1979; LUDLOW, 1976). They enable favorable transpiration ratios (T/P) to occur, which are connected with characteristic ranges of minimal stomatal resistance (R) and expected CO_2 fixation rates (P), see Table 1. Recent research has focused on screening for the different photosynthetic fixation types within the plant kingdom and on correlating the photosynthetic option with the water relations of the plants. Determination of $\partial^{13}C$-isotope discrimination rates proves to be an efficient method for distinguishing between predominant pathways (OSMOND and ZIEGLER, 1975). In addition, the ratio of deuterium to hydrogen in dried plant material reflects plant water relations and is different in C_3, C_4, and CAM species (ZIEGLER et al., 1976).

Preferential occurrence of CAM and C_3 plants under dry and hot climatic conditions is shown for different geographical regions: Indian arid zones (SANKHLA et al., 1975), northern Sahara (WINTER et al., 1976), Israel and Sinai (WINTER and TROUGHTON, 1978), North America (TEERI and STOWE, 1976; STOWE and TEERI, 1978), and Kenya (TIESZEN et al., 1979). Some CAM plants exhibit high photosynthetic flexibility with respect to their prevailing mode of C-fixation (OSMOND, 1975). Dark fixation is stimulated by water and salinity stress effects (WINTER and LÜTTGE, 1976; OSMOND et al., 1976), by photoperiodic conditions (NOLBORCZYK et al., 1975; ZABKA and CHATURVEDI, 1975), and by temperature (TROUGHTON and CARD, 1975). In its natural environment, *Mesembryanthemum crystallinum* shows seasonal shift from C_3 photosynthesis to CAM in response to habitat dryness (WINTER et al., 1978; BLOOM and TROUGHTON, 1979). Tissue $\partial^{13}C$-values in *Opuntia inermis* (eastern Australia) vary according to the habitat water availability, which indicates that control of CAM is environmentally influenced (OSMOND, 1975). This was found to be the case also in alpine species of *Sempervivum* (OSMOND et al., 1975), and in *Welwitschia mirabilis* in the Namib desert (SCHULZE et al., 1976). However, a strict relationship between habitat water conditions and metabolism does not always occur, as shown by many examples demonstrating specific behavior of single taxa (HANSCOM and TING, 1978). Alpine *Sedum* and *Saxifraga* species, which are potential CAM plants, under natural conditions exhibit very little or no CO_2 dark fixation (OSMOND et al., 1975). The same is true for *Sedum acre* (KLUGE, 1977). On the other hand, there are species in which CAM metabolism seems to be obligatorily fixed. *Opuntia phaeacantha* and *Yucca baccata* exhibit CAM exclusively despite significant differences in soil water status along an altitudinal gradient (SZAREK and TROUGHTON, 1976). Mexican Crassulaceae of very dry habitats behave similarly. They do not reduce dark CO_2 fixation even under moist greenhouse conditions (RUNDEL et al., 1979). Members of the Cactaceae and Crassulaceae families prefer different habitats with respect to evaporation and precipitation conditions in North America (TEERI et al., 1978). *Frerea indica*, an Indian Asclepiadaceae, performs

as a leafless stem succulent CAM metabolism during the dry season and produces
C_3 leaves during the moist monsoon time (LANGE and ZUBER, 1977).

6. Habitat Water Relations and Plant Performance

The dependence of plant existence on the specific water relations
of the habitat (cf. also chapters Experimental Ecology in the preceding volumes, Prog. Bot.) has been analyzed, especially for trees
in order to select useful ecotypes for forestry purposes (conifers:
ZOBEL, 1974; UNTERSCHEUTZ et al., 1974; HELLKVIST and PARSBY, 1977;
broadleaved trees: BRAUN, 1977; DAVIES and KOZLOWSKI, 1977), and for
desert and semidesert plants, for Mediterranean-type sclerophyllous
shrubs, and for grassland species and associations.

In an investigation of woody desert perennials STRAIN (1978) was able
to distinguish several features advantageous to drought-evading and
drought-enduring species. In general (1) the former show a higher
photosynthetic potential, but cease to photosynthesize already at
higher water potentials. Other characteristics peculiar to some of
the investigated desert shrubs are (2) seasonal variations in metabolism associated with morphological dimorphism (3) the essential
photosynthetic contribution of chlorophyllous stems to the net productivity of the plant (4) genecological differentiations at species
rank in adaptation to drought and adverse temperature conditions and
(5) physiological acclimations to desert conditions. Transpiration
behavior and gas-exchange in desert species are described by STOCKER
(1974a,b,c) and BATANOUNY (1974). KAPPEN et al. (1976) were able to
correlate water relations and net photosynthesis in *Hammada scoparia*
with its tendency to occur mainly in runnels and loessial plains of
the Negev desert.

Seasonal water use patterns of plants recorded by BRANSON et al.
(1976), HALVORSON and PATTEN (1974), CAMPBELL and HARRIS (1977) and
SHARMA (1976b) were found to correlate with microhabitat conditions
of semi-desert stands. DE PUIT and CALDWELL (1975a,b) found a characteristic gas-exchange for each of several life-forms representative
of cool shrub-steppe communities; BARBOUR et al. (1974) compared the
water relation parameters of *Larrea* proveniences under controlled conditions.

Various species in several chaparral formations and associated communities are affected differently by water stress (POOLE and MILLER,
1975; MOONEY et al., 1975). Morphological parameters can be correlated with drought-adaptation (MORROW and MOONEY, 1974; KRAUSE and KUM-
MEROW, 1977). Also in the xerothermic regions of the old world water
potential determinations have been correlated with vegetation distribution patterns (macchia, tomillares, phrygana: VARDAR and BÜTÜN,
1974; MERINO et al., 1976; MARGARIS, 1977; steppe regions of the
southern alps: FLORINETH, 1974).

Grassland species constituting meadow associations were studied comparatively under various soil water regimes (GLOSER, 1977; REDMANN,
1976). Water use in whole meadow communities is described by RYCH-
NOVSKA (1976). The pair *Dactylis glomerata/Deschampsia cespitosa* shows a
very distinct adaptation to relatively dry and wet sites, respectively (RAHMAN, 1976; ASHENDEN et al., 1975; DAVY and TAYLOR, 1974). In
mixed stands pubescent races of *Danthonia sericea* were unable to survive in wet areas, while glabrous ones were outcompeted on well-

drained upland sites (QUINN, 1975). *Arrhenatherum elatius* is vulnerable
to desiccation in juvenile stages, if growth is retarded by a deficiency of P and N; in *Festuca ovina* this combination is not detrimental
(GRIME and CURTIS, 1976). Such an interaction between water- and nutrient stress may be responsible for some vegetation distribution
patterns (see also LAUENROTH et al., 1978). In considering the occurrence of xeromorphic structures in plants adapted either to water
shortage or nutrient deficiency SMALL (1973) suggested the following:
Since physiological properties associated with xeromorphy may allow
survival in either a water- or nutrient stress situation, they may
enable a species originally adapted to one stress, to extend its occurrence to habitats characterized by the other factor. This may be
meaningful for an understanding of evolutionary processes.

Different N-fertilization heavily influences the competition for moisture between purple nutsedge *(Cyperus rotundus)* and rice (OKAFOR and DE
DATTA, 1976). In a study of the dynamics of co-occurrence of successional annuals in Illinois old-fields PICKETT and BAZZAZ (1976) found
a competitive divergence along a moisture gradient. The complex processes in vegetation replacement resulting from overgrazing under
drought conditions in the Sahel zone are analyzed by BREMAN and CISSE
(1977; see also: MENSCHING, 1978). Good fodder grasses become replaced
by unpalatable species, especially by the short-lived legume *Zornia
glochidiata*, a plant which does not protect the soil surface with cover
during the dry season, thus giving way to erosion.

The water availability in arctic permafrost soil ecosystems is discussed by RYDEN (1976). Experimental work on gas-exchange in tundra
species has been done by JOHNSON and CALDWELL (1975) and EHLERINGER
and MILLER (1975).

MILLER et al. (1975) measured water potentials and stomatal conductances in mangrove plants. MORROW and NICKERSON (1973) determined the
salinity necessary for the existence of *Rhizophora mangle* and *Avicennia
germinans*. The former requires freely exchanged seawater, while the
other prefers lagoons where root-bathing waters show a higher salt
content than seawater.

Waterlogging and Flooding. Different aspects of plant life in anaerobic environments are thoroughly covered by a voluminous book edited by HOOK and
CRAWFORD (1978). Therefore, only a few outstanding original articles will be
cited here. Biochemically oriented investigations deal with the respiration
peculiarities under anoxia (e.g., LAMBERS, 1976: flood-tolerant/-intolerant
Senecio species) and ethylene production (KAWASE, 1974, 1976; JACKSON and
CAMPBELL, 1975, 1976; JACKSON et al., 1978). JONES (1975a,b) studied the influences of waterlogging on ion uptake. There are not only metabolic adaptations to waterlogging and flooding (BASKIN and BASKIN, 1976), but also adaptations in root morphology (GILL, 1975; WAMPLE and REID, 1975; LAMONT, 1976;
STELZER and LÄUCHLI, 1977). The minimizing of ethanol production in seeds subjected to anaerobic conditions is based on a metabolic process similar to that
found in flood-tolerant roots (CRAWFORD, 1977). The combined effects of inunda-
tion and salinity on plant vitality are investigated by ROZEMA and BLOM (1977)
in the flood-tolerant glycophyte, *Agrostis stolonifera* (see also: AHMAD and
WAINWRIGHT, 1977) and the flood-intolerant halophyte, *Juncus gerardi*. NASR et
al. (1977) did similar work in the fruit trees, plum and peach. The consequences of waterlogging conditions for the gas-exchange of anoxia-intolerant species, normally resulting in a complete closure of stomata (PEREIRA and KOZLOWSKI,
1977a), are described by REGEHR et al. (1975), SZLOVÁK (1975), and PHUNG and
KNIPLING (1976).

Irrigation and Agricultural Questions. Of the articles surveyed for this review almost 150 are concerned with irrigation practices and about one third of these are based on studies performed in developing countries. Their emphasis is on water and fertilizer use efficiency, developmental and morphological features and crop quality resulting from a respective irrigation regime. A breakthrough in the field of irrigation with saline waters cannot be reported. In all attempts (e.g., BINGHAM et al., 1974; SHMUELI, 1975) smaller yields resulted from higher salinity levels in the irrigation water. Proposals for scheduling irrigations are presented, e.g., by GRIMES and DICKENS (1974), SINGH (1977). Comparisons of different irrigation practices are made, e.g., by KAUSHAL and PATHAK (1977: sprinkler/border irrig.), BUCKS et al. (1974: trickle/furrow irrig.), DOSS (1974: fog/surface irrig.). KOHL and WRIGHT (1974) found only trifling changes in the microclimate due to sprinkler irrigation.

Many studies of stand microclimates attempt to deal with evapotranspiration (more than 60 titles). For rapid field measurements of evapotranspiration REICOSKY and PETERS (1977) constructed a portable chamber; HARI et al. (1975) described a field method using integrals of Δ wet/dry thermocouple readings; and HEILMAN et al. (1976) inserted remotely sensed canopy temperatures into energy-balance calculations. SHUTTLEWORTH (1976) proposes a theoretical description of vegetation-atmosphere interactions; such interrelationships are also studied by PAHLSSON (1974) and STIGTER (1977). Some titles will be cited as examples of precipitation (SCHNOCK, 1973; STIGTER, 1976) and interception studies (ROWE, 1975; BULTOT et al., 1976; RUTTER et al., 1975). The available water in soils is monitored in different ways, extending from simple irrgation/ evaporation balances (GUMBS and BYAM, 1976) to the use of satellite data (HEILMAN et al., 1977). For soil water/soil nutrient interrelationships see e.g., DUNHAM and NYE (1976), DAKSHINAMURTI (1976), SLAVOV (1976). WOOLHISER (1975) in a discussion of watersheds underlines the importance of models as aides in grasping the complexity of the environment.

7. Water Relations in Poikilohydric Plants

Algae. In unicellular algae, interest in osmoregulation continues (KAUSS, 1978). Carbohydrates (GRANT and WALSBY, 1977) or their reduction products, especially mannitol (KIRST, 1975, 1977) and glycerol (BOROWITZKA et al., 1977) play an important role, but also inorganic ions (BROWN and HELLEBUST, 1978). In *Codium* and *Valonia* the ionic relations are responsible for a constant turgor value (2.3 bar and about 1.5 bar, respectively): BISSON and GUTKNECHT, 1975. In *Valonia* K^+ is actively transported into the vacuole while Cl^- is taken along passively as the counter-ion (HASTINGS and GUTKNECHT, 1976); in *Codium*, on the other hand, a Cl^- pump brings about an active chloride transport (BISSON and GUTKNECHT, 1977). Algal turgor regulation also depends on changes in cell wall permeability and elasticity (RABINOWITCH et al., 1975; ZIMMERMANN et al., 1976). SCHOBERT (1974, 1977b,c) investigates the proline accumulation resulting from water stress in diatoms. Desiccation effects on cyanophycean crusts are described by BROCK (1975a: *Microcoleus*) and RODGERS (1977: *Nostoc*). Akinetes of *Anabaena cylindrica* are more drought-tolerant than vegetative cells (YAMAMOTO, 1975). Habitat zonation in *Fucales* as determined by differences in water relations is discussed by DORGELO (1976) and STROMGREN (1976). MOEBUS et al. (1974) observed a loss of organic carbon in these intertidal brown algae after rehydration.

Fungi. The osmotic relations of *Saccharomyces cerevisiae* are dealt with e.g., by ROSE (1975) and ARNOLD and LACY (1977). Several papers analyze survival and viability of parasitic fungi propagules grown on mediums having different water potentials. COTTER (1977) investigated the effect of osmotic potential on spore germination in the slime mold *Dictyostelium*. ORTH (1976) and MERT and DIZBAY

(1977) studied the limits of substrate water activity for conidiospore germination in *Aspergillus* species. Some interest exists in the water relations of root symbiontic organisms and their effects on symbiosis. Thus, optimal soil moisture conditions for the activity of root nodules have been investigated for soybean by PANKHURST and SPRENT (1975, 1976) and TU and HIETKAMP (1977), for *Vigna unguiculata* and *Dolichos lablab* by HABISH and MAHDI (1976), for pea by MINCHIN and PATE (1975). HUANG et al. (1975) show experimentally that soil flooding as well as desiccation below -19 bar restrict N_2-fixation in soybean. MEJSTRIK (1976) and BANNISTER and NORTON (1974) analyze the productivity of mycorhizal plants in peat bogs and heathlands, respectively.

Lichens. The importance of poikilohydrous life rhythm for lichens and bryophytes is emphasized by BERNER (1976). According to ARMSTRONG (1976) the wetting frequency necessary for optimal radial growth in saxicolous lichens is species-dependent. FARRAR and SMITH (1976) distinguish four steps in the saturation process following rewetting of lichen thalli: the rewetting CO_2 burst, the resaturation respiration, a short period of solute loss from the thallus, and the final recovery of normal metabolism. Continuous water saturation is noxious for the thalli (FARRAR, 1976). Under water stress, on the other hand, lichens can positively photosynthesize up to -56 to -307 bar water potential (the separated alga only between -7 and -146 bar): BROCK (1975b). LARSON (1977) describes a nondestructive method for measurements of lichen moisture content in situ. Metabolic processes which depend on the water status of the lichens are investigated by KERSHAW (1974: nitrogenase activity) and FEIGE (1975: photosynthesis in the marine *Lichina pygmaea*). LARSON and KERSHAW (1976) show that evaporation depends on the lichen morphology, and KERSHAW and FIELD (1975) report steep gradients in temperature and humidity in a *Cladonia alpestris* mat.

Bryophytes. Physiological processes which occur during desiccation in mosses are investigated very intensively. In the commonly studied *Tortula ruralis* in particular the fate of polysomes and the protein synthesis are followed (e.g., DHINDSA and BEWLEY, 1976; TUCKER et al., 1975). Whereas in this xerophytic species polysomes remain stable during desiccation, an irreversible loss of polysomes on desiccation occurs in the aquatic moss *Hygrohypnum luridum* (BEWLEY, 1974). Similar results report also KROCHKO et al. (1978: *Cratoneuron filicinum*). NOERR (1974) and DILKS and PROCTOR (1976a) emphasize seasonal differences in drought tolerance in mosses. Gas exchange of mosses under different water regimes is measured by DILKS and PROCTOR (1974, 1976b), PETERSON and MAYO (1975), BUSBY and WHITFIELD (1978). GUPTA (1977) compares the dependence of photosynthesis on water content in two liverworts, the drought-resistant *Porella platyphylla*, and the drought-sensitive *Scapania undulata*.

Poikilohydrous Cormophytes. GAFF (1977) enlarges the number of desiccation-tolerant vascular plants known from South Africa by 36 species (see also GAFF and CHURCHILL, 1976: *Borya* in Australia). GRUPCHE and GRUPCHE (1976) speculate that already in tertiary times the poikilohydric Balcanic *Ramondia* species (Gesneriaceae) may have evolved the ability to live in anabiosis adapting their metabolism to a rock cliff habitat. *Notholaena parryi*, a xeric fern of the western Colorado desert, occupies a microhabitat protected from extreme environmental conditions by rock outcroppings (NOBEL, 1978). The explanation for the drought-tolerance in poikilohydric plants may lie in several protoplasmatic peculiarities (GENKEL and LEVINA, 1975: for thallophytic species) and in structurally modified mitochondria and chloroplasts (GAFF et al., 1976; WELLBURN and WELLBURN, 1976). The water imbibition in skeletons of *Anastatica hierochuntica* has been studied by FRIEDMAN et al. (1978).

Acknowledgement. We gratefully acknowledge the great help of Mrs. LAURA TENHUNEN in preparing the English version of the manuscript.

References

ACKERSON, R.C., KRIEG, D.R., HARING, C.L., CHANG, N.: Crop. Sci. 17, 81-84 (1977). - ACOCK, B.: Aust. J. Plant Physiol. 2, 253-264 (1975). - AHMAD, I., WAINWRIGHT, S.J.: New Phytol. 79, 605-612 (1977). - ALBERTE, R.S., FISCUS, E.L., NAYLOR, A.W.: Plant Physiol. 55, 317-321 (1975). - ALBERTE, R.S., THORNBER, J.P., FISCUS, E.L.: Plant Physiol. 59, 351-353 (1977). - ALEKSEENKO, L.N.: Bot. Zh. 60, 1740-1749 (1975). - ALEKSEEVA, V.Y., VELIKANOV, G.A., GORDON, L.K., BICHURINA, A.A.: Fiziol. Rast. 24, 496-499 (1977). - ALI-ZADE, M.A., ABDULLAEV, F.I.: Izv. Akad. Nauk Az. SSR Biol. Nauki 3, 33-37 (1974). - ALLEN, L.H.Jr., BOOTE, K.J., HAMMOND, L.C.: Soil Crop Sci. Soc. Fla. Proc. 35, 42-46 (1976). - ANDERSON, J.E., KREITH, F.: Plant Soil 49, 161-173 (1978). - ANDREWS, P., COLLINS, W.J., STERN, W.R.: Aust. J. Agric. Res. 28, 301-308 (1977). - ARAD, S., RICHMOND, A.E.: Plant Physiol. 57, 656-658 (1976). - ARMSTRONG, R.A.: New Phytol. 77, 719-724 (1976). - ARNOLD, W.N., LACY, J.S.: J. Bacteriol. 131, 564-571 (1977). - ASHENDEN, T.W., STEWART, W.S., WILLIAMS, W.: J. Ecol. 63, 97-108 (1975). - ASTON, M.J.: Aust. J. Plant Physiol. 3, 489-502 (1976); - Aust. J. Plant Physiol. 5, 211-218 (1978).

BALLARD, L.A.T.: Aust. J. Plant Physiol. 3, 465-470 (1976). - BALLARD, L.A.T., NELSON, S.O., BUCHWALD, T., STETSON, L.E.: Seed Sci. Technol. 4, 257-274 (1976). - BANNISTER, P., NORTON, W.M.: New Phytol. 73, 81-89 (1974). - BARBOUR, M.G., DIAZ, D.V., BREIDENBACH, R.W.: Ecology 55, 1199-1215 (1974). - BARON, F.J.: Am. J. Bot. 65, 804-810 (1978). - BARTHLOTT, W., CAPESIUS, I.: Z. Pflanzenphysiol. 72, 443-455 (1974). - BASKIN, C.C., BASKIN, J.M.: Oecologia 17, 11-16 (1974). - BASKIN, J.M., BASKIN, C.C.: J. Chem. Ecol. 2, 441-447 (1976). - BASSIRI, A., SIONIT, N.: Physiol. Plant. 34, 226-229 (1975). - BASSIRI, A., KHOSH-KHUI, M., ROUHANI, I.: J. Agric. Sci. 88, 95-100 (1977). - BATANOUNY, K.H.: Flora (Jena) 163, 1-6 (1974). - BAUGHN, J.W., TANNER, C.B.: Crop Sci. 16, 184-190 (1976a); - Crop Sci. 16, 181-184 (1976b). - BAVEL, C.H.M. VAN, AHMED, J.: Ecol. Mod. 2, 189-212 (1976). - BEADLE, C.L., JARVIS, P.G.: Physiol. Plant. 41, 7-13 (1977). - BEADLE, C.L., TURNER, N.C., JARVIS, P.G.: Physiol. Plant. 43, 160-165 (1978). - BEARDSELL, M.F., COHEN, D.: Plant Physiol. 56, 207-212 (1975). - BERNER, L.: Rev. Bryol. 42, 857-865 (1976). - BEWLEY, J.D.: Can. J. Bot. 52, 423-427 (1974). - BIERHUIZEN, J.F.: Irrigation and water use efficiency, 421-431. In: Water and Plant Life, eds. O.L. LANGE, L. KAPPEN, E.-D. SCHULZE, Ecological Studies, Vol. 19. Berlin-Heidelberg-New York: Springer 1976. - BINGHAM, F.T., MAHLER, R.J., PARRA, J., STOLZY, L.H.: Soil Sci. 117, 369-377 (1974). - BISSON, M.A., GUTKNECHT, J.: J. Membr. Biol. 24, 183-200 (1975); - J. Membr. Biol. 37, 85-98 (1977). - BLEKHMAN, G.I.: Fiziol. Rast. 24, 507-512 (1977). - BLOOM, A.J., TROUGHTON, J.H.: Oecologia 38, 35-43 (1979). - BLUM, A., EBERCON, A.: Crop Sci. 16, 428-431 (1976). - BOGGESS, S.F., STEWART, C.R., ASPINALL, D., PALEG, L.G.: Plant Physiol. 58, 398-401 (1976). - BOKHARI, U.G., SINGH, J.S., SMITH, F.M.: J. Appl. Ecol. 12, 153-164 (1975). - BOLE, J.B.: Plant Soil 46, 297-307 (1977). - BOROWITZKA, L.J., KESSLY, D.S., BROWN, A.D.: Arch. Microbiol. 113, 131-138 (1977). - BOYER, J.S.: Planta 117, 187-207 (1974); - Philos. Trans. R. Soc. London 273, 501-512 (1976). - BRANDLE, J.R., HINCKLEY, T.M., BROWN, G.N.: Physiol. Plant. 40, 1-5 (1977). - BRANSON, F.A., MILLER, R.F., MCQUEEN, I.S.: Ecology 57, 1104-1124 (1976). - BRAUN, H.J.: Allg. Forst Jagdztg. 147, 163-168 (1976); - Z. Pflanzenphysiol. 84, 459-462 (1977). - BRAVDO, B.-A.: Physiol. Plant. 41, 36-41 (1977). - BREMAN, H., CISSE, A.M.: Oecologia 28, 301-315 (1977). - BROCK, T.D.: J. Phycol. 11, 316-320 (1975a); - Planta 124, 13-24 (1975b). - BROGÅRDH, T.: Physiol. Plant. 35, 303-309 (1975). - BROGÅRDH, T., JOHNSSON, A.: Physiol. Plant. 35,

115-125 (1975). - BROWN, G.N., BIXBY, J.A., MELCAREK, P.K., HINCKLEY, T.M., ROGERS, R.: Cryobiology 14, 94-99 (1977). - BROWN, L.M., HELLE-BUST, J.A.: Can. J. Bot. 56, 408-412 (1978). - BROWN, R.W.: Agron. J. 68, 432-434 (1976). - BROWN, R.W., MC DONOUGH, W.T.: Plant Soil 48, 5-10 (1977). - BRUN, L.J., WORCESTER, B.K.: Agron. J. 67, 586-589 (1975). - BUCKS, D.A., ERIE, L.J., FRENCH, O.F.: Agron. J. 66, 53-57 (1974). - BULTOT, F., DUPRIEZ, G.L., BODEUX, A.: J. Hydrol. 31, 381-392 (1976). - BUNCE, J.A.: J. Exp. Bot. 28, 156-161 (1977a); - Plant Physiol. 59, 348-350 (1977b). - BUNCE, J.A., MILLER, L.N.: Can. J. Bot. 54, 2457-2464 (1976). - BUSBY, J.R., WHITFIELD, D.W.A.: Can. J. Bot. 56, 1551-1558 (1978). - BUSSCHER, W.J., FRITTON, D.D.: Soil Sci. 125, 1-6 (1978). - BYLINSKA, E.: Monogr. Bot. 50, 5-59 (1975). - BYRNE, G.F., BEGG, J.E., HANSEN, G.K.: Agric. Meteorol. 18, 21-25 (1977).

CALDWELL, M.M.: Root extension and water absorption, 63-85. In: Water and Plant Life, eds. O.L. LANGE, L. KAPPEN, E.-D. SCHULZE, Ecological Studies, Vol. 19. Berlin-Heidelberg-New York: Springer 1976. - CAMACHO-B, S.E., HALL, A.E., KAUFMANN, M.R.: Plant Physiol. 54, 169-172 (1974). - CAMPBELL, G.S.: An Introduction to Environmental Biophysics. 159 p. New York-Heidelberg-Berlin: Springer 1977. - CAMPBELL, G.S., HARRIS, G.A.: Ecology 58, 652-659 (1977). - CAPESIUS, I., BARTHLOTT, W.: Z. Pflanzenphysiol. 75, 436-448 (1975). - ČERMÁK, J., KUČERA, J., PENKA, M.: Biol. Plant. 18, 105-110 (1976a). - ČERMÁK, J., PALAT, M., PENKA, M.: Biol. Plant. 18, 111-118 (1976b). - CHATTERTON, N.J., HANNA, W.W., POWELL, J.B., LEE, D.R.: Can. J. Plant Sci. 55, 641-644 (1975). - CHEN, P.M., LI, P.H.: Plant Physiol. 59, 240-243 (1977). - CHEUNG, Y.N.S., TYREE, M.T., DAINTY, J.: Can. J. Bot. 53, 1342-1346 (1975); -Can. J. Bot. 54, 758-765 (1976). - CHU, T.M., ASPINALL, D., PALEG, L.G.: Aust. J. Plant Physiol. 3, 503-512 (1976). - CHUNG, H.-H., KRAMER, P.J.: Can. J. For. Res. 5, 229-235 (1975). - CLARKSON, N.M., RUSSELL, J.S.: Aust. J. Agric. Res. 27, 227-234 (1976). - CLOUGH, B.F., MILTHORPE, F.L.: Aust. J. Plant Physiol. 2, 291-300 (1975). - COLLATZ, J., FERRAR, P.J., SLATYER, R.O.: Oecologia 23, 95-105 (1976). - COLLINS, J.C., KERRIGAN, A.P.: New Phytol. 73, 309-314 (1974). - COLLIS-GEORGE, N., MELVILLE, M.D.: Aust. J. Soil Res. 13, 141-158 (1975). - COTTER, D.A.: Can. J. Microbiol. 23, 1170-1177 (1977). - COWAN, I.R.: Adv. Bot. Res. 4, 117-228 (1977); - Water use in higher plants, 71-107. In: Water: Planets, Plants and People, ed. A.K. MC INTYRE. Canberra: Australian Academy of Science 1978. - COWAN, I.R., FARQUHAR, G.D.: Stomatal function in relation to leaf metabolism and environment, 471-505. In: Integration of Activity in the Higher Plant, ed. D.H. JENNINGS, Society for Experimental Biology Symposium Nr. XXXI. Cambridge: University Press 1977. - CRAFTS, A.S.: Fiziol. Rast. 8, 563-573 (1976). - CRAM, W.J.: Negative feedback regulation of transport in cells. The maintenance of turgor, volume and nutrient supply, 284-316. In: Encyclopedia of Plant Physiology, New Series, Vol. II, Part A, eds. U. LÜTTGE, M.G. PITMAN. Berlin-Heidelberg-New York: Springer 1976. - CRAWFORD, R.M.M.: New Phytol. 79, 511-518 (1977). - CREVECOEUR, M., DELTOUR, R., BRONCHART, R.: Planta 132, 31-41 (1976). - CUTLER, J.M., RAINS, D.W., LOOMIS, R.S.: Physiol. Plant. 40, 255-260 (1977).

DA, S., HUBAC, C., VARTANIAN, H.: Can. J. Bot. 55, 1236-1245 (1977). DAINTY, J.: Water relations of plant cells, 12-35. In: Encyclopedia of Plant Physiology, New Series, Vol. II, Part A, eds. U. LÜTTGE, M.G. PITMAN. Berlin-Heidelberg-New York: Springer 1976. - DAKSHINA-MURTI, C.: J. Nucl. Agric. Biol. 5, 10-14 (1976). - DALTON, F.N., RAATS, P.A.C., GARDNER, W.R.: Agron. J. 67, 334-339 (1975). - DAMPTEY, H.B., ASPINALL, D.: Ann. Bot. (London) 40, 23-35 (1976). - DAVENPORT, D.C., URIU, K., HAGAN, R.M.: J. Exp. Bot. 25, 410-419 (1974). -

DAVIES, W.J.: Crop Sci. 17, 735-740 (1977); - J. Exp. Bot. 29, 175-182 (1978). - DAVIES, W.J., KOZLOWSKI, T.T.: For. Sci. 21, 191-195 (1975); - Plant Soil 46, 435-444 (1977). - DAVIES, W.J., GILL, K., HALLIDAY, G.: Ann. Bot. (London) 42, 1149-1154 (1978). - DAVIS, S.D., VAN BAVEL, C.H.M., Mc CREE, K.J.: Crop Sci. 17, 640-645 (1977). - DAVY, A.J., TAYLOR, K.: J. Ecol. 62, 367-378 (1974). - DAYANANDAN, P., KAUFMAN, P.B.: Am. J. Bot. 62, 221-231 (1975). - DELL'AQUILA, A., SAVINO, G., DE LEO, P.: Plant Cell Physiol. 19, 349-354 (1978). - DHINDSA, R.S., BEWLEY, J.D.: Science 191, 181-182 (1976); - Proc. Natl. Acad. Sci. U.S.A. 75, 842-846 (1978). - DHINDSA, R.S., CLELAND, R.E.: Plant Physiol. 55, 782-785 (1975). - DILKS, T.J.K., PROCTOR, M.C.F.: J. Bryol. 8, 97-116 (1974); - J. Bryol. 9, 239-247 (1976a); - J. Bryol. 9, 249-264 (1976b). - DITTRICH, P., RASCHKE, K.: Planta 134, 77-81 (1977a); - Planta 134, 83-90 (1977b). - DITTRICH, P., MAYER, M., MEUSEL, M.: Planta 144, 305-309 (1979). - DIXON, M.A., THOMPSON, R.G., FENSOM, D.S.: Can. J. For. Res. 8, 73-80 (1978). - DOLINER, L.H., JOLLIFFE, P.A.: Oecologica 38, 23-34 (1979). - DORGELO, J.: Hydrobiol. Bull. 10, 115-122 (1976). - DOSS, B.D.: Agron. J. 66, 105-107 (1974). - DUBBE, D.R., FARQUHAR, G.D., RASCHKE, K.: Plant Physiol. 62, 413-417 (1978). - DUBE, P.A., STEVENSON, K.R., THURTELL, G.W., NEUMANN, H.H.: Can. J. Plant Sci. 55, 941-948 (1975). - DÜRING, H.: Angew. Bot. 50, 61-70 (1976). - DUMBROFF, E.B., BREWER, W.L.J.: Z. Pflanzenphysiol. 81, 167-172 (1977). - DUNHAM, R.J., NYE, P.H.: J. Appl. Ecol. 13, 967-984 (1976).

EDWARDS, M.: Plant Physiol. 58, 237-239 (1976). - EDWARDS, M., MEIDNER, H.: J. Exp. Bot. 28, 669-677 (1977); - J. Exp. Bot. 29, 771-780 (1978). - EDWARDS, M., MEIDNER, H., SHERIFF, D.W.: J. Exp. Bot. 27, 163-171 (1976). - EHLERINGER, J.R., MILLER, P.C.: Ecology 56, 370-380 (1975). - EL AOUNI, M.H.: Photosynthetica 10, 403-410 (1976). - ELIAS, P.: Biológia 30, 771-779 (1975).

FAIZ, S.M.A., WEATHERLEY, P.E.: New Phytol. 78, 337-347 (1977). - FARQUHAR, G.D.: Aust. J. Plant Physiol. 5, 787-800 (1978). - FARQUHAR, G.D., RASCHKE, K.: Plant Physiol. 61, 1000-1005 (1978). - FARRAR, J.F.: New Phytol. 77, 93-103 (1976). - FARRAR, J.F., SMITH, D.C.: New Phytol. 77, 115-125 (1976). - FEDDES, R.A., KOWALIK, P., KOLINSKA-MALINKA, K., ZARADNY, H.: J. Hydrol. 31, 13-26 (1976). - FEIGE, G.B.: Z. Pflanzenphysiol. 77, 1-15 (1975). - FELLOWS, R.J., BOYER, J.S.: Planta 132, 229-239 (1976). - FENTON, R., MANSFIELD, T.A., WELLBURN, A.R.: J. Exp. Bot. 27, 1206-1214 (1976). - FERERES, E., ACEVEDO, E., HENDERSON, D.W., HSIAO, T.C.: Physiol. Plant. 44, 261-267 (1978). - FERRIER, J.M., DAINTY, J.: Can. J. Bot. 55, 858-866 (1977); - Can. J. Bot. 56, 22-26 (1978). - FISCHER, R.A., TURNER, N.C.: Annu. Rev. Plant Physiol. 29, 277-317 (1978). - FISCUS, E.L.: Plant Physiol. 55, 917-922 (1975); - J. Exp. Bot. 28, 71-77 (1977). - FLORINETH, F.: Oecol. Plant. 9, 295-314 (1974). - FREEMAN, T.P., DUYSEN, M.E.: Protoplasma 83, 131-146 (1975). - FRIEDMAN, J. GUNDERMAN, N., ELLIS, M.: Oecologia 32, 289-301 (1978). - FROMMHOLD, I.: Biol. Zentralbl. 94, 63-73 (1975).

GAFF, D.F.: Oecologia 31, 95-109 (1977). - GAFF, D.F., CHURCHILL, D.M.: Aust. J. Bot. 24, 209-224 (1976). - GAFF, D.F., ZEE, S.-Y., O'BRIEN, T.P.: Aust. J. Bot. 24, 225-236 (1976). - GARRETT, P.W., ZAHNER, R.: Ecology 54, 1328-1334 (1973). - GENEROZOVA, I.P.: Fiziol. Rast. 23, 921-927 (1976). - GENKEL, P.A., LEVINA, V.V.: Fiziol. Rast. 22, 583-586 (1975). - GENKEL, P.A., SATAROVA, N.A., BLEKHMAN, G.I., TVORUS, E.K.: Fiziol. Rast. 21, 113-120 (1974). - GILL, C.J.: Flora (Jena) 164, 85-98 (1975). - GINSBURG, H., GINZBURG, B.Z.: J. Exp. Bot. 25, 28-35 (1974). - GIORDANO, R., SALLEO, A., SALLEO, S., WANDERLINGH, F.: Can. J. Bot. 56, 333-338 (1978). - GLINKA, Z.: Plant Physiol. 59, 933-935 (1977). - GLOSER, J.: Prirodoved. Pr. Ustavu Cesk. Akad. Ved. Brne

11, 1-36 (1977). - GRACE, J.: Plant Response to Wind. 204 p. Experi-
mental Botany, Vol. 13. London-New York: Academic Press 1977. -
GRANIER, A.: Ann. Sci. For. 34, 17-46 (1977). - GRANT, N.G., WALSBY,
A.E.: J. Exp. Bot. 28, 409-415 (1977). - GREENWAY, H.: Aust. J. Plant
Physiol. 1, 247-257 (1974). - GRIME, J.P., CURTIS, A.V.: J. Ecol. 64,
975-988 (1976). - GRIMES, D.W., DICKENS, W.L.: Agron. J. 66, 403-404
(1974). - GRUPCHE, L., GRUPCHE, R.: Bot. Zh. 61, 1454-1458 (1976). -
GULYAEV, B.I.: Fiziol. Biokhim. Kult. Rast. 9, 520-526 (1977). -
GUMBS, F.A., BYAM, L.: Trop. Agric. 53, 31-40 (1976). - GUPTA, R.K.:
Aust. J. Bot. 25, 363-365 (1977).

HAAS, N.F.: Z. Pflanzenphysiol. 75, 427-435 (1975). - HABISH, H.A.,
MAHDI, A.A.: J. Agric. Sci. 86, 553-560 (1976). - HALL, A.E., CAMACHO-
B, S.E., KAUFMANN, M.R.: Physiol. Plant. 33, 62-65 (1975). - HALL,
A.E., SCHULZE, E.-D., LANGE, O.L.: Current perspectives of steady-
state stomatal responses to environment, 169-188. In: Water and Plant
Life, eds. O.L. LANGE, L. KAPPEN, E.-D. SCHULZE, Ecological Studies,
Vol. 19. Berlin-Heidelberg-New York: Springer 1976. - HALVORSON, W.L.,
PATTEN, D.T.: Ecology 55, 173-177 (1974). - HAMMEL, H.T., SCHOLANDER,
P.F.: Osmosis and Tensile Solvent. 133 p. Berlin-Heidelberg-New York:
Springer 1976. - HANOWER, P., BRZOZOWSKA, J.: Phytochemistry 14, 1691-
1694 (1975). - HANSCOM, Z.III, TING, I.P.: Plant Physiol. 61, 327-330
(1978). - HANSON, A.D., NELSEN, C.E.: Plant Physiol. 62, 305-312
(1978). - HARI, P., SMOLANDER, H., LUUKKANEN, O.: J. Exp. Bot. 26,
675-678 (1975). - HASTINGS, D.F., GUTKNECHT, J.: J. Membr. Biol. 28,
263-275 (1976). - HEGARTY, T.W.: J. Exp. Bot. 26, 203-211 (1975). -
HEGARTY, T.W., ROSS, H.A.: Ann. Bot. (London) 42, 1223-1226 (1978). -
HEILMAN, J.L., KANEMASU, E.T., ROSENBERG, N.J., BLAD, B.L.: Remote
Sensing Environ. 5, 137-145 (1976). - HEILMAN, J.L., KANEMASU, E.T.,
BAGLEY, J.O., RASMUSSEN, V.P.: Remote Sensing Environ. 6, 315-326
(1977). - HELLKVIST, J., PARSBY, J.: Physiol. Plant. 41, 211-216
(1977). - HERKELRATH, W.N., MILLER, E.E., GARDNER, W.R.: Soil Sci.
Soc. Am. J. 41, 1033-1038 (1977a); - Soil Sci. Soc. Am. J. 41, 1039-
1043 (1977b). - HILLEL, D., TALPAZ, H.: Soil Sci. 121, 307-312 (1976).
- HILLEL, D., VAN BEEK, C.G.E.M., TALPAZ, H.: Soil Sci. 120, 385-399
(1975). - HINCKLEY, T.M., BRUCKERHOFF, D.N.: Can. J. Bot. 53, 62-72
(1975). - HINCKLEY, T.M., LASSOIE, J.P., RUNNING, S.W.: Temporal and
Spatial Variations in the Water Status of Forest Trees. 72 p. Forest
Science Monograph 20. Washington, D.C.: Society of American Foresters
1978. - HOFFMAN, G.J., HALL, A.E.: Agron. J. 68, 872-875 (1976). -
HOGON, M.W., CHAN, C.W.: Aust. J. Exp. Agric. Anim. Husb. 17, 86-89
(1977). - HONG, S.G., SUCOFF, E.: Plant Physiol. 57, 230-236 (1976).
- HOOK, D.D., CRAWFORD, R.M.M. (eds.): Plant Life in Anaerobic Environ-
ments. 564 p. Ann Arbor: Ann Arbor Science 1978. - HSIAO, T.C.: Stoma-
tal ion transport, 195-221. In: Encyclopedia of Plant Physiology,
New Series, Vol. II, Part B, eds. U. LÜTTGE, M.G. PITMAN. Berlin-
Heidelberg-New York: Springer 1976. - HSIAO, T.C., ACEVEDO, E.,
FERERES, E., HENDERSON, D.W.: Philos. Trans. R. Soc. London 273, 479-
500 (1976a). - HSIAO, T.C., FERERES, E., ACEVEDO, E., HENDERSON, D.W.:
Water stress and dynamics of growth and yield of crop plants, 281-305.
In: Water and Plant Life, eds. O.L. LANGE, L. KAPPEN, E.-D. SCHULZE,
Ecological Studies, Vol. 19. Berlin-Heidelberg-New York: Springer
1976b. - HUANG, C.-Y., BOYER, J.S., VANDERHOEF, L.N.: Plant Physiol.
56, 222-227 (1975). - HUCK, M.G., KLEPPER, B.: Agron. J. 69, 593-597
(1977). - HÜSKEN, D., STEUDLE, E., ZIMMERMANN, U.: Plant Physiol. 61,
158-163 (1978).

IKE, I.F., THURTELL, G.W., STEVENSON, K.R.: Can. J. Bot. 56, 1638-1641
(1978). - IMBAMBA, S.K., TIESZEN, L.L.: Physiol. Plant. 39, 311-316
(1977). - ITAI, C., BENZIONI, A.: Water stress and hormonal response,
225-242. In: Water and Plant Life, eds. O.L. LANGE, L. KAPPEN, E.-D.

SCHULZE. Ecological Studies, Vol. 19. Berlin-Heidelberg-New York: Springer 1976. - ITAI, C., MEIDNER, H.: J. Exp. Bot. 29, 765-770 (1978). - ITAI, C., WEYERS, J.D.B., HILLMAN, J.R., MEIDNER, H., WILLMER, C.: Nature (London) 271, 652-653 (1978).

JACKSON, M.B., CAMPBELL, D.J.: New Phytol. 74, 397-406 (1975); - New Phytol. 76, 21-29 (1976). - JACKSON, M.B., GALES, K., CAMPBELL, D.J.: J. Exp. Bot. 29, 183-194 (1978). - JAEGER, H.-J., MEYER, H.R.: Oecologia 30, 83- 96 (1977). - JARVIS, P.G.: Water transfer in plants, 369-394. In: Heat and Mass Transfer in the Environment of Vegetation, ed. D.A. DE VRIES. Washington D.C.: Scripta 1975; - Philos. Trans. R. Soc. London 273, 593-610 (1976). - JARVIS, P.G., JAMES, G.B., LANDSBERG, J.J.: Coniferous forest, 171- 240. In: Vegetation and the Atmosphere, ed. J.L. MONTEITH, Vol. 2. London - New York: Academic Press 1976. - JENSEN, C.R.: Acta Agric. Scand. 26, 196-202 (1976). - JOHNSON, D.A., BROWN, R.W.: Crop Sci. 17, 507-510 (1977). - JOHNSON, D.A., CALDWELL, M.M.: Oecologia 21, 93-108 (1975); - Physiol. Plant. 36, 271-278 (1976). - JOHNSON, R.P.C.: Planta 136, 187-194 (1977). - JOHNSON, R.R., MOSS, D.N.: Crop Sci. 16, 697-701 (1976). - JOHNSON, R.R., WILLMER, C.M., MOSS, D.N.: Crop Sci. 15, 217-221 (1975). - JONES, M.M., TURNER, N.C.: Plant Physiol. 61, 122-126 (1978). - JONES, R.: J. Ecol. 63, 109-116 (1975a); - J. Ecol. 63, 859-866 (1975b). - JORDAN, C.F., KLINE, J.R.: J. Appl. Ecol. 14, 853-860 (1977). - JORDAN, W.R., BROWN, K.W., THOMAS, J.C.: Plant Physiol. 56, 595-599 (1975).

KAPPEN, L., OERTLI, J.J., LANGE, O.L., SCHULZE, E.-D., EVENARI, M., BUSCHBOM, U.: Oecologia 21, 175-192 (1975). - KAPPEN, L., LANGE, O.L., SCHULZE, E.-D., EVENARI, M., BUSCHBOM, U.: Oecologia 23, 323-334 (1976). - KASSAM, A.H., ELSTON, J.F.: Ann. Bot. (London) 40, 669-679 (1976). - KASSAM, A.H., KOWAL, J.M.: Agric. Meteorol. 15, 333-342 (1975). - KAUFMANN, M.R., ECKARD, A.N.: For. Sci. 23, 27-33 (1977). - KAUSHAL, M.P., PATHAK, B.S.: Indian J. Agric. Sci. 47, 240-244 (1977). - KAUSS, H.: Biochemistry of osmotic regulation, 119-140. In: Plant Biochemistry II, ed. D.H. NORTHCOTE, International Review of Biochemistry, Vol. 13. Baltimore: University Park Press 1977; - Osmotic regulation in algae. Prog. Phytochem. 5, 1-27 (1978). - KAWASE, M.: Physiol. Plant. 31, 29-38 (1974); - Physiol. Plant. 36, 236-241 (1976). - KECK, R.W., BOYER, J.S.: Plant Physiol. 53, 474-479 (1974). - KENNEDY, R.A.: Z. Pflanzenphysiol. 83, 11-24 (1977). - KERSHAW, K.A.: Can. J. Bot. 52, 1423-1427 (1974). - KERSHAW, K.A., FIELD, G.F.: Can. J. Bot. 53, 2614-2620 (1975). - KIRK, C.A. VAN, RASCHKE, K.: Plant Physiol. 61, 474-475 (1978a); - Plant Physiol. 61, 361-364 (1978b). - KIRST, G.O.: Z. Pflanzenphysiol. 76, 316-325 (1975); - Planta 135, 69-75 (1977). - KLUGE, M.: Oecologia 29, 77-83 (1977). - KLUGE, M., TING, I.P.: Crassulacean Acid Metabolism, 209 p. Ecological Studies, Vol. 30. Berlin-Heidelberg-New York: Springer 1978. - KOHL, R.A., KOLAR, J.J.: Agron. J. 68, 536-538 (1976). - KOHL, R.A., WRIGHT, J.L.: Agron. J. 66, 85-88 (1974). - KONDO, N., SUGAHARA, K.: Plant Cell Physiol. 19, 365-373 (1978). - KOZLOWSKI, T.T. (ed.): Water Deficits and Plant Growth. IV. Soil Water Measurement, Plant Response, and Breeding for Drought Resistance. 383 p. London - New York: Academic Press 1976. - KRAUSE, D., KUMMEROW, J.: Oecol. Plant. 12, 133-148 (1977). - KREEB, K.: Methoden der Pflanzenökologie. 235 p. Jena: Fischer 1977. - KRIEDEMANN, P.E., LOVEYS, B.R., DOWNTON, W.J.S.: Aust. J. Plant Physiol. 2, 553-567 (1975). - KROCHKO, J.E., BEWLEY, J.D., PACEY, J.: J. Exp. Bot. 29, 905-917 (1978). - KUČERA, J., ČERMÁK, J., PENKA, M.: Biol. Plant. 19, 413-420 (1977). - KURKOVA, E.B., MOTORINA, M.V.: Fiziol. Rast. 21, 40-44 (1974). - KYRIAKOPOULOS, E., LARCHER, W.: Z. Pflanzenphysiol. 77, 268-271 (1976). - KYRIAKOPOULOS, E., RICHTER, H.: Z. Pflanzenphysiol. 82, 14-27 (1977).

LAMBERS, H.: Physiol. Plant. 37, 117-122 (1976). - LAMONT, B.: Aust.
J. Bot. 24, 691-702 (1976). - LANCASTER, J.E., MANN, J.D., PORTER,
N.G.: J. Exp. Bot. 28, 184-191 (1977). - LANDSBERG, J.J., BLANCHARD,
T.W., WARRIT, B.: J. Exp. Bot. 27, 579-596 (1976). - LANGE, O.L.,
MEDINA, E.: Oecologia 40, 357-363 (1979). - LANGE, O.L., ZUBER, M.:
Oecologia 31, 67-72 (1977). - LANGE, O.L., SCHULZE, E.-D., KAPPEN,
L., BUSCHBOM, U., EVENARI, M.: Photosynthesis of desert plants as
influenced by internal and external factors, 121-143. In: Perspectives
of Biophysical Ecology, eds. D.M. GATES, R.B. SCHMERL. Ecological
Studies, Vol. 12. New York-Heidelberg-Berlin: Springer 1975a; -
Adaptations of desert lichens to drought and extreme temperatures,
20-37. In: Environmental Physiology of Desert Organisms, ed. N.F.
HADLEY. Stroudsburg, Pennsylvania: Dowden, Hutchinson and Ross, Inc.
1975b. - LANGE, O.L., KAPPEN, L., SCHULZE, E.-D. (eds.): Water and
Plant Life. 536 p. Ecological Studies, Vol. 19. Berlin-Heidelberg-
New York: Springer 1976. - LARQUÉ-SAAVEDRA, A.: Physiol. Plant. 43,
126-128 (1978). - LARSON, D.W.: J. Ecol. 65, 135-146 (1977). - LARSON,
D.W., KERSHAW, K.A.: Can. J. Bot. 54, 2061-2073 (1976). - LASSOIE,
J.P., SCOTT, D.R.M., FRITSCHEN, L.J.: For. Sci. 23, 377-390 (1977). -
LAUENROTH, W.K., DODD, J.L., SIMS, P.L.: Oecologia 36, 211-222 (1978).
- LAWLOR, D.W.: Photosynthetica 10, 378-387 (1976). - LAWLOR, D.W.,
FOCK, H.: Planta 126, 247-258 (1975); - J. Exp. Bot 28, 320-328
(1977a); - J. Exp. Bot. 28, 329-337 (1977b). - LAWLOR, D.W., MILFORD,
G.F.J.: J. Exp. Bot. 26, 657-665 (1975). - LEYTON, L.: Fluid Behaviour
in Biological Systems. Oxford: Clarendon Press, Oxford University
Press 1975. - LINDSTROM, M.J., PAPENDICK, R.I., KOEHLER, F.E.: Agron.
J. 68, 137-141 (1976). - LIVSHIN, A.Z., SAMYGIN, G.A.: Fiziol. Rast.
21, 126-134 (1974). - LÖSCH, R.: Oecologia 29, 85-97 (1977); - Ber.
Dtsch. Bot. Ges. 91, 645-656 (1978); - Oecologia 39, 229-238 (1979).
- LÖSCH, R., BRESSEL, C.: Flora (Jena) 168, 109-120 (1979). - LÖSCH,
R., SCHENK, B.: J. Exp. Bot. 29, 781-787 (1978). - LOVEYS, B.R.:
Physiol. Plant. 40, 6-10 (1977). - LUDLOW, M.M.: Ecophysiology of C4
grasses, 364-386. In: Water and Plant Life, eds. O.L. LANGE, L. KAPPEN,
E.-D. SCHULZE. Ecological Studies, Vol. 19. Berlin-Heidelberg-New York:
Springer 1976. - LUDLOW, M.M., NG, T.T.: Aust. J. Plant Physiol. 4,
263-272 (1977). - LÜTTGE, U., KLUGE, M., BALL, E.: Plant Physiol. 56,
613-616 (1975). - LÜTTGE, U., BALL, E., GREENWAY, H.: Plant Physiol.
60, 521-523 (1977). - LURIE, S.: Plant Sci. Lett. 10, 219-223 (1977);
- Planta 140, 245-249 (1978).

MAIER-MAERCKER, U.: Z. Pflanzenphysiol. 91, 25-43 (1979a); - Z.
Pflanzenphysiol. 91, 157-172 (1979b); - Z. Pflanzenphysiol. 91, 225-
238 (1979c); - Z. Pflanzenphysiol. 91, 239-254 (1979d). - MALI,
P.C., MEHTA, S.L.: Phytochemistry 16, 643-646 (1977a); - Phytochem-
istry 16, 1355-1358 (1977b). - MANNING, C.E., MILLER, D.G., TEARE,
I.D.: J. Am. Soc. Hortic. Sci. 102, 756-760 (1977). - MANOHAR, M.S.:
Z. Pflanzenphysiol. 84, 227-236 (1977). - MANOHAR, M.S., MATHUR, M.K.:
Seed Res. (New Delhi) 3, 94-101 (1975). - MANSFIELD, T.A., WELLBURN,
A.R., MOREIRA, T.J.S.: Philos. Trans. R. Soc. London 284, 471-482
(1978). - MARBACH, I., MAYER, A.M.: Plant Physiol. 56, 93-96 (1975).
- MARC, J., PALMER, J.H.: Physiol. Plant. 36, 101-104 (1976). -
MARCHAND, P.J.: Rhodora 77, 53-63 (1975). - MARGARIS, N.S.: Biol.
Plant. 19, 442-447 (1977). - MATTEUCCI, S.D.: Phyton 33, 51-62 (1975).
- MAXWELL, J.O., REDMANN, R.E.: Oecologia 35, 277-284 (1978). -
MCDONOUGH, W.T.: Phyton 32, 107-112 (1974). - MCINTYRE, A.K. (ed.):
Water: Planets, Plants and People. 182 p. Canberra: Australian Academy
of Science 1978. - MCINTYRE, G.I.: Can. J. Bot. 54, 2747-2754 (1976).
- MCMICHAEL, B.L., ELMORE, C.D.: Crop Sci. 17, 905-908 (1977). -
MCNEIL, D.L.: Aust. J. Plant Physiol. 3, 311-324 (1976). - MEDERSKI,
H.J., CHEN, L.H., CURRY, R.B.: Plant Physiol. 55, 589-593 (1975). -
MEHKRI, A.A., SASHIDHAR, V.R., UDAYKUMAR, M., SASTRY, K.S.K.: Indian

J. Plant Physiol. 20, 50-55 (1977). - MEIDNER, H.: J. Exp. Bot. 26, 666-673 (1975); - J. Exp. Bot. 27, 172-174 (1976a); - J. Exp. Bot. 27, 691-694 (1976b). - MEIDNER, H., EDWARDS, M.: J. Exp. Bot. 26, 319-330 (1975). - MEIDNER, H., SHERIFF, D.W.: Water and Plants. 148 p. Glasgow - London: Blackie 1976. - MEIRI, A., PLAUT, Z., SHIMSHI, D.: Physiol. Plant. 35, 72-76 (1975). - MEJSTRIK, V.: Quaest. Geobiol. 16, 99-177 (1976). - MENSCHING, H.: Umschau 78, 99-106 (1978). - MERINO, J., NOVO, F.G., DIAZ, M.S.: Oecol. Plant. 11, 1-12 (1976). - MERT, H.H., DIZBAY, M.: Mycopathol. 61, 125-127 (1977). - MILBURN, J.A., MCLAUGHLIN, M.E.: New Phytol. 73, 861-871 (1974). - MILLAR, B.D.: J. Exp. Bot. 25, 1070-1084 (1974). - MILLAR, B.D., HANSEN, G.K.: Ann. Bot. (London) 39, 915-920 (1975). - MILLER, N.A., ASHBY, W.C.: Bot. Gaz. 139, 211-214 (1978). - MILLER, P.C., HOM, J., POOLE, D.K.: Oecol. Plant. 10, 355-367 (1975). - MINCHIN, F.R., PATE, J.S.: J. Exp. Bot. 26, 60-69 (1975). - MOEBUS, K., JOHNSON, K.M., SIEBURTH, J.M.: Mar. Biol. 26, 127-134 (1974). - MOHANTY, P., BOYER, J.S.: Plant Physiol. 57, 704-709 (1976). - MOLZ, F.J.: Water Resour. Res. 12, 805-808 (1976). - MOLZ, F.J., BOYER, J.S.: Plant Physiol. 62, 423-429 (1978). - MONTEITH, J.L. (ed.): Vegetation and the Atmosphere. 278 p. Principles, Vol. I. London - New York: Academic Press 1975; - Vegetation and the Atmosphere. 439 p. Case Studies, Vol. II. London - New York: Academic Press 1976. - MOONEY, H.A., HARRISON, A.T., MORROW, P.A.: Oecologia 19, 293-302 (1975). - MOONEY, H.A., BJÖRKMAN, O., COLLATZ, G.J.: Carnegie Inst. Yearb. 76, 328-334 (1977). - MORROW, P.A., MOONEY, H.A.: Oecologia 15, 205-222 (1974). - MORROW, L., NICKERSON, N.H.: Rhodora 75, 102-106 (1973). - MOZHAEVA, L.V., PIL'SHCHIKOVA, N.V.: Izv. Timiryazev S. Kh. Akad. 6, 3-11 (1976).

NAGARAJAH, S.: Ann. Bot. (London) 42, 1141-1147 (1978). - NASR, T.A., EL-AZAB, E.M., EL-SHURAFA, M.Y.: Sci. Hortic. 7, 225-236 (1977). - NEILSEN, R.E., JARVIS, P.G.: J. Appl. Ecol. 12, 879-891 (1975). - NELSON, C.E., SAFIR, G.R., HANSON, A.D.: Plant Physiol. 61, 131-133 (1978). - NELSON, S.D., MAYO, J.M.: Can. J. Bot. 55, 489-495 (1977). - NELSON, S.H., HWANG, K.E.: Can. J. Plant Sci. 56, 563-566 (1976). - NEUMANN, H.H., THURTELL, G.W., STEVENSON, K.R.: Can. J. Plant Sci. 54, 175-184 (1974). - NEWMAN, E.I.: Philos. Trans. R. Soc. London 273, 463-478 (1976a); - Plant Physiol. 57, 738-739 (1976b). - NNYAMAH, J.U., BLACK, T.A.: Soil Sci. Soc. Am. J. 41, 972-979 (1977). - NOBEL, P.S.: Oecologia 31, 293-309 (1978). - NOBEL, P.S., LONGSTRETH, D.J., HARTSOCK, T.L.: Physiol. Plant. 44, 97-101 (1978). - NOERR, M.: Flora (Jena) 163, 371-387 (1974). - NOLBORCZYK, E., LACROIX, L.L., HILL, R.D.: Can. J. Bot. 53, 1132-1138 (1975). - NULSEN, R.A., THURTELL, G.W., STEVENSON, K.R.: Agron. J. 69, 951-954 (1977).

OERTLI, J.J.: The states of water in the plant - theoretical consideration, 19-31. In: Ecological Studies, Vol. 19, eds. O.L. LANGE, L. KAPPEN, E.-D. SCHULZE. Berlin-Heidelberg-New York: Springer 1976a; - The soil-plant-atmosphere continuum, 32-41. In: Water and Plant Life, eds. O.L. LANGE, L. KAPPEN, E.-D. SCHULZE, Ecological Studies, Vol. 19. Berlin-Heidelberg-New York: Springer 1976b. - OERTLI, J.J., ACEVES-NAVARRO, E., STOLZY, L.H.: Soil Sci. 119, 162-166 (1975). - OGAWA, T., ISHIKAWA, H., SHIMADA, K., SHIBATA, K.: Planta 142, 61-65 (1978). - OKAFOR, L.I., DE DATTA, S.K.: Weed Sci. 24, 43-46 (1976). - OLOFINBOBA, M.O., KOZLOWSKI, T.T., MARSHALL, P.E.: Plant Soil 40, 619-635 (1974). - OOMES, M.J.M., ELBERSE, W.T.: J. Ecol. 64, 745-755 (1976). - ORTH, R.: Lebensm. Wiss. Technol. 9, 156-159 (1976). - OSMOND, C.B.: Environmental control of photosynthetic options in Crassulacean plants, 311-321. In: Environmental and Biological Control of Photosynthesis, ed. R. MARCELLE. The Hague: Junk 1975; - Annu. Rev. Plant Physiol. 29, 379-414 (1978). - OSMOND, C.B., ZIEGLER, H.: Naturw. Rundsch. 28, 323-328 (1975). - OSMOND, C.B., ZIEGLER, H., STICHLER, W.,

TRIMBORN, P.: Oecologia 18, 209-218 (1975). - OSMOND, C.B., BENDER, M.M., BURRIS, R.H.: Aust. J. Plant Physiol. 3, 787-799 (1976). - OSMOND, C.B., LUDLOW, M.M., DAVIS, R.L., COWAN, I.R., POWLES, S.B., WINTER, K.: Oecologia 41, 65-76 (1979). - O'TOOLE, J.C., OZBUN, J.L., WALLACE, D.H.: Physiol. Plant. 40, 111-114 (1977). - OUTLAW, W.H. Jr., KENNEDY, J.: Plant Physiol. 62, 648-652 (1978).

PAHLSSON, L.: Oikos 25, 176-186 (1974). - PALFI, G., KOVES, E., NEHEZ, R.: Növenytermeles 23, 219-228 (1974a). - PALFI, G., KOVES, E., BITO, M., SEBESTYEN, R.: Phyton 32, 121-128 (1974b). - PALLAS, J.E.Jr., SAMISH, Y.B., WILLMER, C.M.: Plant Physiol. 53, 907-911 (1974). - PALZKILL, D.A., TIBBITTS, T.W.: Plant Physiol. 60, 854-856 (1977). - PANKHURST, C.E., SPRENT, J.I.: J. Exp. Bot. 26, 287-304 (1975); - J. Exp. Bot. 27, 1-9 (1976). - PARRISH, D.J., LEOPOLD, A.C.: Plant Physiol. 59, 1111-1115 (1977). - PARSONS, L.R.: Plant Physiol. 62, 64-70 (1978). - PARSONS, L.R., KRAMER, P.J.: Physiol. Plant. 30, 19-23 (1974). - PATIL, B.B., RAJAT DE: Plant Physiol. 57, 941-943 (1976). - PEARSON, C.J.: Aust. J. Plant Physiol. 2, 85-89 (1975). - PEGELOW, E.J.Jr., BUXTON, D.R., BRIGGS, R.E., MURAMOTO, H., GENSLER, W.G.: Crop Sci. 17, 1-4 (1977). - PEMADASA, M.A.: Ann. Bot. (London) 41, 969-976 (1977). - PENNY, M.G., BOWLING, D.J.F.: Planta 119, 17-25 (1974); - Planta 122, 209-212 (1975). - PENNY, M.G., KELDAY, L.S., BOWLING, D.J.F.: Planta 130, 291-294 (1976). - PEREIRA, J.S., KOZLOW-SKI, T.T.: Physiol. Plant. 41, 184-192 (1977a); - Can. J. For. Res. 7, 145-153 (1977b). - PETERSON, W.L., MAYO, J.M.: Can. J. Bot. 53, 2897-2900 (1975). - PHUNG, H.T., KNIPLING, E.B.: Hortscience 11, 131-133 (1976). - PICKETT, S.T.A., BAZZAZ, F.A.: Ecology 57, 169-176 (1976). - PLAUT, Z.: Physiol. Plant. 30, 212-217 (1974). - POCHARD, E., SERIEYS, H.: Ann. Amelior. Plant. 24, 243-268 (1974). - POLJAKOFF-MAYBER, A., GALE, J. (eds.): Plants in Saline Environments. 213 p. Ecological Studies, Vol. 15. Berlin-Heidelberg-New York: Springer 1975. - PONOMARENKO, S.F.: Izd. Akad. Nauk SSR Biol. Nauki 4, 512-523 (1975). - POOLE, D.K., MILLER, P.C.: Ecology 56, 1118-1128 (1975). - POSPIŠILOVÁ, J., SOLÁROVÁ, J.: Water-in-Plants Bibliography. The Hague: Junk, Vol. 1. 88 p. 1977; - Vol. 2. 129 p. 1978; - Vol. 3. 111 p. 1979. - POSPIŠILOVÁ, J., ZIMA, J., ŠESTÁK, Z.: Biol. Plant. 18, 473-479 (1976). - POWELL, D.B.B., BLANCHARD, T.W.: J. Exp. Bot. 27, 597-607 (1976). - POWLES, S.B., OSMOND, C.B.: Aust. J. Plant Physiol. 5, 619-629 (1978). - PRIEHRADNY, S.: Biológia 31, 545-554 (1976). - PRIHAR, S.S., SINGH, P., GAJRI, P.R.: Agron. J. 67, 369-373 (1975). - PROKOF'EV, A.A., RYBALOVA, B.A., KATS, K.M.: Fiziol. Rast. 21, 108-112 (1974). - PUIT, E.J. DE, CALDWELL, M.M.: J. Ecol. 63, 835-858 (1975a); - Am. J. Bot. 62, 954-961 (1975b).

QUARRIE, S.A., JONES, H.G.: J. Exp. Bot. 28, 192-204 (1977). - QUINN, J.A.: Am. J. Bot. 62, 884-891 (1975).

RABINOWITCH, S., GROVER, N.B., GINZBURG, B.Z.: J. Membr. Biol. 22, 211-230 (1975). - RAGHAVENDRA, A.S., DAS, V.S.R.: J. Exp. Bot. 28, 480-483 (1977). - RAGHAVENDRA, A.S., RAO, I.M., DAS, V.S.R.: Z. Pflanzenphysiol. 80, 36-42 (1976). - RAHMAN, M.S.: J. Ecol. 64, 449-462 (1976). - RAI, V.K., BAPAT, C.M.: Geobios (Jodhpur) 4, 231-234 (1977). - RAJAGOPAL, V., BALASUBRAMANIAN, V., SINHA, S.K.: Physiol. Plant. 40, 69-71 (1977). - RAKHIMOV, G.T.: Uzb. Biol. Zh. 19, 18-19 (1975). - RAKHMANINA, A.T.: Bot. Zh. 61, 1757-1761 (1976). - RASCHKE, K.: Planta 125, 243-259 (1975); - Philos. Trans. R. Soc. London 273, 551-560 (1976); - Movements of stomata, 383-441. In: Encyclopedia of Plant Physiology, New Series, Vol. VII, eds. W. HAUPT, M.E. FEINLEIB. Berlin-Heidelberg-New York: Springer 1979. - RASCHKE, K., DITTRICH, P.: Planta 134, 69-75 (1977). - RASCHKE, K., SCHNABL, H.: Plant Physiol. 62, 84-87 (1978). - RASCHKE, K., ZEEVAART, J.A.D.: Plant

Physiol. 58, 169-174 (1976). - RASCHKE, K., PIERCE, M., POPIELA, C.C.:
Plant Physiol. 57, 115-121 (1976). - RASCHKE, K., HANEBUTH, W.F.,
FARQUHAR, G.D.: Planta 139, 73-77 (1978). - RAWSON, H.M., BEGG, J.E.,
WOODWARD, R.G.: Planta 134, 5-10 (1977). - REDMANN, R.E.: Oecologia
23, 283-295 (1976). - REGEHR, D.L., BAZZAZ, F.A., BOGGESS, W.R.:
Photosynthetica 9, 52-61 (1975). - REICOSKY, D.C., PETERS, D.B.:
Agron. J. 69, 729-732 (1977). - REICOSKY, D.C., RITCHIE, J.T.: Soil
Sci. Soc. Am. J. 40, 293-297 (1976). - RHODES, P.R., MATSUDA, K.:
Plant Physiol. 58, 631-635 (1976). - RICHARDS, D.: Ann. Bot. (London)
41, 279-281 (1977). - RICHTER, H.: Phyton 18, 29-41 (1977); - J. Exp.
Bot. 29, 1197-1203 (1978a); - J. Exp. Bot. 29, 277-280 (1978b). -
ROA, R.L., PICKARD, W.F.: J. Exp. Bot. 26, 469-475 (1975). - ROBERTS,
J.: J. Exp. Bot. 28, 751-767 (1977). - ROBERTS, S.W., KNOERR, K.R.:
Oecologia 28, 191-202 (1977); - Plant Physiol. 61, 311-313 (1978). -
RODGERS, G.A.: Plant Soil 46, 671-674 (1977). - ROSE, C.W., BYRNE,
G.F., HANSEN, G.K.: Agric. Meteorol. 16, 171-184 (1976). - ROSE, D.:
J. Appl. Bacteriol. 38, 169-176 (1975). - ROWE, L.K.: N. Z. J. For.
Sci. 5, 45-61 (1975). - ROZEMA, J., BLOM, B.: J. Ecol. 65, 213-222
(1977).-RUNDEL, P.W., RUNDEL, J.A., ZIEGLER, H., STICHLER, W.:
Oecologia 38, 45-50 (1979). - RUNECKLES, V.C., ROSEN, P.M.: Can. J.
Bot. 55, 193-197 (1977). - RUTTER, A.J., MORTON, A.J., ROBINS, P.C.:
J. Appl. Ecol. 12, 367-380 (1975). - RYCHNOVSKA, M.: Folia Geobot.
Phytotaxon. 11, 427-432 (1976). - RYDEN, B.E.: Nord. Hydrol. 7, 73-80
(1976).

SAMUILOV, F.D., NIKIFOROVA, V.I., NIKIFOROV, E.A.: Fiziol. Rast. 23,
567-572 (1976). - SANDS, R., CORRELL, R.L.: Physiol. Plant. 37, 293-
297 (1976). - SANKHLA, N., ZIEGLER, H., VYAS, O.P., STICHLER, W.,
TRIMBORN, P.: Oecologia 21, 123-129 (1975). - SANTO, A.V. DE, ALFANI,
A., DE LUCA, P.: Ann. Bot. (London) 40, 391-394 (1976). - SANTO, A.V.
DE, DE LUCA, P., ALFANI, A.: G. Bot. Ital. 111, 195-210 (1977). -
SATO, S., ASAHI, T.: Plant Physiol. 56, 816-820 (1975). - SAVINO, G.,
DELL'AQUILA, A., DE LEO, P.: Boll. Soc. Ital. Biol. Sper. 52, 1187-
1192 (1976). - SCHNABL, H.: Planta 144, 95-100 (1978). - SCHNABL, H.,
ZIEGLER, H.: Planta 136, 37-43 (1977). - SCHNOCK, G.: Oecol. Plant. 8,
17-23 (1973). - SCHOBERT, B.: Z. Pflanzenphysiol. 74, 106-120 (1974);
- J. Theor. Biol. 68, 17-26 (1977a); - Z. Pflanzenphysiol. 85, 463-
470 (1977b); - Z. Pflanzenphysiol. 85, 451-462 (1977c). - SCHRÖDER,
C.A.: Bot. Gaz. 136, 94-98 (1975). - SCHULZE, E.-D., HALL, A.E.:
Short-term and long-term effects of drought on steady-state and time-
integrated plant processes: transpiration, carbon dioxide assimila-
tion, biomass production and seed yield. In: Physiological Processes
Limiting Plant Productivity, ed. C.B. JOHNSON. Borough Green: Butter-
worth 1979. - SCHULZE, E.-D., KÜPPERS, M.: Planta 146, 319-326 (1979).
- SCHULZE, E.-D., LANGE, O.L., KAPPEN, L., BUSCHBOM, U., EVENARI, M.:
Planta 110, 29-42 (1973). - SCHULZE, E.-D., LANGE, O.L., EVENARI, M.,
KAPPEN, L., BUSCHBOM, U.: Oecologia 17, 159-170 (1974). - SCHULZE, E.-D.,
LANGE, O.L., KAPPEN, L., EVENARI, M., BUSCHBOM, U.: Oecologia 18, 219-
233 (1975a). - SCHULZE, E.-D., LANGE, O.L., EVENARI, M., KAPPEN, L.,
BUSCHBOM, U.: Oecologia 19, 303-314 (1975b). - SCHULZE, E.-D., ZIEGLER,
H., STICHLER, W.: Oecologia 24, 323-334 (1976). - SEN, D.N., CHAWAN,
D.D.: Biol. Plant. 17, 198-201 (1975). - SEN, S., OSBORNE, D.J.:
J. Exp. Bot. 25, 1010-1019 (1974). - SEPASKHAH, A.R.: Agron. J. 69,
894-896 (1977). - SHALHEVET, J., MAAS, E.V., HOFFMAN, G.J., OGATA, G.:
Physiol. Plant. 38, 224-232 (1976). - SHARMA, M.L.: Agron. J. 68, 390-
394 (1976a); - Aust. J. Ecol. 1, 249-258 (1976b). - SHAW, C.B., GIFFORD,
G.F.: J. Range Manage. 28, 377-379 (1975). - SHEIN, E.V.: Vestn. Mosk.
Univ. Biol. Pochvoved. 29, 117-119 (1974). - SHEPHERD, W.: J. Exp. Bot.
26, 465-468 (1975). - SHERIFF, D.W.: J. Exp. Bot. 27, 175-183 (1976);
- Ann. Bot. (London) 41, 1081-1082 (1977a); - Ann. Bot. (London) 41,
1083-1084 (1977b); - Ann. Bot. (London) 43, 157-171 (1979). -
SHERIFF, D.W., KAYE, P.E.: Z. Pflanzenphysiol. 83, 463-466 (1977a);

- Ann. Bot. (London) 41, 653-655 (1977b). - SHERIFF, D.W., MEIDNER,
H.: J. Exp. Bot. 26, 897-902 (1975). - SHIRAZI, G.A., STONE, J.F.,
CROY, L.I., TODD, G.W.: Physiol. Plant. 33, 214-128 (1975). -
SHIRAZI, G.A., STONE, J.F., TODD, G.W.: J. Exp. Bot. 27, 610-618
(1976a). - SHIRAZI, G.A., STONE, J.F., BACON, C.M.: J. Exp. Bot. 27,
619-633 (1976b). - SHMUELI, M.: Hortscience 10, 506-509 (1975). -
SHONE, M.G.T., WOOD, A.V.: J. Exp. Bot. 28, 872-885 (1977). - SHUTTLE-
WORTH, W.J.: Boundary Layer Meteorol. 10, 273-302 (1976). - SILVIUS,
J.E., JOHNSON, R.R., PETERS, D.B.: Crop Sci. 17, 713-716 (1977). -
SINGH, D.P.: Indian J. Plant Physiol. 20, 63-68 (1977). - SIONIT, N.: J.
Agric. Sci. 89, 663-666 (1977). - SIONIT, N., KRAMER, P.J.: Plant
Physiol. 58, 537-540 (1976); - Agron. J. 69, 274-278 (1977). - SKAAR,
H., JOHNSSON, A.: Physiol. Plant. 43, 390-396 (1978). - SLAVOV, N.:
Fiziol. Rast. 2, 49-57 (1976). - SMALL, E.: Bot. Not. 126, 534-539
(1973). - SMITH, W.K., NOBEL, P.S.: Ecology 58, 1033-1043 (1977a);
- J. Exp. Bot. 28, 169-183 (1977b); - Am. J. Bot. 65, 429-432 (1978).
- SOLNTSEVA, O.N., SUTULOVA, V.I.: Ekologiya 6, 86-89 (1975). -
SORIANO, A., GINZO, H.D.: Agric. Meteorol. 15, 273-284 (1975). -
SRIVASTAVA, A.K., SINGH, G.: Indian J. Ecol. 2, 132-138 (1975). -
SRIVASTAVA, H.S., JOLLIFFE, P.A., RUNECKLES, V.C.: Can. J. Bot. 53,
475-482 (1975). - STELZER, R., LÄUCHLI, A.: Z. Pflanzenphysiol. 84,
95-108 (1977). - STERNE, R.E., KAUFMANN, M.R., ZENTMYER, G.A.:
Physiol. Plant. 41, 1-6 (1977). - STEUDLE, E., ZIMMERMANN, U., LÜTTGE,
U.: Plant Physiol. 59, 285-289 (1977). - STEVENS, R.A., MARTIN, E.S.:
Nature (London) 265, 331-334 (1977a); - Nature (London) 268, 364-365
(1977b); - Can. J. Bot. 55, 2873-2878 (1977c). - STEWART, C.R.,
BOGGESS, S.F.: Plant Sci. Lett. 8, 147-153 (1977). - STEWART, C.R.,
BOGGESS, S.F., ASPINALL, D., PALEG, L.G.: Plant Physiol. 59, 930-932
(1977). - STIGTER, C.J.: Arch. Meteorol. Geophys. Bioklimatol. Ser. B
Klimatol. Umweltmeteorol. Strahlungsforsch. 24, 95-108 (1976); - Arch.
Meteorol. Geophys. Bioklimatol. Ser. B Klimatol. Umweltmeteorol.
Strahlungsforsch. 24, 349-359 (1977). - STOCKER, O.: Flora (Jena) 163,
46-88 (1974a); - Flora (Jena) 163, 89-142 (1974b); - Flora (Jena) 163,
480-529 (1974c). - STONE, J.E., STONE, E.L.: For. Sci. 21, 53-60
(1975). - STONE, J.F., SHIRAZI, G.A.: Planta 122, 169-178 (1975). -
STONE, L.R., TEARE, I.D., NICKELL, C.D., MAYAKI, W.C.: Agron. J. 68,
677-680 (1976). - STOREY, R., WYN JONES, R.G.: Plant Sci. Lett. 4,
161-168 (1975). - STOREY, R., AHMAD, N., WYN JONES, R.G.: Oecologia
27, 319-332 (1977). - STOWE, L.G., TEERI, J.A.: Am. Nat. 112, 609-623
(1978). - STRAIN, B.R.: Physiological ecology of some woody desert
perennials in the southwestern United States, 27-39. In: Environmental
Physiology and Ecology of Plants, eds. D.N. SEN, R.P. BANSAL. Dehra
Dun: Bishen Singh Mahendra Pal Singh Publishers 1978. - STROMGREN, T.:
Sarsia 61, 47-53 (1976). - STUART, K.L., COKE, L.B.: Planta 122, 307-
310 (1975). - STURGES, D.L.: Am. Midl. Nat. 98, 257-274 (1977). -
SZAREK, S.R., TING, I.P.: Photosynthetic efficiency of CAM plants in
relation to C_3 and C_4 plants, 289-297. In: Environmental and Biological
Control of Photosynthesis, ed. R. MARCELLE. The Hague: Junk 1975. -
SZAREK, S.R., TROUGHTON, J.H.: Plant Physiol. 58, 367-370 (1976). -
SZLOVÁK, S.: Acta Bot. Acad. Sci. Hung. 21, 167-174 (1975).

TALBOT, A.J.B., TYREE, M.T., DAINTY, J.: Can. J. Bot. 53, 784-788
(1975). - TALHA, M., LARSEN, P.: Physiol. Plant. 33, 66-70 (1975). -
TAYLOR, G.E.Jr.: New Phytol. 80, 523-534 (1978). - TAYLOR, H.M.,
KLEPPER, B.: Soil Sci. 120, 57-67 (1975). - TEERI, J.A., STOWE, L.G.:
Oecologia 23, 1-12 (1976). - TEERI, J.A., STOWE, L.G., MURAWSKI, D.A.:
Can. J. Bot. 56, 1750-1758 (1978). - TENHUNEN, J.D., GATES, D.M.:
Light intensity and leaf temperature as determining factors in diffu-
sion resistance, 213-225. In: Perspectives of Biophysical Ecology,
eds. D.M. GATES, R.B. SCHMERL. Ecological Studies, Vol. 12. Berlin-
Heidelberg-New York: Springer 1975. - TESHA, A.J., KUMAR, D.: Oeco-

logia 28, 377-382 (1977). - THORPE, N., BRADY, C.J., MILTHORPE, F.L.:
Aust. J. Plant Physiol. 5, 485-493 (1978). - TIESZEN, L.L., SENYIMBA,
M.M., IMBAMBA, S.K., TROUGHTON, J.H.: Oecologia 37, 337-350 (1979). -
TINKER, P.B.: Philos. Trans. R. Soc. London 273, 445-461 (1976). -
TORSSELL, B.W.R.: J. Appl. Ecol. 13, 943-953 (1976). - TRAVIS, A.J.,
MANSFIELD, T.A.: New Phytol. 78, 541-546 (1977). - TROUGHTON, J.H.,
CARD, K.A.: Planta 123, 185-190 (1975). - TU, C.M., HIETKAMP, G.:
Commun. Soil Sci. Plant Anal. 8, 81-86 (1977). - TUCKER, E.B.,
COSTERTON, J.W., BEWLEY, J.D.: Can. J. Bot. 53, 94-101 (1975). -
TUNSTALL, B.R., CONNOR, D.J.: Aust. J. Plant Physiol. 2, 489-499
(1975). - TURNER, N.C.: J. Exp. Bot. 27, 1085-1092 (1976). - TURNER,
N.C., HEICHEL, G.H.: New Phytol. 78, 71-81 (1977). - TYREE, M.T.:
Can. J. Bot. 54, 2738-2746 (1976). - TYREE, M.T., CHEUNG, Y.N.S.,
MACGREGOR, M.E., TALBOT, A.J.B.: Can. J. Bot. 56, 635-647 (1978).

UNTERSCHEUTZ, P., RUETZ, W.F., GEPPERT, R.R., FERRELL, W.K.: Physiol.
Plant. 32, 214-221 (1974).

VAADIA, Y.: Philos. Trans. R. Soc. London 273, 513-522 (1976). -
VÁCLAVÍK, J.: Biol. Plant. 16, 389-394 (1974); - Biol. Plant. 17,
411-415 (1975). - VARDAR, Y., BÜTÜN, G.: Ber. Dtsch. Bot. Ges. 87,
581-588 (1974). - VIEIRA DA SILVA, J.: Water stress, ultrastructure
and enzymatic activity, 207-224. In: Water and Plant Life, eds. O.L.
LANGE, L. KAPPEN, E.-D. SCHULZE. Ecological Studies, Vol. 19. Berlin-
Heidelberg-New York: Springer 1976. - VOLODARSKII, N.I., BYSTRYKH,
E.E.: Fiziol. Rast. 23, 497-501 (1976). - VORA, A.B., PATEL, H.C.,
VYAS, A.V., PATEL, B.R., PATEL, J.A.: Ann. Arid Zone 14, 229-234
(1975).

WAGGONER, P.E., PARLANGE, J.-Y.: Plant Physiol. 57, 153-156 (1976).
- WALTON, D.C., GALSON, E., HARRISON, M.A.: Planta 133, 145-148
(1977). - WAMPLE, R.L., REID, D.M.: Planta 127, 263-270 (1975). -
WARING, R.H., RUNNING, S.W.: Water uptake, storage and transpiration
by conifers: a physiological model, 189-202. In: Water and Plant
Life, eds. O.L. LANGE, L. KAPPEN, E.-D. SCHULZE. Ecological Studies,
Vol. 19. Berlin-Heidelberg-New York: Springer 1976. - WEATHERLEY,
P.E.: Philos. Trans. R. Soc. London 273, 435-444 (1976). - WEAVER,
P.L.: Caribb. J. Sci. 15, 21-30 (1975). - WEBSTER, B.D., LEOPOLD,
A.C.: Am. J. Bot. 64, 1286-1293 (1977). - WELLBURN, F.A.M., WELLBURN,
A.R.: Bot. J. Linn. Soc. 72, 51-54 (1976). - WENKERT, W., LEMON, E.R.,
SINCLAIR, T.R.: Ann. Bot. (London) 42, 295-307 (1978). - WEST, D.W.,
GAFF, D.F.: Planta 129, 15-18 (1976). - WEYERS, J.D.B., HILLMAN, J.R.:
Planta 144, 167-172 (1979). - WIEBE, H.H., PROSSER, R.J.: Plant
Physiol. 59, 256-258 (1977). - WILLIAMS, J.: J. Exp. Bot. 25, 669-674
(1974); - J. Exp. Bot. 27, 121-124 (1976). - WILLMER, C.M., RUTTER,
J.C.: Nature (London) 269, 327-328 (1977). - WINTER, K., LÜTTGE, U.:
Balance between C_3 and CAM pathway of photosynthesis, 323-334. In:
Water and Plant Life, eds. O.L. LANGE, L. KAPPEN, E.-D. SCHULZE.
Ecological Studies, Vol. 19. Berlin-Heidelberg-New York: Springer
1976. - WINTER, K., TROUGHTON, J.H.: Flora (Jena) 167, 1-34 (1978). -
WINTER, K., TROUGHTON, J.H., CARD, K.A.: Oecologia 25, 115-123 (1976).
- WINTER, K., LÜTTGE, U., WINTER, E., TROUGHTON, J.H.: Oecologia 34,
225-237 (1978). - WONG, S.C., COWAN, I.R., FARQUHAR, G.D.: Plant
Physiol. 62, 670-674 (1978). - WOOLHISER, D.A.: J. Environ. Qual. 4,
17-21 (1975). - WRENCH, P., WRIGHT, L., BRADY, C.J., HINDE, R.W.:
Aust. J. Plant Physiol. 4, 703-712 (1977). - WYN JONES, R.G., STOREY,
R., LEIGH, R.A., AHMAD, N., POLLARD, A.: A hypothesis on cytoplasmic
osmoregulation, 121-136. In: Regulation of Cell Membrane Activities
in Plants, eds. E. MARRÈ, O. CIFERRI. Amsterdam-Oxford-New York:
Elsevier/North-Holland Biomedical Press 1977a. - WYN JONES, R.G.,
STOREY, R., POLLARD, A.: Ionic and osmotic regulation in plants
particularly halophytes, 537-544. In: Échanges Ioniques Transmembra-

naires chez les Végétaux (Transmembrane Ionic Exchanges in Plants),
eds. M. THELLIER, A. MONNIER, M. DEMARTY, J. DAINTY, Colloque du
C.N.R.S. No. 258. Paris-Rouen: Centre National de la Recherche Scienti-
fique et Publications de l'Université de Rouen 1977b.

YAMAMOTO, Y.: Plant Cell Physiol. 16, 749-752 (1975). - YOUNG, R.H.,
GARNSEY, S.M.: J. Am. Soc. Hortic. Sci. 102, 751-756 (1977).

ZABKA, G.G., CHATURVEDI, S.N.: Plant Physiol. 55, 532-535 (1975). -
ZANSTRA, P.E., HAGENZIEKER, F.: Plant Soil 48, 347-368 (1977). -
ZEIGER, E., HEPLER, P.K.: Science 196, 887-889 (1977). - ZIEGLER, H.:
Flüssigkeitsströme in Pflanzen, 561-577. In: Biophysik, eds. W. HOPPE,
W. LOHMANN, H. MARKL, H. ZIEGLER. Berlin-Heidelberg-New York: Springer
1977. - ZIEGLER, H., OSMOND, C.B., STICHLER, W., TRIMBORN, P.: Planta
128, 85-92 (1976). - ZIMMERMANN, U.: Cell turgor pressure regulation
and turgor pressure-mediated transport processes, 117-154. In: Integra-
tion of Activity in the Higher Plant, ed. D.H. JENNINGS, Society for
Experimental Biology Symposium, Number XXXI. Cambridge: University
Press 1977; - Annu. Rev. Plant Physiol. 29, 121-148 (1978). - ZIMMER-
MANN, U., STEUDLE, E.: Adv. Bot. Res. 6, 45-117 (1978). - ZIMMERMANN,
U., STEUDLE, E., LELKES, P.I.: Plant Physiol. 58, 608-613 (1976). -
ZLOTNIKOVA, I.F., GUNAR, I.I., PANICHKIN, L.A.: Izv. Timiryazevsk.
Skh. Akad. 2, 10-16 (1977a). - ZLOTNIKOVA, I.F., PANICHKIN, L.A.,
GUNAR, I.I.: Izv. Timiryazevsk. Skh. Akad. 1, 20-24 (1977b). - ZOBEL,
D.B.: Bot. Gaz. 135, 200-210 (1974).

Professor Dr. OTTO L. LANGE and
Dr. RAINER LÖSCH
Lehrstuhl für Botanik II
der Universität Würzburg
Mittlerer Dallenbergweg 64
D 8700 Würzburg

II. Mineral Metabolism

Absorption and Translocation of Mineral Ions in Higher Plants

By ANDRÉ LÄUCHLI

This review discusses the progress in the understanding of the absorption and translocation of mineral ions in higher plants since about 1976. The emphasis is on mineral ions belonging to the macronutrients with the readily absorbed nitrate and ammonium ions excluded. Ion transport in storage tissues is also beyond the scope of this review; interested readers are referred to recent reviews by POOLE (1976) and VAN STEVENINCK (1978).

Various aspects of the subject of this review have been covered in the Encyclopedia of Plant Physiology, new series, Vol. 2 A and B (LÜTTGE and PITMAN, 1976a,b), in two monographs by BOWLING (1976) and BAKER (1978), respectively, and in three symposia (WARDLAW and PASSIOURA, 1976; THELLIER et al., 1977; MARRÉ and CIFERRI, 1977).

1. Absorption by the Root

a) Carrier Concept, Membrane ATPase, and Genetic Control

The low permeability of the plasmalemma of root cells to many kinds of mineral ions has led to the theory that ion transport through this membrane is mediated by ion-selective carrier molecules. Although nobody has been successful as yet in isolating such ion carriers from the plasmalemma of root cells, kinetic and other evidence strongly indicates the existence of carriers as membrane constituents and their involvement in ion absorption (EPSTEIN, 1976). Recent research has been directed toward regulation of carrier-mediated ion transport. During accumulation of K^+, uptake of this ion decreases with increasing internal K^+ content (e.g., GLASS, 1977). The influx of K^+ into barley roots is related to internal K^+ concentration. The change in influx rate with varying internal concentrations may be caused by allosteric inhibition of influx with the internal K^+ functioning as allosteric effector (GLASS, 1975, 1976, 1978). According to this model, the carrier might process a single binding site for external K^+ and four allosteric binding sites for internal K^+. High internal K^+ concentrations would saturate the allosteric binding sites and thus inhibit influx of K^+ by causing conformational changes in the carrier which reduce the K^+ affinity of the external binding site. Other carrier models also involve allosteric regulation (JENSEN and PETTERSSON, 1978) or else no special feedback mechanisms (BANGE, 1978).

There is mounting evidence that a plasmalemma-bound ATPase may function as an alkali cation carrier or at least represent the energy-transducing agent between ATP and alkali cation transport (HODGES, 1976; BOWLING, 1976; LEONARD and HOTCHKISS, 1976). The ATPase of plasma membrane vesicles from oat roots utilizes Mg-ATP as substrate (BALKE and HODGES, 1975) and its activity is further stimulated by monovalent cations, in particular K^+ or Rb^+ (HODGES, 1976). In addition, the specificity of this ATPase toward the alkali cations is similar to the specificity of

alkali cation influx into oat roots (SZE and HODGES, 1977). In these
studies, however, the identification of the plasma membrane vesicles
was based on phosphotungstic acid-chromic acid staining, the specificity of which has been doubted, however, by HALL and FLOWERS (1976).

> Comparative studies on the divalent cation requirement of the plasma membrane
> ATPase from roots of various plant species appear to show interesting ecological
> implications. As described above, the enzyme from roots of oat, which grows well
> on acid, low-Ca soils, requires Mg. In the Ca-requiring wheat, however, the di-
> valent cation stimulation of the ATPase was Ca > Mn > Mg (KÄHR et al., 1977).
> A Ca-sensitive ATPase was also detected in the root of the calcicole *Vicia faba*,
> while it was absent in the root of the calcifuge *Lupinus luteus* (MONESTIEZ-
> LORENZINI et al., 1976).

Strong support for the carrier concept of ion absorption comes from
evidence that membrane transport is controlled genetically and involves de novo synthesis or activation of proteins or enzymes (LÄUCHLI,
1976a). Unfortunately, many of the recently described mineral nutrition mutants cannot be considered transport mutants (e.g., EPSTEIN,
1978). An exception may have been discovered with the wilty mutant
Capsicum annuum, scabrous diminutive. Roots of this mutant absorbed less
Rb^+ but more Na^+ than those of the normal genotype, and the efflux of
Na^+ was lower in the mutant roots (TAL and BENZIONI, 1977; BENZIONI
and TAL, 1978). A possible interpretation is the impairment of the K^+
selective carrier and an effect on the Na^+ efflux mechanism, but changes in membrane permeability to K^+ and Na^+ could also be important. The
soybean varieties *Lee* and *Jackson* show dramatic inheritable differences
in the absorption and translocation of Cl^- and also of Na^+ (LÄUCHLI
and WIENEKE, 1978, 1979; WIENEKE and LÄUCHLI, 1979). Some of the ion
transport mutants discovered earlier have been discussed by LÄUCHLI
(1976a).

b) K^+/Na^+-Selectivity

The well-known selectivity of plant roots for K^+ over Na^+ is probably
due to selective K^+ influx at the plasmalemma in the low range of K^+
concentrations and to Na^+ efflux at the plasmalemma (cf. LÄUCHLI and
PFLÜGER, 1979). Further studies by JESCHKE (1977a,b,c) confirmed that
external K^+ effectively stimulates Na^+ efflux from barley roots via
$K^+ - Na^+$ exchange at the plasmalemma. Furthermore, Na^+ influx at the
tonoplast also contributes to the high K^+/Na^+ ratio of the cytoplasm,
and is possibly linked to K^+ efflux from the vacuole. Accumulation of
Na^+ in vacuoles is probably most important in halophytes (FLOWERS et
al., 1977). In tomato plants, the ability of the roots to accumulate
Na^+ depends on an adequate supply of K^+ (BESFORD, 1978). The salt-
tolerant tomato *Lycopersicon cheesmanii*, however, discriminates very
little between transport of K^+ and Na^+ and consequently is able to
accumulate large concentrations of Na^+ in the leaves (RUSH and EPSTEIN,
1976).

c) Active and Passive Ion Absorption

Ion absorption by plant roots and the significance of active and passive components in this important process have been discussed extensively by PITMAN (1976a). Measurements of membrane potentials (Ψ_{vo})
and of permeability coefficients are important in determining the active or passive transport of particular ions. Values of Ψ_{vo} for root
cells are generally large and negative, but the measurement and the
interpretation of Ψ in multicellular tissues must be considered cau-

tiously (ANDERSON, 1976a; MERTZ and HIGINBOTHAM, 1976). Estimates of permeability coefficients for monovalent ions yielded values for K^+ which are greater than for Na^+ and Cl^- (PITMAN, 1976a). The relative passive transport of Rb^+, Na^+ and Cl^- in plasma membrane vesicles of oat roots was about $1.0 : 0.50 : 0.18$, respectively (SZE and HODGES, 1976). By combining all these data, PITMAN (1976a) concluded that K^+ and Cl^- are absorbed actively at the plasmalemma into the cells, in agreement with EPSTEIN's (1976) view and with flux data by MACKLON (1975a), and that Na^+ is actively extruded. As regards active K^+ absorption, it is considered to be the governing transport process only at low external K^+ concentrations, whereas a passive component of K^+ flux may become more prominent in the higher concentration range (PITMAN, 1976a). However, there is not unanimous agreement on the latter point (EPSTEIN, 1976).

Almost general agreement has now been reached that an active H^+ extrusion pump operates at the plasmalemma of root cells, which is powered by a plasmalemma-bound ATPase and drives K^+ inward (PITMAN et al., 1977a; ANDERSON et al., 1977; DEJAEGERE and NEIRINCKX, 1978; POOLE, 1978). It is still controversial, however, whether K^+ is carried by the H^+ pump itself, or whether a separate carrier for K^+ is involved (cf. POOLE, 1978), as envisioned by RATNER and JACOBY (1976). The membrane ATPase appears to be also directly responsible for active Na^+ efflux, but only indirectly involved in active Cl^- influx which may be mediated by an anion carrier that brings about the exchange of Cl^- and of other anions for OH^- (cf. LÄUCHLI and PFLÜGER, 1979).

Absorption of divalent cations seems to be metabolically dependent (PITMAN, 1976a). There is, however, a paucity of available data. In fact, some evidence was recently published suggesting active extrusion of Mg^{2+} (MACKLON and SIM, 1976) and Ca^{2+} (MACKLON, 1975b; GROSS and MARMÉ, 1978) across the plasmalemma. GROSS and MARMÉ (1978) determined Ca^{2+} uptake into plasma membrane vesicles to be ATP-dependent and active; since these vesicles were inside-out (MARMÉ, personal communication), these authors were in fact looking at active Ca^{2+} efflux.

Iron absorption has recently been reviewed by BROWN (1978). H^+ extrusion from the root and reduction of Fe^{3+} to Fe^{2+} at the cell surface is clearly involved, but the possible involvement of a Fe^{2+} carrier in the absorption of Fe by the root is debated. Microautoradiographic studies by CLARKSON and SANDERSON (1978) showed that labeled Fe appeared readily in internal root tissues only in zones of high rates of translocation to the shoot (about 1 to 4 cm from the tip in barley roots); in other zones most of the Fe appeared bound at the root periphery. KANNAN and KEPPEL (1976) suggested that lateral transport of Fe through the root was via the symplast, but CLARKSON and SANDERSON (1978) considered this transport process to be determined by metabolic factors unrelated to root anatomy.

Absorption of phosphate is complex, due to the number of ionic species available (PITMAN, 1976a). Researchers in New Zealand are particularly interested in studies of phosphate absorption, because there is widespread P deficiency in the soils of this country. The roots of white clover, a common pasture plant in New Zealand, appear to absorb phosphate with the aid of an electrogenic phosphate pump which is dependent on the carbohydrate supply from the shoot (BOWLING and DUNLOP, 1978). This phosphate pump could be driven by a pH gradient across the plasmalemma (DUNLOP and BOWLING, 1978).

d) Regulation in the Intact Plant

A number of exogenous and endogenous factors appear to be involved in the regulation of ion absorption by the root of an intact plant. Some of these phenomena are not apparent in experiments with excised roots. The rate of absorption of an ion usually changes during ontogenesis of the plant; the rate commonly declines at the later stages of development as demonstrated for wheat (JENSEN, 1978). On the other hand, the total concentration of K^+ and Na^+ in the root vacuoles of intact plants can be independent of the external total concentration, and the K^+ concentration in the shoot can be nearly constant and independent of the relative growth rate (PITMAN, 1976b). Our knowledge of the mechanisms whereby an intact plant brings about regulation of ion transport processes is still very sketchy.

The contribution of transpirational water flow to ion absorption and translocation is an aspect of the ion relations of intact plants that is still unresolved (BOWLING, 1976). MARKHART (1978) showed increased total solute translocation with increased water flow through the root of four species. He suggested that this effect was due to solvent drag, in accord with coupled solute/solvent transport in roots. A thorough review of the possible coupling of ion fluxes to water flux has been presented by ZIMMERMANN and STEUDLE (1978). BOWLING (1976), however, proposed that high water fluxes (indicating high transpiration rates) increase the active ion transport across the membranes of the outer root cells.

CRAM (1976) suggested the possibility of negative feedback regulation, yet experimental evidence for it is not conclusive. Clear ist, however, that the shoot is important in supplying energy for active ion absorption to the root by sugar transport in the phloem (PITMAN, 1976a). The metabolic component of the membrane potential in root cells is eliminated when the energy supply from the shoot is blocked (GRAHAM and BOWLING, 1977). In addition, the shoot may transmit information of still unknown nature to the root, and the activity of certain phytohormones in root and shoot may also be involved in regulation (PITMAN and CRAM, 1977). We are, however, far from a detailed understanding of the root/shoot integration of ion absorption and transport.

2. Lateral Transport Into the Xylem

a) Ion Localization and Structural Studies of the Root

There is widespread interest in applying methods of ion localization to the study of lateral transport into the xylem. The two methods most commonly used in recent years are based on electron probe X-ray microanalysis and on the use of ion-specific microelectrodes. Some of these methods, their advantages, and their pitfalls were reviewed by VAN STEVENINCK and VAN STEVENINCK (1978), BOWLING (1976), and LÄUCHLI and PFLÜGER (1979). Thus far, no general agreement exists as to the suitability of these methods for analysis of the pattern of ion distribution. Nonetheless, the present technology of X-ray microanalysis provides for separate analysis of cytoplasm and vacuole of root cells in situ without any chemical treatment of the specimen (PITMAN and LÄUCHLI, 1979).

X-ray fluorescence was used for in situ elemental mapping (MURRAY and SPURR, 1975), but this method allows only a resolution on the tissue level. The microlocalization of boron, which is not possible by X-ray microanalysis, was achieved by means of a specific nuclear reaction (MARTINI et al., 1977).

JESCHKE and STELTER (1976) chose an interesting new approach by measuring ion profiles along the axis of single barley roots in 0.5 mm sections with flameless atomic absorption spectroscopy and relating them to the pattern of cellular differentiation. They found high K^+/Na^+-ratios in the cytoplasm, and much higher K^+ concentration in the cytoplasm than in the vacuole of low-salt barley roots, in qualitative agreement with the X-ray microanalyses of PITMAN and LÄUCHLI (1979).

Transport of ions through the root is affected by root structure. ROBARDS and JACKSON (1976) summarized the present interest in this subject with special reference to the role of endodermal development and plasmodesmata. After the Casparian strip of the endodermis is formed, which occurs 13 to 16 mm from the tip of corn roots (DUPONT and LEONARD, 1977), solutes no longer have access to the stele via the apoplast (cf. LÄUCHLI, 1976b), but can move laterally only through the symplast involving plasmodesmata (cf. ROBARDS and CLARKSON, 1976; VAN STEVENINCK, 1976). The differentiation of the endodermis differs greatly among various plant species. In the salt- and flooding-tolerant *Puccinellia*, a double endodermis is formed at later stages of development with passage cells located in radial positions behind each other (STELZER and LÄUCHLI, 1977). Deposits of Si in the inner tangential wall of the tertiary-state endodermal cells are widespread in the Poaceae (SANGSTER, 1978); however, the significance of this in radial ion transport is unknown.

Transfer cells were discovered recently at various locations in the root. In *Atriplex hastata* they occur in the epidermis and apparently function in selective K^+ absorption from the saline environment (KRAMER et al., 1978). Xylem and phloem parenchyma in *Hieracium* roots develop transfer cells (LETVENUK and PETERSON, 1976); their initiation may be dependent on the nutritional status of the plant. Such xylem parenchyma transfer cells are pronounced in the proximal region of the root of soybean (LÄUCHLI et al., 1974) and bean (KRAMER et al., 1977), and are able to reabsorb Na^+ from the xylem sap in exchange for K^+ (LÄUCHLI, 1976c). Yet the development of wall ingrowths is not an absolute prerequisite for Na^+/K^+ exchange at the xylem parenchyma cells, for these cells feature only simple wall developments in corn roots (YEO et al., 1977). A general review on the possible role of transfer cells in solute transport was presented by GUNNING (1977).

b) Pathways and Mechanisms

The pathways and mechanisms by which ions travel laterally through the root into the xylem have always attracted the attention of transport physiologists. Yet, current experimental evidence and hypotheses are as conflicting as ever. The various views are discussed in a number of reviews (ANDERSON, 1976b; BOWLING, 1976; LÄUCHLI, 1976b,c; PITMAN, 1977). Here, only a selection of areas of research are highlighted.

It was stated already that the symplast is an important ion pathway, particularly across the endodermis. Not withstanding the significance of apoplasmic transport across the cortex (LÄUCHLI, 1976b), the question was studied as to whether all the major ions can move through the symplast. Movement of Ca^{2+} is reduced heavily in regions of the root where the endodermis is suberized, implying restricted Ca^{2+} transport in the symplast (FERGUSON and CLARKSON, 1975; FERGUSON, 1979). The same conclusion holds for Mg^{2+} (FERGUSON and CLARKSON, 1976), though the ratio of Ca^{2+}/Mg^{2+} in the xylem exudate of corn roots was higher than the ratio of these ions absorbed by the root, indicating preferential absorption of Mg^{2+} by the root cells from the apoplast (FERGUSON, 1978). RAVEN (1977) estimated the concentration of free Ca^{2+} in the ground cytoplasm to be less than

5 mM and, hence, the low Ca^{2+} transport capacity of the symplast is considered
to be the consequence of the cytoplasmic nature of this transport pathway. Since
the Na^+ concentration in the ground cytoplasm appears to be also low (RAVEN,
1977), the exclusion of Na^+ from the shoot of *Puccinellia* (STELZER and LÄUCHLI,
1978) might be explained in terms of low Na^+ transport capacity of the symplast.

Whether lateral ion transport into the xylem is governed by a single
or two membranes is still a principal controversy. Based on electrochemical data BOWLING (1976) put a single active transport step at a
membrane located at the outer surface of the root, but his measurements were criticized for methodological reasons (ANDERSON, 1976a).
Similarly the mathematical model by FISCUS and KRAMER (1975) is based
on a two-compartment system consisting of the outer medium and the
xylem separated by a single membrane of undefined nature. FISCUS (1977)
and MARKHART (1978) have generated experimental results that are consistent with the predictions of the model, suggesting that under rea-
sonably high transpiration rates a root can functionally be considered
a two-component system. MARKHART et al. (1979a) have also demonstrated
that a membrane may be the principal resistance to mass flow through
the root. On the other hand, a variety of experimental approaches suggest that two membranes may play a role in lateral ion transport, and
ion transport into the xylem vessels is different in kind from ion absorption by the root (LÄUCHLI, 1976c; PITMAN, 1977).

Correlated transport and structural investigations showed that ion
transport proceeds into fully mature vessels, implying that it is due
to release from the xylem parenchyma cells (LÄUCHLI et al., 1978).
Active K^+ and passive Cl^- transport into the vessels, in addition to
active influx at the plasmalemma of the outer tissues, was deduced
from electrochemical gradients between the xylem vessels and the surrounding stelar parenchyma (DAVIS and HIGINBOTHAM, 1976). Indirect
support for two active steps came from LEONARD and HOTCHKISS (1978),
who demonstrated K^+-stimulated ATPase activity in plasma membrane
fractions from isolated cortex and stele of corn roots, and from
WINTER-SLUITER et al. (1977) who presented cytochemical evidence for
this enzyme at the plasmalemma of xylem parenchyma cells. [It is important to verify cytochemical tests for ATPase activity by X-ray
microanalysis (VAN STEVENINCK, 1979)]. HANSON's (1978) provocative
view of applying the chemiosmotic hypothesis to lateral ion transport
could be reconciled with these data. The pH gradient postulated by
HANSON (1978) may be created by ATP hydrolysis, therefore ion release
from the xylem parenchyma cells would be energy-linked. Furthermore,
ion release to the xylem can be separated experimentally from absorption by the root by compounds that block synthesis of functional protein (WILDES et al., 1976; PITMAN et al., 1977b) and by phytohormones
such as ABA and cytokinins (PITMAN, 1977). ABA appears to inhibit ion
transport into the xylem of barley and sunflower roots by acting on
ion transport and not be reducing water flow (PITMAN and WELLFARE,
1978; ERLANDSSON et al., 1978). In excised bean roots, however, ABA
caused an increase in volume flow and in ion transport to the xylem
(KARMOKER and VAN STEVENINCK, 1978). On the other hand, both absorption and translocation of K^+ were inhibited by ABA in intact bean
seedlings (KARMOKER and VAN STEVENINCK, 1979). In marked contrast to
all the quoted results MARKHART et al. (1979b) found a reduction in
the hydraulic conductivity of soybean roots due to ABA. The apparent
confusion in reconciling the data on effects of ABA is unresolved but
may be caused in part by differences in environmental conditions,
nutritional status of the root systems, and also by differential physiological behavior of the various plant species. Nonetheless, the
speculation drawn from some of these results that ion release to the
xylem might involve a specific carrier protein which turns over rapidly (WILDES et al., 1976) requires further investigation.

3. Translocation in Aerial Organs

Translocation of ions in xylem and phloem, leading to import and export in leaves and other aerial organs, has been reviewed extensively by PATE (1975, 1976). Only a few relevant points will be mentioned here.

There is continued interest in the translocation of some inorganic ions through the phloem. Some debate still exists to what extent Ca^{2+} moves in the phloem. Analyses of the phloem sap show only small amounts of Ca^{2+} (PATE, 1976). Although PENOT et al. (1976) consider the possibility of Ca^{2+} export from leaves through the phloem, there is overwhelming evidence that this is not significant, and that translocation of foliar applied Ca^{2+} might be by means of a reverse xylem flow (HANGER, 1979). Considerable amounts of foliar applied Na^+ (bean plants) may be exported from the leaf via the phloem; the direction of this phloem translocation is almost exclusively toward the root, where it moves out of the phloem from the proximal region of the root into the surrounding solution (MARSCHNER and OSSENBERG-NEUHAUS, 1976). When several species were compared, retranslocation through the phloem of Na^+ and Cl^- applied to leaves varied greatly, and there were no correlations between the rates of translocation of Na^+ and Cl^-, respectively (LESSANI and MARSCHNER, 1978).

Studies of the ionic relations of leaf cells were mostly concerned with the energetics of ion absorption. The absorption of K^+ and Cl^- in cells of greening barley leaves is correlated with ATP levels (LÜTTGE and BALL, 1976), and the energy for ion absorption appears to be derived from ATP (JOHANSEN and LÜTTGE, 1975), synthesized by oxidative or photosynthetic phosphorylation (MACDONALD et al., 1975). Active phosphate absorption is also not obligatorily linked to a specific pathway of energy metabolism (ULLRICH-EBERIUS et al., 1976). The special aspect of ion transport in guard cells and its implications in stomatal movement has been discussed in detail (HSIAO, 1976; LÄUCHLI and PFLÜGER, 1979), and will not be reviewed here.

Attempts have been made to localize ions in leaves of halophytes (the technical problems inherent to such studies were covered above, 2.a.). In leaves of *Salicornia pacifica*, Na^+ and Cl^- were high in both palisade and spongy cells (WEBER et al., 1977); intracellularly, Cl^- appears to be localized mainly in the vacuoles (HESS et al., 1975). Na^+ and Cl^- also seems sequestered mainly in the vacuole of leaf cells of *Suaeda maritima* (HARVEY et al., 1976, 1978). Moreover, the chloroplasts of salt loaded plants can also accommodate some of the accumulated Na^+ and Cl^- (HARVEY and FLOWERS, 1978), while the cytoplasm has a low ion content (HARVEY et al., 1978), in agreement with the expectation from enzymic considerations.

Translocation of ions to developing fruits and seeds is a rather neglected area of investigation, with only a few exceptions mentioned here. PATE and HOCKING (1978) made a thorough investigation of mineral translocation to the fruit of *Lupinus albus* and arrived at the conclusion that most of the P, K, S and micronutrients moved to the fruit by vascular transport; the latter transport did not adequately account for Mg, Ca and Mn in the fruit. Vascular transport of the majority of ions (but not of Ca!) was through the phloem. The problem of Ca translocation to fruits has received additional attention. This nutrient is probably translocated to fruits through the xylem (MIX and MARSCHNER, 1976a; WIENEKE, 1979), but at a low rate compared to the rate of fruit growth which leads to a decline in Ca content during development of the fruit (MIX and MARSCHNER, 1976b). In bean fruits this decreasing Ca content seems to hold only for the pod, while there is continuing transport of Ca to the developing seeds, possibly by apoplasmic transport via the seedcoat (MIX and MARSCHNER, 1976c).

Acknowledgement. I am grateful to Dr. Albert H. Markhart III for his critical review of this manuscript.

References

ANDERSON, W.P.: The electrophysiology of higher plant roots, 125-135.
In: Transport and Transfer Processes in Plants, eds. I.F. WARDLAW,
J.B. PASSIOURA. London, New York: Academic Press 1976a; - Transport
through roots, 129-156. In: Encyclopedia of Plant Physiology, New
Series, eds. U. LÜTTGE, M.G. PITMAN, Vol. II, Part B. Berlin-Heidel-
berg-New York: Springer 1976b. - ANDERSON, W.P., WILLCOCKS, D.A.,
WRIGHT, B.J.: J. Exp. Bot. 28, 894-901 (1977).

BAKER, D.A.: Transport Phenomena in Plants. 80 p. London: Chapman and
Hall 1978. - BALKE, N.E., HODGES, T.K.: Plant Physiol. 55, 83-86
(1975). - BANGE, G.G.J.: Acta Bot. Neerl. 27, 183-198 (1978). -
BENZIONI, A., TAL, M.: J. Exp. Bot. 29, 879-884 (1978). - BESFORD,
R.T.: Plant Soil 50, 399-409 (1978). - BOWLING, D.J.F.: Uptake of
Ions by Plant Roots. 212 p. London: Chapman and Hall 1976. - BOWLING,
D.J.F., DUNLOP, J.: J. Exp. Bot. 29, 1139-1146 (1978). - BROWN, J.C.:
Plant Cell Environ. 1, 249-257 (1978).

CLARKSON, D.T., SANDERSON, J.: Plant Physiol. 61, 731-736 (1978). -
CRAM, W.J.: The regulation of nutrient uptake by cells and roots,
113-124. In: Transport and Transfer Processes in Plants, eds. I.F.
WARDLAW, J.B. PASSIOURA. London, New York: Academic Press 1976.

DAVIS, R.F., HIGINBOTHAM, N.: Plant Physiol. 57, 129-136 (1976). -
DEJAEGERE, R., NEIRINCKX, L.: Z. Pflanzenphysiol. 89, 129-140 (1978).
- DUNLOP, J., BOWLING, D.J.F.: J. Exp. Bot. 29, 1147-1153 (1978). -
DUPONT, F.M., LEONARD, R.T.: Protoplasma 91, 315-323 (1977).

EPSTEIN, E.: Kinetics of ion transport and the carrier concept,
70-94. In: Encyclopedia of Plant Physiology, New Series, eds. U.
LÜTTGE, M.G. PITMAN, Vol. II, Part B. Berlin-Heidelberg-New York:
Springer 1976; - Plant Physiol. 62, 582-585 (1978). - ERLANDSSON, G.,
PETTERSSON, S., SVENSSON, S.-B.: Physiol. Plant. 43, 380-384 (1978).

FERGUSON, I.B.: Aust. J. Plant Physiol. 5, 433-442 (1978); - Commun.
Soil Sci. Plant Anal. 10, 217-224 (1979). - FERGUSON, I.B., CLARKSON,
D.T.: New Phytol. 75, 69-79 (1975); - Planta 128, 267-269 (1976). -
FISCUS, E.L.: Plant Physiol. 59, 1013-1020 (1977). - FISCUS, E.L.,
KRAMER, P.J.: Proc. Natl. Acad. Sci. USA 72, 3114-3118 (1975). -
FLOWERS, T.J., TROKE, P.F., YEO, A.R.: Annu. Rev. Plant Physiol. 28,
89-121 (1977).

GLASS, A.D.M.: Plant Physiol. 56, 377-380 (1975); - Plant Physiol. 58,
33-37 (1976); - Aust. J. Plant Physiol. 4, 313-318 (1977); - Can. J.
Bot. 56, 1759-1764 (1978). - GRAHAM, R.D., BOWLING, D.J.F.: J. Exp.
Bot. 28, 886-893 (1977). - GROSS, J., MARMÉ, D.: Proc. Natl. Acad.
Sci. USA 75, 1232-1236 (1978). - GUNNING, B.E.S.: Sci. Prog. London
64, 539-568 (1977).

HALL, J.L., FLOWERS, T.J.: J. Exp. Bot. 27, 658-671 (1976). - HANGER,
B.C.: Commun. Soil Sci. Plant Anal. 10, 171-193 (1979). - HANSON,
J.B.: Plant Physiol. 62, 402-405 (1978). - HARVEY, D.M.R., FLOWERS,
T.J.: Protoplasma 97, 337-349 (1978). - HARVEY, D.M.R., FLOWERS, T.J.,
HALL, J.L.: New Phytol. 77, 319-323 (1976); - Intracellular localiza-

tion of ions in the halophyte *Suaeda maritima*, 248-249. In: Inaugural
Meeting Federation of European Societies of Plant Physiology, Edinburgh 1978. - HESS, W.M., HANSEN, D.J., WEBER, D.J.: Can. J. Bot. 53,
1176-1187 (1975). - HODGES, T.K.: ATPases associated with membranes
of plant cells, 260-283. In: Encyclopedia of Plant Physiology, New
Series, eds. U. LÜTTGE, M.G. PITMAN, Vol. II, Part A. Berlin-Heidelberg-New York: Springer 1976. - HSIAO, T.C.: Stomatal ion transport,
195-221. In: Encyclopedia of Plant Physiology, New Series, eds.
U. LÜTTGE, M.G. PITMAN, Vol. II, Part B. Berlin-Heidelberg-New York:
Springer 1976.

JENSEN, P.: Physiol. Plant. 43, 129-135 (1978). - JENSEN, P., PETERS-
SON, S.: Physiol. Plant. 42, 207-213 (1978). - JESCHKE, W.D.: Z. Pflanzenphysiol. 84, 247-264 (1977a); - J. Exp. Bot. 28, 1289-1305 (1977b);
- K$^+$-Na$^+$ selectivity in roots, localization of selective fluxes and
their regulation, 63-78. In: Regulation of Cell Membrane Activities
in Plants, eds. E. MARRÉ, O. CIFERRI. Amsterdam: Elsevier/North-Hol-
land Biomedical Press 1977c. - JESCHKE, W.D., STELTER, W.: Planta 128,
107-112 (1976). - JOHANSEN, C., LÜTTGE, U.: Aust. J. Plant Physiol. 2,
471-479 (1975).

KÄHR, M., BERVAES, J., KYLIN, A., KUIPER, P.J.C.: Influence of mineral
nutrition on ATPase activities and relation of saturated to unsatur-
ated fatty acids in roots of wheat and oats, 213-217. In: Transmem-
brane Ionic Exchanges in Plants, eds. M. THELLIER, A. MONNIER, M.
DEMARTY, J. DAINTY. Paris: Editions CNRS 1977. - KANNAN, S., KEPPEL,
H.: Z. Pflanzenphysiol. 79, 132-142 (1976). - KARMOKER, J.L., VAN
STEVENINCK, R.F.M.: Planta 141, 37-43 (1978); - Physiol. Plant (1979).
- KRAMER, D., LÄUCHLI, A., YEO, A.R., GULLASCH, J.: Ann. Bot. 41,
1031-1040 (1977). - KRAMER, D., ANDERSON, W.P., PRESTON, J.: Aust.
J. Plant Physiol. 5, 739-747 (1978).

LÄUCHLI, A.: Genotypic variation in transport, 372-393. In: Encyclopedia of Plant Physiology, New Series, eds. U. LÜTTGE, M.G. PITMAN,
Vol. II, Part B. Berlin-Heidelberg-New York: Springer 1976a; - Apoplasmic transport in tissues, 3-34. In: Encyclopedia of Plant Physiology, New Series, eds. U. LÜTTGE, M.G. PITMAN, Vol. II, Part B.
Berlin-Heidelberg-New York: Springer 1976b; - Symplasmic transport
and ion release to the xylem, 101-112. In: Transport and Transfer
Processes in Plants, eds. I.F. WARDLAW, J.B. PASSIOURA. London, New
York: Academic Press 1976c. - LÄUCHLI, A., PFLÜGER, R.: Potassium
transport through plant cell membranes and metabolic role of potas-
sium in plants. In: Potassium Research-Review and Trends, eds. P.A.
GETHING, A. VON PETER. Bern: International Potash Institute 1979. -
LÄUCHLI, A., WIENEKE, J.: Salt relations of soybean mutants differ-
ing in salt tolerance: distribution of ions and localization by X-ray
microanalysis, 275-282. In: Plant Nutrition 1978, eds. A.R. FERGUSON,
R.L. BIELESKI, I.B. FERGUSON. Wellington: Government Printer 1978; -
Z. Pflanzenernaehr. Bodenkd. 142, 3-13 (1979). - LÄUCHLI, A., KRAMER,
D., STELZER, R.: Ultrastructure and ion localization in xylem paren-
chyma cells of roots, 363-371. In: Membrane Transport in Plants, eds.
U. ZIMMERMANN, J. DAINTY. Berlin-Heidelberg-New York: Springer 1974.
- LÄUCHLI, A., PITMAN, M.G., LÜTTGE, U., KRAMER, D., BALL, E.: Plant
Cell Environ. 1, 217-223 (1978). - LEONARD, R.T., HOTCHKISS, C.W.:
Plant Physiol. 58, 331-335 (1976); - Plant Physiol. 61, 175-179
(1978). - LESSANI, H., MARSCHNER, H.: Aust. J. Plant Physiol. 5,
27-37 (1978). - LETVENUK, L.J., PETERSON, R.L.: Can. J. Bot. 54,
1458-1471 (1976). - LÜTTGE, U., BALL, E.: Z. Pflanzenphysiol. 80,
50-59 (1976). - LÜTTGE, U., PITMAN, M.G.: Encyclopedia of Plant Physiology, New Series, Vol. II, Part A. 400 p. Berlin-Heidelberg-New
York: Springer 1976a; - Encyclopedia of Plant Physiology, New Series,

Vol. II, Part B. 456 p. Berlin-Heidelberg-New York: Springer 1976b.

MACDONALD, I.R., MACKLON, A.E.S., MACLEOD, R.W.G.: Plant Physiol. 56, 699-702 (1975). - MACKLON, A.E.S.: Planta 122, 109-130 (1975a); - Planta 122, 131-141 (1975b). - MACKLON, A.E.S., SIM, A.: Planta 128, 5-9 (1976). - MARKHART III, A.H.: The coupled uptake of solute and solvent by plant root systems under simulated high transpiration rates. Diss. Duke Univ., Durham, North Carolina (1978). - MARKHART III, A.H., FISCUS, E.L., KRAMER, P.J., NAYLOR, A.W.: Plant Physiol. (1979a); - Plant Physiol. (1979b). - MARRÉ, E., CIFERRI, O., eds.: Regulation of Cell Membrane Activities in Plants. p 322. Amsterdam: Amsterdam: Elsevier/North-Holland Biomedical Press 1977. - MARSCHNER, H., OSSENBERG-NEUHAUS, H.: Z. Pflanzenernaehr. Bodenkd. 139, 129-142 (1976). - MARTINI, F., STELZ, T., DUVAL, A., HARTMANN, A., WISSOCQ, J.C., THELLIER, M.: Rôle physiologique, mesure de flux unidirection- nels et microlocalisation du bore chez les végétaux, 477-481. In: Transmembrane Ionic Exchanges in Plants, eds. M. THELLIER, A. MONNIER, M. DEMARTY, J. DAINTY. Paris: Editions CNRS 1977. - MERTZ JR., S.M., HIGINBOTHAM, N.: Plant Physiol. 57, 123-128 (1976). - MIX, G.P., MARSCHNER, H.: Z. Pflanzenernaehr. Bodenkd. 139, 551-563 (1976a); - Z. Pflanzenernaehr. Bodenkd. 139, 537-549 (1976b); - Z. Pflanzen- physiol. 80, 354-366 (1976c). - MONESTIEZ-LORENZINI, M., LAMANT, A., CONVERT, M.: Physiol. Veg. 14, 757-766 (1976). - MURRAY, S.A., SPURR, A.R.: Commun. Soil. Sci. Plant Anal. 6, 139-145 (1975).

PATE, J.S.: Exchange of solutes between phloem and xylem and circula- tion in the whole plant, 451-473. In: Encyclopedia of Plant Physiology, New Series, eds. M.H. ZIMMERMANN, J.A. MILBURN, Vol. I. Berlin-Heidel- berg-New York: Springer 1975; - Nutrients and metabolites of fluids recovered from xylem and phloem: significance in relation to long- distance transport in plants, 253-281. In: Transport and Transfer Pro- cesses in Plants, eds. I.F. WARDLAW, J.B. PASSIOURA. London, New York: Academic Press 1976. - PATE, J.S., HOCKING, P.J.: Ann. Bot. 42, 911- 921 (1978). - PENOT, M., FLOCH, J.-Y., PENOT, M.: PLANTA 129, 7-14 (1976). - PITMAN, M.G.: Ion uptake by plant roots, 95-128. In: Ency- clopedia of Plant Physiology, New Series, eds. U. LÜTTGE, M.G. PITMAN, Vol. II, Part B. Berlin-Heidelberg-New York: Springer 1976a; - Nutrient uptake by roots and transport to the xylem: uptake processes, 85-99. In: Transport and Transfer Processes in Plants, eds. I.F. WARDLAW, J.B. PASSIOURA. London, New York: Academic Press 1976b; - Annu. Rev. Plant Physiol. 28, 71-88 (1977). - PITMAN, M.G., CRAM, W.J.: Regula- tion of ion content in whole plants, 391-424. In: Symp. Soc. Exp. Biol., ed. D.H. JENNINGS, No. XXXI. Cambridge: University Press 1977. - PITMAN, M.G., LÄUCHLI, A.: Plant Physiol. Suppl. 63 (1979). - PITMAN M.G., WELLFARE, D.: J. Exp. Bot. 29, 1125-1138 (1978). - PITMAN, M.G., ANDERSON, W.P., SCHAEFER, N.: H^+ ion transport in plant roots, 147-160 In: Regulation of Cell Membrane Activities in Plants, eds. E. MARRÉ, O. CIFERRI. Amsterdam: Elsevier/North-Holland Biomedical Press 1977a. - PITMAN, M.G., WILDES, R.A., SCHAEFER, N., WELLFARE, D.: Plant Physio 60, 240-246 (1977b). - POOLE, R.J.: Transport in cells of storage tis- sues, 229-248. In: Encyclopedia of Plant Physiology, New Series, eds. U. LÜTTGE, M.G. PITMAN, Vol. II, Part A. Berlin-Heidelberg-New York: Springer 1976; - Annu. Rev. Plant Physiol. 29, 437-460 (1978).

RATNER, A., JACOBY, B.: J. Exp. Bot. 27, 843-852 (1976). - RAVEN, J.A. New Phytol. 79, 465-480 (1977). - ROBARDS, A.W. CLARKSON, D.T.: The role of plasmodesmata in the transport of water and nutrients across roots, 181-201. In: Intercellular Communication in Plants: Studies on Plasmodesmata, eds. B.E.S. GUNNING, A.W. ROBARDS. Berlin-Heidelberg- New York: Springer 1976. - ROBARDS, A.W., JACKSON, S.M.: Root struc- ture and function - an integrated approach, 413-422. In: Perspectives

in Experimental Biology, ed. N. SUNDERLAND, Vol. 2. New York: Pergamon
Press 1976. - RUSH, D.W., EPSTEIN, E.: Plant Physiol. 57, 162-166
(1976).

SANGSTER, A.G.: Am. J. Bot. 65, 929-935 (1978). - STELZER, R., LÄUCHLI,
A.: Z. Pflanzenphysiol. 84, 95-108 (1977); - Z. Pflanzenphysiol. 88,
437-448 (1978). - SZE, H., HODGES, T.K.: Plant Physiol. 58, 304-308
(1976); - Plant Pysiol. 59, 641-646 (1977).

TAL, M., BENZIONI, A.: J. Exp. Bot. 28, 1337-1341 (1977). - THELLIER,
M., MONNIER, A., DEMARTY, M., DAINTY, J., eds.: Transmembrane Ionic
Exchanges in Plants. 607 p. Paris: Editions CNRS 1977.

ULLRICH-EBERIUS, C.I., LÜTTGE, U., NEHER, L.: Z. Pflanzenphysiol. 79,
347-359 (1976).

VAN STEVENINCK, R.F.M.: Cytochemical evidence for ion transport through
plasmodesmata, 131-147. In: Intercellular Communication in Plants:
Studies on Plasmodesmata, eds. B.E.S. GUNNING, A.W. ROBARDS. Berlin-
Heidelberg-New York: Springer 1976; - Control of ion transport in
plant storage tissue slices, 503-542. In: Biochemistry of Wounded
Plant Tissues, ed. G. KAHL. Berlin-New York: Walter de Gruyter 1978;
- Protoplasma 99, 211-220 (1979). - VAN STEVENINCK, R.F.M., VAN
STEVENINCK, M.E.: Ion localization. In: Electron Microscopy and Cyto-
chemistry of Plant Cells, ed. J.L. HALL. Amsterdam: Elsevier/North-
Holland Biomedial Press 1978.

WARDLAW, I.F., PASSIOURA, J.B., eds.: Transport and Transfer Processes
in Plants. 484 p. London, New York: Academic Press 1976. - WEBER, D.J.,
RASMUSSEN, H.P., HESS, W.M.: Can. J. Bot. 55, 1516-1523 (1977). -
WIENEKE, J.: Commun. Soil Sci. Plant Anal. 10, 237-250 (1979). -
WIENEKE, J., LÄUCHLI, A.: Z. Pflanzenernaehr. Bodenkd. 142 (1979). -
WILDES, R.A., PITMAN, M.G., SCHAEFER, N.: Planta 128, 35-40 (1976). -
WINTER-SLUITER, E., LÄUCHLI, A., KRAMER, D.: Plant Physiol. 60,
923-927 (1977).

YEO, A.R., KRAMER, D., LÄUCHLI, A., GULLASCH, J.: J. Exp. Bot. 28,
17-29 (1977).

ZIMMERMANN, U., STEUDLE, E.: Adv. Bot. Res. 6, 45-117 (1978).

Professor Dr. ANDRÉ LÄUCHLI
Botanisches Institut der
Tierärztlichen Hochschule Hannover
Bünteweg 17 d
D 3000 Hannover 71

New address (1.8.79):
Department of Land, Air, and
Water Resources
University of California
Davis, CA 95616/USA

III. Photosynthesis

Structure and Function of the Photosynthetic Membrane

By J. Amesz

1. Introduction

In the time elapsed since the previous review in this series was written (AMESZ, 1977), various advances have been made in the field of photosynthesis research. We shall not try to cover all of these here, since this would render the present review too broad as well as too specialistic, but rather limit ourselves to a more general view of some aspects of structure and function with special emphasis on subjects that were not or only briefly treated in the previous reviews. In order to obtain a more coherent picture we shall not confine the discussion to very recent findings only, but older experiments and concepts that are still pertinent to our present state of knowledge will be discussed as well. Additional information may be found in recent reviews (e.g., SAUER, 1978, 1979; BLANKENSHIP and PARSON, 1978; BARBER, 1978, 1979a; AMESZ and VAN GORKOM, 1978; TREBST, 1978) and in several books and conference proceedings that have appeared during the last few years (PACKER et al., 1977; HALL et al., 1978; CIBA, 1979; CLAYTON and SISTROM, 1978; BARBER, 1977; OLSON and HIND, 1977; TREBST and AVRON, 1977; METZNER, 1978).

2. Membranes, Photochemistry, and Electron Transport

a) Location of Photosynthetic Membranes

There is abundant evidence that all the components that are essential for the photochemical processes of photosynthesis and for the associated electron transport are localized in the photosynthetic membrane. Studies from various laboratories and based on various techniques are gradually providing more specific information regarding the location of these components inside the membrane and in relation to each other. These studies are partly based on biochemical approaches, as, for example, immunochemical studies and techniques to isolate various membrane components, partly physical and biophysical, and based on studies of such phenomena and processes as electrochromism, fluorescence, resonance Raman scattering, linear and circular dichroism, electron and nuclear magnetic resonance, and luminescence.

In eukaryotic cells (algae and higher plants) the photosynthetic membranes are located within the chloroplast. In higher plants one may distinguish the so-called stroma and grana regions. The stroma contains among other things the enzymes of the Calvin cycle. The grana are stacks of closed membrane systems, the so-called thylakoids. Some of these thylakoids extend into the stroma region and form the so-called stroma lamellae. There is evidence (WEHRMEYER, 1964) that the inside of thylakoids and stroma lamellae may form one continuous phase within the chloroplast.

In blue-green algae the photosynthetic membranes are more or less dispersed throughout the cell. In photosynthetic bacteria they are part of the cytoplasmic membrane. The photosynthetic pigments and electron transfer components of purple photosynthetic bacteria appear to be mainly confined to invaginations of the cytoplasmic membrane (REMSEN, 1978). Upon disruption of the cells by sonication or other mechanical methods, photosynthetically active vesicles can be obtained that are usually known as chromatophores.

b) Photochemistry and Electron Transport

The first step in the process of photosynthesis consists of the absorption of light by the photosynthetic pigments. The ubiquitous pigment in algae (including blue-green algae) and higher plants is chlorophyll a. In addition to this there are carotenoids, chlorophyll b (as in green algae and higher plants), chlorophyll c (as in brown algae and diatoms), and phycobiliproteins, which occur in red and blue-green algae. Action spectra of oxygen evolution and other photosynthetic reactions show that light energy absorbed by all of these pigments is used, with various degrees of efficiency, for photosynthesis.

Only a small fraction of the photosynthetic pigments (roughly 1%) is directly involved in the so-called primary photochemical reaction. These are the so-called reaction center pigments, which are chlorophyll a molecules; the excitation energy of the remaining molecules, the so-called antenna chlorophylls and accessory pigments, is transferred to the reaction center.

Two chemically distinct types of reaction center exist in algae and higher plants, belonging to photochemical systems 1 and 2, respectively. The photochemical reactions that occur in these reaction centers and the associated electron transport reactions have been discussed extensively in previous reviews (AMESZ, 1973, 1975, 1977). Therefore, it may suffice to recall here that the primary electron donors of systems 1 and 2, P-700 and P-680, are probably chlorophyll a dimers, which upon excitation are converted to their cation radicals $P-700^+$ and $P-680^+$, and transfer an electron to an acceptor molecule. The two photochemical systems are connected by a series of electron transfer reactions involving plastoquinone, cytochrome f, and plastocyanin. The reduction of $P-680^+$ ultimately involves the oxidation of water, yielding oxygen; the electrons produced at the acceptor side of system 1, together with ATP formed in a reaction coupled to electron transport, are eventually used to drive CO_2 fixation.

The primary donor of purple bacteria, a bacteriochlorophyll dimer, is called P-870 or P-890 in species that contain bacteriochlorophyll a and P-985 in the few species known that contain bacteriochlorophyll b as only chlorophyll pigment. In green bacteria the primary electron donor is known as P-840. In all these instances the number denotes the principal wavelength of bleaching upon conversion to the corresponding cation radical. Various cytochromes function as electron donors for the oxidized dimer (DUTTON and PRINCE, 1978). In purple bacteria it has been well established that the "primary" acceptor is a quinone-iron complex. By flash spectroscopy supplemented by other methods it has been found that bacteriopheophytin acts as an even earlier electron acceptor, the so-called intermediate acceptor (see previous review, AMESZ, 1977).

c) Location of the Electron Transport Components in the Photosynthetic Membrane

There is evidence of various nature that the electron transport components of photosynthesis are assymetrically distributed across the photosynthetic membrane, in such a way that the primary donors and acceptors of systems 1 and 2 and of bacterial photosynthesis are located near opposite sides of the membrane. P-700 and P-680 are thought to be located on the inside, the acceptors on the outside of the thylakoid, and an analogous situation applies to purple bacteria. Much of the evidence for this arrangement, which is also of direct importance for the mechanism of phosphorylation according to the well-known concept of Mitchell (see e.g., ROTTENBERG, 1977; JAGENDORF, 1977) is reviewed and discussed by TREBST (1974, 1978).

Direct evidence for the location of the electron transport components can be obtained from proton translocation studies and from measurements of electrical fields and of the membrane potential generated by primary and secondary electron transport as reflected by electrochronism of photosynthetic pigments and reaction center components (see previous review, AMESZ, 1977, and JUNGE, 1977). For chloroplasts, these studies have so far only given support for the concept of vectorial electron transport in general, but no clear-cut and detailed information about the location of the individual components has been obtained as yet in this way. In part, this is undoubtedly due to the complexity of the system. As an example, we may refer to the uncertain location of plastocyanin. Recent experiments with antibodies (SCHMID et al., 1975; BÖGER, 1978) and with chemical modifiers (SELMAN et al., 1978) suggest that plastocyanin would be located near the external surface of the thylakoid, or at least be accessible from the outside, whereas the Mitchell concept, as well as the potential and pH measurements, favor an internal location. For chromatophores of purple bacteria more detailed schemes have been prepared, partly based on electrochromism, that are discussed by DUTTON and PRINCE (1978) and HOFF (1979).

With improved theories of electron tunneling (POTASEK and HOPFIELD, 1977) rates of electron transfer may be used to estimate the distances between reaction partners, including the primary photochemical reactants (HOFF, 1979). Another recent technique is based on measurements of paramagnetic dipole interaction of gadolinium ions with membrane components. Since the interaction is distance-dependent, it provides information about the location of these components with respect to the membrane surface (see DUTTON and PRINCE, 1978).

3. Pigment-Protein Complexes

a) Algae and Higher Plants

From work in several laboratories it has gradually become apparent that the photosynthetic pigments are not incorporated in the lipid layer, but that they are contained in pigment-protein complexes that are (partially) submerged in the membrane. In oxygen-producing organisms three types of such complex may be discerned: the light-harvesting, the system 1, and the system 2 complexes.

Much of the presently available information about the characteristics of the various pigment complexes is based on detergent fractionation

of higher plant chloroplasts (see reviews by THORNBER, 1975, and by
ANDERSON, 1975). About 50% of the total chlorophyll content of these
chloroplasts seem to be contained in the so-called *light-harvesting chlorophyll a/b protein complex*. In the older literature the complex is also
referred to by various other names, such as CP II or (pigment-protein)
complex II (THORNBER, 1975), but these names are to be avoided because
they lead to confusion with the system 2 reaction center complex. The
complex contains about equal amounts of chlorophyll a and b. The molar
ratio of these two pigments is 1.2-1.3 according to some workers
(DUNKLEY and ANDERSON, 1979; RÉMY et al., 1977; ARNTZEN and DITTO,
1976; BURKE et al., 1978); 1.0 according to others (THORNBER et al.,
1977; KUNG and THORNBER, 1971; VAN METTER, 1977). In addition, there
are smaller amounts of carotenoid. The complex has been obtained from
various species of higher plants, including spinach, tobacco, pea,
barley, and maize, from green algae, and from *Euglena* (THORNBER, 1975).
Resonance Raman spectroscopy of the isolated light-harvesting complex
from spinach (and also of the system 1 complex to be mentioned below)
yielded spectra that were very similar to those of intact membrane
systems, indicating that the structural characteristics of the complexes, as "seen" by the chlorophyll molecules, remain essentially intact during the isolation procedure (LUTZ, 1977; LUTZ et al., 1979).
Studies on fluorescence polarization and circular dichroism by VAN
METTER (1977) indicate that each subunit of the complex contains three
chlorophyll b molecules which together form a trimer with strong exciton coupling, surrounded by three chlorophyll a molecules which are
further apart from each other and which are not strongly coupled. One
of these appears to absorb at longer wavelength (677 nm) then the
other two (670 nm). Energy transfer from chlorophyll b to chlorophyll a
could proceed very efficiently in such an arrangement, whereas the
chlorophyll a molecules would provide a lattice for energy transfer
along the pigment-protein complex as a whole (KNOX and VAN METTER,
1979). The complex appears to contain 8 to 10 subunits, with a total
molecular weight of about 300 kD (THORNBER et al., 1977; THORNBER,
1979; cf. DUNKLEY and ANDERSON, 1979).

Light-harvesting pigment complexes have also been obtained from algae that do
not contain chlorophyll b. Among these are the so-called phycobilisomes of red
and blue-green algae that are discussed in another section of this review, and that
are separate particles, attached to but not part of the photosynthetic membrane.
Pigment-protein complexes containing chlorophyll c in addition to chlorophyll a
were isolated from several species of marine brown algae (ANDERSON and BARRETT,
1979; BARRETT and ANDERSON, 1977). Two types of complex were obtained. One is
a fucoxanthin-chlorophyll a/c$_2$ protein, the other contains chlorophylls a, c,
and c$_2$, but no fucoxanthin. Only fluorescence emitted by chlorophyll a was observed in these preparations, indicating that energy transfer from chlorophyll c
to chlorophyll a is quite efficient.

Peridinin-chlorophyll a protein complexes have been obtained from dinoflagel-
ates (HAXO et al., 1976; PRÉZELIN and HAXO, 1976; SONG et al., 1976; SIEGELMAN
et al., 1977). These complexes have no chlorophyll c and are soluble in water,
which suggests that they are only loosely bound to the membrane.

Chlorophyll a protein complexes (*complex I*) that are enriched in P-700,
and are apparently part of system 1, have been obtained by detergent
treatment of a variety of organisms, including higher plants, green,
brown, red and blue-green algae (THORNBER, 1975; THORNBER et al., 1977;
VERNON et al., 1971; WESSELS et al., 1973; BROWN, 1976; ANDERSON and
BARRETT, 1979; BENGIS and NELSON, 1975). These preparations contain
typically one P-700 per 40 chlorophyll a molecules, some β-carotene
(SHIOZAWA et al., 1974) and, if sufficiently purified, very little
chlorophyll b. In spinach they account for about 30% of the total
chlorophyll (THORNBER, 1979; THORNBER et al., 1977). By special treat-

ments preparations of lower chlorophyll-P-700 ratio have been obtained. However, these appear to be particles that have lost some of their
native chlorophyll by the preparation method, rather than a "purer"
and more "basic" subunit of photosystem 1 (IKEGAMI and KATOH, 1975;
BENGIS and NELSON, 1977; ALBERTE and THORNBER, 1978). The properties
of complex I are remarkably independent of the source of material
(THORNBER et al., 1977). This is especially striking with respect to
the amino-acid composition, which is quite similar for complex I derived from widely different organisms (THORNBER, 1975) and indicates
that this complex evolved in an early phase of the evolution of photosynthetic organisms.

> By means of sodium dodecyl sulfate electrophoresis six different polypeptide
> subunits have been obtained from complex I of spinach (NELSON and NOTSANI,
> 1977). The largest of these have a molecular weight of 70 kD (see also WESSELS,
> 1978) and contain chlorophyll a. The complex probably contains two of these
> subunits, associated with one P-700 molecule and one of each of the other units
> which vary in weight between 25 and about 8 kD. Together, these units would give
> a particle weight of about 225 kD; lower particle weights have been reported by
> Thornber and coworkers (THORNBER et al., 1977; MARKWELL et al., 1978). A signi-
> ficantly lower molecular weight of 48 kD for the major subunit was reported by
> THORNBER et al. (1977) for complex I from the blue-green alga *Phormidium luridum*.
> At present it is not clear if the difference is due to the different source of
> complex I or to differences in the experimental methods used by the two groups.

The absorption spectrum of complex I appears to consist of several
overlapping chlorophyll bands between about 650 and 700 nm, but the
number of these bands and their exact location is difficult to determine with certainty (SHIOZAWA et al., 1974; RÉMY et al., 1977; THORNBER
et al., 1977). The complexity of the absorption spectrum indicates that
the structure of the complex may be less simple than that inferred for
the light-harvesting chlorophyll a/b complex or for light-harvesting
complexes of bacteria. The linear dichroism spectrum of electrically
oriented samples appeared to be less complicated than the total absorption spectrum; it showed a narrow peak at 686 nm, suggesting a
relatively high degree of orientation for chlorophylls absorbing in
this region (GAGLIANO et al., 1979).

> The arrangement of the chlorophyll molecules that constitute the P-700 dimer
> has been studied in immobilized complex I particles by means of photoselection
> studies using polarized flashes to oxidize P-700 and polarized measuring light
> to measure the resulting absorption changes. The results did not allow a unique
> interpretation, but, if supplemented by other evidence, may provide useful in-
> formation about the structure of the reaction center (JUNGE et al., 1977; JUNGE
> and SCHAFFERNICHT, 1979). This also applies to measurements of linear dichroism
> with oriented chloroplasts (BRETON, 1977).

Preparations that show high *system 2* activity have been prepared by
VERNON et al. (1971) and by WESSELS et al. (1973). These preparations
contain only little chlorophyll b and may represent a system 2 reaction center pigment-protein complex analogous to complex I. They show
a light-induced reduction of cytochrome b-559 (KE et al., 1972) (which
is present in relatively high amount), and relatively large absorbance
changes at 675 and 430 nm upon illumination, which are presumably due
to oxidation of P-680 (VAN GORKOM et al., 1975). From the size of these
signals and from those attributed to the electron acceptor (Q) of system 2, it was estimated that the preparations of WESSELS et al. (1973)
contained one reaction center of system 2 per 20 to 70 chlorophyll
molecules (VAN GORKOM et al., 1975). So far these preparations have
only been obtained from spinach (VERNON et al., 1971; WESSELS et al.,
1973), from certain mutants of tobacco and barley (THIELEN and VAN

GORKOM, unpublished results), and from the blue-green alga *Synechococcus cedrorum* (NEWMAN and SHERMAN, 1978). The preparation technique is quite laborious and has a low yield, and it is difficult to obtain quantitatively reproducible results. Presumably for these reasons, the complex has not been well-defined biochemically so far.

Electron micrographs obtained by the freeze-fracture technique show that in green algae and higher plants the membranes from the stacked (grana) regions contain particles that are different from those observed in unstacked (stroma) lamellae (REMY, 1969; SANE et al., 1970; PHUNG NHU HUNG et al., 1970; GOODENOUGH and STAEHELIN, 1971; STAEHELIN and ARNTZEN, 1979). The particles that occur in the grana regions, the so-called EF_s particles, are observed in two sizes: 100-140 and 140-180 Å, whereas the particles that are seen in the stroma region, the EF_u particles, have an average diameter of 110 Å (STAEHELIN, 1976). As reviewed by ANDERSON (1975), the occurrence of the EF_s particles is correlated with that of the light-harvesting complex. The EF_s particles are thought to consist of photosystem 2 complexes combined with varying amounts of the light-harvesting complex. The EF_u particles may represent system 1, which is known to occur in excess to system 2 in the stroma lamellae (SANE et al., 1970).

Studies of fluorescence induction by MELIS and HOMANN (1976) and RIJGERSBERG et al. (1979a) indicate that there are two different "types" of reaction center of system 2, with different cross sections for light absorption, i.e., different numbers of associated light-harvesting chlorophylls. It is conceivable that these may belong to system 2 complexes combined with different numbers of light-harvesting pigment complexes as discussed above; the action spectra for fluorescence indeed show different relative contributions of chlorophyll b, as would be expected in this case (HORTON and CROZE, 1979). From measurements of light-induced changes in ultraviolet absorption it may be concluded that the primary acceptor Q is a plastoquinone molecule in both types of reaction centers (MELIS and DUYSENS, 1979); however, the oxidation-reduction potentials of these plastoquinones appear to be different (MELIS, 1978; HORTON and CROZE, 1979).

> There is evidence that the light-harvesting complex is involved in grana formation (STAEHELIN et al., 1977): The synthesis of the light-harvesting complex appears to occur in parallel with the appearance of grana in greening leaves (ARMOND et al., 1976, 1977), and mutants or algae that do not contain the light-harvesting chlorophyll a/b complex show little or no stacking (ANDERSON, 1975). Studies of cation-regulated stacking and unstacking in isolated chloroplasts similarly suggest that the EF_s particles are involved in this process (STAEHELIN, 1976; STAEHELIN and ARNTZEN, 1979). Interestingly, although the effect of cation concentration on energy distribution (HOMANN, 1969; MURATA, 1971; BARBER, 1979b; see review by WILLIAMS, 1977) appears to be correlated with the presence of the light-harvesting complex (DAVIS et al., 1976; BURKE et al., 1978; see also BARBER, 1979b), the different time scales at which these phenomena occur in response to a change in ion concentration indicate that stacking or unstacking by itself is not the major controlling factor in this process.

b) Emission Spectra of Chlorophyll

At room temperature, the main emission band of chlorophyll a in algae and higher plants is located at about 685 nm. There is various evidence that most of the emission comes from chlorophyll associated with system 2, both from studies with intact cells (DUYSENS and SWEERS, 1963) and with subcellular fractions (BOARDMAN et al., 1966).

Although fluorescence studies have provided a wealth of information about the photochemistry of system 2 (see e.g., review by LAVOREL and ETIENNE, 1977), at room temperature most of these studies were based on measurement of the intensity rather than on the spectral distribution of fluorescence, since the spectrum is rather featureless and does not provide much information about the organization and energy conversion processes of the photosynthetic apparatus.

Upon cooling the emission spectrum of intact cells and of isolated chloroplasts increases strongly in intensity, and long-wave emission bands develop that are absent or too weak to be observed at room temperature. Three major emission bands have been observed at low temperature in most organisms (including higher plants, green, red, and blue-green algae), located at 685, 695, and 710-740 nm, respectively (KOK, 1963; MURATA et al., 1966; CHO and GOVINDJEE, 1970; GOEDHEER, 1972). In this section we will confine ourselves to green algae and chloroplasts of higher plants; some experiments with red and blue-green algae will be discussed elsewhere in this review.

Cooling below about 200 K gives rise to a strong increase in long-wave fluorescence, and causes the development of a strong emission band at 710-740 nm (KOK, 1963; MURATA et al., 1966). At about 100 K, the band at 695 nm appears, which band grows strongly upon further cooling (Fig. 1), followed by an increase in the emission intensity of the band at 685 nm below about 40-50 K (CHO et al., 1966; CHO and GOVINDJEE, 1970; RIJGERSBERG et al., 1979a,b; BUTLER, 1979). A detailed explanation of this complicated temperature dependence is not yet available, but it may at least in part be explained by the assumption that upon cooling certain long-wave forms of chlorophyll a become unable to transfer their energy to other chlorophyll forms. Thus, energy transfer to the reaction centers is interrupted and the energy is emitted as fluorescence instead. In agreement with this, the relative increase in fluorescence was strongest when measured under conditions where the reaction center traps are open (RIJGERSBERG and AMESZ, 1978).

It is generally assumed that the emission at 710-740 nm comes from system 1, as its intensity is little affected by the trapping state of the system 2 reaction centers (MURATA, 1968; GOVINDJEE and YANG, 1966), especially at very low temperature (RIJGERSBERG et al., 1979b). The band is relatively weak in purified system 1 pigment-protein complex (BROWN, 1976; VACEK et al., 1977; RIJGERSBERG et al., 1979b), suggesting that it is due to an aggregated chlorophyll which is dissociated upon purification. The bands at 685 and 695 nm appear to be associated with system 2. The last one may again be due to aggregated chlorophyll; the band at 685 nm may correspond to the emission band emitted by the isolated system 2 pigment-protein complex at about this wavelength (WESSELS et al., 1973; RIJGERSBERG et al., 1979b). Experiments with a barley mutant indicate that the light-harvesting complex emits in a weak band near 680 nm (RIJGERSBERG et al., 1979b). At 5 K, the amplitude of this band seems to be unaffected by the intensities of the 685 and 695 nm bands (RIJGERSBERG et al., 1979b), which indicates that energy transfer from the light-harvesting complex to system 2 is irreversible at this temperature.

c) Bacteria

Pigment-protein complexes that have a function similar to those described above for oxygen-evolving organisms were also obtained from photosynthetic bacteria. The so-called reaction center complexes that have been isolated from various nonsulfur purple bacteria and recently

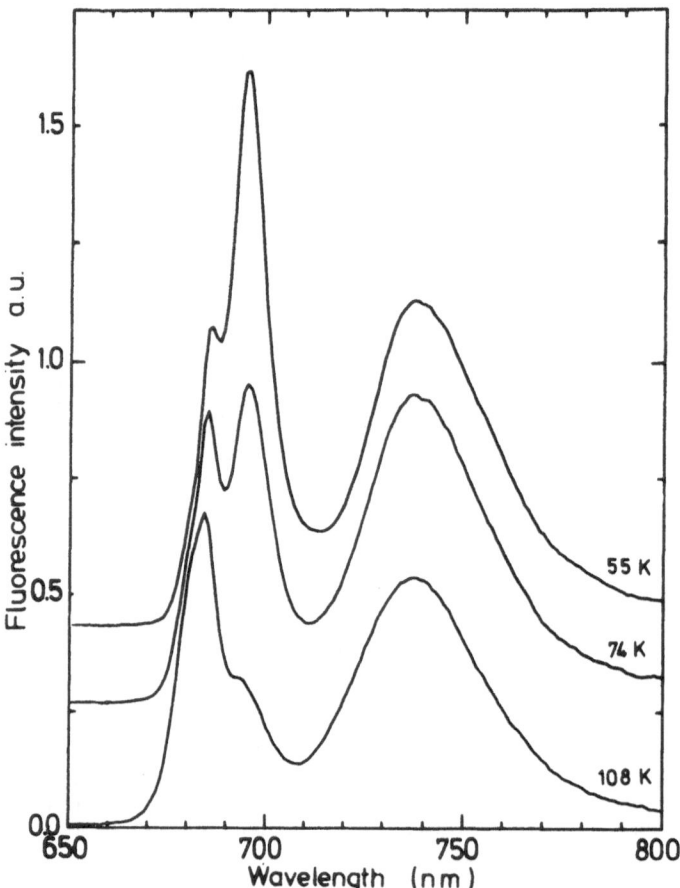

Fig. 1. Fluorescence emission spectra of spinach chloroplasts at three different temperatures. Note the strong temperature dependence of the band at 695 nm, which is barely visible at 108 K, but develops upon cooling into the highest peak of the spectrum. In the temperature range shown here, the bands near 685 and 740 nm show little temperature dependence. Emission from the light-harvesting complex at 680 nm is visible as a weak shoulder on the left flank of the 685 nm band (RIJGERSBERG et al., 1979b)

also from *Chromatium vinosum* (LIN and THORNBER, 1975; TIEDE et al., 1976; ROMIJN and AMESZ, 1977) were discussed in a previous review (AMESZ, 1975; see also reviews by GINGRAS, 1978, FEHER and OKAMURA, 1978, and DUTTON et al., 1978). In contrast to the complexes obtained from algae and higher plants, these complexes contain only the pigments that belong to the reaction center proper: if sufficiently purified, they are free of antenna bacteriochlorophylls. This type of reaction center particle has not yet been isolated from green bacteria; the "best" preparation obtained so far has a molecular weight of about 350 kD and contains about 35 bacteriochlorophylls per reaction center (SWARTHOFF, unpublished experiments).

Light-harvesting complexes have been obtained from purple bacteria and from the green bacterium *Prosthecochloris aestuarii*. The complex from *Prosthecochloris* is a water-soluble bacteriochlorophyll a protein that was first isolated by OLSON and ROMANO already in 1962. It has a molecular weight of 145 kD and consists of three identical subunits each

containing 7 bacteriochlorophylls (THORNBER and OLSON, 1968; FENNA et
al., 1974). So far, it is the only pigment-protein complex that has
been fully characterized by X-ray crystallography (FENNA and MATTHEWS,
1977; FENNA et al., 1977).

The light-harvesting complexes of purple bacteria have been isolated
and studied by various groups (CLAYTON and CLAYTON, 1972; OELZE and
GOLECKI, 1975; VAN DER REST et al., 1974; SAUER and AUSTIN, 1978;
FEICK and DREWS, 1978; COGDELL and CROFTS, 1978; COGDELL and THORNBER,
1979). The basic units of these complexes appear to be relatively
small (10-20 kD) and to contain three (sometimes two) bacteriochlorophylls a and one carotenoid molecule. Several of these subunits together are thought to constitute the native complex located in the
photosynthetic membrane. The best-known complexes contain bacteriochlorophylls absorbing at 850 and 800 nm (COGDELL and THORNBER, 1979).
Absorption and circular dichroism spectra indicate that the structure
of these is not basically altered during the isolation (COGDELL and
CROFTS, 1978), and that there is strong exciton interaction between
the two bacteriochlorophylls absorbing at 850 nm, but not between these
two and the third one, if present (SAUER and AUSTIN, 1978). A second
type of complex, and the only one in, for example, *Rhodospirillum rubrum*,
absorbs maximally near 890 nm. Here the basic unit seems to contain
two bacteriochlorophylls and one carotenoid (COGDELL and THORNBER,
1979; FEICK and DREWS, 1978). Carotenoid-protein complexes have also
been isolated (e.g., SCHWENKER and GINGRAS, 1973).

d) Phycobilisomes and "Chlorobium Vesicles"

At least two important exceptions exist to the general rule that photosynthetic pigments are contained in the photosynthetic membrane: the
phycobilisomes and the "chlorobium vesicles". Phycobilisomes are large
aggregates of phycobili-proteins that occur in red and blue-green algae
(see reviews by GLAZER, 1977; BOGORAD, 1975; GANTT, 1979). In electron
micrographs they are observed as particles of a few hundred Å diameter
that are attached in a regular array to the photosynthetic membrane
(GANTT and CONTI, 1966, 1969; GANTT et al., 1968). Interestingly,
phycobilisomes were not observed in Cryptomonads; in these algae the
phycobilins are probably dispersed in the intrathylakoidal space (GANTT
et al., 1971). The early studies of EMERSON and LEWIS (1942), DUYSENS
(1952) and HAXO and BLINKS (1950) have shown that light energy absorbed by phycobilins is used with high efficiency for photosynthesis; it
was subsequently found that this excitation energy is mainly used to
bring about photosystem 2 photochemistry (DUYSENS and AMESZ, 1962;
AMESZ and DUYSENS, 1962).

Phycocyanins and allophycocyanins are always present in phycobilisomes (GANTT
et al., 1979). Four forms of allophycocyanin have been isolated from blue-green
algae (ZILINSKAS et al., 1978; GLAZER and BRYANT, 1975). Three of these absorb
near 650 nm and show emission bands at 660, 662, and 673 nm, respectively. The
fourth one, called allophycocyanin B, occurs in smaller amounts than the other
three pigments, it absorbs at 618 and 678 and emits at about 680 nm (GLAZER and
BRYANT, 1975). A similar pigment was obtained from the red alga *Porphyridium
cruentum* (LEY et al., 1977). In addition to the phycocyanins a large proportion
of the total pigment content in phycobilisomes may be accounted for by phycoerythrin (GANTT and LIPSCHULTZ, 1974), as is usually the case in red algae.
Phycobilisomes obtained from the blue-green alga *Anabaena cylindrica* contained
small amounts of chlorophyll and carotenoids, but these pigments may represent
membrane pigments that remain attached to the phycobilisomes during purification (MIMURO and FUJITA, 1977; see also GANTT et al., 1979).

The most extensively studied phycobilisomes are those of the red alga
Porphyridium cruentum (GANTT and LIPSCHULZ, 1974; GANTT et al., 1976,
1977). Combined immunochemical and dissociation studies indicate that
the phycoerythrins are located on the outside and the phycocyanins on
the inside. The allophycocyanins are situated near the membrane. Such
a structure would minimize energy losses, since light energy absorbed
by phycoerythrins and phycocyanins is funneled into allophycocyanin
and hence transferred to membrane chlorophylls (GRABOWSKI and GANTT,
1978a,b). Energy transfer along the sequence phycoerythrin-phycocyanin-
allophycocyanin-chlorophyll a could be observed directly by measurement of the fluorescence rise times of these pigments after a 6 ps
laser flash (PORTER et al., 1978; SEARLE et al., 1978). Back transfer
from allophycocyanin to the other pigments is largely prevented (see
also WANG and MYERS, 1977) by the infavorable overlap between emission
and absorption spectra.

The bands at 695 and 710-730 nm in the fluorescence spectra of red and
blue-green algae at low temperature are, like the corresponding bands
in gree algae, attributed to chlorophyll a belonging to systems 2 and
1, respectively (MURATA et al., 1966; GOEDHEER, 1972; SHERMAN and
CUNNINGHAM, 1979). The emission band at 685 nm is probably at least
partly due to allophycocyanin, since a prominent emission band at
680-685 nm is also observed in the emission spectrum of isolated phycobilisomes of various species (GANTT et al., 1976, 1977, 1979). It
is, however, difficult to assess how much of the emission at 685 nm
upon excitation of the phycobiliproteins is actually due to chlorophyll a: removal of phycobilisomes from membrane vesicles of *Anabaena
variabilis* resulted in a drastic lowering of the emission at both 685
and 695 nm (KATOH and GANTT, 1979) but this effect is certainly to a
large part due to removal of most of the antenna of system 2 (i.e.,
the phycobilisome). Upon direct excitation of chlorophyll a with blue
light almost all fluorescence is emitted in the 720 band by *Porphyridium*
(MURATA et al., 1966; GANTT et al., 1977; LEY and BUTLER, 1977), which
indicated that the amount of chlorophyll a associated with the photosystem 2 pigment-protein complex is much smaller than that associated
with system 1 in this organism.

Chlorobium vesicles are oval-shaped bodies of several hundred to more
than a thousand Å large, first seen by COHEN-BAZIRE (1963), that occur
in green bacteria (see review by PIERSON and CASTENHOLZ, 1978). They
contain the accessory pigment chlorobium chlorophyll (bacteriochlorophyll c, d, or e) (CRUDEN and STANIER, 1970), which accounts for roughly 90% of the total bacteriochlorophyll present in these bacteria, and
little or no bacteriochlorophyll a (BOYCE et al., 1977). They are located adjacent to the cytoplasmic membrane and account for 10%-25% of
the total volume of the bacterial cell (PIERSON and CASTENHOLZ, 1978).
Their internal structure appears to be rather complicated (STAEHELIN
et al., 1978), and it is clear that they are not vesicles in the normal sense of the word, which are made up of a membrane enclosing an
inner aqueous phase. Therefore, the name chlorosomes has been proposed
(STAEHELIN et al., 1978). The outer "membrane" of the chlorobium vesicle, which is 30 Å thick (COHEN-BAZIRE and SISTROM, 1966; PIERSON
and CASTENHOLZ, 1974) may consist of a protein layer (OLSON et al.,
1977). In *Chlorobium*, all the protein of the vesicle (one-third of the
dry weight) would then be accounted for by this "membrane", the remaining constituents being chlorobium chlorophyll (28%), other lipids (16%)
and carbohydrate (15%) (CRUDEN and STANIER, 1970). Thus, with a particle weight of more than 30,000 kD, there are about 10,000 chlorobium
chlorophylls per vesicle which means that there is one vesicle per
about 10 reaction centers in *Chlorobium*. The efficiency of energy transfer from chlorobium chlorophyll to bacteriochlorophyll a was estimated

to be 70%-80% (OLSON and SYBESMA, 1963). Emission spectra of fluorescence and of luminescence indicate that back transfer does also occur,
but with lower efficiency (CLAYTON, 1965).

References

ALBERTE, R.S., THORNBER, J.P.: FEBS Lett. 91, 126-130 (1978). - AMESZ,
J.: Fortschr. Bot. 35, 89-102 (1973); - Prog. Bot. 37, 107-120 (1975);
- Prog. Bot. 39, 48-61 (1977). - AMESZ, J., DUYSENS, L.N.M.: Biochim.
Biophys. Acta 64, 261-278 (1962). - AMESZ, J., VAN GORKOM , H.J.:
Annu. Rev. Plant Physiol. 29, 47-66 (1978). - ANDERSON, J.M.: Biochim.
Biophys. Acta 416, 195-235 (1975). - ANDERSON, J.M., BARRETT, J.:
Chlorophyll-protein complexes of brown algae: P700 reaction center
and light-harvesting complexes, 81-96. In: CIBA FOUNDATION Symp. 61,
New Series: Chlorophyll Organization and Energy Transfer in Photosyn-
thesis. Amsterdam: Excerpta Medica 1979. - ARMOND, P.A., ARNTZEN, C.J.,
BRIANTAIS, J.-M., VERNOTTE, C.: Arch. Biochem. Biophys. 175, 54-63
(1976). - ARMOND, P.A., STAEHELIN, L.A., ARNTZEN, C.J.: J. Cell Biol.
73, 400-418 (1977). - ARNTZEN, C.J., DITTO, C.L.: Biochim. Biophys.
Acta 449, 259-274 (1976).

BARBER, J.: (ed.) Primary Processes of Photosynthesis. 516 pp. Amsterdam: Elsevier 1977; - Rep. Prog. Phys. 41, 1157-1199 (1978); - Photochem. Photobiol. 29, 203-207 (1979a); - Energy transfer and its dependence on membrane properties, 283-298. In: CIBA FOUNDATION Symp.
61, New Series: Chlorophyll Organization and Energy Transfer in Photosynthesis. Amsterdam: Excerpta Medica 1979b. - BARRETT, J., ANDERSON,
J.M.: Plant Sci. Lett. 9, 275-283 (1977). - BENGIS, C., NELSON, N.:
J. Biol. Chem. 250, 2783-2788 (1975); - J. Biol. Chem. 252, 4564-4569
(1977). - BLANKENSHIP, R.E., PARSON, W.W.: Annu. Rev. Biochem. 47,
635-653 (1978). - BOARDMAN, N.K., THORNE, S.W., ANDERSON, J.M.: Proc.
Natl. Acad. Sci. USA 56, 586-593 (1966). - BÖGER, P.: Some properties
of plastocyanin and its function in algal photosynthesis, 755-764. In:
Proc. 4th Int. Congr. Photosynthesis, eds. D.O. HALL, J. COOMBS, T.W.
GOODWIN. London: The Biochemical Society 1978. - BOGORAD, L.: Annu.
Rev. Plant Physiol. 26, 369-401 (1975). - BOYCE, C.O.L., OYEWOLE,
S.H., FULLER, R.C.: Localization of photosynthetic reaction center in
Chlorobium limicola, 365. In: Chlorophyll-Proteins, Reaction Centers and
Photosynthetic Membranes, eds. J.M. OLSON, G. HIND. Brookhaven Symp.
Biol., No. 28. Upton: Biol. Dep. Brookhaven Natl. Lab. 1977. - BRETON,
J.: Biochim. Biophys. Acta 459, 66-75 (1977). - BROWN, J.S.: Carnegie
Inst. Yearb. 75, 460-465 (1976). - BURKE, J.J., DITTO, C.L., ARNTZEN,
C.J.: Arch. Biochem. Biophys. 187, 252-263 (1978). - BUTLER, W.L.:
Tripartite und bipartite models of the photochemical apparatus of
photosynthesis, 237-252. In: CIBA FOUNDATION Symp. 61, New Series:
Chlorophyll Organization and Energy Transfer in Photosynthesis. Amster-
dam: Excerpta Medica 1979.

CHO, F., GOVINDJEE: Biochim. Biophys. Acta 216, 139-150 (1970). - CHO,
F., SPENCER, J., GOVINDJEE: Biochim. Biophys. Acta 126, 174-176 (1966).
- CIBA FOUNDATION Symp., New Series: Chlorophyll Organization and
Energy Transfer in Photosynthesis. 374 pp. Amsterdam: Excerpta Medica
1979. - CLAYTON, R.K.: J. Gen. Physiol. 48, 633-646 (1965). - CLAYTON,
R.K., CLAYTON, B.J.: Biochim. Biophys. Acta 283, 492-504 (1972). -
CLAYTON, R.K., SISTROM, W.R. (eds.): The Photosynthetic Bacteria.
946 pp. New York: Plenum Press 1978. - COGDELL, R.J., CROFTS, A.J.:
Biochim. Biophys. Acta 502, 409-416 (1978). - COGDELL, R.J., THORNBER,
J.P.: The preparation and characterization of different types of light-

harvesting pigment-protein complexes from some purple bacteria, 61-73.
In: CIBA FOUNDATION Symp. 61, New Series: Chlorophyll Organization
and Energy Transfer in Photosynthesis. Amsterdam: Excerpta Medica
1979. - COHEN-BAZIRE, G.: The photosynthetic apparatus of procaryotic
organisms, 65-90. In: Bacterial Photosynthesis, eds. H. GEST, A. SAN
PIETRO, L.P. VERNON. Yellow Springs: The Antioch Press 1963. - COHEN-
BAZIRE, G., SISTROM, W.R.: The procaryotic photosynthetic apparatus,
313-341. In: The Chlorophylls, eds. L.P. VERNON, G.R. SEELY. London:
Academic Press 1966. - CRUDEN, D.L., STANIER, R.Y.: Arch. Mikrobiol.
72, 115-134 (1970).

DAVIS, C.J., ARMOND, P.A., CROSS, E.L., ARNTZEN, C.J.: Arch. Biochem.
Biophys. 175, 64-70 (1976). - DUNKLEY, P.R., ANDERSON, J.M.: Biochim.
Biophys. Acta 545, 174-187 (1979). - DUTTON, P.L., PRINCE, R.C.:
Reaction-center-driven cytochrome interactions in electron and proton
translocation and energy coupling, 525-570. In: The Photosynthetic
Bacteria, eds. R.K. CLAYTON, W.R. SISTROM. New York: Plenum Press
1978. - DUTTON, P.L., PRINCE, R.C., TIEDE, D.M.: Photochem. Photobiol.
28, 939-949 (1978). - DUYSENS, L.N.M.: Transfer of Excitation Energy
in Photosynthetis. 96 pp. Thesis Univ. Utrecht 1952. - DUYSENS, L.N.M.,
AMESZ, J.: Biochim. Biophys. Acta 64, 243-260 (1962). - DUYSENS,
L.N.M., SWEERS, H.E.: Mechanism of two photochemical reactions in algae as studied by means of fluorescence, 353-372. In: Studies on Micro-
algae and Photosynthetic Bacteria, Special Issue of Plant and Cell
Physiol. Tokyo: Jpn. Soc. Plant Physiol. Univ. Tokyo Press 1963.

EMERSON, R., LEWIS, C.M.: J. Gen. Physiol. 25, 579-594 (1942).

FEHER, G., OKAMURA, M.Y.: Chemical composition and properties of re-
action centers, 349-386. In: The Photosynthetic Bacteria, eds. R.K.
CLAYTON, W.R. SISTROM. New York: Plenum Press 1978. - FEICK, R.,
DREWS, G.: Biochim. Biophys. Acta 501, 499-513 (1978). - FENNA, R.E.,
MATTHEWS, B.W.: Structure of a bacteriochlorophyll a-protein from
Prosthecochloris aestuarii, 170-182. In: Chlorophyll-Proteins, Reaction
Centers and Photosynthetic Membranes, eds. J.M. OLSON, G. HIND. Brook-
haven Symp. Biol., No. 28. Upton: Biol. Dep. Brookhaven Natl. Lab.
1977. - FENNA, R.E., MATTHEWS, B.W., OLSON, J.M., SHAW, E.K.: J. Mol.
Biol. 84, 231-240 (1974). - FENNA, R.E., TEN EYCK, L.F., MATTHEWS,
B.W.: Biochem. Biophys. Res. Commun. 75, 751-755 (1977).

GAGLIANO, A.G., GEACINTOV, N.E., BRETON, J., ACKER, S., RÉMY, R.:
Photochem. Photobiol. 29, 415-418 (1979). - GANTT, E.: Structure and
function of phycobilisomes: Light-harvesting pigment complexes in red
and blue-green algae. In: International Review of Cytology, eds. G.
BOURNE, J.F. DANIELLI. New York: Academic Press 1979, in press. -
GANTT, E., CONTI, S.F.: J. Cell Biol. 29, 423-434 (1966); - J. Bac-
teriol. 97, 1486-1493 (1969). - GANTT, E., LIPSCHULZ, C.A.: Bio-
chemistry 13, 2960-2966 (1974). - GANTT, E., EDWARDS, M.R., CONTI,
S.F.: J. Phycol. 4, 65-71 (1968). - GANTT, E., EDWARDS, M.R., PROVA-
SOLI, L.: J. Cell Biol. 48, 280-290 (1971). - GANTT, E., LIPSCHULZ,
C.A., ZILINSKAS, B.A.: Biochim. Biophys. Acta 430, 375-388 (1976);
- Phycobilisomes in relation to the thylakoid membranes, 347-357. In:
Chlorophyll-Proteins, Reaction Centers and Photosynthetic Membranes,
eds. J.M. OLSON, G. HIND. Brookhaven Symp. Biol., No. 28. Upton:
Biol. Dep. Brookhaven Natl. Lab. 1977. - GANTT, E., LIPSCHULZ, C.A.,
GRABOWSKI, J., ZIMMERMAN, B.K.: Plant Physiol. 63, 615-620 (1979). -
GINGRAS, G.: A comparative review of photochemical reaction centre
preparations from photosynthetic bacteria, 119-131. In: The Photosyn-
thetic Bacteria, eds. R.K. CLAYTON, W.R. SISTROM. New York: Plenum
Press 1978. - GLAZER, A.N.: Mol. Cell. Biochem. 18, 125-140 (1977).
- GLAZER, A.N., BRYANT, D.A.: Arch. Microbiol. 104, 15-22 (1975). -

GOEDHEER, J.C.: Annu. Rev. Plant Physiol. 23, 87-112 (1972). - GOOD-
ENOUGH, U.W., STAEHELIN, L.A.: J. Cell Biol. 48, 594-619 (1971). -
GOVINDJEE, YANG, L.: J. Gen. Physiol. 49, 763-780 (1966). - GRABOWSKI,
J., GANTT, E.: Photochem. Photobiol. 28, 39-45 (1978a); - Photochem.
Photobiol. 28, 47-54 (1978b).

HALL, D.O., COOMBS, J., GOODWIN, T.W. (eds.): Proc. 4th International
Congress Photosynthesis. 827 pp. London: The Biochemical Society 1978.
- HAXO, F.T., BLINKS, L.R.: J. Gen. Physiol. 33, 389-422 (1950). -
HAXO, F.T., KYCIA, J.H., SOMERS, G.F., BENNETT, A., SIEGELMAN, H.W.:
Plant Physiol. 57, 297-303 (1976). - HOFF, A.J.: Photooxidation of
the reaction center chlorophylls. In: Light Reactions in Photosyn-
thesis, ed. F.K. FONG. Berlin-Heidelberg-New York: Springer 1979,
in press. - HOMANN, P.H.: Plant Physiol. 44, 932-936 (1969). - HORTON,
P., CROZE, E.: Biochim. Biophys. Acta 545, 188-201 (1979).

IKEGAMI, I., KATOH, S.: Biochim. Biophys. Acta 376, 588-592 (1975).

JAGENDORF, A.T.: Photophosphorylation, 307-337. In: Photosynthesis I,
Photosynthetic Electron Transport and Photophosphorylation, ed. A.
TREBST, M. AVRON. Berlin-Heidelberg-New York: Springer 1977. - JUNGE,
W.: Physical aspects of light harvesting, electron transport and electrochemical potential generation in photosynthesis, 59-93. In: Photosynthesis I, Photosynthetic Electron Transport and Photophosphorylation, ed. A. TREBST, M. AVRON. Berlin-Heidelberg-New York: Springer
1977. - JUNGE, W., SCHAFFERNICHT, H.: The field of possible structures
for the chlorophyll a dimer in photosystem I of green plants deline-
ated by polarized photochemistry, 127-142. In: CIBA FOUNDATION Symp.
61, New Series: Chlorophyll Organization and Energy Transfer in Photosynthesis. Amsterdam: Excerpta Medica 1979. - JUNGE, W., SCHAFFERNICHT,
H., NELSON, N.: Biochim. Biophys. Acta 462, 73-85 (1977).

KATOH, T., GANTT, E.: Biochim. Biophys. Acta 546, 383-393 (1979). - KE, B.,
VERNON, L.P., CHANEY, T.H.: Biochim. Biophys. Acta 256, 345-357 (1972).
- KNOX, R.S., VAN METTER, R.L.: Fluorescence of light-harvesting
chlorophyll a/b-protein complexes: implications for the photosynthetic
unit, 177-186. In: CIBA FOUNDATION Symp. 61, New Series: Chlorophyll
Organization and Energy Transfer in Photosynthesis. Amsterdam: Excerpta
Medica 1979. - KOK, B.: Fluorescence studies, 45-53. In: Photosynthetic
Mechanisms of Green Plants. Washington: Natl. Acad. Sci.-Natl. Res.
Council 1963. - KUNG, S.D., THORNBER, J.P.: Biochim. Biophys. Acta
253, 285-289 (1971).

LAVOREL, J., ETIENNE, A.-L.: In vivo chlorophyll fluorescence, 203-
268. In: Primary Processes of Photosynthesis, ed. J. BARBER. Amsterdam:
Elsevier 1977. - LEY, A.C., BUTLER, W.L.: The distribution of excita-
tion energy between photosystem I and photosystem II in *Porphyridium
cruentum*, 33-46. In: Photosynthetic Organelles, Structure and Function,
Special Issue of Plant and Cell Physiol., no. 3, eds. S. MIYACHI, S.
KATOH, Y. FUJITA, K. SHIBATA. Tokyo: Jpn. Soc. Plant Physiol.-Center
Acad. Publ. Japan 1977. - LEY, A.C., BUTLER, W.L., BRYANT, D.A.,
GLAZER, A.N.: Plant Physiol. 59, 974-980 (1977). - LIN, L., THORNBER,
J.P.: Photochem. Photobiol. 22, 37-40 (1975). - LUTZ, M.: Biochim.
Biophys. Acta 460, 408-430 (1977). - LUTZ, M., BROWN, J.S., RÉMY, R.:
Resonance Raman spectroscopy of chlorophyll-protein complexes, 105-
121. In: CIBA FOUNDATION Symp. 61, New Series: Chlorophyll Organiza-
tion and Energy Transfer in Photosynthesis. Amsterdam: Excerpta Medica
1979.

MARKWELL, J.P., REINMAN, S., THORNBER, J.P.: Arch. Biochem. Biophys.
190, 136-141 (1978). - MELIS, A.: FEBS Lett. 95, 202-206 (1978). -

MELIS, A., DUYSENS, L.N.M.: Photochem. Photobiol. 29, 373-382 (1979). - MELIS, A., HOMANN, P.H.: Photochem. Photobiol. 23, 343-350 (1976). - METZNER, H. (ed.): Photosynthetic Oxygen Evolution. 532 pp. New York: Academic Press 1978. - MIMURO, M., FUJITA, Y.: A native phycobilin aggregate separated from the blue-green algae *Anabaena cylindrica* and *Plectonema boryanum*, 23-31. In: Photosynthetic Organelles, Structure and Function, Special Issue of Plant and Cell Physiol. no. 3, eds. S. MIYACHI, S. KATOH, Y. FUJITA, K. SHIBATA. Tokyo: Jpn. Soc. Plant Physiol. - Center Acad. Publ. Japan 1977. - MURATA, N.: Biochim. Biophys. Acta 162, 106-121 (1968); - Biochim. Biophys. Acta 245, 365-372 (1971). - MURATA, N., NISHIMURA, M., TAKAMIYA, A.: Biochim. Biophys. Acta 126, 234-243 (1966).

NELSON, N., NOTSANI, B.-E.: Function and organization of individual polypeptides in chloroplast photosystem I reaction center, 233-244. In: Bioenergetics of Membranes, eds. L. PACKER, G.C. PAPAGEORGIOU, A. TREBST. Amsterdam: Elsevier-North Holland 1977. - NEWMAN, P.J., SHERMAN, L.A.: Biochim. Biophys. Acta 503, 343-361 (1978).

OELZE, J., GOLECKI, J.R.: Arch. Microbiol. 102, 59-64 (1975). - OLSON, J.M., HIND, G. (eds.): Chlorophyll-Proteins, Reaction Centers and Photosynthetic Membranes. Brookhaven Symp. Biol., No. 28. 385 pp. Upton: Biol. Dep. Brookhaven Natl. Lab. 1977. - OLSON, J.M., ROMANO, C.A.: Biochim. Biophys. Acta 59, 726-728 (1962). - OLSON, J.M., SYBESMA, C.: Energy transfer and cytochrome oxidation in green bacteria, 413-422. In: Bacterial Photosynthesis, eds. H. GEST, A. SAN PIETRO, L.P. VERNON. Yellow Springs: The Antioch Press 1963. - OLSON, J.M., PRINCE, R.C., BRUNE, D.C.: Reaction center complexes from green bacteria, 238-245. In: Chlorophyll-Proteins, Reaction Centers and Photosynthetic Membranes, eds. J.M. OLSON, G. HIND. Brookhaven Symp. Biol., No. 28. Upton: Biol. Dep. Brookhaven Natl. Lab. 1977.

PACKER, L., PAPAGEORGIOU, G.C., TREBST, A. (eds.): Bioenergetics of Membranes. 538 pp. Amsterdam: Elsevier-North Holland 1977. - PHUNG NHU HUNG, S., LACOURLY, A., SARDA, C.: Z. Pflanzenphysiol. 62, 1-16 (1970). - PIERSON, B.K., CASTENHOLZ, R.W.: Arch. Mikrobiol. 100, 5-24 (1974); - Photosynthetic apparatus and cell apparatus of the green bacteria, 179-197. In: The Photosynthetic Bacteria, eds. R.K. CLAYTON, W.R. SISTROM. New York: Plenum Press 1978. - PORTER, G., TREDWELL, C.J., SEARLE, G.F.W., BARBER, J.: Biochim. Biophys. Acta 501, 232-245 (1978). - POTASEK, M.J., HOPFIELD, J.J.: Proc. Natl. Acad. Sci. USA 74, 229-233 (1977). - PRÉZELIN, B.B., HAXO, F.T.: Planta 128, 133-141 (1976).

REMSEN, C.C.: Comparative subcellular architecture of photosynthetic bacteria, 31-60. In: The Photosynthetic Bacteria, eds. R.K. CLAYTON, W.R. SISTROM. New York: Plenum Press 1978. - RÉMY, R.: C.R. Acad. Sci. Paris 268d, 3057-3060 (1969). - RÉMY, R., HOARAU, J., LECLERC, J.C.: Photochem. Photobiol. 26, 151-158 (1977). - RIJGERSBERG, C.P., AMESZ, J.: Biochim. Biophys. Acta 502, 152-160 (1978). - RIJGERSBERG, C.P., MELIS, A., AMESZ, J., SWAGER, J.A.: Quenching of chlorophyll fluorescence and photochemical activity of chloroplasts at low temperature, 305-318. In: CIBA FOUNDATION Symp. 61, New Series: Chlorophyll Organization and Energy Transfer in Photosynthesis. Amsterdam: Excerpta Medica 1979a. - RIJGERSBERG, C.P., AMESZ, J., THIELEN, A.P.G.M., SWAGER, J.A.: Biochim. Biophys. Acta 545, 473-482 (1979b). - ROMIJN, J.C., AMESZ, J.: Biochim. Biophys. Acta 461, 327-338 (1977). - ROTTENBERG, H.: Proton and ion transport across the thylakoid membranes, 338-349. In: Photosynthesis I, Photosynthetic Electron Transport and Photophosphorylation, ed. A. TREBST, M. AVRON. Berlin-Heidelberg-New York: Springer 1977.

SANE, P.V., GOODCHILD, D.J., PARK, R.B.: Biochim. Biophys. Acta 216, 162-178 (1970). - SAUER, K.: Acc. Chem. Res. 11, 257-264 (1978); - Annu. Rev. Phys. Chem. (1979) in press. - SAUER, K., AUSTIN, L.A.: Biochemistry 17, 2011-2019 (1978). - SCHMID, G.H., RADUNZ, A., MENKE, W.: Z. Naturforsch. 30c, 201-212 (1975). - SCHWENKER, U., GINGRAS, G.: Biochem. Biophys. Res. Commun. 51, 94-99 (1973). - SEARLE, G.F.W., BARBER, J., PORTER, G., TREDWELL, C.J.: Biochim. Biophys. Acta 501, 246-256 (1978). - SELMAN, B.R., SMITH, D.D., VOEGELI, K.K., JOHNSON, G., DILLEY, R.A.: Chloroplast membrane sidedness. Location of plasto-cyanin determined by chemical modifiers, 793-798. In: Proc. 4th Inter-national Congress Photosynthesis, eds. D.O. HALL, J. COOMBS, T.W. GOODWIN. London: The Biochemical Society 1978. - SHERMAN, L.A., CUN-NINGHAM, J.: Plant Sci. Lett. 14, 121-131 (1979). - SHIOZAWA, J.A., ALBERTE, R.S., THORNBER, J.P.: Arch. Biochem. Biophys. 165, 388-397 (1974). - SIEGELMAN, H.W., KYCIA, J.H., HAXO, F.T.: Peridinin-chloro-phyll a-proteins of dinoflagellate algae, 162-169. In: Chlorophyll-Proteins, Reaction Centers and Photosynthetic Membranes, eds. J.M. OLSON, G. HIND. Brookhaven Symp. Biol., No. 28. Upton: Biol. Dep. Brookhaven Natl. Lab. 1977. - SONG, P.S., KOKA, P., PRÉZELIN, B.B., HAXO, F.T.: Biochemistry 15, 4422-4427 (1976). - STAEHELIN, L.A.: J. Cell Biol. 71, 136-158 (1976). - STAEHELIN, L.A., ARNTZEN, C.J.: Effects of ions and gravity forces on the supramolecular organization and excitation energy distribution in chloroplast membranes, 147-169. In: CIBA FOUNDATION Symp. 61, New Series: Chlorophyll Organization and Energy Transfer in Photosynthesis. Amsterdam: Excerpta Medica 1979. - STAEHELIN, L.A., ARMOND, P.A., MILLER, K.R.: Chloroplast mem-brane organization at the supramolecular level and its functional implications, 278-315. In: Chlorophyll-Proteins, Reaction Centers and Photosynthetic Membranes, eds. J.M. OLSON, G. HIND. Brookhaven Symp. Biol., No. 28. Upton: Biol. Dep. Brookhaven Natl. Lab. 1977. - STAEHELIN, L.A., GOLECKI, J.R., FULLER, R.C., DREWS, G.: Arch. Micro-biol. 119, 269-277 (1978).

THORNBER, J.P.: Annu. Rev. Plant Physiol. 26, 127-158 (1975); - Gen-eral discussion: grosser structure of chlorophyll-protein complexes, 351-354. In: CIBA FOUNDATION Symp. 61, New Series: Chlorophyll Organi-zation and Energy Transfer in Photosynthesis. Amsterdam: Excerpta Medica 1979. - THORNBER, J.P., OLSON, J.M.: Biochemistry 7, 2242-2249 (1968). - THORNBER, J.P., ALBERTE, R.S., HUNTER, F.A., SHIOZAWA, J.A., KAN, K.-S.: The organization of chlorophyll in the plant photosynthetic unit, 132-148. In: Chlorophyll-Proteins, Reaction Centers and Photo-synthetic Membranes, eds. J.M. OLSON, G. HIND. Brookhaven Symp. Biol., No. 28. Upton: Biol. Dep. Brookhaven Natl. Lab. 1977. - TIEDE, D.M., PRINCE, R.C., DUTTON, P.L.: Biochim. Biophys. Acta 449, 447-468 (1976). - TREBST, A.: Annu. Rev. Plant Physiol. 25, 423-458 (1974); - Organi-zation of the photosynthetic electron transport system of chloroplasts in the thylakoid membrane, 84-95. In: 29. Colloquium der Gesellschaft für Biologische Chemie, Mosbach/Baden, Energy Conservation in Biological Membranes, eds. G. SCHAFER, M. KLINGENBERG. Berlin-Heidelberg-New York: Springer 1978. - TREBST, A., AVRON, M. (eds.): Photosynthesis I. Photo-synthetic Electron Transport and Photophosphorylation. 730 pp. Berlin-Heidelberg-New York: Springer 1977.

VACEK, K., WONG, D., GOVINDJEE: Photochem. Photobiol. 26, 269-276 (1977). - VAN DER REST, M., NOËL, H., GINGRAS, G.: Arch. Biochem. Bio-phys. 164, 285-292 (1974). - VAN GORKOM, H.J., PULLES, M.P.J., WESSELS, J.S.C.: Biochim. Biophys. Acta 408, 331-339 (1975). - VAN METTER, R.L.: Biochim. Biophys. Acta 462, 642-658 (1977). - VERNON, L.P., SHAW, E.R., OGAWA, T., RAVEED, D.: Photochem. Photobiol. 14, 343-357 (1971).

WANG, R.T., MYERS, J.: Reverse energy transfer from chlorophyll to phycobilin in *Anacystis nidulans*, 3-7. In: Photosynthetic Organelles,

Structure and Function, Special Issue of Plant and Cell Physiol., no. 3, eds. S. MIYACHI, S. KATOH, Y. FUJITA, K. SHIBATA. Tokyo Jpn. Soc. Plant Physiol. - Center Acad. Publ. Japan 1977. - WEHRMEYER, W.: Planta 62, 272-293 (1964). - WESSELS, J.S.C.: Biochim. Biophys. Acta 503, 78-93 (1978). - WESSELS, J.S.C., VAN ALPHEN-VAN WAVEREN, O., VOORN, G.: Biochim. Biophys. Acta 292, 741-752 (1973). - WILLIAMS, W.P.: The two photosystems and their interactions, 99-147. In: Primary Processes of Photosynthesis, ed. J. BARBER. Amsterdam: Elsevier 1977.

ZILINSKAS, B.A., ZIMMERMAN, B.K., GANTT, E.: Photochem. Photobiol. 27, 587-595 (1978).

Dr. J. AMESZ
Department of Biophysics
Huygens Laboratory of the
State University
P.O. Box 9504
NL 2300 RA Leiden

IVa. Metabolism of Inorganic Nitrogen Compounds

By Erich Kessler

1. Dinitrogen Reduction

Within the period of time covered by this review, i.e., 1976-1978, again many reviews, symposia, and books on nitrogen fixation have appeared. Some of them are of a more or less general nature (BOTHE, 1976; BRILL, 1977; SHANMUGAM et al., 1978; and the symposia edited by NEWTON and NYMAN, 1976, and NEWTON et al., 1977); others stress mainly biological (HARDY and SILVER, 1977), biochemical (MORTENSON, 1978a; WINTER and BURRIS, 1976; ZUMFT, 1976a,b), or chemical aspects (CHATT, 1977; CHATT et al., 1978). Problems concerned with agriculture and with the utilization of nitrogen fixation have been treated by EVANS and BARBER (1977), HARDY and GIBSON (1977), HOLLAENDER (1977), POSTGATE (1977), and STEWART (1976).

a) Dinitrogen Reduction by Free-Living Organisms

Most of the basic knowledge on the biochemical mechanism of nitrogen fixation was derived from work with *Clostridium* and *Azotobacter* (cf. Progr. Bot. 38, 108). Recent studies with the nitrogenases from other organisms have led to essentially similar results. The MoFe proteins from *Corynebacterium autotrophicum, Bacillus polymyxa,* and *Rhodospirillum rubrum* have molecular weights of 215,000-232,000 daltons, and the Fe proteins of 55,000-73,000 daltons (BERNDT et al., 1978; EMERICH and BURRIS, 1978a; LUDDEN and BURRIS, 1978; NORDLUND et al., 1978).

The further analysis of the MoFe protein yielded values of 2 Mo, 23-33 Fe, and 19-22 atoms of sulfide per molecule. Most authors now agree that the MoFe protein is a tetramer which is composed of two different types of subunit (BERNDT et al., 1978; EMERICH and BURRIS, 1978a; KENNEDY et al., 1976; LUNDELL and HOWARD, 1978; NORDLUND et al., 1978; SWISHER et al., 1977; ZIMMERMANN et al., 1978). An apparently identical cofactor of low molecular weight, which contains 8 Fe and 6 atoms of acid-labile sulfide per Mo, has been isolated from the nitrogenases of *Azotobacter, Clostridium, Klebsiella,* and other bacteria (SHAH and BRILL, 1977; PIENKOS et al., 1977; RAWLINGS et al., 1978). ZUMFT (1978) obtained Mo compounds related to thiomolybdate anions from the MoFe protein of *Clostridium.*

Analyses of the Fe protein indicated the presence of 3-4 Fe and 2-4 atoms of sulfide per molecule (BERNDT et al., 1978; EMERICH and BURRIS, 1978a; NORDLUND et al., 1978; TANAKA et al., 1977). There are two subunits which were claimed to be identical (BERNDT et al., 1978; NORDLUND et al., 1978) or different (LUDDEN and BURRIS, 1978). TANAKA et al. (1977) obtained the complete amino-acid sequence of 273 residues per monomer of the Fe protein from *Clostridium.* The Fe proteins from *Rhodospirillum rubrum* and *Spirillum lipoferum* require activation by an activating factor (LUDDEN and BURRIS, 1976, 1978; LUDDEN et al.,1978).

The hydrolysis of ATP during dinitrogen reduction is coupled to the electron transfer from the Fe protein to the MoFe protein. The former protein (also designated as nitrogenase reductase) serves as a specific reducing agent for the latter, the true nitrogenase. (EADY et al., 1978; HAGEMAN and BURRIS, 1978). Studies by THORNELEY et al. (1978) with nitrogenase from *Klebsiella* suggest the existence of an enzyme-bound intermediate of nitrogen fixation which yields hydrazine upon hydrolysis. Different substrates of nitrogenase, i.e., N_2, C_2H_2, and H^+, seem to be reduced by different conformational forms of the enzyme (THORNELEY and EADY, 1977; RENNIE et al., 1978; cf. also APTE et al., 1978a, for the blue-green alga *Anabaena*).

EMERICH and BURRIS (1978b) studied the complementary functioning of the nitrogenase proteins from 8 different organisms. 45 out of 56 possible heterologous combinations of MoFe protein and Fe protein were catalytically active in the reduction of C_2H_2.

The generation of reducing equivalents for the nitrogenase of *Azotobacter* takes place in the cytoplasmic membrane and is mediated by NADH-flavodoxin oxidoreductase. The flavodoxin is reduced to its hydroquinone which has a redox potential low enough for the reduction of nitrogenase (HAAKER and VEEGER, 1977). In *Klebsiella*, 29 mol of ATP are required for the fixation of one mol N_2 (HILL, 1976).

Nitrogenase has been known for a long time to reduce protons to dihydrogen. On the other hand, most, if not all, nitrogen-fixing organisms possess also an uptake hydrogenase (review: MORTENSON, 1978b), which is capable of scavenging and recycling at least part of the H_2 evolved by the nitrogenase. This reaction serves to provide electrons and energy for nitrogen reduction, and to protect the nitrogenase by means of an oxyhydrogen reaction against inactivation by oxygen in heterotrophic bacteria, photosynthetic bacteria, and blue-green algae (L.A. SMITH et al., 1976; WALKER and YATES, 1978; GOGOTOV, 1978; KELLEY et al., 1977; BOTHE et al., 1977a,b; TEL-OR et al., 1977).

In addition to ammonia, or a product of ammonia assimilation via glutamine synthetase (cf. Progr. Bot. 38, 109), also L-amino acids repress the biosynthesis of nitrogenase in *Klebsiella* (SHANMUGAM and MORANDI, 1976). L-methionine-DL-sulfoximine, an inhibitor of glutamine synthetase, which derepresses nitrogenase in *Rhodospirillum* even in the presence of NH_4^+, leads to a light-dependent excretion of ammonia into the culture medium. This reaction represents a utilization of solar energy for the production of ammonia from N_2 (WEARE and SHANMUGAM, 1976).

The activity of nitrogenase from *Clostridium* is regulated by the ratio MgADP : MgATP (DAVIS and ORME-JOHNSON, 1976). NH_4^+, glutamine, asparagine, and urea produce an immediate, reversible inactivation of the light-dependent reduction of C_2H_2 in *Rhodopseudomonas* (ZUMFT and CASTILLO, 1978).

The genes for nitrogen fixation (*nif* genes) can be transferred into other bacteria by means of conjugation or transduction (cf. Progr. Bot. 38, 109). Plasmids carrying *nif* genes from *Klebsiella pneumoniae* were introduced into *Salmonella typhimurium* (POSTGATE and KRISHNAPILLAI, 1977) and into a *nif⁻* mutant of *Azotobacter vinelandii* (CANNON and POSTGATE, 1976). In the latter case, *Azotobacter* was able to express under aerobic conditions the genes from an anaerobic nitrogen fixer.

Nitrogenase has been shown to be present in the hydrogen bacteria *Corynebacterium autotrophicum* and *Mycobacterium flavum* (BERNDT et al., 1976; DE BONT and LEIJTEN, 1976; KALININSKAYA and NOZHEVNIKOVA, 1977) and in *Thiobacillus ferrooxidans* (MACKINTOSH, 1978).

Development and biochemical activities of the heterocysts of nitrogen-
fixing blue-green algae, or cyanobacteria, are still in the center of
interest (cf. Progr. Bot. 36, 100; review: HASELKORN, 1978). Isolated
heterocysts from *Anabaena* were found to contain 60%-90% of the nitrogenase activity of whole filaments (PETERSON and BURRIS, 1976; PETER-
SON and WOLK, 1978; TEL-OR and STEWART, 1977). In agreement with these
results, mutants without heterocysts are unable to fix nitrogen (PADHY
and SINGH, 1978; SINGH, 1976). The presence of cellulose in the cell
walls of heterocysts could be confirmed by GRANHALL (1976).

The electron transport for dinitrogen reduction in heterocysts of
Anabaena proceeds from glucose-6-phosphate via glucose-6-phosphate dehydrogenase and ferredoxin-NADP oxidoreductase to ferredoxin (APTE et
al., 1978b; LOCKAU et al., 1978). Cyclic photophosphorylation, or oxidative phosphorylation in the dark, provide the ATP (BOTTOMLEY and
STEWART, 1977; RIPPKA and STANIER, 1978).

In the absence of Mo, or in the presence of W, *Plectonema boryanum* syn-
thesizes both proteins of nitrogenase; an addition of Mo restores the
activity of the MoFe protein, even after application of chloramphenicol
(NAGATANI and HASELKORN, 1978). Certain mutants of *Nostoc muscorum*, however, are able to grow in the presence of W or Cr, and they require
these elements instead of Mo for growth with N_2 or nitrate (SINGH et
al., 1978).

Nitrogen fixation in a marine *Oscillatoria* (= *Trichodesmium*) is associated
with special cells localized in the center of colonies which consist
of several hundred trichomes. These cells have a lower concentration
of pigments and they are unable to reduce CO_2 or to evolve O_2 in photosynthesis. This observation seems to explain the occurrence under aerobic conditions of N_2 reduction without heterocysts (CARPENTER and
PRICE, 1976). RIPPKA and WATERBURY (1977) found anaerobic nitrogen
fixation in many genera of nonheterocystous blue-green algae, i.e.,
*Gloeothece, Synechococcus, Dermocarpa, Xenococcus, Myxosarcina, Chroococcidiopsis,
Pleurocapsa, Oscillatoria, Pseudanabaena, Lyngbya, Plectonema,* and *Phormidium*. Only
Gloeothece showed nitrogenase activity also in aerobic conditions.

b) Symbiotic Dinitrogen Reduction

Symbiotic nitrogen fixation has been the subject of a book (NUTMAN, 1976) and
a review (QUISPEL, 1977).

The rather spectacular developments which led to the discovery of dinitrogen reduction in asymbiotic *Rhizobium* have been described in the
previous review (Progr. Bot. 38, 110). Not only cell or callus cultures of leguminous or nonleguminous plants, but also intact seedlings
of *Petunia* are able to induce nitrogenase activity in free *Rhizobium*
(HESS and GÖTZ, 1977). The induction of the enzyme requires the presence of a pentose (e.g., arabinose, ribose, or xylose) and an organic
acid (e.g., succinate, pyruvate, fumarate, malate, or oxoglutarate) as
carbon sources (CHILD and KURZ, 1978; GIBSON et al., 1976). In addition, a nitrogen compound has to be provided (e.g., glutamine, glutamate, asparagine, urea, or NH_4^+: GIBSON et al., 1976). The optimum concentration of oxygen ranges from 0.1%-2% O_2 (KEISTER and EVANS, 1976;
WERNER and STRIPF, 1978). The development of nitrogenase activity in
Rhizobium is inhibited by chloramphenicol and rifampicin; it involves
de novo protein synthesis (WERNER, 1978). The assimilation of the newly fixed N_2 follows the glutamine synthetase-glutamate synthase pathway (BERGENSEN and TURNER, 1978; cf. also MEEKS et al., 1978).

SKOTNICKI and ROLFE (1978) observed a transfer of the genes for nitrogen fixation from *Rhizobium* to *Escherichia coli*. MAIER et al. (1978) and PAGE (1978) achieved the transformation with DNA isolated from *Rhizobium* of a *nif⁻* mutant of *Azotobacter* to *nif⁺*. The transformed *Azotobacter* mutant was not only able to reduce dinitrogen, but had also the ability for nodulation (MAIER et al., 1978).

The ATP-dependent evolution of H_2 by the nitrogenase of root nodules leads to considerable losses of reductant and energy (SCHUBERT and EVANS, 1976). The presence in *Rhizobium* of a hydrogenase, however, allows a recapture and recycling of dihydrogen and increases the efficiency of dinitrogen reduction (SCHUBERT et al., 1978; review: DIXON, 1978).

Nitrate has been known to decrease nitrogen fixation in root nodules. The effective agent seems to be the nitrite produced by nitrate reduction. An addition of nitrite results in a rapid inhibition of nitrogenase (KAMBERGER, 1977; MAGALHÃES et al., 1978; RIGAUD, 1976; RIGAUD and PUPPO, 1977).

Studies on the localization and genetics of leghemoglobin synthesis indicate that it takes place in the cytoplasm of the host cells and is regulated by genes of the legume plant (SIDLOI-LUMBROSO et al., 1978; VERMA and BAL, 1976). Bacteroid enzymes, however, seem to be responsible for the synthesis of the heme component (NADLER and AVISSAR, 1977).

The loose associations or true symbioses of the roots of higher plants (mainly grasses) with nitrogen-fixing *Spirillum lipoferum* have attracted much attention (e.g., KUMARI et al., 1976; NAYAK and RAO, 1977; cf. Progr. Bot. **38**, 111; review: NEYRA and DÖBEREINER, 1977). Also other nitrogen-fixing bacteria, e.g., *Bacillus macerans*, *B. polymyxa*, and *Enterobacter cloacae*, have been found in association with the rhizosphere of grasses (BARBER and EVANS, 1976; NELSON et al., 1976). Work on the effect of *Spirillum lipoferum* on the yield of maize and sorghum, however, has led to contradictory results (increase: R.L. SMITH et al., 1976; no effect: ALBRECHT et al., 1977; BARBER et al., 1976).

The endophytic actinomycetes from root nodules of *Alnus, Hippophaë, Shepherdia, Myrica,* and *Comptonia* have been isolated (CALLAHAM et al., 1978; VAN STRATEN et al., 1977). They are able to reduce C_2H_2. In the case of *Comptonia*, a reinfection leading to the formation of active nodules was possible (CALLAHAM et al., 1978). *Trevoa trinervis* (Rhamnaceae) from Chile was found to possess root nodules capable of nitrogen fixation (RUNDEL and NEEL, 1978). The nitrogen-fixing member of the Ulmaceae (cf. Progr. Bot. **36**, 101) was identified as *Parasponia parviflora* rather than *Trema aspera*. Its root nodules, which contain *Rhizobium*, are similar in structure to those of other nonlegumes (AKKERMANS et al., 1978; TRINICK and GALBRAITH, 1976).

Azotobacter cells, introduced into the mycelium of the mycorrhizal fungus *Rhizopogon* by means of a treatment with polyethylene glycol, retain the ability for nitrogen fixation and make the fungus independent of combined nitrogen (GILES and WHITEHEAD, 1976, 1977).

GRANHALL and HOFSTEN (1976) found a symbiosis of *Sphagnum* with *Noctoc*. The nitrogen-fixing blue-green algae are localized in the water-filled, porous hyaline cells of the moss.

Bacteria capable of dinitrogen reduction were isolated from the guts of termites (FRENCH et al., 1976; POTRIKUS and BREZŇAK, 1977). They could be identified as *Citrobacter freundii* and *Enterobacter agglomerans* and are important for the nitrogen economy of the termites.

2. Nitrate Reduction

a) Assimilatory Nitrate Reduction

The biochemistry of nitrate assimilation has been reviewed by LOSADA (1976) and ZUMFT (1976a).

The heme component of the NAD(P)H-nitrate reductase of eukaryotic plants (cf. Progr. Bot. 36, 101; 38, 111) has been identified as cyto-chrome b-557 (GEWITZ et al., 1978; GUERRERO and GUTIERREZ, 1977; NOTTON et al., 1977; PAN and NASON, 1978). The nitrate reductase from *Neurospora* seems to consist of two equal, cytochrome-containing subunits (molecu-lar weight 115,000 daltons) linked together by a Mo cofactor of low molecular weight (PAN and NASON, 1978). MENDEL and MÜLLER (1978) were able to reconstitute in vitro (and in vivo by protoplast fusion: GLIMELIUS et al., 1978) an active nitrate reductase from two types of tobacco mutants lacking the apoprotein and the Mo cofactor, respec-tively. Similar results were obtained with the apoprotein from Mo-de-ficient spinach and the Mo cofactor (RUCKLIDGE et al., 1976).

Prokaryotic bacteria and blue-green algae, on the other hand, contain a ferredoxin-nitrate reductase (CANDAU et al., 1976). Cell-free prepa-rations from *Anacystis* and *Nostoc* are able to reduce, with their ferre-doxin-dependent nitrate and nitrite reductases, nitrate to ammonia in the light (CANDAU et al., 1976; MANZANO et al., 1976; ORTEGA et al., 1976).

The activity of nitrate reductase is subject to regulation by reductive inactivation and oxidative activation (cf. Progr. Bot. 36, 102; 38, 111). The reduced, inactive form of the enzyme from spinach and *Chlorella* can be reactivated within seconds by illumination with white or blue, but not red, light; FAD accelerates this photoreactivation (APARICIO et al., 1976; ROLDÁN et al., 1978). After growth in blue light, the level of nitrate reductase in maize leaves is 1.5- to 3-fold higher than after growth in red light (JONES and SHEARD, 1977).

The work by VENNESLAND's group on the regulation of nitrate reductase activity in vivo by cyanide (cf. Progr. Bot. 38, 112; SOLOMONSON and SPEHAR, 1977) has led to the discovery of amino acids, especially his-tidine, as precursors of HCN in *Chlorella* and higher plants (GEWITZ et al., 1976). This reaction is catalyzed by a D-amino acid oxidase (PISTORIUS et al., 1977).

In addition to compounds of low molecular weight, which are involved in the regulation of nitrate reductase activity, proteinaceous in-activating factors for nitrate reductase have been isolated from *Neuro-spora* and from leaves and roots of higher plants (JOLLY and TOLBERT, 1978; SORGER et al., 1978; WALLACE, 1978; YAMAYA and OHIRA, 1976, 1978).

Green plants, including blue-green algae, possess a ferredoxin-nitrite reductase (cf. Progr. Bot. 36, 102; 38, 112). The enzyme from spinach contains 3 Fe; one of them is in the siroheme and the other two con-stitute, with two atoms of acid-labile sulfide, an iron-sulfur center (VEGA and KAMIN, 1977).

For fungi and bacteria, on the other hand, a NAD(P)H-nitrite reductase is characteristic. The enzyme from *Neurospora* has a siroheme prosthetic group (GREENBAUM et al., 1978). The nitrite reductase from *Escherichia coli*, however, does not contain any heme (COLEMAN et al., 1978a); an enzyme-NH$_2$OH complex seems to act as an intermediate in the reduction of nitrite to ammonia (COLEMAN et al., 1978b).

The reduction of nitrite in green plants takes place in the chloroplasts and is closely coupled to the photosynthetic electron transport system in the light (cf. Progr. Bot. 36, 102; 38, 112). The reduction of nitrate to nitrite, which occurs in the cytoplasm, is dependent upon NAD(P)H produced by respiration or by the oxidation of intermediates of photosynthetic CO$_2$ reduction. The oxidation of glyceraldehyde-3-phosphate, dihydroxyacetone phosphate, malate, and isocitrate is the main source of NADH (MANN et al., 1978; NEYRA and HAGEMAN, 1976, 1978; RATHNAM, 1978; SAWHNEY et al., 1978).

The assimilation of the ammonia produced by nitrate reduction is catalyzed by glutamine synthetase and glutamate synthase (BAUER et al., 1977; SKOKUT et al., 1978; STEWART and RHODES, 1976).

In leaves of C$_4$ plants, both nitrate reductase and nitrite reductase are localized predominantly in the mesophyll cells (HAREL et al., 1977; NEYRA and HAGEMAN, 1978; RATHNAM and EDWARDS, 1976), in contrast to an earlier report (cf. Progr. Bot. 36, 102).

Nitrate and Mo are required for the synthesis of nitrate reductase. In rose cell cultures, actinomycin D, cycloheximide, and puromycin inhibit only the induction by nitrate and not by Mo. These results indicate that the induction by nitrate involves the mRNA-dependent synthesis of the apoprotein, whereas the activation by Mo of the enzyme is independent of protein synthesis (JONES et al., 1978). CHOUDARY and RAO (1976) were able to separate by means of actinomycin D and cycloheximide the transcription and translation during the induction of nitrate reductase in *Candida*. The presence of nitrate during transcription, but not during translation, is necessary for the biosynthesis of the enzyme to occur. When cells of green algae grown with NH$_4^+$ are transferred into nitrate medium, the synthesis of nitrate reductase is not inhibited by 6-methylpurine. This result suggests that NH$_4^+$-grown algae contain a mRNA for the synthesis of nitrate reductase, and that the control takes place after transcription (HIPKIN and SYRETT, 1977). In *Neurospora*, however, the mRNA for nitrate reductase seems to have a half-life of only 8.5 min (PREMAKUMAR et al., 1978).

SOSA et al. (1978) (cf. also NICHOLS et al., 1978) obtained mutants of *Chlamydomonas* which were unable to grow with nitrate. Some of them have NADH-diaphorase activity but lack the terminal nitrate reductase; others are without NADH-diaphorase activity but contain a functional FMNH$_2$-nitrate reductase. Another group of mutants lacks both activities; finally, there are some strains which are unable to grow with nitrate, although both activities of the nitrate reductase are present.

The carrier-mediated uptake of nitrate is an ATP-dependent process (BUTZ and JACKSON, 1977; EISELE and ULLRICH, 1977; OHMORI et al., 1977; RAO and RAINS, 1976). Ammonia inhibits nitrate uptake in some organisms (OHMORI et al., 1977; RAO and RAINS, 1976; SERRA et al., 1978), but was found to be ineffective in others (TOPINKA, 1978; cf. Progr. Bot. 38, 113). A release into the medium of OH$^-$ with a ratio 1:1 is coupled with the uptake of NO$_3^-$ (EISELE and ULLRICH, 1977; LÖPPERT et al., 1977; RAVEN and JAYASURIYA, 1977).

b) Dissimilatory Nitrate Reduction

Denitrification has been reviewed by DELWICHE and BRYAN (1976) and, with special emphasis on its ecological aspects, by FOCHT and VERSTRAETE (1977).

The respiratory nitrate reductase from *Escherichia coli* is a tetramer with a molecular weight of 880,000 daltons. The monomer consists of subunits of about 150,000 and 60,000 daltons (CLEGG, 1976; DE MOSS, 1977; LUND and DE MOSS, 1976). The monomer contains 12 nonheme Fe, 12 atoms of acid-labile sulfide, 24 cysteine residues, and 1 Mo; there is neither flavin nor cytochrome in the enzyme (LUND and DE MOSS, 1976). Similar enzymes were obtained from *Propionibacterium acidipropionici* and *Proteus mirabilis* (KANEKO and ISHIMOTO, 1978; OLTMANN et al., 1976). A soluble nitrate reductase from *Clostridium perfringens*, which is involved in nitrate fermentation, has a molecular weight of 90,000 daltons. It contains Mo, Fe, and sulfide, does not have subunits, and requires ferredoxin as electron donor (SEKI-CHIBA and ISHIMOTO, 1977).

The dissimilatory nitrite reductase from *Rhodopseudomonas sphaeroides* is soluble, has a molecular weight of 80,000 daltons, with two identical subunits, and contains 2 Cu (SAWADA et al., 1978). A membrane-bound, sulfide-linked nitrite reductase from *Thiobacillus denitrificans*, on the other hand, has a molecular weight of 120,000 daltons and contains cytochromes (SAWHNEY and NICHOLAS, 1978). Both enzymes reduce nitrite to NO.

Acetylene is a specific inhibitor of the reduction of N_2O to N_2 during denitrification (BALDERSTON et al., 1976; YOSHINARI and KNOWLES, 1976). In the presence of C_2H_2, *Pseudomonas* and *Micrococcus* reduce nitrate and nitrite only to N_2O.

The ability for denitrification has been observed in the nitrogen-fixing bacterium *Spirillum lipoferum* (NEYRA et al., 1977) and in a strain of the photosynthetic bacterium *Rhodopseudomonas sphaeroides* (SATOH, 1977; SAWADA et al., 1978).

3. Nitrification

A review on the ecology of nitrification has been published by FOCHT and VERSTRAETE (1977).

Autotrophic growth, coupled to the oxidation of ammonia, has been found in *Nitrosovibrio tenuis* (HARMS et al., 1976), in a halophilic *Nitrosococcus* (KOOPS et al., 1976), and in a thermophilic *Nitrosomonas* from hot springs (GOLOVACHEVA, 1976).

Acetylene and methane are inhibitors of the oxidation of ammonia by *Nitrosomonas* (HYNES and KNOWLES, 1978; SUZUKI et al., 1976). The soluble hydroxylamine oxidoreductase from *Nitrosomonas*, which oxidizes NH_2OH to nitrite, contains cytochromes; its only metal component is Fe (HOOPER et al., 1978).

The nitrite oxidase of *Nitrobacter* is repressed under heterotrophic growth conditions and induced in the presence of nitrite (STEINMÜLLER and BOCK, 1977). The oxidation of nitrite by cytochrome c in particles from *Nitrobacter* is accompanied by a translocation of protons (COBLEY, 1976a,b).

Heterotrophic nitrification of ammonia to nitrite occurs in *Methylococcus* and other methane-oxidizing bacteria (DALTON, 1977; ROMANOVSKAYA et al., 1977). An oxidation of nitrite to nitrate was observed in excised pea roots (SAHULKA and LISÁ, 1978).

References

AKKERMANS, A.D.L., ABDULKADIR, S., TRINICK, M.J.: Nature (London) 274, 190 (1978). - ALBRECHT, S.L., OKON, Y., BURRIS, R.H.: Plant Physiol. 60, 528-531 (1977). - APARICIO, P.J., ROLDÁN, J.M., CALERO, F.: Biochem. Biophys. Res. Commun. 70, 1071-1077 (1976). - APTE, S.K., DAVID, K.A.V., THOMAS, J.: Biochem. Biophys. Res. Commun. 83, 1157-1163 (1978a). - APTE, S.K., ROWELL, P., STEWART, W.D.P.: Proc. R. Soc. Lond. B 200, 1-25 (1978b).

BALDERSTON, W.L., SHERR, B., PAYNE, W.J.: Appl. Environ. Microbiol. 31, 504-508 (1976). - BARBER, L.E., EVANS, H.J.: Can. J. Microbiol. 22, 254-260 (1976). - BARBER, L.E., TJEPKEMA, J.D., RUSSELL, S.A., EVANS, H.J.: Appl. Environ. Microbiol. 32, 108-113 (1976). - BAUER, A., URQUHART, A.A., JOY, K.W.: Plant Physiol. 59, 915-919 (1977). - BERGERSEN, F.J., TURNER, G.L.: Biochim. Biophys. Acta 538, 406-416 (1978). - BERNDT, H., OSTWAL, K.-P., LALUCAT, J., SCHUMANN, C., MAYER, F., SCHLEGEL, H.G.: Arch. Microbiol. 108, 17-26 (1976). - BERNDT, H., LOWE, D.J., YATES, M.G.: Eur. J. Biochem. 86, 133-142 (1978). - BOTHE, H.: Naturwiss. Rundsch. 29, 316-324 (1976). - BOTHE, H., TENNIGKEIT, J., EISBRENNER, G.: Arch. Microbiol. 114, 43-49 (1977a). - BOTHE, H., TENNIGKEIT, J., EISBRENNER, G., YATES, M.G.: Planta 133, 237-242 (1977b). - BOTTOMLEY, P.J., STEWART, W.D.P.: New Phytol. 79, 625-638 (1977). - BRILL, W.J.: Sci. Am. 236, 68-81 (1977). - BUTZ, R.G., JACKSON, W.A.: Phytochemistry 16, 409-417 (1977).

CALLAHAM, D., DEL TREDICI, P., TORREY, J.G.: Science 199, 899-902 (1978). - CANDAU, P., MANZANO, C., LOSADA, M.: Nature (London) 262, 715-717 (1976). - CANNON, F.C., POSTGATE, J.R.: Nature (London) 260, 271-272 (1976). - CARPENTER, E.J., PRICE, C.C.: Science 191, 1278-1280 (1976). - CHATT, J.: Philos. Trans. R. Soc. Lond. B 281, 243-248 (1977). - CHATT, J., DILWORTH, J.R., RICHARDS, R.L.: Chem. Rev. 78, 589-625 (1978). - CHILD, J.J., KURZ, W.G.W.: Can. J. Microbiol. 24, 143-148 (1978). - CHOUDARY, V.P., RAO, G.R.: Biochem. Biophys. Res. Commun. 72, 598-602 (1976). - CLEGG, R.A.: Biochem. J. 153, 533-541 (1976). - COBLEY, J.G.: Biochem. J. 156, 481-491 (1976a); - Biochem. J. 156, 493-498 (1976b). - COLEMAN, K.J., CORNISH-BOWDEN, A., COLE, J.A.: Biochem. J. 175, 483-493 (1978a); - Biochem. J. 175, 495-499 (1978b).

DALTON, H.: Arch. Microbiol. 114, 273-279 (1977). - DAVIS, L.C., ORME-JOHNSON, W.H.: Biochim. Biophys. Acta 452, 42-58 (1976). - DE BONT, J.A.M., LEIJTEN, M.W.M.: Arch. Microbiol. 107, 235-240 (1976). - DELWICHE, C.C., BRYAN, B.A.: Annu. Rev. Microbiol. 30, 241-262 (1976). - DE MOSS, J.A.: J. Biol. Chem. 252, 1696-1701 (1977). - DIXON, R.O.D.: Biochimie 60, 233-236 (1978).

EADY, R.R., LOWE, D.J., THORNELEY, R.N.F.: FEBS Lett. 95, 211-213 (1978). - EISELE, R., ULLRICH, W.R.: Plant Physiol. 59, 18-21 (1977). - EMERICH, D.W., BURRIS, R.H.: Biochim. Biophys. Acta 536, 172-183 (1978a); - J. Bacteriol. 134, 936-943 (1978b). - EVANS, H.J., BARBER, L.E.: Science 197, 332-339 (1977).

FOCHT, D.D., VERSTRAETE, W.: Adv. Microbial Ecol. 1, 135-214 (1977). -

FRENCH, J.R.J., TURNER, G.L., BRADBURY, J.F.: J. Gen. Microbiol. 95, 202-206 (1976).

GEWITZ, H.-S., PISTORIUS, E.K., VOSS, H., VENNESLAND, B.: Planta 131, 149-153 (1976). - GEWITZ, H.-S., PIEFKE, J., VENNESLAND, B.: Planta 141, 323-328 (1978). - GIBSON, A.H., SCOWCROFT, W.R., CHILD, J.J., PAGAN, J.D.: Arch. Microbiol. 108, 45-54 (1976). - GILES, K.L., WHITEHEAD, H.: Science 193, 1125-1126 (1976); - Plant Sci. Lett. 10, 367-372 (1977). - GLIMELIUS, K., ERIKSSON, T., GRAFE, R., MÜLLER, A.J.: Physiol. Plant. 44, 273-277 (1978). - GOGOTOV, I.N.: Biochimie 60, 267-275 (1978). - GOLOVACHEVA, R.S.: Mikrobiologija 45, 377-379 (1976). - GRANHALL, U.: Physiol. Plant. 38, 208-216 (1976). - GRANHALL, U., v. HOFSTEN, A.: Physiol. Plant. 36, 88-94 (1976). - GREENBAUM, P., PRODOUZ, K.N., GARRETT, R.H.: Biochim. Biophys. Acta 526, 52-64 (1978). - GUERRERO, M.G., GUTIERREZ, M.: Biochim. Biophys. Acta 482, 272-285 (1977).

HAAKER, H., VEEGER, C.: Eur. J. Biochem. 77, 1-10 (1977). - HAGEMAN, R.V., BURRIS, R.H.: Proc. Natl. Acad. Sci. USA 75, 2699-2702 (1978). - HARDY, R.W.F., GIBSON, A.H. (eds.): A Treatise on Dinitrogen Fixation. Section IV: Agronomy and Ecology. New York: Wiley-Interscience 1977. - HARDY, R.W.F., SILVER, W.S. (eds.): A Treatise on Dinitrogen Fixation. Section III: Biology. New York: Wiley-Interscience 1977. - HAREL, E., LEA, P.J., MIFLIN, B.J.: Planta 134, 195-200 (1977). - HARMS, H., KOOPS, H.-P., WEHRMANN, H.: Arch. Microbiol. 108, 105-111 (1976). - HASELKORN, R.: Annu. Rev. Plant Physiol. 29, 319-344 (1978). - HESS, D., GÖTZ, E.-M.: Z. Pflanzenphysiol. 85, 185-188 (1977). - HILL, S.: J. Gen. Microbiol. 95, 297-312 (1976). - HIPKIN, C.R., SYRETT, P.J.: J. Exp. Bot. 28, 1270-1277 (1977). - HOLLAENDER, A.: Genetic Engineering for Nitrogen Fixation. New York and London: Plenum Press 1977. - HOOPER, A.B., MAXWELL, P.C., TERRY, K.R.: Biochem. 17, 2984-2989 (1978). - HYNES, R.K., KNOWLES, R.: FEMS Microbiol. Lett. 4, 319-321 (1978).

JOLLY, S.O., TOLBERT, N.E.: Plant Physiol. 62, 197-203 (1978). - JONES, R.W., SHEARD, R.W.: Plant Sci. Lett. 8, 305-311 (1977). - JONES, R.W., ABBOTT, A.J., HEWITT, E.J., BEST, G.R., WATSON, E.F.: Planta 141, 183-189 (1978).

KALININSKAYA, T.A., NOZHEVNIKOVA, A.N.: Izv. Akad. Nauk SSSR Ser. Biol. 201-208 (1977). - KAMBERGER, W.: Arch. Microbiol. 115, 103-108 (1977). - KANEKO, M., ISHIMOTO, M.: J. Biochem. 83, 191-200 (1978). - KEISTER, D.L., EVANS, W.R.: J. Bacteriol. 127, 149-153 (1976). - KELLEY, B.C., MEYER, C.M., GANDY, C., VIGNAIS, P.M.: FEBS Lett. 81, 281-285 (1977). - KENNEDY, C., EADY, R.R., KONDOROSI, E., REKOSH, D.K.: Biochem. J. 155, 383-389 (1976). - KOOPS, H.-P., HARMS, H., WEHRMANN, H.: Arch. Microbiol. 107, 277-282 (1976). - KUMARI, M.L., KAVIMANDAN, S.K., RAO, N.S.S.: Ind. J. Exp. Biol. 14, 638-639 (1976).

LOCKAU, W., PETERSON, R.B., WOLK, C.P., BURRIS, R.H.: Biochim. Biophys. Acta 502, 298-308 (1978). - LÖPPERT, H.G., KRONBERGER, W., KANDELER, R.: Coll. Int. CNRS 258, 283-288 (1977). - LOSADA, M.: J. Mol. Catal. 1, 245-264 (1976). - LUDDEN, P.W., BURRIS, R.H.: Science 194, 424-426 (1976); - Biochem. J. 175, 251-259 (1978). - LUDDEN, P.W., OKON, Y., BURRIS, R.H.: Biochem. J. 173, 1001-1003 (1978). - LUND, K., DE MOSS, J.A.: J. Biol. Chem. 251, 2207-2216 (1976). - LUNDELL, D.J., HOWARD, J.B.: J. Biol. Chem. 253, 3422-3426 (1978).

MACINTOSH, M.E.: J. Gen. Microbiol. 105, 215-218 (1978). - MAGALHÃES, L.M.S., NEYRA, C.A., DÖBEREINER, J.: Arch. Microbiol. 117, 247-252 (1978). - MAIER, R.J., BISHOP, P.E., BRILL, W.J.: J. Bacteriol. 134,

1199-1201 (1978). - MANN, A.F., HUCKLESBY, D.P., HEWITT, E.J.: Planta 140, 261-263 (1978). - MANZANO, C., CANDAU, P., GOMEZ-MORENO, C., RELIMPIO, A.M., LOSADA, M.: Mol. Cell. Biochem. 10, 161-169 (1976). - MEEKS, J.C., WOLK, C.P., SCHILLING, N., SHAFFER, P.W., AVISSAR, Y., CHIEN, W.-S.: Plant Physiol. 61, 980-983 (1978). - MENDEL, R.R., MÜLLER, A.J.: Mol. Gen. Genet. 161, 77-80 (1978). - MORTENSON, L.E.: Curr. Top. Cell. Regul. 13, 179-232 (1978a); - Biochimie 60, 219-223 (1978b).

NADLER, K.D., AVISSAR, Y.J.: Plant Physiol. 60, 433-436 (1977). - NAGATANI, H.H., HASELKORN, R.: J. Bacteriol. 134, 597-605 (1978). - NAYAK, D.N., RAO, V.R.: Arch. Microbiol. 115, 359-360 (1977). - NELSON, A.D., BARBER, L.E., TJEPKEMA, J., RUSSELL, S.A., POWELSON, R., EVANS, H.J., SEIDLER, R.J.: Can. J. Microbiol. 22, 523-530 (1976). - NEWTON, W.E., NYMAN, C.J. (eds.): Proceedings of the 1st international Symposium on Nitrogen Fixation, Vol. 1,2. Pullman: Washington State Univ. Press 1976. - NEWTON, W., POSTGATE, J.R., RODRIGUEZ-BARRUECO, C. (eds.): Recent Developments in Nitrogen Fixation. New York: Academic Press 1977. - NEYRA, C.A., DÖBEREINER, J.: Adv. Agron. 29, 1-38 (1977). - NEYRA, C.A., HAGEMAN, R.H.: Plant Physiol. 58, 726-730 (1976); - Plant Physiol. 62, 618-621 (1978). - NEYRA, C.A., DÖBEREINER, J., LALANDE, R., KNOWLES, R.: Can. J. Microbiol. 23, 300-305 (1977). - NICHOLS, G.L., SHEHATA, S.A.M., SYRETT, P.J.: J. Gen. Microbiol. 108, 79-88 (1978). - NORDLUND, S., ERIKSSON, U., BALTSCHEFFSKY, H.: Biochim. Biophys. Acta 504, 248-254 (1978). - NOTTON, B.A., FIDO, R.J., HEWITT, E.J.: Plant Sci. Lett. 8, 165-170 (1977). - NUTMAN, P.S. (ed.): Symbiotic Nitrogen Fixation in Plants. Int. Biol. Prog. Ser., Vol. 7. Cambridge: Univ. Press 1976.

OHMORI, M., OHMORI, K., STROTMANN, H.: Arch. Microbiol. 114, 225-229 (1977). - OLTMANN, L.F., REIJNDERS, W.N.M., STOUTHAMER, A.H.: Arch. Microbiol. 111, 25-35 (1976). - ORTEGA, T., CASTILLO, F., CÁRDENAS, J.: Biochem. Biophys. Res. Commun. 71, 885-891 (1976).

PADHY, R.N., SINGH, P.K.: Mol. Gen. Genet. 162, 203-211 (1978). - PAGE, W.J.: Can. J. Microbiol. 24, 209-214 (1978). - PAN, S.-S., NASON, A.: Biochim. Biophys. Acta 523, 297-313 (1978). - PETERSON, R.B., BURRIS, R.H.: Arch. Microbiol. 108, 35-40 (1976). - PETERSON, R.B., WOLK, C.P.: Proc. Natl. Acad. Sci. USA 75, 6271-6275 (1978). - PIENKOS, P.T., SHAH, V.K., BRILL, W.J.: Proc. Natl. Acad. Sci. USA 74, 5468-5471 (1977). - PISTORIUS, E.K., GEWITZ, H.-S., VOSS, H., VENNESLAND, B.: Biochim. Biophys. Acta 481, 384-394 (1977). - POSTGATE, J.R.: Philos. Trans. R. Soc. Lond. B 281, 249-260 (1977). - POSTGATE, J.R., KRISHNAPILLAI, V.: J. Gen. Microbiol. 98, 379-385 (1977). - POTRIKUS, C.J., BREZNAK, J.A.: Appl. Environ. Microbiol. 33, 392-399 (1977). - PREMAKUMAR, R., SORGER, G.J., GOODEN, D.: Biochim. Biophys. Acta 519, 275-278 (1978).

QUISPEL, A.: Ber. Dtsch. Bot. Ges. 90, 397-410 (1977).

RAO, K.P., RAINS, D.W.: Plant Physiol. 57, 55-58 (1976). - RATHNAM, C.K.M.: Plant Physiol. 62, 220-223 (1978). - RATHNAM, C.K.M., EDWARDS, G.E.: Plant Physiol. 57, 881-885 (1976). - RAVEN, J.A., JAYASURIYA, H.D.: Coll. Int. CNRS 258, 299-305 (1977). - RAWLINGS, J., SHAH, V.K., CHISNELL, J.R., BRILL, W.J., ZIMMERMANN, R., MÜNCK, E., ORME-JOHNSON, W.H.: J. Biol. Chem. 253, 1001-1004 (1978). - RENNIE, R.J., FUNNELL, A., SMITH, B.E.: FEBS Lett. 91, 158-161 (1978). - RIGAUD, J.: Physiol. Veg. 14, 297-308 (1976). - RIGAUD, J., PUPPO, A.: Biochim. Biophys. Acta 497, 702-706 (1977). - RIPPKA, R., STANIER, R.Y.: J. Gen. Microbiol. 105, 83-94 (1978). - RIPPKA, R., WATERBURY, J.B.: FEMS Microbiol. Lett. 2, 83-86 (1977). - ROLDÁN, J.M., CALERO, F., APARICIO, P.J.:

Z. Pflanzenphysiol. 90, 467-474 (1978). - ROMANOVSKAYA, V.A., SHUROVA,
Z.P., YURCHENKO, V.V., TKACHUK, L.V., MALASHENKO, Y.R.: Mikrobiologija
46, 66-70 (1977). - RUCKLIDGE, G., NOTTON, B., HEWITT, E.: Biochem.
Soc. Trans. 4, 77-80 (1976). - RUNDEL, P.W., NEEL, J.W.: Flora 167,
127-132 (1978).

SAHULKA, J., LISÁ, L.: Biol. Plant. 20, 359-367 (1978). - SATOH, T.:
Arch. Microbiol. 115, 293-298 (1977). - SAWADA, E., SATOH, T., KITA-
MURA, H.: Plant Cell Physiol. 19, 1339-1351 (1978). - SAWHNEY, V.,
NICHOLAS, D.J.D.: J. Gen. Microbiol. 106, 119-128 (1978). - SAWHNEY,
S.K., NAIK, M.S., NICHOLAS, D.J.D.: Biochem. Biophys. Res. Commun. 81,
1209-1216 (1978). - SCHUBERT, K.R., EVANS, H.J.: Proc. Natl. Acad.
Sci. USA 73, 1207-1211 (1976). - SCHUBERT, K.R., JENNINGS, N.T.,
EVANS, H.J.: Plant Physiol. 61, 398-401 (1978). - SEKI-CHIBA, S.,
ISHIMOTO, M.: J. Biochem. 82, 1663-1671 (1977). - SERRA, J.L., LLAMA,
M.J., CADENAS, E.: Plant Physiol. 62, 991-994 (1978). - SHAH, V.K.,
BRILL, W.J.: Proc. Natl. Acad. Sci. USA 74, 3249-3253 (1977). -
SHANMUGAM, K.T., MORANDI, C.: Biochim. Biophys. Acta 437, 322-332
(1976). - SHANMUGAM, K.T., O'GARA, F., ANDERSEN, K., VALENTINE, R.C.:
Annu. Rev. Plant Physiol. 29, 263-276 (1978). - SIDLOI-LUMBROSO, R.,
KLEIMAN, L., SCHULMAN, H.M.: Nature (London) 273, 558-560 (1978). -
SINGH, H.N., VAISHAMPAYAN, A., SINGH, R.K.: Biochem. Biophys. Res.
Commun. 81, 67-74 (1978). - SINGH, P.K.: Z. Allg. Mikrobiol. 16,
453-463 (1976). - SKOKUT, T.A., WOLK, C.P., THOMAS, J., MEEKS, J.C.,
SHAFFER, P.W., CHIEN, W.-S.: Plant Physiol. 62, 299-304 (1978). -
SKOTNICKI, M.L., ROLFE, B.G.: J. Bacteriol. 133, 518-526 (1978). -
SMITH, L.A., HILL, S., YATES, M.G.: Nature (London) 262, 209-210
(1976). - SMITH, R.L., BOUTON, J.H., SCHANK, S.C., QUESENBERRY, K.H.,
TYLER, M.E., MILAM, J.R., GASKINS, M.H., LITTELL, R.C.: Science 193,
1003-1005 (1976). - SOLOMONSON, L.P., SPEHAR, A.M.: Nature (London) 265,
373-375 (1977). - SORGER, G.J., PREMAKUMAR, R., GOODEN, D.: Biochim.
Biophys. Acta 540, 33-47 (1978). - SOSA, F.M., ORTEGA, T., BAREA,
J.L.: Plant Sci. Lett. 11, 51-58 (1978). - STEINMÜLLER, W., BOCK, E.:
Arch. Microbiol. 115, 51-54 (1977). - STEWART, G.R., RHODES, D.:
FEBS Lett. 64, 296-299 (1976). - STEWART, W.D.P.: Philos. Trans. R.
Soc. Lond. B 274, 341-358 (1976). - STRATEN, J. VAN, AKKERMANS, A.D.L.,
ROELOFSEN, W.: Nature (London) 266, 257-258 (1977). - SUZUKI, I.,
KWOK, S.-C., DULAR, U.: FEBS Lett. 72, 117-120 (1976). - SWISHER, R.H.,
LANDT, M.L., REITHEL, F.J.: Biochem. J. 163, 427-432 (1977).

TANAKA, M., HANIU, M., YASUNOBU, K.T., MORTENSON, L.E.: J. Biol. Chem.
252, 7093-7100 (1977). - TEL-OR, E., STEWART, W.D.P.: Proc. R. Soc.
Lond. B 198, 61-86 (1977). - TEL-OR, E., LUIJK, L.W., PACKER, L.:
FEBS Lett. 78, 49-52 (1977). - THORNELEY, R.N.F., EADY, R.R.: Biochem.
J. 167, 457-461 (1977). - THORNELEY, R.N.F., EADY, R.R., LOWE, D.J.:
Nature (London) 272, 557-558 (1978). - TOPINKA, J.A.: J. Phycol. 14,
241-247 (1978). - TRINICK, M.J., GALBRAITH, J.: Arch. Microbiol. 108,
159-166 (1976).

VEGA, J.M., KAMIN, H.: J. Biol. Chem. 252, 896-909 (1977). - VERMA,
D.P.S., BAL, A.K.: Proc. Natl. Acad. Sci. USA 73, 3843-3847 (1976).

WALKER, C.C., YATES, M.G.: Biochimie 60, 225-231 (1978). - WALLACE,
W.: Biochim. Biophys. Acta 524, 418-427 (1978). - WEARE, N.M., SHAN-
MUGAM, K.T.: Arch. Microbiol. 110, 207-213 (1976). - WERNER, D.:
Z. Naturforsch. 33c, 859-862 (1978). - WERNER, D., STRIPF, R.: Z. Na-
turforsch. 33c, 245-252 (1978). - WINTER, H.C., BURRIS, R.H.: Annu.
Rev. Biochem. 45, 409-426 (1976).

YAMAYA, T., OHIRA, K.: Plant Cell Physiol. 17, 633-641 (1976); - Plant
Cell Physiol. 19, 211-220 (1978). - YOSHINARI, T., KNOWLES, R.:

Biochem. Biophys. Res. Commun. 69, 705-710 (1976).

ZIMMERMANN, R., MÜNCK, E., BRILL, W.J., SHAH, V.K., HENZL, M.T.,
RAWLINGS, J., ORME-JOHNSON, W.H.: Biochim. Biophys. Acta 537, 185-207
(1978). - ZUMFT, W.G.: Naturwissenschaften 63, 457-464 (1976a); -
Struct. Bonding (Berlin) 29, 1-65 (1976b); - Eur. J. Biochem. 91,
345-350 (1978). - ZUMFT, W.G., CASTILLO, F.: Arch. Microbiol. 117,
53-60 (1978).

Professor Dr. ERICH KESSLER
Institut für Botanik und Pharmazeutische
Biologie der Universität
Schloßgarten 4
D 8520 Erlangen

IVb. Metabolism of Organic N-Compounds

Ammonium Assimilation and Amino Acid Metabolism

By Thomas Hartmann

The physiology and biochemistry of plant amino acid metabolism has been reviewed by MIFLIN and LEA (1977); a text-book covering the literature up to
1973 has been published by BEEVERS (1976).

1. Ammonium Assimilation

For many years the reductive amination of 2-oxoglutarate catalyzed by the enzyme glutamate dehydrogenase (GDH) has been considered as the main reaction
of ammonia assimilation in plants. Now there is convincing evidence (cf. Prog.
Bot. 38, 118) that most of the inorganic nitrogen available to a plant is incorporated into the amide-amino group of glutamine via the enzyme glutamine
synthetase (GS). Subsequently this nitrogen can be transferred to the 2-posi-
tion of 2-oxoglutarate yielding two moles of glutamate. This amido-transfer
reaction is catalyzed by the enzyme glutamate synthase (GOGAT; glutamine:
2-oxoglutarate amino transferase NAD(P)H oxidizing (EC 2.6.1.53) or reduced
feredoxin oxidizing (EC 1.4.7.1)). The coupled action of the two enzymes (the
GS/GOGAT-pathway) generates both glutamine, the essential substrate of the
various amide transfer reactions, as well as glutamate which supplies the
amino acid metabolism with α-amino-nitrogen. The GS/GOGAT system has been established as an obligatory component of the different pathways of inorganic
nitrogen assimilation known in phototrophic and heterotrophic plant tissues.
Furthermore it is involved in the process of organic nitrogen interconversion
taking place especially in developing and germinating seeds during synthesis
and mobilization of nitrogenous storage compounds. As a consequence, the function of plant GDH as an assimilatory enzyme has become obscure. It may operate
in special situations of elevated intracellular ammonium levels auxiliary to
GS to prevent ammonia toxification.

The current state of ammonium assimilation has been reviewed by MIFLIN
and LEA (1976, 1977), the metabolic detoxification of ammonia in plant
tissues by GIVAN (1979).

a) Ammonium Assimilation in Phototrophic Tissues

The common way to trace the pathway of NH_4^+ assimilation in whole plants
or plant tissues is to follow the specific incorporation of ^{15}N-label-
ed precursors. In recent studies the introduction of inhibitors of the
assimilatory enzymes has proved very valuable to discriminate between
the GS/GOGAT- and the GDH-pathway. Most frequently methionine sulfoximine, a potent inhibitor of GS (TATE and MEISTER, 1973), and azaserine
(O-diazoacetyl-L-serine), an inhibitor of GOGAT and other glutamine-
amido transfer reactions (LEA and NORRIS, 1976), are used. For instance
when $^{15}NO_3^-$ is fed to photosynthesizing leaves of *Datura stramonium* the
highest accumulation of ^{15}N occurred in glutamine, but the highest ^{15}N
enrichment was found in glutamate, thus suggesting the operation of

both GS/GOGAT and GDH (LEWIS and PROBYN, 1978). However, when the same experiment was carried out in the presence of methionine sulfoximine, which does not interfere with GDH, the supply of newly reduced ^{15}N to glutamate and other amino acids was completely restricted. The observed persistance of a large ^{14}N-glutamine pool not easily available to glutamate synthesis explains the above-mentioned lower ^{15}N-enrichment in glutamine than in glutamate. The results indicate the perhaps exclusive role of the GS/GOGAT pathway in NH_4^+ assimilation of *Datura* leaves. The same conclusion was reached in similar studies with spinach mesophyll cells (ITO et al., 1978).

GOGAT has been purified from *Vicia faba* leaves (WALLSGROVE and MIFLIN, 1977; WALLSGROVE et al., 1977). It has a molecular weight of 145,000 and does not appear to contain nonhem iron. The enzyme is essentially specific for glutamine (K_M 330 µmol/l), 2-oxo-glutarate (K_M 150 µmol/l) and ferredoxine (K_M 2 µmol/l). Localization studies using protoplasts obtained from pea mesophyll cells revealed that GOGAT is located solely in the chloroplasts, whereas GS occurs in the chloroplasts as well as in the cytosol (WALLSGROVE et al., 1979). RATHNAM and EDWARDS (1976) found that at least 74% of GS activity of maize mesophyll cells is of plastid origin. In a continuation of this work HAREL et al. (1977) established that GS and GOGAT occur in the mesophyll as well as in the bundle sheet tissue, whereas nitrate and nitrite reductase appeared to be restricted to the mesophyll, the proposed locus of nitrate reduction (MAYNE et al., 1971). GOGAT is already active in etiolated wheat leaves and increases in activity during illumination (NICKLISCH et al., 1977).

Further evidence that the light-dependent NH_4^+ assimilation in intact chloroplasts occurs via the GS/GOGAT-pathway comes from experiments in which the reductant dependent step has been coupled to O_2-evolution. Oxygen evolution by intact illuminated chloroplasts is dependent on glutamine and 2-oxoglutarate (ANDERSON and DONE, 1977a) or for complete assimilation on NH_4^+, 2-oxoglutarate, ADP, PPi, and Mg^{2+} (ANDERSON and DONE, 1977b). Since both reactions are strongly inhibited by azaserine, a participation of photosynthetically linked GDH could be ruled out.

The regulation of NH_4^+ assimilation has been studied with *Lemna minor* grown on different nitrogen sources. RHODES et al. (1976) observed a small but significant reduction in level of GS and GOGAT and an increase in NAD-dependent GDH in *Lemna* grown in the presence of high (up to 10 mmol/l) NH_4^+ concentrations. Since under these conditions the tissue concentrations of NH_4^+ are in the range of 10-30 mmol/l participation of GDH in NH_4^+ assimilation is suggested (STEWART and RHODES, 1977a). On the contrary in the presence of NO_3^-, which obviously is the more physiological substrate, NH_4^+ was found to proceed exclusively via the GS/GOGAT pathway (STEWART and RHODES, 1976). A rapid inactivation of GS was observed in *Lemna* plants transferred in the dark, concomitantly the intracellular level of glutamine decreased and that of glutamate increased. The effects are reversed by re-illumination (STEWART and RHODES, 1977b). The changes in GS activity are apparently caused by a reversible inactivation. In vitro reactivation is temperature-dependent and requires ATP, glutamate, Mg^{2+}, and NADPH. Presumably this mechanism serves as a means of limiting the production of glutamine at the expense of ATP and perhaps glutamate which cannot be replenished in the absence of light. The mechanisms by which nitrogen assimilation in the chloroplast may be controlled have excellently been discussed in detail by MIFLIN (1977).

The pathway of nitrogen assimilation in the N_2-fixing blue-green alga
Anabaena cylindrica has further been studied using the ^{13}N-labeling technique (see Prog. Bot. 38, 123). It has been established that the NH_4^+
produced by the algal nitrogenase is exclusively incorporated into
organic nitrogenous compounds via the GS/GOGAT pathway (WOLK et al.,
1976; MEEKS et al., 1977). The whole process represents an interesting example of cellular compartmentation. Glutamine is synthesized
in the heterocysts by GS which is metabolically coupled to nitrogenase. Glutamine in turn is translocated to the vegetative cells where
glutamate is formed via GOGAT which seems not to occur in the heterocysts (THOMAS et al., 1977).

b) Ammonium Assimilation in Heterotrophic Tissues

Further attention has been drawn to the pathway of nitrogen assimilation in plant heterotrophic tissues and its compartmentation. Apparently the intracellular distribution of the enzymes involved in nitrogen
assimilation corresponds to the pattern observed in green tissues. In
root apices of young peas nitrate reductase was found in the cytosol,
nitrite reductase and GOGAT are restricted to the plastids, and GS
shares a location between cytosol and plastid; GDH was only detected
in the mitochondria (EMES and FOWLER, 1979a). In contrast WASHITANI
and SATO (1977a) identified NADP-dependent GDH in a plastid fraction
from cultured tobacco cells accompanied by GS and GOGAT (WASHITANI
and SATO, 1977b). The plastid localization of at least two reductive
enzymes involved in heterotrophic nitrogen assimilation raises the
question for the supply with reducing equivalents. Recently JESSUP
and FOWLER (1977) observed an increased pentose phosphate pathway
carbon flux in actively nitrogen-assimilating sycamore cell cultures.
Subsequently EMES and FOWLER (1979b) established that in addition to
a location within the cytosol all the enzymes of the pentose phosphate
pathway are also present in the plastids of pea root apices. Therefore
a generation of NADPH within the plastid may supply at least nitrite
reductase. With GOGAT the situation is somewhat more complex since
its coenzyme requirement is still conflicting. Although most enzymes
from heterotrophic tissues are active with both nicotinamide nucleotides, they seem to prefer NAD (CHIU and SHARGOOL, 1979; SODEK and
DA SILVA, 1977; STOREY and REPORTER, 1978; BEEVERS and STOREY, 1976).
Furthermore GOGAT activity in rice roots is essentially dependent on
ferredoxin (ARIMA, 1978), whereas the enzyme of lupin root nodules
is NADH-specific (BOLAND and BENNY, 1977). DOUGALL and BLOCH (1976)
presented evidence for two GOGATs in carrot cell cultures, each specific for one pyridine nucleotide.

The pathway of nitrogen assimilation in tobacco cell cultures has been
elegantly followed using ^{13}N-labeled precursors (SKOKUT et al., 1978).
The kinetics of labeling and pulse chase studies with [13]NO_3 and [13]NH_4^+
showed that ^{13}N entered glutamate mainly via glutamine. Methionine
sulfoximine inhibits the incorporation of [13]NH_4^+ from [13]NO_3^- into glutamate more strongly than it inhibits the incorporation into glutamine,
indicating that the assimilation may occur solely via the GS/GOGAT
pathway. Only when [13]NH_4^+ is fed to the cultures may a minor portion
of ^{13}N be incorporated into glutamate via GDH. In soybean cell cultures CHIU and SHARGOOL (1979) observed a positive correlation between
growth and the levels of GS and GOGAT, but not GDH in response to variations in the NO_3^- or NH_4^+ content of the medium.

Several lines of evidence have established that in symbiotic N_2-fixa-
tion of legume root nodules the NH_4^+ produced by the bacteroid nitrogenase is solely assimilated into organic nitrogen via the GS/GOGAT

system of the host plant. GS has been purified from the plant cytosol of soybean root nodules (McPARLAND et al., 1976). The enzyme accounts for 2% of the total cytosolic protein; it has a molecular weight of 376,000 and is composed of eight subunits arranged in two sets of planar tetramers. GOGAT has been isolated from the cytosol of *Lupinus angustifolia* nodules (BOLAND and BENNY, 1977). It consists of a single polypeptide chain with a molecular weight of 235,000 which is considerably higher than that of the chloroplast enzyme. It is specific for NADH (K_M 1.3 µmol/l) and inactive with ferredoxin. Both enzymes are present in symbiotic systems of a great number of legumes (BOLAND et al., 1978). Plant GS may account for more than 95% of total nodule GS activity (PLANQUE et al., 1977). In nodule bacteroids (BISHOP et al., 1976) and free-living rhizobia induced for their N_2-fixation system (O'GARA and SHANMUGAM, 1976), GS (and GOGAT) appeared to be repressed. As a consequence most newly fixed N_2 is excreted from the cells as NH_4^+. Glutamate auxotrophs of *Rhizobium* lacking GOGAT activity were still effective in symbiotic N_2-fixation, whereas mutants lacking GS were not (KONDOROSI et al., 1977). Rhizobial GS appears to be involved in both the regulation of nitrogenase activity and in shutting off the assimilation of NH_4^+ as part of the overall shift in nodule metabolism from bacterial to plant cells (LUDWIG and SIGNER, 1977). Finally SCOTT et al. (1976) detected a glutamine-dependent asparagine synthetase in the plant cytosol of lupin nodules. The enzyme level increased rapidly during nodule development. This enzyme in cooperation with GS, GOGAT and a glutamate-oxaloacetate amino transferase produces asparagine, the final product of nodule nitrogen metabolism, which is known to be the predominant amino acid translocated to the shoot.

2. Biosynthesis and Metabolism of Asparagine

Glutamine-dependent asparagine synthase has been purified from germinating cotyledons of *Lupinus luteus* (ROGNES, 1975) and *L. albus* (LEA and FOWDEN, 1975a). It catalyzes the reaction: L-aspartate + L-glutamine + ATP $\rightleftharpoons$ L-asparagine + L-glutamate + AMP + PPi.

The enzyme occurs in the cytosol; it has a low K_M for glutamine (0.16 and 0.04 mmol/l, respectively) and a high K_m for NH_4^+ (approx. 2 mmol/l) which at least in vitro can replace glutamine as a substrate. Since asparagine shares the same cell compartment with GS, which has a much higher affinity for NH_4^+ (K_M approx. 0.02 mmol/l), a direct incorporation of NH_4^+ into asparagine seems unlikely. This idea could be proven by STEWART, who demonstrated that in soybean leaves NH_4^+- but not glutamine-stimulated asparagine synthesis is inhibited in the presence of methionine sulfoximine. Similar conclusions were reached from tracer studies with rice roots (KANAMORI and MATSUMOTO, 1975) and the kinetic studies of BAUER et al. (1977a).

Although asparagine is one of the major N-translocates and low molecular N-storage constituents in plants, our knowledge of its metabolism is still limited. Three possible enzymatic routes of asparagine breakdown have been discussed by LEA and FOWDEN (1975b): (a) conversion of asparagine to aspartate and NH_4^+ by asparaginase (b) transamination to 2-oxosuccinamate (c) asparagine: 2-oxoglutarate amido transamination, a reaction which would be analogous to that of GOGAT which, however, has not been substantiated so far (see MIFLIN and LEA, 1977). Asparaginase has been purified from maturing lupin seeds (LEA et al., 1978). Although this enzyme has a high K_M for asparagine (12.2 mmol/l), it might operate efficiently in vivo because the growing seed is supplied

by fairly high concentrations of asparagine (30 mmol/l calculated for
the phloem) (ATKINS et al., 1975). Furthermore feeding of asparagine-
[14]C,-^{15}N-amide to lupins has shown that the amide-N is metabolized in
a different way to the carbon skeleton; a result which would be consistent with asparaginase action. From ^{15}N labeling studies BAUER et
al. (1977a) suggested that α-transamination plays an important role in
the distribution of nitrogen from asparagine in growing pea leaves.
Subsequently LLOYD and JOY (1978) identified 2-hydroxysuccinamate as
a major intermediate which is formed rapidly from 2-oxosuccinamate,
the product of asparagine transamination. Hydroxysuccinamate accumulates in leaves but is also metabolized particularly in the dark. Again
in tracer experiments with ^{15}N-labeling in growing pea leaves BAUER et
al. (1977b) observed a previously unrecognized direct transfer of as-
paragine-N into glutamine, which did not appear to involve free NH_4^+.

3. Interconversion of Amide-N and Amino-N in Developing Seeds

[14]C tracer studies have established that the biosynthetic pathway of
protein amino acids is operative in developing seeds (SODEK, 1976;
MACNICOL, 1977). In many plants glutamine and asparagine are the major
N-compounds supplied to the developing seeds via the phloem translocation. ^{15}N-tracer studies have shown that the amide-N is readily incorporated into the amino groups of the various amino acids which are
needed for the biosynthesis of seed storage proteins (ATKINS et al.,
1975; LEWIS, 1975; LEWIS and PATE, 1973). The suggestion of LEWIS that
GOGAT may be responsible for the transfer of glutamine amide-N into
amino-N has now been verified for developing legume seeds (BEEVERS
and STOREY, 1976; STOREY and REPORTER, 1978) and maize endosperm
(SODEK and DA SILVA, 1977). STOREY and BEEVERS (1978) studied the
changes in protein and amino acid content as well as GS and GOGAT activity in the leaf and subtended fruit of peas during seed development. The high GS activity in the supply organs (leaf and pod) could
be related to the re-assimilation of nitrogen released during the
break-down of leaf and pod proteins (STOREY and BEEVERS, 1977). Thus
the nitrogen is channeled into the glutamine, which in turn is translocated to the developing cotyledons and subsequently transferred to
2-amino-N via cotyledonary GOGAT which reaches its maximal activity
during the phase of embryo growth and protein synthesis.

4. Glycine, Serine

Glycine and serine formation in photosynthesizing plant tissues is
thought to proceed mainly via the glycolate pathway during photorespiration. In this pathway two glycine molecules are converted to serine,
CO_2 and NH_4^+. The reaction takes place in leaf mitochondria. The NH_4^+
released may be reassimilated by cytosolic GS rather than mitochondrial
GDH (KEYS et al., 1978). SERVAITES and ORGEN (1977) postulated an alternative pathway of serine formation in soybean leaf cells, since
they found that about 50% of serine synthesis remained although the
glycolate pathway was completely inhibited. Subsequently LARSSON and
ALBERTSSON (1979) established that spinach chloroplasts contain all
enzymes required for the conversion of 3-phosphoglycerate into serine.
The presence of the "phosphorylated pathway of serine biosynthesis"
in chloroplasts is further indicated by the detection of phosphoserine
and phosphohydroxypyruvate in chloroplasts (CHAPMAN and LEECH, 1976;
DALEY and BIDWELL, 1977).

5. Ornithine, Arginine, Proline

It is now well documented that higher plants as bacteria synthesize
ornithine from glutamate via N-acetylated intermediates. N-Acetyla-
tion may occur either by acetyl-CoA-glutamate-transacetylation or
acetylornithine-glutamate-transacetylation (McKAY and SHARGOOL, 1977;
MORRIS and THOMPSON, 1977). The first reaction is needed to provide
de novo formation of acetylated intermediates, the second would promote ornithine/arginine formation while preserving the energy required
for acetylation. Arginine is a potent feed-back inhibitor of acetyl-
CoA-glutamate-transacetylase as well as acetylglutamate phosphokinase
(MORRIS and THOMPSON, 1977).

Localization studies starting with protoplasts obtained from soybean
cell cultures revealed that glutamine-dependent carbamoylphophate
synthetase and ornithine carbamoyl transferase are localized in plastids, whereas argininosuccinate synthase and argininosuccinate-lyase
appeared to be cytosolic (SHARGOOL et al., 1978). By contrast GLENN
and MARETZKI (1977) reported on distinct cytosolic and mitochondrial
isoenzymes of ornithine transcarbamoylase. However, the authors did
not examine their "mitochondrial 10,000 g pellet" obtained by differential centrifugation for plastid contamination.

Proline can be synthesized from ornithine by transamination (LU and
MAZELIS, 1975) and reduction of the resulting pyrroline-5-carboxylate
(MILER and STEWART, 1976). A complete oxidative degradation of proline was found to proceed in plant mitochondria (BOGGESS and KOEPPE,
1978).

6. Lysine, Threonine, Methionine

Although some of the enzymatic steps in lysine biosynthesis still require elucidation, there is no doubt that in higher plants the three
amino acids are synthesized via the same branched sequence of reactions as in microorganisms, except for one deviation: in plants O-
phosphohomoserine rather than homoserine is the branchpoint for methionine synthesis. O-Phosphohomoserine has been established as the dominant physiological substrate for cystathionine biosynthesis via trans-
sulfuration (DATKO et al., 1974) or direct sulfhydration (DATKO et al.,
1977). For lysine biosynthesis at least dihydrodipicolinate synthase
(CHESHIRE and MIFLIN, 1975; MAZELIS et al., 1977), the first enzyme
in the pathway, and diaminopimelate decarboxylase (MAZELIS and CREVE-
LING, 1978), the final enzyme, could be isolated from plants. The
presence of these two key enzymes in plants and the effective co-regu-
lation of the lysine and threonine/methionine pathways (see below) indicate that the diaminopimelate pathway is perhaps the exclusive pathway of lysine synthesis in plants. The 2-aminoadipate pathway, often
discussed as an alternative route for lysine synthesis in plants (Prog.
Bot. 36, 91) could not be confirmed. Studies with labeled 2-aminoadipate
and saccharopine revealed no evidence for a preferential incorporation
into lysine (MOLLER, 1976a,b).

A number of studies have dealt with the regulation of the biosynthesis
of the aspartate derived amino acids. Similarly as in microorganisms
(Prog. Bot. 32, 82) a great diversity of individual patterns of feedback control was found to exist in higher plants. In carrot roots and
cell cultures a strictly additive control of aspartate kinase by lysine

and threonine was observed (DAVIES and MIFLIN, 1977). Subsequently
lysine- and threonine-sensitive isoenzymes of aspartate kinase were
isolated (SAKANO and KOMAMINE, 1978; DAVIES and MIFLIN, 1978) and in
addition single species of lysine-sensitive dihydropicolinate synthase
and threonine-sensitive homoserine dehydrogenase (MATTHEWS and WIDHOLM,
1978). By contrast only single species of aspartate kinase which are
dominantly regulated by lysine have been isolated from barley (AARNES,
1977; SHEWRY and MIFLIN, 1977) and maize (CHESHIRE and MIFLIN, 1975;
HENKE and WAHNBAECK, 1977). In both organisms homoserine dehydrogenase
exists in two or more threonine-sensitive and threonine-insensitive
molecular forms (AARNES, 1977; MATTHEWS et al., 1975). BRIGHT et al.
(1978a) observed that addition of lysine plus threonine causes a synergistic growth inhibition of isolated wheat and barley embryos, which
is specifically relieved by methionine, homocysteine, and homoserine.
Whereas threonine synthesis is inhibited almost completely in the presence of threonine alone or threonine plus lysine, methionine synthesis
is inhibited only in the presence of both endproducts (BRIGHT et al.,
1978b). The authors suggest that the synergistic effect may be caused
by homoserine dehydrogenase becoming more sensitive to threonine inhibition as its substrate aspartate semialdehyde is depleted by lysine
inhibition of aspartate kinase. Methionine synthesis itself is sensi-
tively controlled by an interesting compensatory mechanism (MADISON
and THOMPSON, 1976; THOEN et al., 1978): Threonine synthase is strongly activated by S-adenosylmethionine (a product of methionine synthesis). Thus in the presence of sufficient methionine O-phosphohomoserine
is channeled into threonine/isoleucine synthesis. In the absence of S-
adenosylmethionine threonine synthesis is almost completely restricted
(AARNES, 1978).

Recent evidence indicates that chloroplasts are the major sites of the
synthesis of the aspartate-family amino acids (MILLS and WILSON, 1978).
Several enzymes of the pathway including homoserine dehydrogenase
(BRYAN et al., 1977) and diaminopimelate decarboxylase (MAZELIS et al.,
1976) have been found in chloroplasts.

7. Aromatic Amino Acids

There is increasing evidence that the shikimi acid pathway which has
been well established in higher plants (Prog. Bot. 34, 159) takes place
at least partially in the chloroplasts. BICKEL and SCHULTZ (1979) observed an incorporation of ^{14}C from $^{14}CO_2$ and ^{14}C-shikimate into the
aromatic amino acids in suspensions of intact spinach chloroplasts.
Furthermore phenylalanine and tyrosine were found to exert endproduct
control over their own rates of synthesis, whereas tryptophan controls
the synthesis of all three amino acids. All enzymes required for the
conversion of chorismate into tryptophane were found in pea etioplasts
(GROSSE, 1976). Shikimate dehydrogenase occurs predominantly associated with plastids in various green and nongreen plant tissues (FEIER-
ABEND and BRASSEL, 1977).

References

AARNES, H.: Plant Sci. Lett. 9, 137-145 (1977); - Planta 140, 185-192
(1978). - ANDERSON, J.W., DONE, J.: Plant Physiol. 60, 354-359 (1977a);
- Plant Physiol. 60, 504-508 (1977b). - ARIMA, Y.: Plant Cell Physiol.

19, 955-961 (1978). - ATKINS, C.A., PATE, J.S., SHARKEY, P.J.: Plant
Physiol. 56, 807-812 (1975).

BAUER, A., URQUHART, A.A., JOY, K.W.: Plant Physiol. 59, 915-919
(1977a). - BAUER, A., JOY, K.W., URQUHART, A.A.: Plant Physiol. 59,
920-924 (1977b). - BEEVERS, L.: Nitrogen Metabolism in Plants. 333 p.
London: Arnold 1976. - BEEVERS, L., STOREY, R.: Plant Physiol. 57,
862-866 (1976). - BICKEL, H., SCHULTZ, G.: Phytochemistry 18, 498-
499 (1979). - BISHOP, P.E.: GUEVARA, J.G., ENGELKE, J.A., EVANS, H.J.:
Plant Physiol. 57, 542-546 (1976). - BOGGESS, S.F., KOEPPE, D.E.: Plant
Physiol. 62, 22-25 (1978). - BOLAND, M.J., BENNY, A.G.: Eur. J. Bio-
chem. 79, 355-362 (1977). - BOLAND, M.J., FORDYCE, A.M., GREENWOOD,
R.M.: Aust. J. Plant. Physiol. 5, 553-559 (1978). - BRIGHT, S.W.J.,
WOOD, E.A., MIFLIN, B.J.: Planta 139, 113-117 (1978a). - BRIGHT,
S.W.J., SHEWRY, P.R., MIFLIN, B.J.: Planta 139, 119-125 (1978b). -
BRYAN, J.K., LISSIK, E.A., MATTHEWS, B.F.: Plant Physiol. 59, 673-
679 (1977).

CHAPMAN, D.J., LEECH, R.M.: FEBS Lett. 68, 160-164 (1976). - CHESHIRE,
R.M., MIFLIN, B.J.: Phytochemistry 14, 695-698 (1975). - CHIU, J.Y.,
SHARGOOL, P.D.: Plant Physiol. 63, 409-415 (1979).

DALEY, L.S., BIDWELL, R.G.S.: Plant Physiol. 60, 109-114 (1977). -
DATKO, A.H., GIOVANELLI, J., MUDD, S.H.: J. Biol. Chem. 249, 1139-
1155 (1974). - DATKO, A.H., MUDD, S.H., GIOVANELLI, J.: J. Biol. Chem.
252, 3436-3445 (1977). - DAVIES, H.M., MIFLIN, B.J.: Plant Sci. Lett.
9, 323-332 (1977); - Plant Physiol. 62, 536-541 (1978). - DOUGALL,
D.K., BLOCH, J.: Can J. Bot. 54, 2924-2927 (1976).

EMES, M.J., FOWLER, M.W.: Planta 144, 249-253 (1979a); - Planta 145,
287-292 (1979b).

FEIERABEND, J., BRASSEL, D.: Z. Pflanzenphysiol. 82, 334-346 (1977).

GIVAN, C.V.: Phytochemistry 18, 375-382 (1979). - GLENN, E., MARETZKI,
A.: Plant Physiol. 60, 122-126 (1977). - GROSSE, W.: Z. Pflanzenphysiol. 80, 463-468 (1976).

HAREL, E., LEA, P.J., MIFLIN, B.J.: Planta 134, 195-200 (1977). -
HENKE, R.R., WAHNBAECK, R.: Biochem. Biophys. Res. Commun. 79, 38-45
(1977).

ITO, O., YONEYAMA, T., KUMAZAWA, K.: Plant Cell Physiol. 19, 1109-
1119 (1978).

JESSUP, W., FOWLER, M.W.: Planta 137, 71-76 (1977).

KANAMORI, T., MATSUMOTO, H.: Z. Pflanzenphysiol. 74, 264-266 (1975).
- KEYS, A.J., BIRD, I.F., CORNELIUS, M.J., LEA, P.J., WALLSGROVE,
R.M., MIFLIN, B.J.: Nature (London) 275, 741-743 (1978). - KONDOROSI,
A., SVA, B.Z., KISS, G.B., DIXON, R.A.: Mol. Gen. Genet. 151, 221-226
(1977).

LARSSON, C., ALBERTSSON, E.: Physiol. Plant 45, 7-10 (1979). - LEA,
P.J., FOWDEN, L.: Proc. R. Soc. London 192, 13-26 (1975a); - Biochem.
Physiol. Pflanz. 168, 3-14 (1975b). - LEA, P.J., NORRIS, R.D.: Phytochemistry 15, 585-595 (1976). - LEA, P.J., FOWDEN, L., MIFLIN, B.J.:
Phytochemistry 17, 217-222 (1978). - LEWIS, O.A.M.: J. Exp. Bot. 26,
361-366 (1975). - LEWIS, O.A.M., PATE, J.S.: J. Exp. Bot. 24, 596-
606 (1973). - LEWIS, O.A.M., PROBYN, T.A.: New Phytol. 81, 519-526
(1978). - LLOYD, D.H., JOY, K.W.: Biochem. Biophys. Res. Commun. 81,

186-192 (1978). - LU, T.-S., MAZELIS, M.: Plant Physiol. 55, 502-506
(1975). - LUDWIG, R.A., SIGNER, E.R.: Nature (London) 167, 245-247
(1977).

MACNICOL, P.K.: Plant Physiol. 60, 344-348 (1977). - MADISON, J.T.,
THOMPSON, J.F.: Biochem. Biophys. Res. Commun. 71, 684-691 (1976). -
MATTHEWS, B.F., WIDHOLM, J.M.: Planta 141, 315-321 (1978). - MATTHEWS,
B.F., GURMAN, A.W., BRYAN, J.K.: Plant Physiol. 55, 991-998 (1975). -
MAYNE, B.C., EDWARDS, G.E., BLACK, C.C.: In: Photosynthesis and photo-
respiration, pp. 361-371. Eds. M.D. HATCH, C.B. OSMOND, R.O. SLAYTER.
New York: Wiley Interscience 1971. - MAZELIS, M., CREVELING, R.K.:
J. Food Biochem. 2, 29-37 (1978). - MAZELIS, M., MIFLIN, B.J., PRATT,
H.M.: FEBS Lett. 64, 197-200 (1976). - MAZELIS, M., WHATLEY, F.R.,
WHATLEY, J.: FEBS Lett. 84, 236-240 (1977). - McKAY, G., SHARGOOL,
P.D.: Plant Sci. Lett. 9, 189-193 (1977). - McPARLAND, R.H., GUEVARA,
J.G., BECKER, R.R., EVANS, H.J.: Biochem. J. 153, 597-606 (1976). -
MEEKS, J.C., WOLK, C.P., THOMAS, J., LOCKAU, W., SHAFFER, P.W., AUSTIN,
S.M., CHIEN, W.-S., GALONSKI, A.: J. Biol. Chem. 252, 7894-7900 (1977).
- MIFLIN, B.J.: In: Regulation of enzyme synthesis and activity in
higher plants, pp. 23-40. ed. H. SMITH. London, New York: Academic
Press 1977. - MIFLIN, B.J., LEA, P.J.: Phytochemistry 15, 873-885
(1976); Annu. Rev. Plant Physiol. 28, 299-329 (1977). - MILER, P.M.,
STEWART, C.R.: Phytochemistry 15, 1855-1857 (1976). - MILLS, W.R.,
WILSON, G.: Planta 142, 153-160 (1978). - MOLLER, B.L.: Phytochemistry
15, 695-696 (1976a); - Plant. Physiol. 57, 687-692 (1976b). - MORRIS,
C.J., THOMPSON, J.F.: Plant Physiol. 59, 684-687 (1977).

NICKLISCH, A., TSENOVA, E.N., HOFFMANN, P.: Biochem. Physiol. Pflanz.
171, 375-384 (1977).

O'GARA, F., SHANMUGAM, K.T.: Biochim. Biophys. Acta 437, 313-321 (1976).

PLANQUE, K., KENNEDY, I.R., DE VRIES, G.E., QUISPEL, A., v. BRUSSEL,
A.A.N.: J. Gen. Microbiol. 102, 95-104 (1977).

RATHNAM, C.K.M., EDWARDS, G.E.: Plant Physiol. 57, 881-883 (1976). -
RHODES, D., RENDON, G.A., STEWART, G.R.: Planta 129, 203-210 (1976).
- ROGNES, S.E.: Phytochemistry 14, 1975-1982 (1975).

SAKANO, K., KOMAMINE, A.: Plant Physiol. 61, 115-118 (1978). - SCOTT,
D.B., FARNDEN, K.J.F., ROBERTSON, J.G.: Nature (London) 263, 703-705
(1976). - SERVAITES, J.C., OGREN, W.K.: Plant Physiol. 60, 461-466
(1977). - SHARGOOL, P.D., STEEVES, T., WEAVER, M., RUSSELL, M.: Can.
J. Biochem. 56, 273-279 et erratum p. 926 (1978). - SHEWRY, P.R.,
MIFLIN, B.J.: Plant Physiol. 59, 69-73 (1977). - SKOKUT, T.A., WOLK,
C.P., THOMAS, J., MEEKS, J.C., SHAFFER, P.W.: Plant Physiol. 62, 299-
304 (1978). - SODEK, L.: Phytochemistry 15, 1903-1906 (1976). - SODEK,
L., DA SILVA, W.J.: Plant Physiol. 60, 602-605 (1977). - STEWART, C.R.:
Plant Sci. Lett. 14, 269-273 (1979). - STEWART, G.R., RHODES, D.:
FEBS Lett. 64, 296-299 (1976); - New Phytol. 79, 257-268 (1977a);
- In: Regulation of enzyme synthesis and activity in higher plants,
pp. 1-22. ed. H. SMITH. London, New York: Academic Press 1977b. -
STOREY, R., BEEVERS, L.: Planta 124, 77-87 (1977); - Plant Physiol.
61, 494-500 (1978). - STOREY, R., REPORTER, M.: Can. J. Bot. 56,
1349-1356 (1978).

TATE, S.S., MEISTER, A.: In: The enzymes of glutamine metabolism,
pp. 77-127. eds. S. PRUSINER, E.R. STADTMAN. London, New York: Academic
Press 1973. - THOEN, A., ROGNES, S.E., AARNES, H.: Plant Sci. Lett. 13,
113-119 (1978). - THOMAS, J., MEEKS, J.C., WOLK, C.P., SHAFFER, P.W.,
AUSTIN, S.M., CHIEN, W.-S.: J. Bacteriol. 129, 1545-1555 (1977).

WALLSGROVE, R.M., MIFLIN, B.J.: Biochem. Soc. Trans. 5, 269-271 (1977).
- WALLSGROVE, R.M., HAREL, E., LEA, P.J., MIFLIN, B.J.: J. Exp. Bot.
28, 588-596 (1977). - WALLSGROVE, R.M., LEA, P.J., MIFLIN, B.J.:
Plant Physiol. 63, 232-236 (1979). - WASHITANI, I., SATO, S.: Plant
Cell Physiol. 18, 117-125 (1977a); - Plant Cell Physiol. 18, 505-512
(1977b). - WOLK, C.P., THOMAS, J., SHAFFER, P.W., AUSTIN, S.M.,
GALONSKI, A.: J. Biol. Chem. 251, 5027-5034 (1976).

Professor Dr. THOMAS HARTMANN
Institut für Pharmazeutische Biologie
der Technischen Universität Braunschweig
D 3300 Braunschweig

V. Secondary Plant Substances

Special Topics of the Phenylpropanoid Metabolism

By Horst Robert Schütte

1. Introduction

In addition to topics which were discussed last year (cf. Prog. Bot. 40, 126ff.) in the following the news of some other phenylpropanoid compounds are summarised.

2. Capsaicinoids, Gingerol, and Related Compounds

Capsaicinoids like capsaicin (I) are the pungent principle in red peppers. On the biosynthesis it has been reported that phenylalanine,

HO—⟨benzene ring⟩—CH$_2$NH–CO(CH$_2$)$_4$–CH=CH–CH⟨CH$_3$, CH$_3$⟩

CH$_3$O

Capsaicin (I)

cinnamate derivatives and vanillylamine are incorporated in vivo into the vanillylamine moiety of capsaicinoids (BENNET and KIRBY, 1968) and that valine is incorporated into the acyl moiety (LEETE and LOUDEN, 1968). Furthermore dihydrocapsaicin, capsaicin, and nordihydrocapsaicin are formed and accumulated in fruits of sweet pepper, *Capsicum annuum* var. *grossum*, when the peduncle was put into the aqueous solution of vanillylamine and isocapric acid and the whole fruit was aged for several days under continuous light (IWAI et al., 1977). Dihydrocapsaicin was also formed by cell-free extracts of the fruits in a reaction mixture containing vanillylamine and isocapric acid.

The capsaicinoid-synthesising enzyme activity in sweet pepper fruits was found in both mitochondrial and microsomal fractions of placenta and pericarp (IWAI et al., 1978). No enzyme activity was detected in seeds. The soluble mitochondrial enzyme of the placenta utilises different branched fatty acids, i.e., 7-methyloctanoic acid, 8-methylnon-trans-6-enoic acid and 9-methyl-decanoic acid, at the same rate as it utilises isocapric acid as a substrate, and yields the corresponding capsaicinoids.

The rhizome of ginger *(Zingiber officinale)* has long been valued for its flavouring qualities. These have been ascribed to phenolic constituents of the oleoresin, particularly the (S)-gingerols (MASADA et al., 1974; CONNELL and SUTHERLAND, 1969) of which (S)-(+)-(6)-gingerol (V) is the most abundant. They are phenylpropane derivatives with a longer side chain.

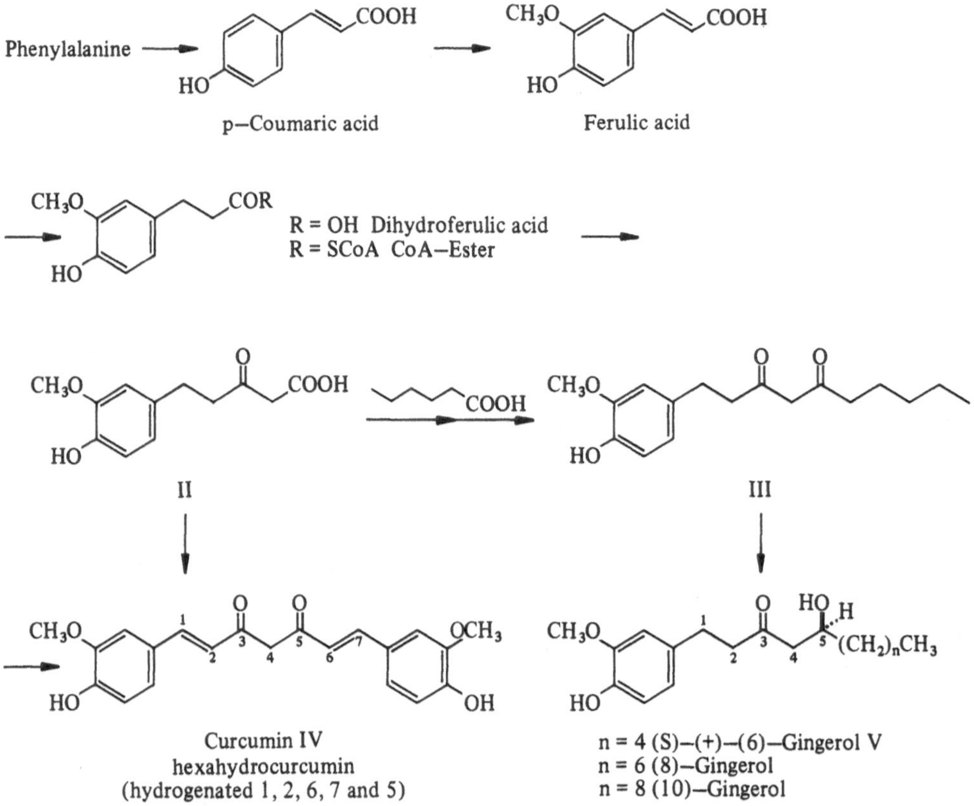

Fig. 1. Biosynthesis of gingerols and curcumin

The biosynthesis of (6)-gingerol (Fig. 1) (DENNIFF and WHITING, 1976) appears to involve cinnamate-acetate condensations and in this case there are similarities to the related biogenesis of stilbenes, diaryl-heptanoids, benzophenones, etc. Phenylalanine, p-coumaric acid and ferulic acid were incorporated specifically as a C_6-C_3-unit into 6-gingerol (V). The other part is derived from acetate. With methyl-labelled acetate the majority of radioactivity was found at C-4 and C-6-C-10. Carboxyl-labelled acetate yielded a gingerol with the major activity at C-5 and C-6-C-10 showing that the C-4-C-10 unit is of ace-tate-malonate origin. Their mode of assembly is indicated by the in-corporation of sodium (1-^{14}C) hexanoate giving all the activity in the C-5-C-10 unit located at C-5. Therefore it is postulated that phenylalanine is elaborated to ferulic acid through p-coumaric acid. Reduction to dihydroferulic acid ensues, followed by condensation with a malonate and hexanoate residue preformed from acetate-malonate per-haps via a biological Claisen reaction. In this case an acetate-malonate condenses with two other acids. The (6)-gingerdione (III) would then be reduced at C-5 to give natural (6)-gingerol (V).

The reaction of an active zingerone (like II) with a second cinnamic acid part has been postulated for the synthesis of curcumin (IV), the pigment of *Curcuma longa* rhizome (ROUGHLEY and WHITING, 1973) and hexa-hydrocurcumin which has been identified also as a component of ginger oleoresin (Fig. 1).

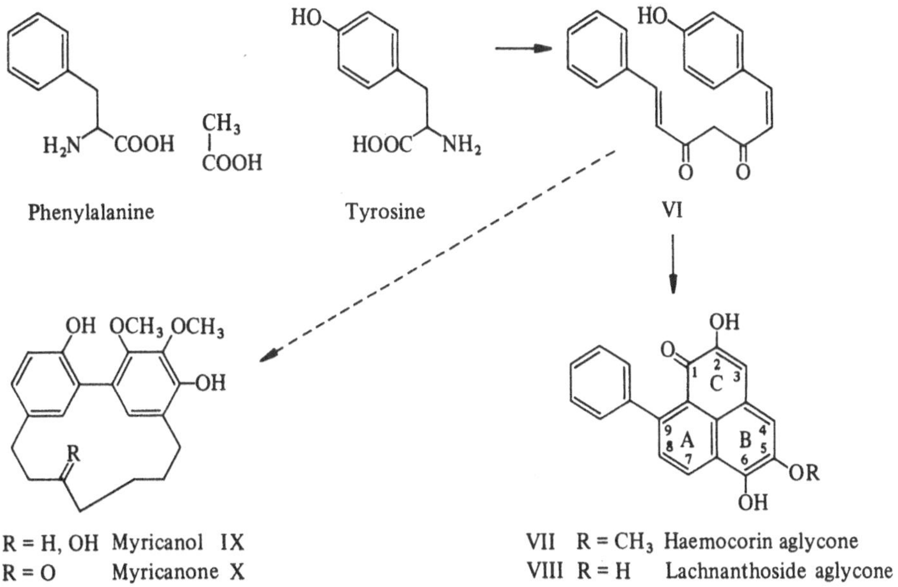

Phenylalanine   Tyrosine   VI

R = H, OH Myricanol IX   VII R = CH₃ Haemocorin aglycone
R = O   Myricanone X   VIII R = H   Lachnanthoside aglycone

Fig. 2. Formation of phenylphenalenone and related compounds

The m,m-bridged biphenyls myricanol (Fig. 2; IX) and myricanone (X)
(BEGLEY et al., 1971) from *Myrica nagi* are also closely related in
structure to the diarylheptanoids like curcumin (IV). To this type
of structure belong also the haemocorin aglycone (VII) of *Haemodorum
corymbosum* (Haemodoraceae) and 2,5,6-trihydroxy-9-phenylphenalenone
(VIII), the aglycone of lachnanthoside, which is the major pigment
isolated from the root system of *Lachnanthes tinctoria* (Haemodoraceae;
"paint root") (EDWARDS and WEISS, 1974). 9-Phenylphenalenones appear
to be characteristic of the family Haemodoraceae. It was suggested
(THOMAS, 1961) that the biosynthesis of the 9-phenylphenalenones
might proceed through a sequence involving the unique combination of
one mol each of phenylalanine and tyrosine, and one mol of acetic
acid (Fig. 2); oxidative cyclisation of the intermediate biarylhepta-
noid (VI) (BAZAN et al., 1977) would lead directly to the natural products. Biosynthetic studies with *L. tinctoria* and *Haemodorum corymbosum*
using labelled precursors have established the specific incorporation
of phenylalanine, tyrosine, and acetate into these compounds (EDWARDS
et al., 1972; THOMAS, 1971; HARMON et al., 1977), showing that the 9-
phenyl substituent, and carbons 7, 8, and 9 arise from phenylalanine
and the ring C with carbons 4, 5, and 6 from tyrosine. No transformation of phenylalanine to tyrosine is possible in these plants. These
results are in contrast to the formation of fungal phenalenones, which
are biosynthesised via the acetate-malonate pathway.

3. Xanthones and Benzophenones

Cinnamate-acetate condensations similar to those mentioned for the
formation of gingerols and related compounds are realised during
the biosynthesis of flavonoids, stilbenes, and some xanthones. The
first two groups of natural compounds were reviewed last year (cf.
Prog. Bot. 40, 126ff.). Meanwhile a soluble enzyme preparation of
stilbene synthase could be prepared from rhizome of *Rheum rhaponticum*

which is capable of converting p-coumaroyl-CoA and malonyl-CoA into a 3,5,4'-trihydroxystilbene, resveratrol (RUPPRICH and KINDL, 1978).

It has been suggested that naturally occurring xanthones are biosynthesised via benzophenone-like intermediates derived wholly from polyketide in fungi, and from shikimate-polyketide in higher plants (cf. BIRCH et al., 1976). Thus regarding the biosynthesis of xanthones in *Gentiana lutea*, it has been demonstrated that gentisin (gentisein-7-methyl ether), gentisein (1,3,7-trihydroxyxanthone) and related xanthones are biosynthesised by oxidative coupling of a benzophenone, derived from three malonates for the phloroglucinol ring unit and an intermediate (C_6-C_1) formed by loss of two carbon fragments from phenylalanine (FLOSS and RETTIG, 1964; GUPTA and LEWIS, 1971).

A different pathway is elaborated for mangiferin (XIII), a C-glycosyl-xanthone which is widely distributed in several families (CARPENTER et al., 1969). Its biosynthesis has been studied using the aerial parts of *Anemarrhena asphodeloides* (Liliaceae) (FUJITA and INOUE, 1977) showing an intact incorporation of a C_6-C_3 unit from shikimate according Figure 3. All the feeding experiments indicate that, for example, the 1,3,6,7-tetrahydroxyxanthone can be biosynthesised by the cyclisation of an intermediate derived from p-coumarate and two malonates.

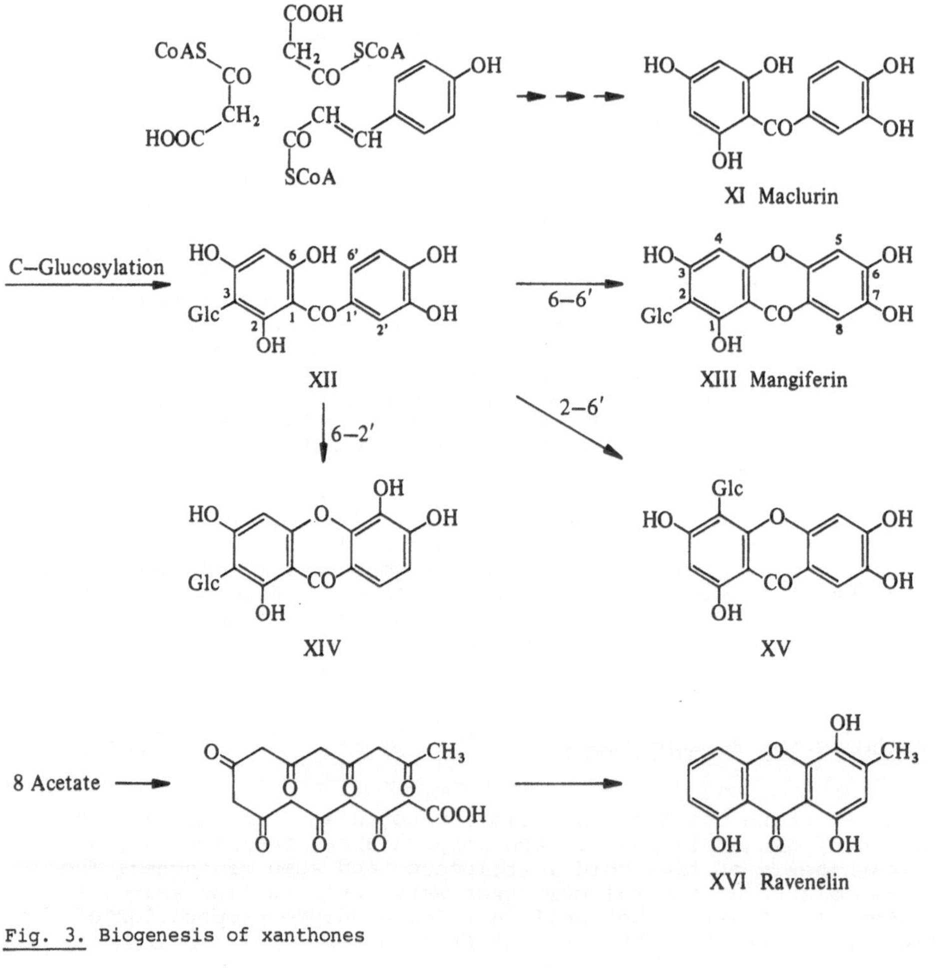

Fig. 3. Biogenesis of xanthones

Furthermore these results show that C-glucosylation occurs at the
benzophenone stage of maclurin (XI) prior to the formation of the
xanthone nucleus and that mangiferin (XIII) would be biosynthesised
via 3-C-glucosylmaclurin (XII) as a possible intermediate. When XII
is converted to C-glucosylxanthone by ring closure, four isomeric C-
glucosylxanthones could be formed (ARITOMI and KAWASAKI, 1970), mangiferin (XIII) and isomangiferin (XV) besides 1,3,5,6-tetrahydroxy-
xanthose-2-C-glucoside (XIV) and the corresponding 4-C-glucoside. XIV
and its 5-methyl ether (irisxanthone) were isolated from the Gentianacea
Canscora decussata (CHOSAL and CHAUDHURI, 1973) and from *Iris florentina*
(ARISAWA et al., 1973) along with XIII respectively. The co-occurrence
of XIII, its 3-methyl ether (homomangiferin) and XV in *Mangifera indica*
has been reported (SALEH and EL-ANSARI, 1975). These facts strongly
support that 3-C-glucosylmaclurin (XII) would be a key intermediate
for the biosynthesis of the above C-glucosylxanthone.

> Contrasting to these shikimate-acetate derived xanthones the fungal xanthones
> like ravenelin (XVI) from *Helminthosporium ravenellii* is formed wholly from
> acetate, also via an oxygenated benzophenone derivative but with a replacement
> of a carbonyl group by an oxygen atom (BIRCH et al., 1976).

4. Coumarins

Coumarins are lactones of cis-o-hydroxycinnamic acids. These benzo-
pyran-2-ones are widely distributed in the plant kingdom. The structures of the different coumarins from natural sources derive by O-
methylation, O-glycosylation and simple or modified O- or C-bound isoprene substituents; many coumarins contain a condensed furan ring
originating from an isoprene unit. The latter structures are distributed especially in the Apiaceae and Rutaceae (GRAY and WATERMAN, 1978).
For the biosynthesis of coumarin itself (BROWN, 1963) cinnamic acid
is hydroxylated in o-position (GESTETNER and CONN, 1974) and glucosylated to the o-coumaric acid glucoside which occurs in all coumarin-
containing plants, trans-cis-isomerisation yields coumarinic acid
glucoside. After hydrolysis by a β-glucosidase, spontaneous cyclisation of coumarinic acid occurs giving coumarin. (Similar pathway as
umbelliferone, only without 7-OH, Fig. 4.)

With exception of coumarin itself the most known coumarins are oxygenated at C-7 like umbelliferone (Fig. 4), the most simple of these derivatives. Its biosynthesis, which has been studied preferably in
Hydrangea macrophylla (BROWN et al., 1964; AUSTIN and MEYERS, 1965a,b;
FLOSS and PAIKERT, 1969), starts from phenylalanine and goes via cinnamic acid to p-coumaric acid, which is hydroxylated to umbellic acid.
After glucosylation to the diglucoside there occurs a transformation
to the cis-cinnamic acid derivative (XVII), which is partly hydrolysed,
giving the lactone ring of umbelliferone. The coumarin itself cannot
be hydroxylated to umbelliferone.

The central role of p-coumaric acid has been substantiated by numerous
feeding experiments. Further substitution to yield simple substituted
coumarins seems to occur mostly after cyclisation. Thus daphnin is produced mainly via p-coumaric acid and the hydroxylation is achieved at
a later step (SATO and HASEGAWA, 1972). But for some coumarins it
seems that hydroxyl group pattern is introduced on the cinnamic acid
step. Thus caffeic acid is the precursor of esculetin (SATO, 1967),
ferulic acid the precursor of scopoletin (STECK, 1967; FRITTIG et al.,

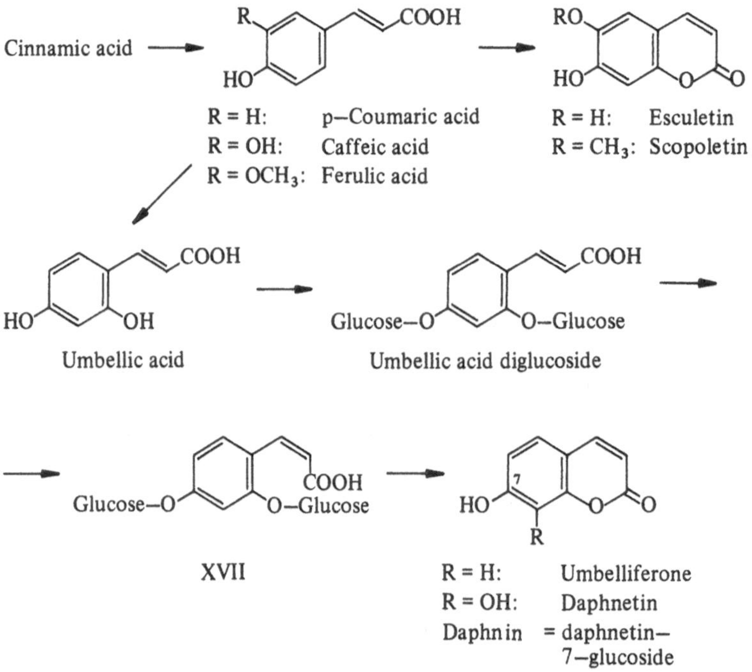

Fig. 4. Biosynthesis of simple coumarins like umbelliferone

1970), and 2-hydroxy-4-methoxy-cinnamic acid the precursor of her-
niarin (7-methoxycoumarin) (BROWN, 1963; Fig. 4).

Many coumarins contain a condensed furan ring. Based on the frequent
co-occurrence of such furanocoumarins with 5'-isopropylfuranocoumarins
and isoprenylated coumarins, it has long been suspected that the furan
ring in these compounds is of isoprenoid origin (ANEJA et al., 1958)
like other aromatic hemiterpene derivatives, e.g., furanochromones,
furanoflavones and furanoquinolines (GRUNDON, 1978). This concept requires that the two extra carbon atoms of the furan ring originate
from C-4 and C-5 of mevalonic acid by isoprenylation of a phenylpropanoid precursor, followed by loss of a 3-carbon fragment. This could
be confirmed by tracer experiments with different plant species including cell cultures, for example, of *Ruta graveolens* (FLOSS and MOTHES,
1964, 1966; FLOSS et al., 1969; BROWN, 1970; FLOSS, 1972; AUSTIN and
BROWN, 1973; KUTNEY et al., 1973a,b,c; OVERTON and PICKEN, 1977). In
particular, umbelliferone, which is, for example, together with some
furanocoumarins a natural constituent of *Pimpinella magna* roots, seems
to occupy a key position in the biogenetic pathway of the furano derivatives (Fig. 5) (FLOSS and PAIKERT, 1969; BROWN, 1970; BROWN et
al., 1970; STECK and BROWN, 1970; AUSTIN and BROWN, 1973). Umbelliferone can be isoprenylated preferably either in 6- or 8-positions,
the two activated ortho positions to the 7-hydroxy group. Such a scheme
would explain plausibly why the furan ring is found attached only to
the 6,7- (linear psoralen-type) and 7,8 (angular angelicin type) positions, the oxygen always being at C-7.

In *Ruta graveolens* the addition of the dimethylallyl unit at C-6 appears to be
specifically controlled by the enzyme dimethylallylphosphate: umbelliferone
transferase, which is found preferably in the chloroplasts (ELLIS and BROWN,
1974; DHILLON and BROWN, 1976). O- and C-8 prenylation presumably being mediated by other, similar, enzyme systems.

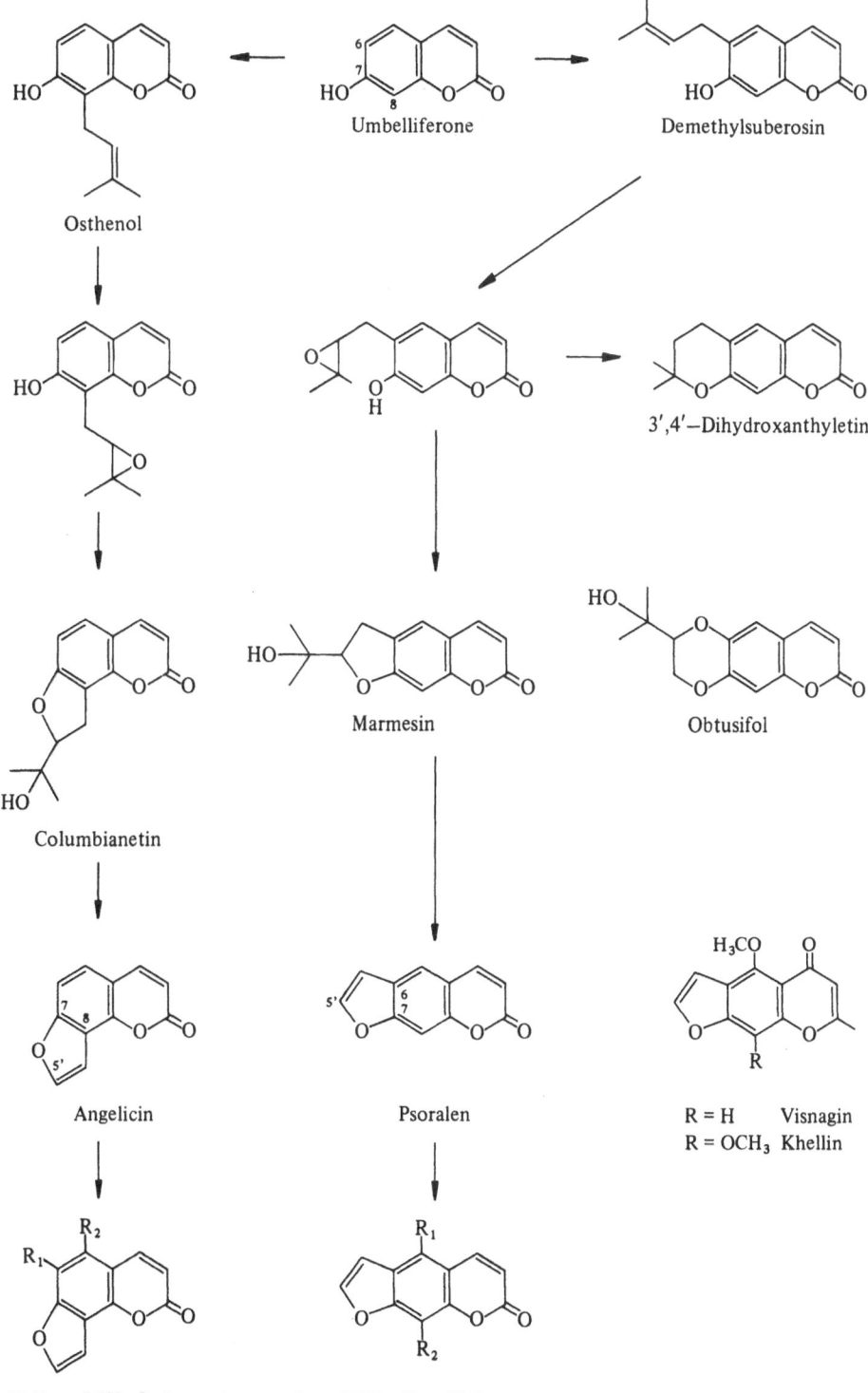

Fig. 5. Biosynthesis of furanocoumarins and related compounds

Secondary modifications of the prenyl side chain may take place via an initial epoxidation of the olefinic double bond, which is controlled by distinct mono-oxygenases (GRUNDON and McCOLL, 1975). The dimethylallyl umbelliferones deme-thyl-suberosin and osthenol and the hydroxyisopropyl dihydrofurans marmesin and columbianetin were shown to be good precursors of the linear furanocouma-rins, e.g., bergapten and psoralen, and of the angular derivatives respective-ly, e.g., angelicin (Fig. 5) (BROWN et al., 1970; STECK and BROWN, 1970, 1971; GAMES and JAMES, 1972; CAPORALE et al., 1972; BROWN and STECK, 1973; AUSTIN and BROWN, 1973). The corresponding dihydrofurano derivatives like dihydro-psoralen seem not to be on the physiological pathway (CAPORALE et al., 1970a; BROWN, 1973).

The further substitution in the form of hydroxylation followed by methylation or O-prenylation occurs after furan ring formation (CAPORALE et al., 1971; DALL'ACQUA et al., 1972, 1975) suggesting that the hy-droxylated angular furanocoumarins originate from angelicin and the linear furanocoumarins, like xanthotoxin, bergapten and isopimpinellin, from psoralen (STECK and BROWN, 1970; BROWN and SAMPATHKUMAR, 1977), but a generalisation is not yet justifiable. For the formation of pyr-anocoumarins, which co-occur in different plants with the furanoiso-mers, a pathway analogous to that for furanocoumarins is in operation e.g., 3',4'-dihydroxanthyletin is yielded from demethylsuberosin (Fig. 5) (STECK and BROWN, 1970). Occasionally the side-chains exhibit more complex elaborations, e.g., obtusifol (Fig. 5) and others.

Besides the 3,3-dimethylallylderivatives, a number of plant products have been isolated containing a C- or O-1,1 dimethylallyl substituent, e.g., rutamarin, which is biosynthesised from umbelliferone involving formation and Claisen rearrangement of a 4-prenyloxycoumarin (XVIII) (Fig. 6). The formation of the dihydrofuran ring and the loss of the 4-hydroxy group occur at later stages (GRUNDON, 1978; DONELLY et al., 1978).

Contrasting to the phenylpropanoid origin of the different coumarins mentioned the isocoumarin asperetin in the fungus *Aspergillus flavus* is biosynthesised from 8 two-carbon units via the acetate-polymalonate pathway (Fig.7) (CATTEL et al., 1973). Also the related chromone system, e.g., in visnagin and khellin (Fig. 5), is of polyketide origin but the furan ring derives from a prenyl group (HARRISON et al., 1971).

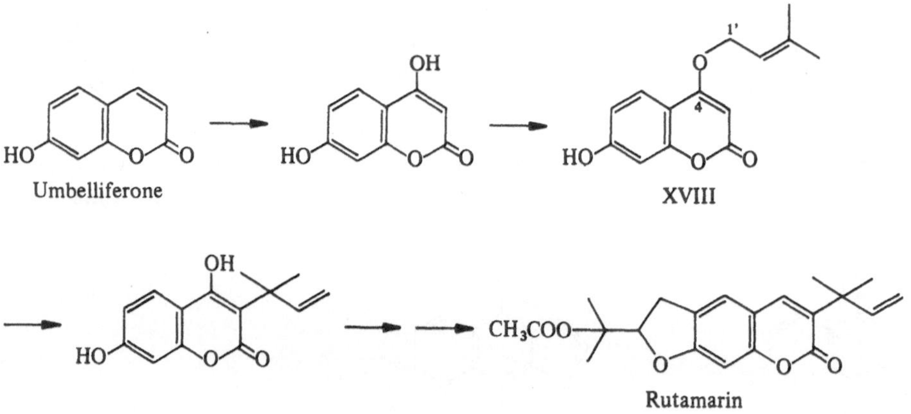

Fig. 6. Biogenesis of rutamarin from umbelliferone

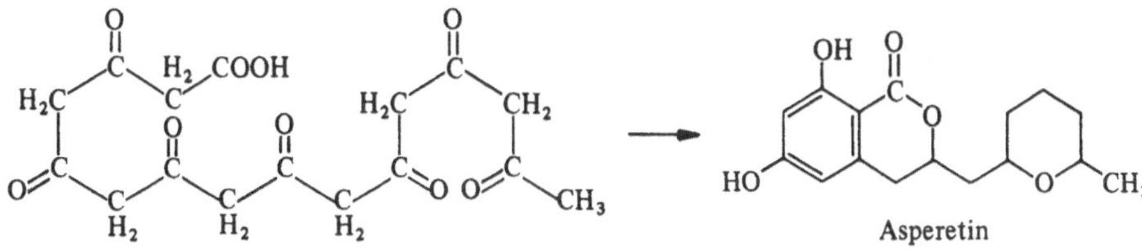

Fig. 7. Formation of asperetin according to polyketid-pathway

5. Gliotoxin

Gliotoxin is a fungal metabolite, for example, of *Trichoderma viride*
(Gliocladium deliquescens). It contains a hydrogenated indole ring which
is derived biosynthetically from phenylalanine including its amino
group (Fig. 8) (SUHADOLNIK and CHENOWETH, 1958; BOSE et al., 1968a,b;
BRANNON et al., 1971; JOHNS et al., 1975) by a process retaining all
the aryl protons in a manner consistent with the involvement of a
2,3-arene oxide intermediate (Bu'LOCK and RYLES, 1970; JOHNS and KIRBY,
1971). Before the incorporation of phenylalanine into gliotoxin this
amino acid undergoes an unusual exchange reaction involving replacement of the pro-R-methylene proton by an external proton with overall retention of configuration (JOHNS et al., 1975; Bu'LOCK et al.,
1972). Cyclo-(phenylalanylseryl) is an intermediate for gliotoxin
biosynthesis (Bu'LOCK and LEIGH, 1975; MacDONALD and SLATER, 1975).
Cyclo-(L-alanyl-L-phenylalanyl) is converted in cultures of *Trichoderma
viride* into the gliotoxin analogue, 3a-deoxygliotoxin, with an efficiency similar to that of the corresponding natural process (KIRBY
and ROBINS, 1976). In a similar manner phenylalanine is incorporated
into acetylaranotin by *Aspergillus tereus* (BRANNON et al., 1971) but in
this case the benzene ring is oxidised additionally. Tryptophan gives
the corresponding sporidesmin A in *Phithomyces chartarum* (Fig. 8) (KIRBY
and VARLEY, 1974).

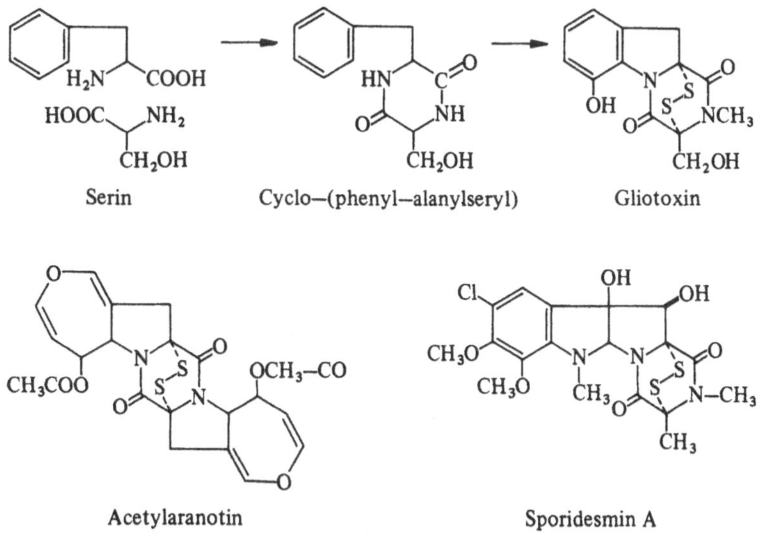

Serin   Cyclo—(phenyl—alanylseryl)   Gliotoxin

Acetylaranotin   Sporidesmin A

Fig. 8. Formation of gliotoxin and related compounds

6. Betalains

Betalains are nitrogenous water-soluble plant pigments. To this group belong the red-violet betacyans and the yellow betaxanthins. They occur exclusively in the families of the Centrospermae (MABRY et al., 1972a; REZNIK, 1975). The occurrence of the betacyans excludes strictly the presence of anthocyans. Caryophyllaceae and Mollugenaceae do not have betacyans but contain anthocyans; all other Centrospermae investigated contain betacyans, but cannot synthesise anthocyans.

Contrasting to this fact flavones co-occur with betacyans and betaxanthins. Both types are characterised by the retained betalamic acid unit in their structures. Biosynthetically the betacyans belong to the phenylpropanoids both for the dihydroindole and for the dihydropyridine unit (Fig. 9) (WYLER et al., 1963; HÖRHAMMER et al., 1964; GARAY and TOWERS, 1966; MILLER et al., 1968; LIEBISCH et al., 1969; SCIUTO et al., 1974). Corresponding experiments with *Opuntia ficusindica* have shown an incorporation of Dopa into the betalamic acid unit of indicaxanthin (MINALE et al., 1965). The formation of the dihydroindole unit via cyclodopa is related to the melanines but also to gliotoxin.

The hydropyridine moiety of these pigments is shown to have originated from 3,4-dihydroxyphenylalanine by an extradiol fission of the aromatic nucleus and subsequent closure to the heterocyclic ring system of the

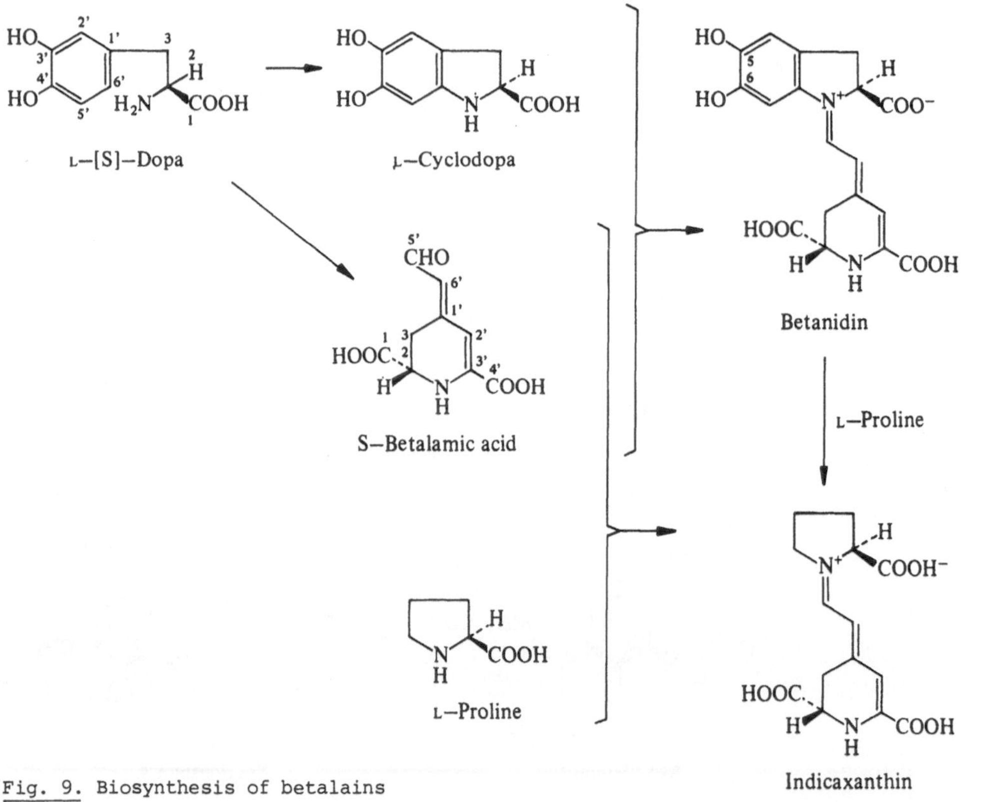

Fig. 9. Biosynthesis of betalains

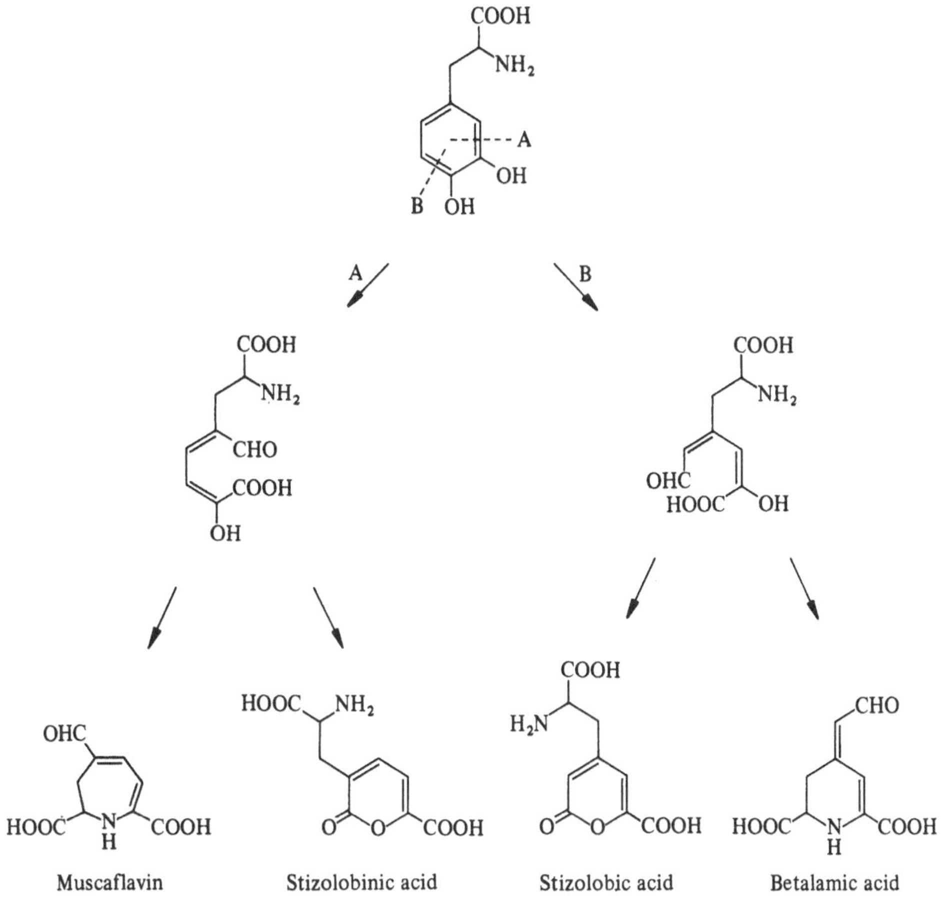

Fig. 10. Ring fission of Dopa for the formation of stizolobic acid and related compounds

betalamic acid (Figs. 9 and 10) (FISCHER and DREIDING, 1972; IMPELLIZ-ZERI and PIATTELLI, 1972; CHANG et al., 1974). The 5'-position of Dopa corresponds to the aldehyde group of betalamic acid. Betalamic acid is a key intermediate for the synthesis of all betacyans and beta-xanthins and has been detected as a naturally occurring pigment in betalain-synthesising plants (KIMLER et al., 1971; MARBY et al., 1972b; REZNIK, 1978). This acid yields betanidin with cyclodopa and indica-xanthin with proline (Fig. 9). While different betalain-containing plants can synthesise these compounds in the dark (WOHLPART and MABRY, 1968), the formation of amaranthin in *Amaranthus* species and of beta-xanthin, e.g., in *Celosia plumosa*, and especially the formation of the dihydropyridine moiety from Dopa is light-controlled by phytochrome and an additional blue light-dependent high energy reaction (PIATELLI et al., 1969; KÖHLER, 1972b, 1973a,b; GIUDICI DE NICOLA et al., 1972a, b, 1973a,b, 1974, 1975a; FRENCH et al., 1973; WOODHEAD and SWAIN, 1974). In the dark the synthesis of betacyan is stimulated by kinetin and cAMP (PIATELLI et al., 1971; GIUDICI DE NICOLA et al., 1972a,b, 1973a, 1974, 1975b); KÖHLER, 1972a,b; RAST et al., 1973). The kinetin effect in *Amaranthus* is the basis for a very sensitive kinetin test (CONRAD, 1974).

7. Stizolobic Acid and Related Compounds

A similar ring fission as for the formation of betalamic acid is suggested for the biosynthesis of stizolobinic acid, α-amino-6-carboxy-2-oxo-2H-pyran-3-propionic acid, and stizolobic acid, α-amino-6-carboxy-2-oxo-2H-pyran-4-propionic acid, α-pyrone-bearing non-protein amino acids (Fig. 10), which are distributed in five genera of Lotoideae (Leguminosae) (HATTORI and KOMAMINE, 1959) and in some species of toxic mushrooms, *Amanita pantherina* and *A. gemmata* (Basidiomycetes) (CHILTON et al., 1974) suggesting that the same 4,5-extra-diol dioxygenase may be present in these unrelated species. In *Stizolobium hassjoo* and in *Mucuna deeringiana (Stizolobium deeringianum)* it has been demonstrated that these two compounds are derived directly from dihydroxyphenylalanine via an extradiol type of ring cleavage reaction followed by recyclisation to form α-pyrone-6-carboxylic acid derivatives (SAITO et al., 1975, 1976; ELLIS, 1976). The stizolobinic acid synthase and stizolobic acid synthase yielding these two compounds from dihydroxyphenylalanine have been purified and characterised (SAITO and KAMAMINE, 1976, 1978). Also muscaflavin (v. ARDENNE et al., 1974) from *Amanita muscaria* is probably a recyclised product of an extradiol ring fission of Dopa (Fig. 10).

References

ANEJA, R., MUKERJEE, S.K., SESHADRI, T.R.: Tetrahedron $\underline{4}$, 256-270 (1958). - ARDENNE, v. R., DÖPP, H., MUSSO, H., STEGLICH, W.: Z. Naturforsch. $\underline{29c}$, 637-639 (1974). - ARISAWA, M., MORITA, N., KONDO, Y., TAKEMOTO, T.: Chem. Pharm. Bull. $\underline{21}$, 2562-2565 (1973). - ARITOMI, M., KAWASAKI, T.: Chem. Pharm. Bull. $\underline{18}$, 2327-2333 (1970). - AUSTIN, D.J., BROWN, S.A.: Phytochemistry $\underline{12}$, 1657-1667 (1973). - AUSTIN, D.J., MEYERS, M.B.: Phytochemistry $\underline{4}$, 245-254 (1965a); - Phytochemistry $\underline{4}$, 255-262 (1965b).

BAZAN, A.C., EDWARDS, J.M., WEISS, U.: Tetrahedron Lett. $\underline{1977}$, 147-150. - BEGLEY, M.J., CAMPBELL, R.V.M., CROMBIE, L., TUCK, B., WHITING, D.A.: J. Chem. Soc. C $\underline{1971}$, 3634-3642. - BENNET, D.J., KIRBY, G.W.: J. Chem. Soc. C $\underline{1968}$, 442-446. - BIRCH, A.J., BALDAS, J., HLUBUCEK, J.R., SIMPSON, T.J., WESTERMAN, P.W.: Chem. Soc. Perkin I $\underline{1976}$, 898-904. - BOSE, A.K., DAS, K.G., FUNKE, P.T., KUGAJEVSKY, I., SHUKLA, O.P., KHANCHANDANI, K.S., SUHADOLNIK, R.J.: J. Am. Chem. Soc. $\underline{90}$, 1038-1041 (1968a). - BOSE, A.K., KHANCHANDANI, K.S., TAVARES, R., FUNKE, P.T.: J. Am. Chem. Soc. $\underline{90}$, 3593-3594 (1968b). - BRANNON, D.R., MABE, J.A., MOLLOY, B.B., DAY, W.A.: Biochem. Biophys. Res. Commun. $\underline{43}$, 588-594 (1971). - BROWN, S.A.: Phytochemistry $\underline{2}$, 137-144 (1963); - Phytochemistry $\underline{9}$, 2471-2475 (1970); - Can. J. Biochem. $\underline{51}$, 965-968 (1973). - BROWN, S.A., SAMPATHKUMAR, S.: Can. J. Biochem. $\underline{55}$, 686-692 (1977). - BROWN, S.A., STECK, W.: Phytochemistry $\underline{12}$, 1315-1324 (1973). - BROWN, S.A., TOWERS, G.H.N., CHEN, D.: Phytochemistry $\underline{3}$, 469-476 (1964). - BROWN, S.A., EL-DEKHAKHNY, M., STECK, W.: Can. J. Biochem. 48, 863-871 (1970). - Bu'LOCK, J.D., LEIGH, C.: Chem. Commun. $\underline{1975}$, 628-629. - Bu'LOCK, J.D., RYLES, A.P.: Chem. Commun. $\underline{1970}$, 1404-1406. - Bu'LOCK, J.D., RYLES, A.P., JOHNS, N., KIRBY, G.W.: Chem. Commun. $\underline{1972}$, 100-101.

CAPORALE, G., DALL'ACQUA, F., MARCIANI, S., CAPOZZI, A.: Z. Naturforsch. $\underline{25b}$, 700 (1970). - CAPROALE, G., DALL'ACQUA, F., CAPOZZI, A., MARCIANI, S., CROCCO, R.: Z. Naturforsch. $\underline{26b}$, 1256-1259 (1971). - CAPORALE, G., DALL'ACQUA, F., MARCIANI, S.: Z. Naturforsch. $\underline{27b}$, 871-872 (1972). -

CARPENTER, I., LOCKSLEY, H.D., SCHEINMANN, F.: Phytochemistry 8,
2013-2026 (1969). - CATTEL, L., GORVE, J.F., SHAW, D.: J. Chem. Soc.
Perkin I 1973, 2626-2629. - CHANG, C., KIMMLER, L., MABRY, T.J.:
Phytochemistry 13, 2771-2775 (1974). - CHILTON, W.S., HSU, C.P.,
ZDYBAK, W.T.: Phytochemistry 13, 1179-1181 (1974). - CONNELL, D.W.,
SUTHERLAND, M.D.: Aust. J. Chem. 22, 1033-1043 (1969). - CONRAD, K.:
Biochem. Physiol. Pflanz. 165, 531-535 (1974).

DALL'ACQUA, F., CAPOZZI, A., MARCIANI, S., CAPORALE, G.: Z. Natur-
forsch. 27b, 813-817 (1972). - DALL'ACQUA, F., INNOCENTI, G., CAPORALE,
G.: Planta Med. 27, 343-348 (1975). - DENNIF, P., WHITING, D.A.: Chem.
Commun. 1976, 711-712. - DHILLON, D.S., BROWN, S.A.: Arch. Biochem.
Biophys. 177, 74-83 (1976). - DONELLY, W.J., GRUNDON, M.F., RAMACHAN-
DRAN, V.N.: Proc. R. Ir. Acad. Sect. B 77B, 443-447 (1977); C.A. 89,
103824 (1978).

EDWARDS, J.M., WEISS, U.: Phytochemistry 13, 1597-1602 (1974). -
EDWARDS, J.M., SCHMIDT, R.C., WEISS, U.: Phytochemistry 11, 1717-1720
(1972). - ELLIS, B.E.: Phytochemistry 15, 489-491 (1976). - ELLIS,
B.E., BROWN, S.A.: Can. J. Biochem. 52, 734-738 (1974).

FISCHER, N., DREIDING, A.S.: Helv. Chim. Acta 55, 649-658 (1972). -
FLOSS, H.G.: Recent Adv. Phytochem. 4, 143-164 (1972). - FLOSS, H.G.,
MOTHES, U.: Z. Naturforsch. 19b, 770-771 (1964); - Phytochemistry 5,
161-169 (1966). - FLOSS, H.G., PAIKERT, H.: Phytochemistry 8, 589-596
(1969). - FLOSS, H.G., RETTIG, A.: Z. Naturforsch. 19b, 1103-1105
(1964). - FLOSS, H.G., GÜNTHER, H., HADWIGER, L.A.: Phytochemistry 8,
585-588 (1969). - FRITTIG, B., HIRTH, L., OURISSON, G.: Phytochemistry
9, 1963-1975 (1970). - FRENCH, C.J., PECKET, R.C., SMITH, H.: Phytochemistry 12, 2887-2891 (1973). - FUJITA, M., INOUE, T.: Tetrahedron
Lett. 1977, 4503-4506.

GAMES, D.E., JAMES, D.H.: Phytochemistry 11, 868-869 (1972). - GARAY,
A.S., TOWERS, G.H.N.: Can. J. Bot. 44, 231-236 (1966). - GESTETNER,
B., CONN, E.E.: Arch. Biochem. Biophys. 163, 617-624 (1974). - GHOSAL,
S., CHAUDHURI, R.K.: Phytochemistry 12, 2035-2038 (1973). - GIUDICI DE
NICOLA, M., PIATELLI, M., CASTROGIOVANNI, V., MOLINA, C.: Phyto-
chemistry 11, 1005-1010 (1972a). - GIUDICI DE NICOLA, M., PIATELLI,
M., CASTROGIOVANNI, V., AMICO, V.: Phytochemistry 11, 1011-1017
(1972b). - GIUDICI DE NICOLA, M., PIATELLI, M., AMICO, V.: Phyto-
chemistry 12, 353-357 (1973a); - Phytochemistry 12, 2163-2166 (1973b).
- GIUDICI DE NICOLA, M., AMICO, V., PIATELLI, M.: Phytochemistry 13,
439-442 (1974). - GIUDICI DE NICOLA, M., AMICO, V., SCIUTO, S.,
PIATELLI, M.: Phytochemistry 14, 479-481 (1975a). - GIUDICI DE NICOLA,
M., AMICO, V., PIATELLI, M.: Phytochemistry 14, 989-991 (1975b). -
GRAY, A.I., WATERMAN, P.G.: Phytochemistry 17, 845-864 (1978). -
GRUNDON, M.F.: Tetrahedron 34, 143-161 (1978). - GRUNDON, M.F., McCOLL,
I.S.: Phytochemistry 14, 143-150 (1975). - GUPTA, P., LEWIS, J.R.:
J. Chem. Soc. (C) 1971, 629-631.

HARMON, A.D., EDWARDS, J.M., HIGHET, R.J.: Tetrahedron Lett. 1977,
4471-4474. - HARRISON, P.G., BAILEY, B.K., STECK, W.: Can. J. Biochem.
49, 964-970 (1971). - HATTORI, S., KOMAMINE, A.: Nature (London) 183,
1116-1117 (1959). - HÖRHAMMER, L., WAGNER, H., FRITZSCHE, W.: Biochem.
Z. 339, 398-400 (1964).

IMPELLIZZERI, G., PIATELLI, M.: Phytochemistry 11, 2499-2502 (1972).
- IWAI, K., SUZUKI, T., LEE, K.-R., KOBASHI, M., OKA, S.: Agric. Biol.
Chem. 41, 1877-1882 (1977). - IWAI, K., LEE, K.-R., KOBASHI, M.,
SUZUKI, T., OKA, S.: Agric. Biol. Chem. 42, 201-202 (1978).

JOHNS, N., KIRBY, G.W.: Chem. Commun. 1971, 163-164. - JOHNS, N., KIRBY, G.W., Bu'LOCK, J.D., RYLES, A.P.: J. Chem. Soc. Perkin I 1975, 383-386.

KIMLER, L., LARSON, R.A., MESSENGER, L., MOORE, J.B., MABRY, T.J.: Chem. Commun. 1971, 1329-1330. - KIRBY, G.W., ROBINS, D.J.: Chem. Commun. 1976, 354-355. - KIRBY, G.W., VARLEY, M.J.: Chem. Commun. 1974, 833-834. - KOEHLER, K.H.: Phytochemistry 11, 127-131 (1972a); - Phytochemistry 11, 133-137 (1972b); - Biol. Zentralbl. 92, 307-336 (1973a); - Pharmazie 28, 18-24 (1973b). - KUTNEY, J.P., VERMA, A.K., YOUNG, R.N.: Tetrahedron 29, 2645-2660 (1973a); - Tetrahedron 29, 2661-2671 (1973b.) - KUTNEY, J.P., SALISBURY, P.J., VERMA, A.K.: Tetrahedron 29, 2673-2681 (1973c).

LEETE, E., LOUDEN, M.C.L.: J. Am. Chem. Soc. 90, 6837-6841 (1968). - LIEBISCH, H.W., MATSCHINER, B., SCHÜTTE, H.R.: Z. Pflanzenphysiol. 61, 269-278 (1969). - LOEWENBERG, J.R.: Phytochemistry 9, 361-366 (1970).

MABRY, T.J., KIMLER, L., CHANG, C., in: Recent Advances in Phytochemistry 5, eds. V.C. RUNECKLES, T.C. TSO, 105-134 (1972a). - MABRY, T.J., KIMLER, L., LARSON, R.A.: Hoppe Seylers Z. Physiol. Chem. 353, 127-128 (1972b). - MacDONALD, J.C., SLATER, G.P.: Can. J. Biochem. 53, 475-478 (1975). - MASADA, Y., INOUE, T., HASHIMOTO, K., FUJIOKA, M., UCHINO, C.: J. Pharm. Soc. Jn. 94, 735-738 (1974). - MILLER, H.E., RÖSLER, H., WOHLPART, A., WYLER, H., WILCOX, M.E., FROHOFER, H., MABRY, T.J., DREIDING, A.S.: Helv. Chim. Acta 51, 1470-1474 (1968). - MINALE, L., PIATELLI, M., NICOLAUS, R.A.: Phytochemistry 4, 593-597 (1965).

OVERTON, K.H., PICKEN, D.J.: Fortschr. Chem. Org. Naturst. 34, 249-298 (1977).

PIATELLI, M., GIUDICI DE NICOLA, M., CASTROGIOVANNI, V.: Phytochemistry 8, 371-736 (1969); - Phytochemistry 10, 289-293 (1971).

RAST, D., SKŘIVANOVÁ, R., BACHOFEN, R.: Phytochemistry 12, 2669-2672 (1973). - REZNIK, H.: Ber. Dtsch. Bot. Ges. 88, 179-190 (1975); - Z. Pflanzenphysiol. 87, 95-102 (1978). - ROUGHLEY, P.J., WHITING, D.A.: J. Chem. Soc. Perkin I 1973, 2379-2388. - RUPPRICH, N., KINDL, H.: Hoppe Seylers Z. Physiol. Chem. 359, 165-172 (1978).

SAITO, K., KOMAMINE, A.: Eur. J. Biochem. 68, 237-243 (1976); - Eur. J. Biochem. 82, 385-392 (1978). - SAITO, K., KOMANINE, A., SENOH, S.: Z. Naturforsch. 30c, 659-662 (1975); - Z. Naturforsch. 31c, 15-17 (1976). - SALEH, N.A.M., EL-ANSARI, M.A.I.: Planta Med. 28, 124-130 (1975). - SATO, M.: Phytochemistry 6, 1363-1373 (1967). - SATO, M., HASEGAWA, M.: Phytochemistry 11, 657-662 (1972). - SCIUTO, S., ORIENTE, G., PIATELLI, M., IMPELLIZERI, G., AMICO, V.: Phytochemistry 13, 947-951 (1974). - STECK, W.: Can. J. Biochem. 45, 1995-2003 (1967). - STECK, W., BROWN, S.A.: Can. J. Biochem. 48, 872-880 (1970); - Can. J. Biochem. 49, 1213-1216 (1971). - SUHADOLNIK, R.J., CHENOWETH, R.G.: J. Am. Chem. Soc. 80, 4391-4392 (1958).

THOMAS, R.: Biochem. J. 78, 807-813 (1961); - Chem. Commun. 1971, 739-740.

WOHLPART, A., MABRY, T.J.: Plant Physiol. <u>43</u>, 457-459 (1968). -
WOODHEAD, S., SWAIN, T.: Phytochemistry <u>13</u>, 953-956 (1974). - WYLER,
H., MABRY, T.J., DREIDING, A.S.: Helv. Chim. Acta <u>46</u>, 1745-1748
(1963).

Professor Dr. HORST-ROBERT SCHÜTTE
Institut für Biochemie der Pflanzen
des Forschungszentrums für
Molekularbiologie und Medizin der
Akademie der Wissenschaften der DDR
Weinberg
DDR 401 Halle (Saale)

VI. Growth

By Nikolaus Amrhein

1. Prefatory Remarks

The topic of this progress report - the biochemistry and physiology of the
gibberellins and cytokinins - was last reviewed in this series two years ago
(Progr. Bot. 39, 101). A computer-aided search of the pertinent literature
that was published in the last three years revealed that nearly 3000 reports
dealt in some regard or other with these two classes of plant hormones. This
number gives a presentiment of the difficulty one finds in integrating these
data into our understanding of the complex hormonal regulation of plant growth
and differentiation. With this difficulty in mind the highly individual and
stimulating account by K.V. THIMANN of "Hormone action in the whole life of
plants" (THIMANN, 1977), is mentioned here first. LETHAM et al. (1978a) edited
a comprehensive treatise on phytohormones and related compounds, which covers
the literature up to 1976 and will certainly become a standard reference for
years to come. A thought-provoking article by TREWAVAS (1976) assists the analysis of some basic questions of plant growth regulator action. Conference proceedings were edited by PILET (1977), SCHÜTTE and GROSS (1977), BOGORAD and
WEIL (1977), and KUDREV et al. (1977). A manual with guidelines on methods of
extraction, purification, and quantitative analysis of plant hormones with
modern analytical procedures was edited by HILLMAN (1978).

2. Gibberellins

Monograph: KRISHNAMOORTHY (1975).

a) Methods of Isolation and Determination

Combined gas chromatography-mass spectrometry (GC-MS) provides a highly sensitive method for the definitive identification of gibberellins
(GAs) and their metabolites (GASKIN and MACMILLAN, 1978). The method
is not directly applicable, however, to the analysis of nonvolatile
GA-conjugates and it requires a minimum level of purity of the sample
(normally in the range of 5%), which is a formidable task to achieve
with extracts from higher plants. As the use of the apparatus should
be entirely dedicated to GA analysis, it makes this procedure accessible to only a very limited number of investigators. Mass fragmento-
graphy was employed by BROWNING and SAUNDERS (1977) in the quantitative analysis of GA's in wheat chloroplasts. CROZIER and REEVE (1977)
and REEVE and CROZIER (1978) outline the use of high-performance liquid chromatography (HPLC) in GA analysis.

GA's differing from each other only by the presence or absence of a
double bond (GA_4/GA_7; GA_5/GA_{20}; GA_1/GA_3) were successfully separated
in the form of their p-nitrobenzyl esters by argentation HPLC (HEFT-
MANN et al., 1978) or thin-layer chromatography (TLC) (HEFTMANN and

SAUNDERS, 1978). MORRIS and ZAERR (1978) used the bromophenacyl esters
in HPLC analysis of Ga's. KÜLLERTZ et al. (1978) described the quantitative gaschromatographic determination of as little as 2 nanograms
of GA by electron capture detection. None of the merely chromatographic
methods can, however, provide for the inherent specificity of GC-MS.
Bioassays can more easily be performed in a plant physiological laboratory, but their usefulness in the quantitative analysis of GA's is
limited by the varying biological potencies of individual GA's. This
was again aptly illustrated by SPONSEL et al. (1977), who determined
the biological activity of six new GA's in six bioassays. MAPELLI and
RAINIERI (1978) introduced a variant of the barley aleurone layer bioassay, in which total protein release in response to GA is measured.
The assay is less sensitive than the classical α-amylase assay. REEVE
and CROZIER (1975) critically discuss practical and theoretical aspects
of five bioassays for GA's.

b) Occurrence

Presently 52 gibberellins have been positively identified in the fungus *Gibberella fujikuroi (Fusarium moniliforme)* and/or higher plants. The
complete set of their structures is given by BEARDER and SPONSEL (1977)
and HEDDEN et al. (1978). From 50 kg of seeds of *Cucurbita pepo* FUKUI et
al. (1977a) isolated 50 mg GA_{39}, 27 mg GA_{48} and 3 mg GA_{49}. All three
GA's belong to the C_{20}-GA's, and are oxygenated at C-12 (for number-
ing of the *ent*-gibberellane skeleton see Fig. 1). GA_{48} and GA_{49} are
epimers with respect to C-12 (FUKUI et al., 1977b). GA_{50} (C_{19}) and GA_{52}
(C_{20}) were isolated from another cucurbitaceous plant, *Lagenaria leucantha*
(FUKUI et al., 1978). A novel nonpolar GA extracted from needles of
Picea sitchensis was tentatively identified as an isomer of GA_9 in which
the exocyclic double-bond between C-16 and C-17 (see Fig. 1) has migrated between positions C-15 and C-16 in the D-ring (LORENZI et al.,
1977). The structures of the 2-hydroxylated GA's, GA_{46} (from seeds of
Marah macrocarpus, formerly *Echinocystis macrocarpa*) and GA_{47} (from *Gibberella
fujikuroi*), were established by BEELEY and MACMILLAN (1976). GA_{51}
(2β-OH-GA_9) is an endogenous constituent of pea seeds (SPONSEL and
MACMILLAN, 1977).

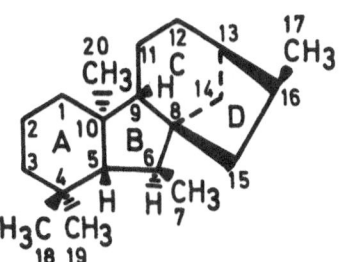

Fig. 1. *ent*-Gibberellane skeleton

While the chemical synthesis of some simpler gibberellins was achieved
earlier, the total synthesis of GA_3, the major GA produced by *Gibberella
fujikuroi*, resisted numerous attempts. The stereospecific total synthesis
of GA_3 has now been reported by COREY et al. (1978a,b) filling a major
gap in synthetic work on GA's. FUJITA et al. (1977) described the total
synthesis of GA_{15} and GA_{37}. Preparation of several radioactively labeled GA's with high specific activity, which is required for metabolic
and binding studies, was reported by YOKOTA et al. (1976) and MUROFUSHI
et al. (1977). GA's doubly-labeled with deuterium and tritium have been
prepared and were successfully employed in metabolic studies (MACMILLAN,

1977; SPONSEL and MACMILLAN, 1977, 1978). The use of the stable heavy isotope is highly advantageous, since it allows to distinguish the deuterated metabolite from the endogenous compound by means of MS-analysis.

AUGIER (1977) compiled an exhaustive list on the tentative identification of GA's and other hormones in algae. Since in all of the cited studies "GA-like" material was characterized only by bioassays, an unambiguous identification of GA's in algae, as well as in lower plants other than *Gibberella fujikuroi*, has not been proven by the time of this writing.

ECKERT et al. (1978) identified GA_1 and GA_3 by MS from *Triticum aestivum* and *Secale cereale* plants and quantified them in the dwarf pea bioassay. Toward the end of axis growth the GA content declined. On the other hand, DATHE et al. (1978) found GA_8, GA_{16}, and GA_{24} in developing *Secale cereale* caryopses. Free GA's were found predominantly in the endosperm, while a GA_{16} conjugate prevailed in the embryo. In developing barley caryopses total GA-like activity was correlated with the rate of dry weight increase, suggesting a function of GA's in the accumulation of assimilates (MOUNLA, 1978). YAMANE et al. (1977) isolated four further GA's (GA_{17}, GA_{20}, GA_{29}, GA_{44}) in addition to the previously known seven GA's from immature *Phaseolus vulgaris* seeds. TURNER et al. (1978) investigated the molecular requirements for gibberellin-like activity in the barley endosperm bioassay with the aid of synthetic analogs of helminthosporic acid, the structure of which resembles the C and D rings of GA's (see Fig. 1) and which is thought to "mimic" a partial GA molecule.

c) Biosynthesis and Metabolism

Reviews: HEDDEN et al. (1978) presented an extensive and authoritative review of this topic. More selected brief reviews are by BEARDER and SPONSEL (1977) and RAPPAPORT and ADAMS (1978).

The main pathway of GA biosynthesis has been established in the fungus *Gibberella fujikuroi*, and much progress has been made in the synthesis of GA's in cell-free systems from higher plants, notably immature seeds from the cucurbitaceous species *Marah macrocarpus* and *Cucurbita maxima*, as well as the legumiosous species *Pisum sativum*. ent-Kaur-16-ene and GA_{12} aldehyde are central intermediates both in the fungal and higher plant systems. Kaurene synthetase, which catalyzes the first committed step in GA biosynthesis, has been partially purified from *Marah macrocarpus* endosperm and characterized (FROST and WEST, 1977). No resolution of activity A (cyclization of geranylgeranyl-pyrophosphate to copalyl-pyrophosphate) and activity B (cyclization of copalyl-pyrophosphate to ent-kaurene) was observed. The two activities differed in a number of properties, however. Copalyl-pyrophosphate generated by the A activity was utilized in preference to exogenous copalyl-pyrophosphate. A decision whether both reactions are catalyzed at the same site, at different sites of the same protein, or of separate proteins was not possible. The molecular weight of the synthetase was in the range of 45,000 daltons, which is ten times less than that of the previously purified enzyme from *Gibberella fujikuroi*. A number of growth retardants, among them Phosfon D and AMO 1618 effectively inhibited activity A at µM concentrations, while higher concentrations were required for inhibition of the B-activity. The potent inhibition of kaurene formation from mevalonate in *Gibberella fujikuroi*, probably at the site of kaurene synthetase A-activity, by the novel growth retardant N,N,N-trimethyl-1-methyl-(2',6',6'-trimethylcyclohex-2'-en-1'-yl) prop-2-enylammonium iodide was used by HEDDEN et al. (1977) to show that only the natural intermediate *ent*-kaur-16-ene, but not a number of isomers, is efficiently incorporated into GA_3.

The view that Phosfon D, AMO 1618 and other agents with growth-retarding activity act exclusively via inhibition of kaurene synthetase, thus depriving the plant of endogenous GA's, has again been challenged by DOUGLAS and PALEG (1978a,b). These authors showed that AMO 1618 inhibited incorporation of ^{14}C-mevalonate into sterols in tobacco seedlings, especially in the stems, and caused accumulation of labeled squalene-2,3-epoxide, the substrate for the cyclizing enzyme in sterol biosynthesis. Experiments with cell-free extracts showed that AMO 1618 inhibits also hydroxymethylglutaryl-CoA reductase, the mevalonate-producing enzyme. DOUGLAS and PALEG argue not quite convincingly that AMO 1618 retards growth through deprivation of sterols that are required for membrane synthesis. The specificity of other growth retardants for inhibition of GA synthesis has also been questioned again. Phosfon-D was shown by LENDZIAN et al. (1978) to uncouple photophosphorylation at 1-100 μM concentrations and to inhibit photosynthetic electron transport at higher concentrations. MÜLLER and SHUPHAN (1976) suggest that CCC affects photosynthetic membranes.

Another potent growth retardant, α-cyclopropyl-α-[p-methoxy-phenyl]-5-pyrimidine methyl alcohol (ancymidol), inhibits the oxidation of kaurene to kaurenol in extracts of *Marah oreganus* (Prog. Bot. $\underline{39}$, 101). The effect was studied in more detail by COOLBAUGH et al. (1978), who showed that ancymidol was an extremely active inhibitor of the cytochrome P-450-dependent oxidation of kaurene, kaurenol, and kaurenal by the microsomal fraction from *Marah macrocarpus* endosperm. Kaurenoic acid oxidation was affected to a much lesser extent. Interestingly, kaurene oxidation in *Gibberella fujikuroi* is inhibited by ancymidol neither in vitro nur in vivo, indicating a significant difference in the oxidative reactions in spite of the identical sequence of intermediates.

CHO et al. (1978) introduced twelve new quarternary ammonium compounds with growth-retarding activity of which N,N,N-trimethyl-1-methyl-3-(3',3',5'-trimethylcyclohexyl)-2-propenylammonium iodide and the 3',3',5',5'-tetramethyl analog were superior in activity to any other growth retardant hitherto known.

> According to KNOTZ et al. (1977) kaurene synthesis from mevalonate is subject to regulation by adenylate energy charge, the site of control being pyrophosphomevalonate decarboxylase. A specific regulation of GA biosynthesis at such an early step in the general isoprenoid pathway seems unlikely, however.

Kaurene synthetase from *Marah macrocarpus* accepts the (R,S)-14,15-epoxide of geranylgeranylpyrophosphate as a substrate and converts it to a mixture of 3α- and 3β-hydroxykaurene (COATES et al., 1976). The possible physiological significance of this reaction in the generation of 3-hydroxylated gibberellins (see below) remains open. Efficient conversion of 3-hydroxylated kaurenoids into GA's by *Gibberella fujikuroi* (LUNNON et al., 1977) demonstrates, however, that the fungal enzymes at least lack sufficient substrate specificity. This is also evident from experiments which prove the conversion of 15α-fluorokaurenoic acid into fluorogibberellins by the fungus (CROSS and ERASMUSON, 1978). Fluorogibberellins are of interest as competitive inhibitors of GA in higher plants (Prog. Bot. $\underline{39}$, 102).

Beyond GA_{12}-aldehyde the pathways leading to the multitude of GA's diverge in *Gibberella fujikuroi* and higher plants, the difference being mainly due to differing capacities to hydroxylate the gibberellane

skeleton (Fig. 1) at positions C-3 and C-13 (HEDDEN et al., 1978).
Extent and position of hydroxylation have a pronounced effect on the
biological activity of GA's and their subsequent metabolism. An important step is the removal of C-20 (Fig. 1) leading to the series of
C_{19} GA's. Chemical oxidative decarboxylation of *ent*-gibberellane-19,20-
dioic acids with lead tetraacetate resulted in the loss of either car-
boxyl group and production of the corresponding lactones (BEARDER and
MACMILLAN, 1976). In vivo studies with *Gibberella fujikuroi* indicate that
C-20 of GA_{13}-7-aldehyde is lost as CO_2 during conversion of the latter
to GA_3 (DOCKERILL et al., 1977) and the same was found for C-20 of *ent*-
kaur-16-ene (DOCKERILL and HANSON, 1978). Both oxygen atoms of the car-
boxyl group (C-19) are retained in this conversion (BEARDER et al.,
1976a; BEARDER and SPONSEL, 1977). It is presently still unknown,
which GA metabolite is the immediate substrate for the reaction leading to removal of C-20, and the elucidation of this question will depend on the availability of cell-free systems which efficiently produce C_{19}-GA's from C_{20}-precursors.

The sequences of the metabolic steps leading from the central intermediate GA_{12}-aldehyde to the plethora of GA's are quite a labyrinth
for the uninitiated observer. As a general rule, in the course of
these conversions, biologically active GA's are formed which are subsequently deactivated by further substitution reactions and/or conjugation. GA-biosynthetic pathways beyond GA_{12}-aldehyde in *Gibberella
fujikuroi* are extensively discussed by HEDDEN et al. (1978) and will
not be considered here, except that MCINNES et al. (1977) found a
second example of 2α-hydroxylation (GA_4 → GA_{47}) in addition to the
known conversion of GA_9 to GA_{40}. The 2α-hydroxylated GA's are active
in a number of bioassays, while 2β-hydroxilated GA's are devoid of
any activity (SPONSEL et al., 1977). In the cell-free system from
Cucurbita maxima the conversion of GA_{12}-aldehyde to GA_{43} via GA_{24} and
and GA_{36} is catalyzed by the 200,000 *g* supernatant and depends on the
presence of Fe^{2+}, while Mn^{2+} is inhibitory (HEDDEN et al., 1978).
Thus, this oxidative system is different from the microsomal system,
which catalyzes the oxidation of *ent*-kaur-16-ene to GA_{12}-aldehyde.
Conversion of GA_9 into a GA_{20} compound by 13-hydroxylation in excised
lettuce hypocotyls is blocked by the iron chelator 2,2'-dipyridyl,
and $FeSO_4$ completely restored the tissue's capacity to metabolize GA_9
(NASH et al., 1978). 13-Hydroxylation of GA_9 to GA_{20} was also observed
by SPONSEL and MACMILLAN (1977) at an early stage of pea seed maturation, while at later stages 2β-hydroxylation (GA_9 → GA_{51}) predominated.
Further metabolism of GA_{20} to GA_{29} (2β-hydroxlation) was studied using
GA_{20} doubly labeled with deuterium and tritium and a very high deuterium content. With this elegant technique the metabolite produced (GA_{29})
could be distinguished from the endogenous compound. While the endogenous GA_{29} content declined with progressing maturity, only low metabolism of exogenously added or endogenously produced radioactive GA_{29}
was observed. In a subsequent report SPONSEL and MACMILLAN (1978) investigated this apparent paradox and found that a novel catabolite, a
2-oxo-GA (Fig. 2), is formed from either $[1\beta,3\alpha - {}^2H_2][1\beta,3\alpha - {}^3H_2]$ GA
or $[2\alpha - {}^2H_1][2\alpha - {}^3H_1]$ GA_{29}, in which the isotope had been lost and was
therefore no longer traceable. As to be expected the catabolite had no
biological activity in three bioassays. The studies described above in
a very abbreviated form show the highly sophisticated level at which
GA metabolism is being studied at present. Stringent criteria that
should be met in such metabolic studies have been restated by HEDDEN
et al. (1978).

Cell-free 13-hydroxylation was observed for the first time in a system from
immature pea seeds by ROPERS et al. (1978). This system rather unexpectedly
produced GA_{12}-alcohol from GA_{12}-aldehyde, a compound that was observed for the
first time in a biological system.

Fig. 2. Catabolite of GA_{29} in Pea Seeds and Seedlings

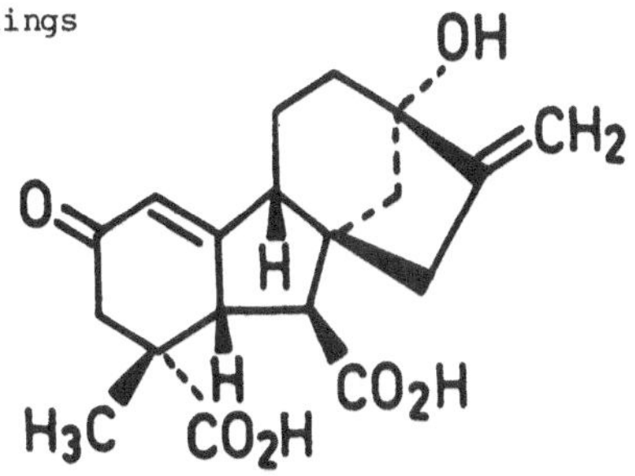

While immature seeds have provided most valuable systems for investigations of GA biosynthesis and metabolism, the function of GA's in
seed maturation is anything but clear. The relationship between GA
metabolism and the effects of applied GA in excised lettuce hypocotyls
has been studied by JONES and coworkers (SILK et al., 1977; STODDART
and JONES, 1977; NASH et al., 1978). GA_1-induced elongation is accompanied by its conversion to polar metabolites which have tentatively been identified as a glucosyl ether and ester. The amount of these
polar metabolites is proportional to the exogenous GA_1 concentration.
Removal of exogenous GA_1 results in a decline of accumulated label to
a constant level in a way which suggests the existence of at least two
GA-accumulating compartments. Growth is sustained after removal of GA_1,
and in this respect GA-induced growth differs from auxin-induced growth.
GA_9, which has no hydroxyl groups, but is also active in the lettuce
bioassay, is rapidly converted to presumably GA_{20} (by 13-hydroxyla-
tion). Inhibition of both growth and GA_{20} formation by 2,2'-bipyridyl
(see above) suggests a causal relationship between GA_9-induced growth
and its metabolism.

A relation between the metabolism of GA_9 and its effect on apical
growth and senescence in peas is also suggested by the work of PROEB-
STING et al. (1978). On the other hand, lack of correlation between
conversion of GA_1 to GA_8 (2β-hydroxlation) and its effect on growth
in dwarf rice seedlings was reported by RAILTON (1978). The relation
between GA_1 metabolism and induction of α-amylase in barley aleurone
layers was investigated by STOLP et al. (1977). Wilting of etiolated
pea seedlings or treatment with abscisic acid (ABA) was shown by
TAYLOR and RAILTON (1977) to be associated with the reduced conversion
of the biologically active GA_{20} into the less active GA_{29} (2β-hydroxy-
lation). These results argue against inhibition by ABA of pea stem
growth via GA inactivation.

Deactivation of GA's may be achieved by 2β-hydroxylation (see above)
and formation of conjugates, preferentially glucosylated derivatives
(HEDDEN et al., 1978). The glucose conjugates are either O-β-D-gluco-
pyranosides (glucose ethers) or O-β-D-glucopyranosylesters. Novel
O-(3)- and O-(13)-glucosylated GA's as well as GA's carrying two glucose moieties both in an ether and ester linkage have been chemically
synthesized (SCHNEIDER et al., 1977a,b). GA -O(3)-^{3}H-glucoside was
synthesized by SCHNEIDER (1978).

The speculation that the glucose ethers may generally represent storage forms of GA's, from which biologically active GA's can be released, is probably untenable because many of the naturally occurring
glucose ethers are those of 2β-hydroxylated GA's, which are biologically inactive anyhow. Reactivation by 2β-dehydroxylation does not
seem feasible. The biological activity of conjugates of active GA's
is probably based on the plants' (or contaminating microorganisms')
capacity to release the GA's from the bound forms. A recent survey of

enzymes from various sources, which are capable of hydrolyzing such conjugates revealed that cellulase from *Aspergillus niger* and helicase from *Helix pomatia* were most efficient (MÜLLER et al., 1978). GA_1-O(3)-glucoside, which was recently isolated also from a natural source (YOKOTA et al., 1978), was especially resistant to enzymatic hydrolysis. Glucosylation of GA's to their esters in maturing bean seeds was found to be predominant in later stages of maturation (YAMANE et al., 1977). During germination and early seedling growth these esters are hydrolyzed (LIEBISCH, 1974) and are, therefore, better candidates for storage forms than the glucosyl ethers.

O(2)-acetyl-GA_3 is a metabolite of *Gibberella fujikuroi* (SCHREIBER et al., 1966), and NADEAU and RAPPAPORT (1974) isolated an amphoteric GA_1-conjugate with probably a peptide side chain from barley aleurone layers. ADAM et al. (1977) synthesized a number of GA-amino acid conjugates linked through a peptide bond, which are analogous to the known amino acid conjugates of indole acetic acid (Prog. Bot. 40, 153). Their natural occurrence is unknown.

Compartmentation of GA synthesis and metabolism in chloroplasts (Prog. Bot. 39, 103) has received further attention. GA_4 and GA_9 were extracted with the detergent Triton X100 from chloroplast membranes of wheat (BROWNING and SAUNDERS, 1977). While methanol was not suited for GA extraction from wheat chloroplasts, RAILTON and RECHAV (1979) successfully employed this solvent for the extraction of GA's from pea chloroplasts. These chloroplasts are capable of hydroxylating GA_9 at the exocyclic C-17 position (Fig. 1) to yield 16,17-dihydro-16,17-dihydroxy GA_9 (RAILTON, 1977a,b) in a reaction, which is also known to occur in *Gibberella fujikuroi* (BEARDER et al., 1976b). RAPPAPORT and ADAMS (1978) added a new facet to the compartmentation studies by the demonstration that vacuoles isolated from barley protoplasts were able to convert GA_1 to GA_8 by 2β-hydroxylation. Further conversion of GA to its glucoside and other products was observed only with protoplasts.

d) Mechanism of Action

PEGG (1976) discusses endogenous gibberellins in healthy and diseased plants. JACOBSEN (1977) examines the regulation of RNA metabolism by plant hormones.

While in spite of many unresolved questions noticeable progress is evident in our understanding of GA biosynthesis and metabolism, this cannot be said of GA action. Work on the effect of GA on the barley aleurone layer (Fortschr. Bot. 35, 132; 36, 131; 105), which has produced so much valuable information in the past, has advanced only little.

Immunohistochemical studies of JONES and CHEN (1976) suggested association of α-amylase in GA-treated cells with the perinuclear region rather than the aleurone grain membrane as had been reported by JACOBSEN and KNOX (1973). Synthesis of α-amylase is thought to depend upon the formation of new endoplasmic reticulum from the nuclear envelope. Release of α-amylase was suggested to occur without participation of secretory organelles.

LOCY and KENDE (1978) reinvestigated the mode of secretion of α-amylase with biochemical methods and found 40%-60% of the intracellular activity in a membrane-bound, latent form, from which it could be released with membrane-disrupting agents. The membrane vesicles are most likely derived from the endoplasmic reticulum (rough and smooth), but an unequivocal identification was not possible after all. The results reinforce the conclusion that the mode of α-amylase release is particulate rather than soluble (Prog. Bot. 39, 106).

Nevertheless, these results do not exclude the possibility of a soluble mode of α-amylase secretion, since merely the synthesis of the
enzyme might be associated with the endoplasmic reticulum (rER), the
vesiculation of which during disruption of the tissue would produce
particulate α-amylase. A putative precursor of α-amylase in wheat
aleurone layers, which is ca. 1500 daltons larger than the excreted
form of the enzyme, was just recently found in a cell-free translational system by OKITA et al. (1979). The authors relate these findings to the signal hypothesis that has been proposed for the transport of proteins across membranes (BLOBEL and DOBBERSTEIN, 1975).

Specific binding of ^{3}H-GA$_1$ to a fraction from aleurone layers enriched
in aleurone grains (ca. 85%) with a single binding site and a dissociation constant of 1.5 µM was reported by JELSEMA et al. (1977a). Specificity of the binding was deduced from the fact that GA$_8$, steviol, indole acetic acid, and kinetin did not compete for binding. Clearly,
more competition studies with other GA's, either active or inactive
in the aleurone layer, are required to demonstrate specificity. Interestingly, abscisic acid, which is known to counteract the effect
of GA in the aleurone layer (Fortschr. Bot. 35, 132), eliminated the
specific binding of GA$_1$. Ca^{2+} is essential for specific binding, which
correlates with the requirement for Ca^{2+} of GA-induced synthesis and
secretion of α-amylase in aleurone layers.

Acid lipase and phospholipase D activities, initially associated with
aleurone grains in wheat aleurone layers, disappeared from this fraction upon incubation of the layers with GA$_3$ (JELSEMA et al., 1977b).
The binding of GA to the aleurone grains might be involved in the initiation of this process. The released enzymes might catalyze lipid
mobilization and turnover, thus providing substrates for membrane proliferation (Fortschr. Bot. 34, 187). Whether the aleurone grain is one
of the primary sites of action of GA in the aleurone layer, as the
authors suggest, remains to be seen.

Studies on the time-course of α-amylase synthesis after addition of
GA$_3$ to aleurone layers has previously shown that most of the amylase
activity appears after 12 h of GA$_3$ treatment (Fortschr. Bot. 35, 133).
While de novo synthesis of α-amylase had unambiguously been proven by
the classical density-labeling experiment with H$_2$^{18}O of FILNER and
VARNER (1967), these experiments did not rule out the possibility that
an inactive α-amylase precursor is synthesized early after addition of
GA$_3$ and is only later converted to the active enzyme. HO and VARNER
(1978) have now eliminated this possibility. They made α-amylase production dependent on exogenously added amino acids by inhibiting pro-
teolysis of endogenous protein reserves with KBrO$_3$ and then fed either
^{12}C- or ^{13}C-amino acids for either 24 h, or for the first or second 12-h
period of 24 h. From the observed density shifts the percentage of α-
amylase synthesized during the first period was 95% and during the
second period 100%. Thus, accumulation of an inactive α-amylase precursor during the first 12-h period can be ruled out. The correlation
between the rate of in vivo production of α-amylase and the level of
translatable mRNA for α-amylase (HIGGINS et al., 1976) and the fact
that the transcription inhibitor cordycepin prevents α-amylase formation only during the first 12 h of GA$_3$ treatment provide evidence that
α-amylase is translated from stable mRNA formed during the initial
12 h of GA$_3$ treatment.

While the available data can be interpreted as being in favor of GA$_3$-
stimulated α-amylase mRNA synthesis (but see JACOBSEN, 1977), it can
by no means be inferred that GA$_3$ interacts directly or in association
with a receptor protein with the transcription process in analogy to

the known mechanism of action of steroid hormones. In this context it is of interest to note that the exclusive action of steroid hormones at the nuclear locus is not at all certain (BAULIEU et al., 1978).

CARLSON (1972) had reported that α-amylase synthesized in aleurone layers in the presence of 5-fluoro-uracil (5-FU) had a decreased thermal stability. He inferred from this result that translation of α-amylase mRNA, into which 5-FU had been incorporated, led to production of α-amylase molecules with altered physical properties and concluded further from experiments, in which aleurone layers were treated with 5-FU before or after addition of GA's, that GA_3 had no effect on α-amylase mRNA synthesis. RODAWAY and KENDE (1978) extensively purified α-amylase from 5-FU-treated aleurone layers and found no difference in its physical properties from those of the enzyme synthesized in the absence of 5-FU. Thus, CARLSON's conclusions do not seem valid. The mechanism by which GA affects mRNA levels remains to be discovered (JACOBSEN, 1977).

> Cells of the aleurone layer are surrounded by thick cell walls consisting mainly of an arabinoxylan. In response to GA_3 a cell-wall-degrading endoxylanase is formed, and xylose and arabinose (free or bound to larger molecules) are released into the medium (DASHEK and CHRISPEELS, 1977). Carboxypeptidase has also to be added to the list of hydrolases that are released from the aleurone layer (SCHROEDER and BURGER, 1978). GA_3 had little effect on the total amount of the enzyme as well as on its release. An attempt to use isolated aleurone cells (spheroplasts) from oat aleurone layers in studies of GA action failed again (EASTWOOD, 1977).

A gibberellin-tannin antagonism (Prog. Bot. $\underline{39}$, 107) was also reported for GA_3-induced α-amylase and phosphatase in barley half grains (JACOBSON and CORCORAN, 1977). As the effect of tannins could be completely overcome by additional GA_3 the authors suggest that tannins should be considered native plant growth inhibitors. The significance of GA_3-enhanced DNA turnover in aleurone cells (TAIZ and STARKS, 1977) remains obscure.

GA's added in relatively high concentrations (0.07-0.3 mM) affect the subcellular development or organelles in the endosperm of germinating castor bean seeds. GA_3 and GA_7 are equally effective in accelerating the appearance and enhancing the total activity of isocitrate lyase, a glyoxysomal marker enzyme (MARRIOTT and NORTHCOTE, 1977). GA_3 failed, however, to enhance the glyoxysomal population as measured by protein and phospholipid determinations (WRIGLEY and LORD, 1977) and to stimulate the phosphatidylcholine-synthesizing capacity of the tissue. GONZALEZ (1978), on the other hand, concluded that GA_3 treatment leads to the assembly of a special class of glyoxysomes with high specific activity of isocitrate lyase. As the effects of GA's in the castor bean endosperm system consist mainly in the acceleration and enhancement of processes that occur also in the absence of exogenous GA, this system is less suitable for studies of GA action than the barley endosperm.

GA-induced growth (Prog. Bot. $\underline{39}$, 105) has been further investigated in a number of systems. STUART et al. (1977) analyzed the growth of excised lettuce hypocotyl sections. Inhibition of elongation growth by light is overcome by GA_3, and GA_3 is equally effective in γ-irradiated or 5-fluorodeoxyuridine-treated sections, in which cell division is nearly eliminated. Thus, GA_3-induced growth is solely attributable to cell alongation. K^+ and Na^+, which have no effect on the growth of the sections in the absence of GA_3, synergistically enhance

the growth response (STUART and JONES, 1977). Not the increased osmotic potential, but rather the increase in cellular extensibility is
responsible for the increased growth rates, which is analogous to
auxin-induced growth. Anion and cation uptake are stoichiometric, and
experiments with anions that do not readily penetrate membranes indicate that anion uptake is required for the salt effect (STUART and
JONES, 1978). The temperature dependence of the GA response is thought
to be attributable to changes in membrane fluidity (STODDART et al.,
1978), but whether this is a direct or indirect effect of GA on membranes is open. A direct interaction of GA_3 with model lipid membranes
had been demonstrated by WOOD and PALEG (1974), but LESHEM and INBAR
(1978) failed to find a GA_3-induced change in the microviscosity of
the mitochondrial membrane. NEUMANN and JANOSSY (1977a) related GA_3-
induced changes in intracellular ion levels of dwarf maize leaves to
hormone-induced membrane permeability changes. While these results
were based on long-term experiments which permit alternative interpretations, it was found in a subsequent study (NEUMANN and JANOSSY,
1977b) that in the same system changes in ion ratios in response to
GA_3 occurred without a lag phase. The physiologically inactive GA_3-
methylester did not produce these changes, but studies with a larger
number of active and inactive GA's are clearly required to establish
the specificity of the observed effect. The significance of the changes in ion rations for the induction of growth remains obscure.

A rapid GA_3-induced increase within a few minutes of the membrane potential difference of dwarf maize coleoptile cells (NELLES, 1977) also
indicates a direct interaction of the hormone with membranes. No distinction between interaction of GA_3 with the lipid or the protein material of the membranes could, however, be made. Kinetic analyses of
GA_3-induced growth of *Pharbitis nil* seedlings by BARENDSE et al. (1978)
revealed in some cases latent periods of less than 20 min, and the
authors argue that this period is too short to allow for de novo nucleic acid and protein synthesis. This is reminiscent of the same argument raised in the interpretation of the kinetics of auxin-induced
growth (Fortschr. Bot. 34, 182). Also related to auxin-induced growth
is the question raised by STUART and JONES (1978) whether the acid-
growth hypothesis (Fortschr. Bot. 34, 182; 36, 127; 38, 152; 40, 154)
might be applicable to GA-induced growth since both GA_3 and auxin evoke
the elongation response by reducing the tissue's cell wall extensibility (see above). Neither GA_3 nor fusicoccin (Prog. Bot. 40, 155) cause
acidification of the medium surrounding the sections, although both
agents cause elongation. In the presence of KCl or NaCl and fusicoccin,
but not GA_3, acidification does occur, however, but apparently unrelated to growth. In a detailed argumentation the authors conclude that
the acid growth hypothesis does not explain either GA_3- or fusicoccin-
induced growth in lettuce hypocotyls.

The involvement of microtubules in GA_3-induced growth (Prog. Bot. 39,
105) was investigated by KEITH and SRIVASTAVA (1978) using both colchicine and lumicolchicine to discriminate between metabolic and micro-
tubule-mediated processes. Both drugs inhibited the respiration of lettuce seedlings, but only colchicine disrupted microtubules and inhibited growth. In wheat coleoptile sections the orientation of microtubules
was related to the differential growth changes elicited by auxin, kinetin, or GA_3 (VOLFOVA et al., 1977). Stimulation of GDP-glucose-dependent
β-glucan synthetase activity in GA_3-treated *Avena* stem segments was found
by MONTAGUE and IKUMA (1978). The increase is obviously related to the
enhancement by GA_3 of hemicellulose and cellulose synthesis in this
system (Prog. Bot. 39, 105). Suppression of GA_3-stimulated growth in
Avena stem segments by IAA, native as well as applied exogenously, excludes the possibility that the GA_3-effect is mediated through IAA

(RAPOPORT et al., 1978) and clearly shows that the response of a tissue to a given hormone is determined by its "competence" (see also discussion of cytokinin action).

A number of less related reports will briefly be discussed for their general interest: GA$_3$ at 0.03 ppm (ca. 10^{-7}M) suppressed the formation of bulbous swellings in the petioles of the water hyacinth *Eichhornia crassipes*, which provide the buoyancy of this floating plant (PIETERSE et al., 1976). Overgrowth of water areas by this harmful weed can thus be reduced. A similarly obnoxious plant is *Potamogeton nodosus*, a heterophyllous species that produces submerged-type leaves without and floating-type leaves with stomates. The development of the latter type was promoted by 1-10 μM abscisic acid, and this effect was overcome by simultaneous exposure to GA$_3$ or kinetin (ANDERSON, 1978).

Marked changes in the phyllotaxis (leaf arrangement) of *Xanthium pennsylvanicum* due to GA3-treatment were observed by MAKSYMOWYCH and ERICKSON (1977). In a series of papers CHAILAKHYAN and KHRYANIN (1978a,b,c; 1979) investigated the effect of growth regulators on the sex expression in hemp *(Cannabis sativa)* and spinach plants. In the presence of roots or exogenous cytokinin pistillate (female) flowers were predominantly produced, while treatment with GA$_3$ resulted in the formation of mainly staminate (male) flowers. Enhancement by GA$_3$ of the male tendency of cucumber floral buds was reported by FUCHS et al. (1977), while LESHEM and OPHIR (1977) found that in carob *(Ceratonia siliqua)* and the date palm *(Phoenix dactylifera)* female growth is correlated with higher endogenous levels of gibberellin-like activity. DAUPHIN et al. (1979) found female sex expression associated with cytokinin levels in *Mercurialis annua*. Any generalization of the hormonal control of sex expression is thus apparently still not possible.

Inhibition of betacyanin synthesis in *Amaranthus caudatus* by GA's was shown to be related to the availability of the precursor tyrosine for pigment production (LALORAYA et al., 1976; STOBART and KINSMAN, 1977).

3. Cytokinins

DEKHUIJZEN (1976) surveyed endogenous cytokinins in healthy and diseased plants. SCHLEE (1975) discussed cytokinins in microorganisms.

a) Analytical Methods

A very useful introduction to analytical chemical procedures for cytokinins was written by HORGAN (1978a).

While gibberellins are strictly classified on the basis of their chemical structure, cytokinins are defined by their capacity to induce cell division in certain excised plant tissues or tissue cultures. As with the gibberellins, it is a formidable task to unambiguously identify and quantify cytokinins from plant sources with (physico)-chemical procedures as compared to the seeming ease of bioassay performance. However, these procedures avoid the ambiguity and mere approximation, which is associated with bioassays, and a number of suggestions can be found in the recent literature directed towards the identification of cytokinins by physicochemical means. New thin-layer chromatographic systems for the separation of natural and synthetic cytokinins are given by PACES and KAMINEK (1978). In addition to the procedures for high-performance liquid chromatographic (HPLC) separation of cytokinins listed in the last review (Prog. Bot. 39, 109), novel HPLC systems were described by KANNANGARA et al. (1978) and HOLLAND et al.

(1978). Gas chromatography (GC) and gas liquid chromatography (GLC) have the major disadvantage that the cytokinins have to be derivatized prior to chromatographical analysis, but in combination with mass spectrometry these procedures allow selective detection and identification of cytokinins and their metabolites at submicrogram levels (MACLEOD et al., 1976; YOUNG, 1977; MORRIS, 1977; DAUPHIN et al., 1979). HASHIZUME et al. (1979) developed a system for cytokinin analysis by GC-MS and selected ion monitoring and prepared deuterated cytokinin standards to measure precise recovery rates. The analysis time, including extraction, was 3 days as compared to several weeks in the tobacco or soybean callus assay. ^{13}C-NMR spectroscopy can be successfully employed in the elucidation of the structure of cytokinins and their glucosides (DUKE and MACLEOD, 1978). However, the requirement for milligram amounts of the material to be analyzed makes this technique hardly suited for studies on natural cytokinins.

A method, the tremendous potential of which in the quantitative analysis of plant constituents is only just emerging is the radioimmunoassay (RIA) (ECKERT, 1976), in which the extreme specificity of the antigen-antibody reaction is exploited. The sensitivity of the assay is mainly dependent on the specific radioactivity of the antigen in question. RIA's for plant hormones have recently been developed for indoleacetic acid (PENGELLY and MEINS, 1977) and abscisic acid (WEILER, 1979). In the latter case, due to the high specificity of the immunoassay, no purification of crude extracts was required, thus permitting the analysis of an extremely large number of samples per day, which hitherto had simply been impssible due to the necessity of highly purifying ABA prior to any physicochemical analysis. A RIA for isopentenyladenosine was developed by KHAN et al. (1977), but in their work it was used to quantitate the isopentenyladenosine content of tRNA's. The specificity of the antigen-antibody reaction is evident from the fact that 2-methylthioisopentenyladenosine hardly crossreacted in this system. This assay was also applied in the estimation of the isopentenyladenosine content of rice embryos by SHIVAKUMAR et al. (1976). Using antibodies to isopentenyladenine MILSTONE et al. (1978) demonstrated that isopentenyladenosine-containing tRNA's can be purified by affinity chromatography on immobilized antibodies. The antibodies reacted equally well with free or tRNA-bound nucleotides. A similar technique involving sepharose-bound anti-GA$_3$ antibodies had earlier been developed by FUCHS and GERTMAN (1974) both for RIA of GA$_3$ and for fractionation of GA's from pea seedlings. Their method lacked, however, the desired sensitivity and specificity. Immunological assays for IAA and GA's with nonradioactive antigens, using either the inhibition of complement fixation (FUCHS and FUCHS, 1969) or the inhibition of inactivation of modified bacteriophages (FUCHS et al., 1971) were also only of limited value for the analysis of these hormones in tissue extracts. The mode of coupling the antigen to the carrier protein (usually serum albumin) is obviously extremely important for the production of antibodies of the desired specificity. To ensure sensitivity it is required to prepare antigens with high specific radioactivity for the competitive binding assay. ^{3}H-N^6-Benzyladenine (SUSSMAN and FIRN, 1976) and ^{3}H-zeatin with specific activities > 10 ci/mmol are available for binding studies. CONSTANTINIDOU et al. (1978) prepared antibodies against N^6-benzyladenosine coupled to bovine serum albumin and determined the affinity constants for a number of natural cytokinins. The N^6-substituent was immunodominant. Immobilized antibodies were used to trap cytokinins from solutions, but while kinetin, N^6-benzyladenine, and isopentenyladenine were efficiently trapped, zeatin was not. A general applicability of these antibodies to either cytokinin extraction or quantitation is at present not conceivable. It is somewhat unfortunate, too, that the unnatural N^6-benzyladenosine was used as antigen, rather than, e.g., the natural zeatin.

b) Occurrence

A brief account of the nature and distribution of cytokinins is given by HORGAN
(1978b). The localization of cytokinins in tRNA and the possible physiological
significance of this are discussed by KAMINEK (1975) and LETHAM and WETTENHALL
(1977).

In addition to N^6-isopentenyladenosine and 2-methylthio-isopentenyl-
adenosine, cytokinins commonly occurring in bacterial tRNA, 2-methyl-
thio-ribosyl-cis-zeatin has been detected with chromatographical procedures in ^{35}S-labeled tRNA of the plant-associated bacteria *Rhizobium
leguminosarum*, *Agrobacterium tumefaciens* and *Corynebacterium fascians* (CHERAYIL
and LIPSETT, 1977). EINSET and SKOOG (1977), based on a more extended
characterization, detected the unmethylthiolated form ribosyl-cis-zeatin in *Corynebacterium fascians* tRNA, while MORRIS and CHAPMAN (1976) reported the isolation of ribosyl-trans-zeatin from *Agrobacterium tumefaciens*
tRNA. In any case, these findings extend previous reports on the occurrence of side-chain hydroxylated cytokinins in bacterial tRNA's (Prog.
Bot. **39**, 108). The finding that only plant-associated bacteria, which
induce growth in their hosts, contain side-chain hydroxylated cytokinins is suggestive of a causal relationship, especially, if the
preliminary report by CLAEYS et al. (1978) that only oncogenic strains
of *Agrobacterium tumefaciens* produce trans-zeatin should be confirmed. While
cis-zeatin previously had been isolated from the culture filtrate of
Corynebacterium fascians (Prog. Bot. **36**, 134), trans-zeatin is secreted
by the virulent *Agrobacterium temefaciens* strain C 58 (KAISS-CHAPMAN and
MORRIS, 1977) along with isopentenyladenine and the riboside of its
methylthiolated derivative.

An investigation of the cytokinin content of *Euglena gracilis* tRNA preparations (SWAMINATHAN and BOCK, 1977; SWAMINATHAN et al., 1977) revealed that of the three major cytokinins identified, isopentenyladenosine is found in cytoplasmic as well as chloroplast tRNA, while
ribosyl-cis-zeatin was mainly restricted to cytoplasmic tRNA and ri-
bosyl-2-methylthio-zeatin was localized exclusively in chloroplast
tRNA. ^{35}SO$_4$$^{2-}$ was not incorporated into the latter component in two
chloroplast bleached mutants, showing that the methiolation reaction
is restricted to the chloroplast. Both the cis- and trans-isomers of
ribosyl-2-methylthio-zeatin had previously been chemically synthesized
and also been isolated from a pea tRNA preparation (VREMAN et al.,
1974). An improved chemical synthesis of ribosyl-2-ethylthio-trans-
zeatin was reported by SUGIYAMA et al. (1978). HECKER et al. (1976)
had tentatively identified 2-methylthio-isopentenyladenosine in the
phenylalanine-accepting tRNA of *Euglena* chloroplasts. In a subsequent
study with spinach leaves and isolated spinach chloroplasts VREMAN
et al. (1978) isolated isopentenyladenosine and the cis-isomer of ribo-
syl-2-methylthio-zeatin from chloroplast tRNA preparations, while
tRNA preparations from leaves yielded, in addition, the cis- (98%) and
trans- (2%) isomers of ribsylzeatin and the trans-isomer of ribosyl-2-
methylthio-zeatin. Thus, in agreement with the results obtained with
Euglena, ribosyl-zeatin is probably exclusively localized in the cytoplasm. Allowing for some overgeneralization, a parallelism between
prokaryotes and chloroplasts with respect to the complement of cyto-
kinin-active nucleotides in their tRNA's is becoming apparent. Fractionation of wheat germ tRNA's and examination of their cytokinin content showed that occurrence of the cytokinins is restricted to those
tRNA species that would be expected to respond to codons beginning
with U (STRUXNESS et al., 1979), but within the U-group of tRNA species only a few actually contained cytokinin modifications. LESTER
et al. (1979) found that two of the six leucine isoaccepting tRNA
species from soybean cotyledons responded to codons beginning with U
and contained cytokinins.

Cytokinins have been isolated from root nodules containing nitrogen-fixing bacteria (Prog. Bot. <u>39</u>, 108) and the mitotic activity of the nodule meristem seems to be related to their occurrence (NEWCOMB et al., 1977; PUPPO and RIGAUD, 1978). In a series of investigations HENSON and WHEELER (1977a,b,c,d,e; WHEELER and HENSON, 1978) found higher cytokinin activity (tentatively identified as *trans*-zeatin and its riboside, zeatin-O-β-D-glucoside, a β-D-glucoside of ribosyl zeatin and the corresponding dihydro-zeatin derivatives) in root nodules of *Alnus glutinosa* and other nonleguminous nodule-bearing angiosperms than in the roots. Radioactivity from $8-^{14}C$-zeatin supplied to the nodules was recovered from all parts of the plant, either as zeatin or its metabolites, but the authors argue that the nodules are not a substantial source of cytokinins for other plant parts.

Most identifications of cytokinins reported in the literature are tentative in nature since they relate the chromatographic behavior of a compound to that of an authentic standard. A number of recent unambiguous identifications is listed in the following: side chain O-β-D-glucosyl zeatin and O-β-D-glucosyl-ribosylzeatin were identified as natural constituents of *Vinca rosea* crown gall cultures as their permethylated derivatives by MS (PETERSON and MILLER, 1977; MORRIS, 1977). The latter compound was identified in a plant system for the first time, while the former was also identified as a zeatin-derived metabolite in lupin seedlings (LETHAM et al., 1977; PARKER et al., 1978). O-Glucosyl-zeatin was less active by an order of magnitude in the soybean callus assay than zeatin itself or ribosylzeatin, but at high ($10^{-4}M$) concentrations the callus had a higher tolerance for the glucosylated cytokinin (VAN STADEN and PAPAPHILIPPOU, 1977). A storage function of the glucoside was indicated (see also next section). Dihydrozeatin-O-βD-glucoside was the major cytokinin in the leaves of decapitated bean plants (WANG et al., 1977), while dihydro-zeatin riboside was a minor cytokinin in the same tissue (WANG and HORGAN, 1978). The N-glucosylated zeatin-7-β-D-glucoside (raphanatin, Prog. Bot. <u>36</u>, 135), which had originally been detected as a metabolite of ^3H-zeatin supplied to derooted radish seedlings, was recognized as an endogenous cytokinin of radish seeds by SUMMONS et al. (1977). Full details of the chemical synthesis of the 7- and 9-glucosides of zeatin and 6-benzylaminopurin are given by COWLEY et al. (1978). The spectrum of zeatin-related compounds was extended by a very unusual compound, named lupinic acid (Fig. 3, structure I), which was isolated as a zeatin metabolite from derooted lupin seedlings (DUKE et al., 1978; PARKER et al., 1978). Elucidation of its structure, confirmed by chemical synthesis, revealed that an alanine moiety is attached to the N-9-position of the purine ring of *trans*-zeatin. HORGAN (1978b) states briefly that soybean callus metabolizes benzyladenine to the corresponding 9-alanyl derivate. A structurally related compound is the spore germination inhibitor of *Dictyostelium discoideum*, termed discadenine (Fig. 3, structure II), in which an α-amino-butyryl moiety if attached to N-3 of the purine ring of isopentenyladenine (ABE et al., 1976) and which has approx. 1% of the activity of isopentenyl-

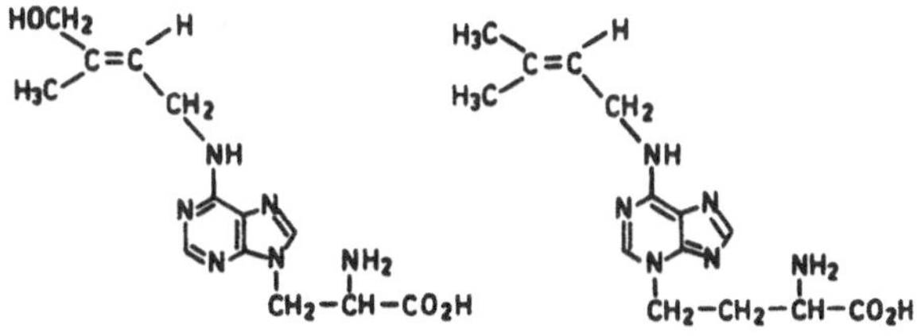

Fig. 3. Structures of Lupinic
Acid (I) and Discadenine (II)   I   II

adenine in the tobacco callus bioassay (NOMURA et al., 1977). Isopentenyladenine itself has also been isolated from the slime mold
(TANAKA et al., 1978). Details of the cell-free biosynthesis of these
compounds will be given in the next section. Further, the conclusive
identification by GC-MS of the moss cytokinin ("bryokinin", Fortschr.
Bot. 35, 157) as isopentenyladenine can be reported (BEUTELMANN and
BAUER, 1977). Our knowledge of the presence and function of cytokinins
in pteridophytes is less advanced, but SCHRAUDOLF and FISCHER (1979),
using the *Amaranthus* biotest, could tentatively identify ispentenyl-
adenine, zeatin, and its riboside in gametophytes of *Anemia phyllitidis*.
Studies related to the tuberization process in shoots and roots, in
which cytokinins are supposed to be involved as the "tuber-forming
stimulus" (VAN STADEN and DIMALLA, 1977), led to the isolation of ri-
bosyl-*cis*-zeatin from potato tubers (MAUK and LANGILLE, 1978) and of
trans-zeatin from radish roots (KOYAMA et al., 1978).

The relation between chemical structure and physiological activity of
cytokinins was discussed in the last review (Prog. Bot. 39, 109). Dihydrozeatin (Fig. 4, structure IV) is optically active, and the natural antipode has the (S)(-)-configuration. Nevertheless, the unnatural
(R)(+)-antipode exhibited a higher biological activity in a number of
bioassays (MATSUBARA et al., 1977). While all previous reports had indicated that a modification of the purine moiety lowered the biological
activity of a cytokinin, MATSUBARA et al. (1978) reported that 6-ben-
zoylamino-1-deaza-purine possessed very high activity in a number of
bioassays equalling or even exceeding that of the highly active zeatin.
The optimum concentration of the "urea-type" cytokinin N-phenyl-N'-
(2-chloro-4-pyridyl)-urea in the tobacco callus bioassay was less than
4×10^{-9}M, while that for diphenylurea was 5×10^{-5}M (TAKAHASHI et al.,
1978). The former compound is also 10 times more active than either
benzyladenine, isopentenyladenine, or zeatin in inducing shoot formation from tobacco callus (OKAMOTO et al., 1978). 8-azido-N^6-benzyl-
adenine was synthesized by SUSSMAN and KENDE (1977) as a potential
photoaffinity reagent for cytokinin receptor sites (see also last
section) and found to be active in two bioassays. The 2-azido derivatives of isopentenyl-adenine and benzyladenine had previously been
synthesized (THEILER et al., 1976). BITTNER et al. (1977) attached
cytokinins covalently to starch and cellulose and demonstrated a sustained release of the hormones. This approach might prove useful for
the control and prolongation of the release of plant hormones. An
alternative would be the application of 9-alkyl-cytokinins, which are
gradually converted in vivo to active cytokinin molecules (LETHAM et
al., 1978b).

c) Biogenesis and Metabolism

The question whether cytokinins are released intact during the turnover of tRNA or whether they are synthesized de novo independent of
tRNA (Prog. Bot. 36, 135; 39, 110) is still not satisfactorily resolved, but most of the available evidence is in favor of the latter
pathway. BURROWS (1978a) examined free and tRNA-bound cytokinins both
in lupin seeds and poplar leaves. tRNA from both sources contained
four cytokinins: ribosyl-*cis*-zeatin (Fig. 4, structure I), 2-methyl-
thioribosyl-zeatin (Fig. 4, structure II; stereochemistry of side
chain not known); isopentenyladenosine (Fig. 4, structure III) and
its 2-methylthio-derivative (Fig. 4, structure IV). N^6-(2-hydroxy-
benzyl)-adenosine (Fig. 4, structure V) and dihydrozeatin (Fig. 4,
structure VI) were the free cytokinins found in poplar and lupin respectively. Since especially N^6-(2-hydroxybenzyl)-adenosine cannot
possibly have its origin by side-chain rearrangement in any of the

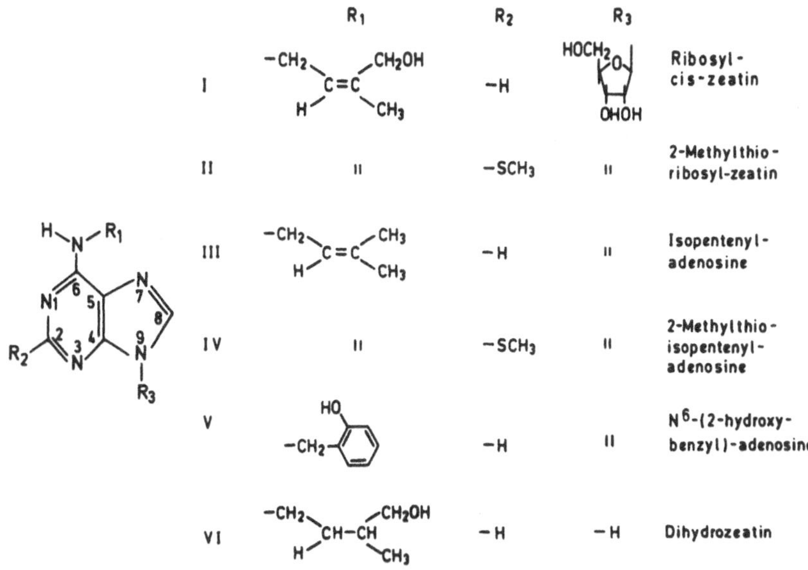

	R₁	R₂	R₃	
I	$-CH_2$ $C=C$ H CH_2OH CH_3	$-H$	$HOCH_2$ ribose (OHOH)	Ribosyl-cis-zeatin
II	"	$-SCH_3$	"	2-Methylthio-ribosyl-zeatin
III	$-CH_2$ $C=C$ H CH_3 CH_3	$-H$	"	Isopentenyl-adenosine
IV	"	$-SCH_3$	"	2-Methylthio-isopentenyl-adenosine
V	$-CH_2$ phenyl-HO	$-H$	"	N^6-(2-hydroxy-benzyl)-adenosine
VI	$-CH_2$ $CH-CH$ H CH_3 CH_2OH	$-H$	$-H$	Dihydrozeatin

Purine ring structure with positions: N_1 6 5 7 8, N_2 2 3 4 9, substituents R_1, R_2, R_3.

Fig. 4. Structures of tRNA-bound Cytokinins (I-IV) and of Free Cytokinins (V-VI) in Lupin Seedlings (V) and Poplar Leaves (VI)

cytokinins present in the tRNA, the author concludes that this finding lends strong support to biosynthesis de novo of the free cytokinin. Subsequently BURROWS (1978b) fed ³H-adenine for 60 min to a cytokinin-autonomous tobacco callus and found incorporation of label into the free cytokinins *trans*-zeatin, ribosyl-*trans*-zeatin, isopentenyladenine (containing the highest radioactivity) and isopentenyladenosine. No discrete peak of radioactivity was associated with the *cis*-isomers of zeatin and its riboside, which are predominant in tRNA both with respect to the *trans*-isomer and to isopentenyladenosine. This result corroborates the author's previous conclusion. tRNA-isopentenyltransferase purified from maize root tips and kernels catalyzed the isopentenylation of tRNA's, oligonucleotides, and, to a very low extent and only at high substrate concentrations, of adenosine (HOLTZ and KLÄMBT, 1978). A physiological significance of the latter reaction seems, therefore, unlikely. An intriguing contribution to the cell-free synthesis of cytokinins was made by TAYA et al. (1978a). During an investigation of the biosynthesis of the spore germination inhibitor discadenine in *Dictyostelium discoideum* (Fig. 3, structure II, see previous section) they showed that 5'-AMP is the actual acceptor molecule for the isopentenyl group and that the overall pathway for isopentenyladenine synthesis follows the sequence

5'-AMP ⟶ isopentenyl-AMP ⟶ isopentenyladenine
        ↑
isopentenylpyrophosphate

It will be very interesting to see if such a reaction can also be detected in higher plants.

HAHN et al. (1976) calculated that tRNA turnover in *Agrobacterium tumefaciens* can account for the release of free cytokinins, but could not exclude an independent biosynthesis for de novo synthesis. Thus, either of the pathways, or both together may be involved in cytokinin biosynthesis. HELBACH et al. (1978) concluded from double-labeling experiments (2-¹⁴C-mevalonate, ³H-methyl-methionine) with *Lactobacillus acidophilus* that the tRNA half life is the same irrespective of whether the tRNA contains cytokinins or not.

While HAHN et al. (1974) concluded from experiments with in vitro cultivated developing peapods that the seeds produce cytokinins independent of the presence of the root system, the reinvestigation of the experiment by KRECHTING et al. (1978) led to the conclusion that pea seeds are incapable of cytokinin synthesis and received the cytokinins in the in vitro experiment most likely from the pod walls. CHEN and PETSCHOW (1978a) demonstrated cytokinin synthesis from ^{14}C-adenine in cultured rootless tobacco plants, which indicates that in any case roots are not the exclusive site of cytokinin biosynthesis. Aseptically cultured excised roots of maize and tomato are capable of cytokinin synthesis (VAN STADEN and SMITH, 1978). Formation in excised roots of O-glucosyl-zeatin, which is not found in roots of intact maize plants, is probably an inactivation mechanism due to inhibited export of zeatin.

Exogenously added radioactive cytokinins undergo extensive metabolism, which varies with the plant species, the plant organ and the age of the tissue (Prog. Bot. 36, 135; 39, 111). The metabolism can be very complex, as is illustrated by the work of LETHAM and coworkers on lupin, in which 16 metabolites have been characterized apart from some minor unidentified metabolites (PARKER et al., 1978). Glucosylation of benzyladenine can occur at the 3-, 7-, and 9-positions of the purine ring, while zeatin can be glucosylated at the 7- and 9-positions of the ring, as well as at the side chain hydroxyl group (Prog. Bot. 39, 111). The unusual 7-glucosylation is neither restricted to molecules with cytokinin activity (LETHAM et al., 1978b) nor is it required for the expression of the biological activity of a cytokinin (LALOUE, 1977). GAWER et al. (1977) and LALOUE et al. (1977) suggest that 7-glucosylation protects a cytokinin against side-chain cleavage, and hence irreversible loss of activity, and that therefore the 7-glucosyl derivatives should be considered storage forms. A similar protection of the side chain of zeatin is afforded by O-glucosylation, which results in a considerable metabolic stability of the O-glucoside (PARKER et al., 1978).

Cell-free synthesis of the 7- and 9-glucosides of benzyladenine by separate enzymes with UDP-glucose as glucosyl donor in radish cotyledon extracts was found for the first time by ENTSCH and LETHAM (1979). The very low activity of the enzymes has so far not allowed to determine their specificity. The cell-free synthesis of lupinic acid (Fig. 3, structure I) requires O-acetylserine as donor for the side chain (MURAKOSHI et al., 1977). The synthesis of the isomeric β-uracil-substituted alanines willardiin and isowillardiin is known to proceed according to the same mechanism (MURAKOSHI et al., 1978), and in an analogous reaction discadenine (Fig. 3, structure II) is synthesized from isopentenyladenine and S-adenosylmethionine (TAYA et al., 1978b). Thus, a very promising start has been made in studying cytokinin conjugation in cell-free systems. Since serine at high concentrations is known to promote leaf senescence and to suppress the senescence-retarding activity of cytokinins, PARKER et al. (1978) give thought to the idea that in the presence of high serine concentrations cytokinins are inactivated by transformation into their 9-alanine derivatives. Increased glucosylation, and hence inactivation (?), of cytokinins occurs in maturing leaves of *Alnus glutinosa* (HENSON, 1978a,b).

Ribosylation of cytokinin bases by adenosine phosphorylase from wheat germ was demonstrated by CHEN and PETSCHOW (1978b) and phosphorylation of the nucleosides by adenosine kinase from the same source by CHEN and

ECKERT (1977). Formation in vivo of cytokinin-ribosyl-5'-mono, di-
and triphosphates from exogenously added cytokinins was observed in
cultured tobacco cells (LALOUE et al., 1977), inflorescence stalks
of *Yucca* (VONK, 1978), and germinating lettuce seeds (MIERNYK and
BLAYDES, 1977; MIERNYK, 1979). Formation of the nucleotide-triphos-
phates is a prerequisite for a possible incorporation of cytokinins
into tRNA. Very low incorporation of exogenous benzyladenine into
tRNA and to a higher degree into rRNA had been reported previously
(Prog. Bot. 39, 110). Now MURAI et al. (1977) have found that tobacco
callus incorporates also kinetin preferentially into rRNA and to a
four times greater extent than benzyladenine. Kinetin in the rRNA preparation was unambiguously identified by GLC-MS and had thus been incorporated as the intact molecule. Kinetin was found in both 18S and
25S rRNA and was associated with oligonucleotides after partial enzymic digestion of the rRNA (MURAI et al., 1978). Whether incorporation
was specific or the result of transcriptional errors or rather of errors in the process of nucleotidyl transfer remains to be seen. Co-
valent linkage of benzyladenine as benzyladenosine-5'-monophosphate
in the total RNA of cultured tobacco cells is also suggested by experiments of JOUANNEAU et al. (1977); unambiguous chemical identification
is, however, lacking and no attempt was made to differentiate between
tRNA and rRNA. Cytokinin-dependent tobacco callus grown on ^{14}C-labeled
N,N'-diphenylurea (a nonpurine cytokinin) contained ribosyl-*cis*-zeatin,
ribosyl-isopentenyl-adenosine and ribosyl-2-methylthio-zeatin in its
tRNA, and radioactivity was not incorporated into any specific ribo-
nucleoside (BURROWS, 1976a). Taking all the available evidence together it seems highly unlikely that the incorporation of cytokinins into
tRNA is related to their mechanism of action. While cytokinins in tRNA
do have a positive effect on codon recognition, thus maintaining codon-
anticoden fidelity, the relevance of their occurrence in tRNA is apparently unrelated to the mode of action of free cytokinins.

d) Mode of Action

The current interest in plant hormone receptors is reflected by the articles
of KENDE and GARDNER (1976), FOX and ERION (1977), VENIS (1977) and LIBBENGA
(1978). Of these only the review of FOX and ERION deals specifically with cytokinin-binding proteins, while auxin-binding sites which are known best at present, are preferentially discussed in the other reviews. General reviews dealing with one aspect or other of cytokinin function are those by BURROWS (1976b),
JACOBSEN (1977), KLÄMBT (1977), LALOUE (1978), and WAREING et al. (1977).

There are two generally acknowledged strategies to study the primary
action of hormones. The "backward"-approach traces a recognized biochemical event, which occurs after addition of the hormone, back to
earlier reactions, until the site and nature of the primary action
are found. This strategy, used by SUTHERLAND and his coworkers
(SUTHERLAND, 1972) led to the development of the second-messenger concept of hormone action. The "forward"-approach is based on the general consensus that a hormone must be "recognized" by a specific receptor, presumably of a macromolecular and proteinaceous nature, and
that hormone-receptor interactions produce a "transformed" receptor
(or hormone-receptor complex) which initiates secondary reactions
leading to the ultimate physiological responses. Elucidation of the
mechanism of steroid hormone action is the example to be cited here.
By definition, a hormone-receptor is a bifunctional element: it must
(1) be able to recognize the hormone (discrimination) and (2) generate
a response which is ultimately manifested in the observed physiological response. Unless both criteria are fulfilled one should call a
cellular component, which interacts with a hormone, a binding-site or

acceptor rather than a receptor. A receptor is characterized by (1) a specificity for the hormone corresponding to the biological specificity of the hormone (2) an affinity, which equals the concentrations of the hormone required in vivo to elicit a response (3) a low capacity (4) its occurrence in the target tissue and (5) noncovalent interaction with the hormone. Many other proteins will exhibit hormone affinity without being receptors, e.g., proteins involved in the synthesis, export, transport, uptake, and catabolism or conjugation of a given hormone. With these precautions in mind one has to admit that the unequivocal identification of a plant hormone receptor has so far not been achieved.

Binding of various molecules with low or high biological cytokinin activity to the cytokinin-binding factor from wheat germ ribosomes (Prog. Bot. $\underline{39}$, 112) corresponded closely with their in vivo activity (FOX and ERION, 1977). Root ribosomes were more effective in binding cytokinins than shoot ribosomes. No biological function of the ribosome-bound or soluble cytokinin-binding sites is presently known. It cannot be excluded that a ribosomal binding site is involved in the recognition of cytokinin-containing tRNA's. Binding of benzyladenine to ribosomes of *Mercurialis annua* has also been reported (CHUNG et al., 1977).

The 4000-5000 dalton cytokinin-binding protein from tobacco leaves (Prog. Bot. $\underline{39}$, 113) has a dissociation constant (K_D) of 4×10^{-5}M for benzyladenine (YOSHIDA and TAKEGAMI, 1977), which is certainly too high to be of physiological significance. The protein bound to the 40S subunits of leaf ribosomes, and binding was enhanced in the presence of 10^{-5}M benzyladenine (TAKEGAMI and YOSHIDA, 1977). The binding protein is, however, not identical with that of FOX and ERION (1977) since the latter has a molecular weight of 93,000 daltons. Also working with wheat germ POLYA and DAVIS (1978) isolated a soluble 180,000-dalton, high-affinity ($K_D = 2 \times 10^{-7}$M) kinetin-binding protein by conventional techniques, which exhibited specificity for cytokinins. The relationship between this protein and that isolated by FOX and ERION (1977, see above) is unknown. Binding of benzyladenine and, to a lesser extent, zeatin to a particulate, nonribosomal fraction from *Funaria hygrometrica* protonema was studied by GARDNER et al. (1978) with ^{3}H-benzyladenine of high specific activity (10 Ci/mmol), which would permit the detection of binding sites with a $K_D < 10^{-8}$M. Abnormal Scatchard plots did not allow the calculation of K_D, however, and were thought to be the result of merely physical adsorption phenomena together with the biologically significant recognition of the hormone. Competitive and rather specific binding of cytokinin to talcum powder (SUSSMAN and KENDE, 1978), must be a warning in the interpretation of hormone-binding studies. Particulate, again probably nonribosomal, cytokinin-binding sites were also found in extracts from cultured tobacco cells (SUSSMAN and KENDE, 1978) exhibiting nonspecific, low-affinity, high-capacity and heat-resistant binding as well as specific, high-affinity, low-capacity and heat-sensitive binding. The high ratio of nonspecific to specific binding is clearly a serious obstacle in the characterization of the cytokinin specific binding (receptor?) sites. The very promising report, dating back to 1970 (MATTHYSSE and ABRAMS, 1970) that in the presence of a cytokinin mediator protein (a good candidate for a receptor!) isolated from pea chromatin, kinetin and zeatin stimulated RNA synthesis with *E. coli* RNA polymerase and pea chromatin or DNA as template has never been followed up. In fact, MENNES et al. (1978) working recently with nuclei isolated from tobacco pith callus tissue were unable to detect any effect of kinetin or auxin on RNA synthesis in the nuclei. Kinetin stimulation of the rate of RNA synthesis prior to the onset of DNA synthesis and cytokinesis was observed by SHININGER and POLLEY (1977) in cultured pea root segments and cortical explants as early as 9 h after addition of the hormone. Both auxin and cytokinin are required in this system for the determined sequence of DNA synthesis, cell division and cytodifferentiation (SIMPSON and TORREY, 1977). While a direct effect of cytokinins on transcription in vitro remains to be confirmed, KLÄMBT (1977) has presented new evidence for a direct effect of cytokinins on in vitro protein synthesis (Prog. Bot. 39, 113) and suggests a quantitative rather than a qualitative effect of the hormones "within the limit of special gene activations". Kinetin-mediated acceleration of protein synthesis (incorporation of ^{14}C-amino acids into nascent polypeptides isolated from polyribosomes) in vivo in the presumable absence of RNA synthesis reported by MAASS and KLÄMBT (1977) would support the concept that cytokinins act at the translational level, and substantial support for this hypothesis comes from careful and detailed studies on the hormone-mediated control of protein synthesis in cultured soybean cells (MUREN and FOSKET, 1977; FOSKET et al., 1977; TEPFER and FOSKET, 1978; FOSKET and TEPFER, 1978): within 15 min of treatment with cytokinin the polyribosome-monoribosome ratio in these cells increased nearly twofold, which is considered evidence for increased protein synthesis. Using the polyribosome-monoribosome ratio as a criterion for the protein-synthesizing capacity of the cells, the authors avoided the ambiguity of ^{14}C-amino acid incorporation experiments, in which effects of cytokinins on amino acid uptake and distribution in intracellular pools are difficult to assess. The cytokinin-induced polyribosome formation is observed without preceding or concurrent alteration in the rate of mRNA and rRNA synthesis and is not suppressed by inhibitors of RNA synthesis. Nonpolyribosomal poly(A)-containing (and therefore, presumably messenger-) RNA, pulse-labeled with ^{3}H-uridine in the absence of cytokinin, is recovered in the polyribosomal fraction after treatment of the cells with zeatin. This is very good evidence indeed, that under the influence of the cytokinin more messenger is made available ("unmasked"?) for protein synthesis. It was shown in addition that cytokinin produces qualitative changes in the spectrum of newly synthesized proteins, and cytokinin-induced cell division can therefore not simply be the result of a general acceleration of protein synthesis. In fact, the efficiency of the protein-synthesizing machinery would remain unaltered by the presence of cytokinin. Cytokinin would rather permit the translation of certain preformed mRNA's coding for proteins that are required for cell division. FOSKET and TEPFER (1978) conclude "that cytokinins are *permissive* factors whose presence is necessary for the completion of a genetically determined program the transcription of which was initiated by other factors". The permissive role of cytokinins is also aptly illustrated by the work of SOSSINKA and FEIERABEND (1978) on the senescence of rye coleoptiles: while kinetin stimulated leaf growth of derooted rye seedlings (DE BOER and FEIERABEND, 1978), it enhanced, rather than retarded, the senescence of the coleoptiles, indicating that the response of the individual target tissue depends on its "competence" and that cytokinin rather allows the completion of a program than initiating a program at the gene level. Apart from the effect of cytokinins on the overall leaf growth there is a much more pronounced effect on the development of chloroplastic enzymes, which is apparent only at higher cytokinin concentrations (FEIERABEND and DE BOER, 1978). According to the results of these authors cytokinin selectively promotes the plastidic ribosome content as well as the plastidic polyribosome-monoribosome ratio, while the effect on cytoplasmic ribosomes was hardly significant. A preferential effect of benzyladenine or kinetin on chloroplast rRNA, as compared to cytoplasmic RNA, in isolated pumpkin cotyledons (MIKULOVICH et al., 1978) and tobacco leaves (GRIERSON et al., 1977) was also observed, but the primary site of cytokinin action is unknown.

That the response of a tissue to an exogenously added cytokinin depends on its competence is also evident from the cytokinin-induced differentiation (bud formation) in the protonema of the moss *Funaria hygrometrica* (Fortschr. Bot. 35, 158). Of the two differentiation stages of the protonema - chloronema and caulonema - only caulonema cells are able to form buds in response to cytokinin. As BOPP and coworkers (ERICHSEN et al., 1977; BOPP et al., 1978; SOOD et al., 1978) have shown, the ability of the caulonema to respond to cytokinin is related to the presence of high molecular weight caulonema specific proteins (CSP), which establish the nature of the caulonema as a target tissue for cytokinin and are considered "reaction partners" (receptors?) for the hormone. This system appears to be very promising for the study of the mechanism of action of cytokinins.

Possible direct interactions of cytokinins with membranes were reported in the last review (Prog. Bot. 39, 112). Further evidence has been presented that cytokinins interfere with solute (amino acids, glucose, nucleosides) transport in the oomycete *Achlya* by rapid interaction with a membrane-bound phsophorylated proteoglycan (GOH and LEJOHN, 1978; STEVENSON and LEJOHN, 1978). As a consequence of this interaction Ca^{2+} and three polyphosphorylated dinucleosides are released from the proteoglycan. The concentrations of cytokinins that are required to produce inhibition appear to be rather high (inhibition constants in the range of 100 μM), and a relation to cytokinin effects in higher plants remains obscure. Effects of cytokinins on the movement of K^+ (SONKA, 1976) and the K^+/Ca^{2+} ratio (GÖRING and MARDANOV, 1976) may be related to an effect on membrane functions. A role of K^+ and Ca^{2+} in cytokinin-mediated phenomena is also indicated by the enhancement by K^+ and Ca^{2+} of the cytokinin-induced expansion of cucumber cotyledons (GREEN and MUIR, 1978) and the synergistic stimulation of auxin-induced ethylene production in mung bean hypocotyls by cytokinins and Ca^{2+} (LAU et al., 1977). The recent demonstration of a Ca^{2+}-dependent modulator protein of NAD kinase in peas (ANDERSON and CORMIER, 1978) suggests that Ca^{2+} transients may function in the regulation of metabolic processes in plants.

References

ABE, M., UCHIYAMA, M., TANAKA, Y., SAITO, H.: Tetrahedron Lett., 3807-3810 (1976). - ADAM, G., LISCHEWSKI, M., SYCH, F.J., ULRICH, A.: Tetrahedron 33, 95-100 (1977). - ANDERSON, L.W.J.: Science 201, 1135-1138 (1978). - ANDERSON, J.M., CORMIER, M.J.: Biochem. Biophys. Res. Commun. 84, 595-602 (1978). - AUGIER, H.: Bot. Mar. 20, 187-203 (1977).

BARENDSE, G.W.M., VAN WIJCK, C.J.A., DE BOO, Th.: Z. Pflanzenphysiol. 88, 413-421 (1978). - BAULIEU, E.E., GODEAU, F., SCHORDERET, M., SCHORDERET-SLATKINE, S.: Nature (London) 275, 593-598 (1978). - BEARDER, J.R., MACMILLAN, J.: J. Chem. Soc. Chem. Commun. (1976). - BEARDER, J.R., SPONSEL, V.M.: Biochem. Soc. Trans. 5, 569-582 (1977). - BEARDER, J.R., MACMILLAN, J., PHINNEY, B.O.: J. Chem. Soc. Chem. Commun., 834-835 (1976a). - BEARDER, J.R., FRYDMAN, V.M., GASKIN, P., HATTON, I.K., HARVEY, W.E., MACMILLAN, J., PHINNEY, B.O.: J. Chem. Soc. Perkin Trans. I, 178-183 (1976b). - BEELEY, L.J., MACMILLAN, J.: J. Chem. Soc. Perkin Trans. I, 1022-1028 (1976). - BEUTELMANN, L., BAUER, L.: Planta 133, 215-217 (1977). - BITTNER, S., PERRY, I., KNOBLER, Y.: Phytochemistry 16, 305-307 (1977). - BLOBEL, G., DOBBERSTEIN, B.: J. Cell Biol. 67, 835-851 (1975). - BOGORAD, L., WEIL, J.H. (eds.): Acides nucléiques et synthèse des protéines chez les végétaux.

714 p. Coll. Int. CNRS No. 261, Strasbourg 1976. Paris: CNRS 1977. -
BOPP, M., ERICHSEN, U., NESSEL, M., KNOOP, B.: Physiol. Plant. 42,
73-78 (1978). - BROWNING, G., SAUNDERS, P.F.: Nature (London) 265,
375-377 (1977). - BURROWS, W.J.: Planta 130, 313-316 (1976a); - Mechanisms of action of cytokinins, 263-273. In: Commentaries in Plant
Science, ed. H. SMITH. Oxford: Pergamon Press 1976b; - Planta 138,
53-57 (1978a); - Biochem. Biophys. Res. Commun. 84, 743-748 (1978b).

CARLSON, P.S.: Nature New Biol. 237, 39-41 (1972). - CHAILAKHYAN,
M., K.Kh., KHRYANIN, V.N.: Planta 138, 181-184 (1978a); - Planta 138,
185-187 (1978b); - Planta 142, 207-2120 (1978c); - Planta 144, 205-
207 (1979). - CHEN, C.M., ECKERT, R.L.: Plant Physiol. 59, 443-447
(1977). - CHEN, C.M., PETSCHOW, B.: Plant Physiology 62, 861-865
(1978a); - Plant Physiol. 62, 871-874 (1978b). - CHERAYIL, J.D.,
LIPSETT, M.N., J. Bacteriol. 131, 741-744 (1977). - CHO, K.Y., SAKURAI,
A., TAKAHASHI, N., TAMURA, S.: Agric. Biol. Chem. 42, 1389-1396 (1978).
- CHUNG, S.R., DURAND, R., DURAND, B.: C. R. Acad. Sci. (Paris) Ser.D.
284, 1417-1420 (1977). - CLAEYS, M., MESSENS, E., VANMONTAGU, M.,
SCHELL, J.: Fresenius Z. Anal. Chem. 290, 125-126 (1978). - COATES,
R.M., CONRADI, R.A., LEY, D.A., AKESON, A., HARADA, J., LEE, S.C.,
WEST, C.A.: J. Am. Chem. Soc. 98, 4659-4661 (1976). - CONSTANTINIDOU,
H.A., STEELE, J.A., KOZLOWSKI, T.T., UPPER, C.D.: Plant Physiol. 62,
968-974 (1978). - COOLBAUGH, R.C., HIRANO, S.S., WEST, C.A.: Plant
Physiol. 62, 571-576 (1978). - COREY, E.J., DANHEISER, R.L., CHANDRA-
SEKARAN, S., SIRET, P., KECK, G.E., GRAS, J.L.: J. Am. Chem. Soc. 100,
8031-8034 (1978a). - COREY, E.J., DANHEISER, R.L., CHANDRASEKARAN,
S., KECK, G.E., GOPALAN, B., LARSEN, S.D., SIRET, P., GRAS, J.L.:
J. Am. Chem. Soc. 100, 8034-8036 (1978b). - COWLEY, D.E., DUKE, C.C.,
LIEPA, A.J., MACLEOD, J.K., LETHAM, D.S.: Aust. J. Chem. 31, 1095-1111
(1978). - CROSS, B.E., ERASMUSON, A.: J. Chem. Soc. Chem. Commun.,
1013-1015 (1978). - CROZIER, A., REEVE, D.R.: The application of high
performance liquid chromatography to the analysis of plant hormones,
67-76, see: PILET, P.E. (ed.) 1977.

DASHEK, W.V., CHRISPEELS, M.J.: Planta 134, 251-256 (1977). - DATHE,
W., SCHNEIDER, G., SEMBDNER, G.: Phytochemistry 17, 963-966 (1978). -
DAUPHIN, B., TELLER, G., DURAND, B.: Planta 144, 113-119 (1979). -
DE BOER, J., FEIERABEND, J.: Planta 142, 67-73 (1978). - DEKHUIJZEN,
H.M.: Endogenous cytokinins in healthy and diseased plants, 526-559.
In: Encyclopedia of Plant Physiology, New Series, Vol. IV. eds. R.
HEITEFUSS, P.H. WILLIAMS. Berlin-Heidelberg-New York: Springer 1976.
- DOCKERILL, B., HANSON, J.R.: Phytochemistry 17, 701-704 (1978). -
DOCKERILL, B., EVANS, R., HANSON, J.R.: J. Chem. Soc. Chem. Commun.,
919-921 (1977). - DOUGLAS, T.J., PALEG, L.G.: Phytochemistry 17, 705-
712 (1978a); - Phytochemistry 17, 713-718 (1978b). - DUKE, C.C.,
MACLEOD, J.K.: Aust. J. Chem. 31, 2219-2223 (1978). - DUKE, C.C.,
MACLEOD, J.K., SUMMONS, R.E., LETHAM, D.S., PARKER, C.W.: Aust. J.
Chem. 31, 1291-1301 (1978).

EASTWOOD, D.: Plant Physiol. 60, 457-459 (1977). - ECKERT, H.G.:
Angew. Chem. 88, 565-574 (1976). - ECKERT, H., SCHILLING, G., PODLESAK,
W., FRANKE, P.: Biochem. Physiol. Pflanz. 172, 475-486 (1978). -
EINSET, J.W., SKOOG, F.K.: Biochem. Biophys. Res. Commun. 79, 1117-
1121 (1977). - ENTSCH, B., LETHAM, D.S.: Plant Sci. Lett. 14, 205-212
(1979). - ERICHSEN, J., KNOOP, B., BOPP, M.: Planta 135, 161-168 (1977).

FEIERABEND, J., DE BOER, J.: Planta 142, 75-82 (1978). - FILNER, P.,
VARNER, J.E.: Proc. Natl. Acad. Sci. USA 58, 1520-1526 (1967). -
FOSKET, D.E., TEPFER, D.A.: In Vitro 14, 63-75 (1978). - FOSKET, D.E.,
VOLK, M.J., GOLDSMITH, M.R.: Plant Physiol. 60, 554-562 (1977). -
FOX, J.E., ERION, J.: Cytokinin-binding proteins in higher plants,

139-146, see PILET, P.E., (ed.) 1977. - FROST, R.G., WEST, C.A.: Plant Physiol. 59, 22-29 (1977). - FUCHS, E., ATSMON, D., HALEVY, A.H.: Plant Cell Physiol. 18, 1193-1201 (1977). - FUCHS, S., FUCHS, Y.: Biochim. Biophys. Acta 192, 528-530 (1969). - FUCHS, S., HAIMOVICH, J., FUCHS, Y.: Eur. J. Biochem. 18, 384-390 (1971). - FUCHS, Y., GERMAN, E.: Plant Cell Physiol. 15, 629-633 (1974). - FUJITA, E., NODE, M., HORI, H.: J. Chem. Soc. Perkin Trans. I, 611-621 (1977). - FUKUI, H., KOSHIMIZU, K., USUDA, S., YAMAZAKI, Y.: Agric. Biol. Chem. 41, 175-180 (1977a). - FUKUI, H., NEMORI, R., KOSHIMIZU, K., YAMAZAKI, Y.: Agric. Biol. Chem. 41, 181-187 (1977b). - FUKUI, H., KOSHIMIZU, K., NEMORI, R.: Agric. Biol. Chem. 42, 1571-1576 (1978).

GARDNER, G., SUSSMAN, M.R., KENDE, H.: Planta 143, 67-73 (1978). - GASKIN, P., MACMILLAN, J.: GC and GC-MS techniques for gibberellins, 79-95, see HILLMAN, J.R. (ed.) 1978. - GAWER, M., LALOUE, M., TERRINE, C., GUERN, J.: Plant Sci. Lett. 8, 267-274 (1977). - GÖRING, H., MARDANOV, A.A.: Biol. Rundsch. 14, 177-189 (1976). - GOH, S.H., LEJOHN, H.B.: Can. J. Biochem. 56, 246-256 (1978). - GONZALEZ, E.: Plant Physiol. 62, 449-453 (1978). - GREEN, J.F., MUIR, R.M.: Physiol. Plant. 43, 213-218 (1978). - GRIERSON, D., CHAMBERS, S.E., PENNIKET, L.P.: Planta 134, 29-34 (1977).

HAHN, H., DE ZACKS, R., KENDE, H.: Naturwissenschaften 61, 170 (1974). - HAHN, H., HEITMANN, I., BLUMBACH, M.: Z. Pflanzenphysiol. 79, 143-153 (1976). - HASHIZUME, T., SUBIYAMA, T., IMURA, M., CORY, H.T., SCOTT, M.F., MCCLOSKEY, J.A.: Anal. Biochem. 92, 111-122 (1979). - HECKER, L.I., UZIEL, M., BARNETT, W.E.: Nucleic Acids Res. 3, 371-380 (1976). - HEDDEN, P., PHINNEY, B.O., MACMILLAN, J., SPONSEL, V.M.: Phytochemistry 16, 1913-1917 (1977). - HEDDEN, P., MACMILLAN, J., PHINNEY, B.O.: Annu. Rev. Plant Physiol. 29, 149-192 (1978). - HEFTMANN, E., SAUNDERS, G.A.: J. Liq. Chromatogr. 1, 333-341 (1978). - HEFTMANN, E., SAUNDERS, G.A., HADDON, W.F.: J. Chromatogr. 156, 71-77 (1978). - HELBACH, M., LEINEWEBER, M., KLÄMBT, D.: Physiol. Plant. 44, 313-314 (1978). - HENSON, I.E.: Z. Pflanzenphysiol. 86, 363-369 (1978a); - J. Exp. Bot. 29, 935-951 (1978b). - HENSON, I.E., WHEELER, C.T.: Z. Pflanzenphysiol. 84, 179-182 (1977a); - J. Exp. Bot. 28, 205-214 (1977b); - J. Exp. Bot. 28, 1076-1086 (1977c); - J. Exp. Bot. 28, 1087-1098 (1977d); - J. Exp. Bot. 28, 1099-1110 (1977e). - HIGGINS, T.J.V., ZWAR, J.A., JACOBSEN, J.V.: Nature (London) 260, 166-169 (1976). - HILLMAN, J.R. (ed.): Isolation of Plant Growth Substances. 157 pp. Cambridge: Cambridge University Press 1978. - HO, T.H.D., VARNER, J.E.: Arch. Biochem. Biophys. 187, 441-446 (1978). - HOLLAND, J.A., MCKERRELL, E.H., FUELL, K.J., BURROWS, W.J.: J. Chromatogr. 166, 545-553 (1978). - HOLTZ, J., KLÄMBT, D.: Hoppe Seylers Z. Physiol. Chem. 359, 89-101 (1978). - HORGAN, R.: Analytical procedures for cytokinins, 97-114, see: HILLMAN, J.R. (ed.) 1978a; - Philos. Trans. R. Soc. Lond. B 284, 439-447 (1978b).

JACOBSEN, J.V.: Annu. Rev. Plant Physiol. 28, 537-564 (1977). - JACOBSEN, J.V., KNOX, R.B.: Planta 112, 213-224 (1973). - JACOBSON, A., CORCORAN, M.R.: Plant Physiol. 59, 129-133 (1977). - JELSEMA, C.L., RUDDAT, M., MORRE, D.J., WILLIAMSON, F.A.: Plant Cell Physiol. 18, 1009-1019 (1977a). - JELSEMA, C.L., MORRE, D.J., RUDDAT, M., TURNER, C.: Bot. Gaz. 138, 138-149 (1977b). - JONES, R.L., CHEN, R.F.: J. Cell Sci. 20, 183-198 (1976). - JOUANNEAU, J.P., GANDAR, J.-C., PEAUD-LENOEL, C.: Plant Sci. Lett. 9, 77-87 (1977).

KAISS-CHAPMAN, R.W., MORRIS, R.O.: Biochem. Biophys. Res. Commun. 76, 453-459 (1977). - KAMINEK, M.: Biol. Rundsch. 13, 137-152 (1975). - KANNANGARA, T., DURLEY, R.C., SIMPSON, G.M.: Physiol. Plant. 44, 295-299 (1978). - KEITH, B., SRIVASTAVA, L.M.: Planta 139, 301-303 (1978).

- KENDE, H., GARDNER, G.: Annu. Rev. Plant Physiol. 27, 267-290 (1976).
- KHAN, S.A., HUMAYUN, M.Z., JACOB, T.M.: Anal. Biochem. 83, 632-635
(1977). - KLÄMBT, D.: Cytokinin and cell metabolism, 154-160, see
PILET, P.E. (ed.) 1977. - KNOTZ, J., COOLBAUGH, R.C., WEST, C.A.:
Plant Physiol. 60, 81-85 (1977). - KOYAMA, S., KAWAI, H., KUMAZAWA,
Z., OGAWA, Y., IWAMURA, H.: Agric. Biol. Chem. 42, 1997-2001 (1978).
- KRECHTING, H.C.J.M., VARGA, A., BRUINSMA, J.: Z. Pflanzenphysiol.
87, 91-93 (1978). - KRISHNAMOORTHY, H.N. (ed.): Gibberellins and Plant
Growth. 356 pp. New Delhi: Wiley Eastern Ltd. 1975. - KUDREV, T.,
IVANOVA, I., KARANOV, E.: Plant Growth Regulators. 769 pp. Proc. 2nd
Int. Symp. Sofia 1975. Sofia: Bulgarian Acad. Sci. 1977. - KÜLLERTZ,
G., ECKERT, H., SCHILLING, G.: Biochem. Physiol. Pflanz. 173, 186-187
(1978).

LALORAYA, M.M., SRIVASTAVA, H.N., GURUPRASAD, K.N.: Planta 128, 275-276
(1976). - LALOUE, M.: Planta 134, 273-275 (1977); - Philos. Trans. R.
Soc. Lond. B 284, 449-457 (1978). - LALOUE, M., TERRINE, C., GUERN, J.:
Plant Physiol. 59, 478-483 (1977). - LAU, O.-L., JOHN, W.W., YANG,
S.F.: Physiol. Plant. 39, 1-3 (1977). - LENDZIAN, K.J., ZIEGLER, H.,
SANKHLA, N.: Planta 141, 199-204 (1978). - LESHEM, Y., OPHIR, D.: Ann.
Bot. 41, 375-379 (1977). - LESHEM, Y.Y., INBAR, M.: J. Exp. Bot. 29,
671-675 (1978). - LESTER, B.R., MORRIS, R.O., CHERRY, J.H.: Plant
Physiol. 63, 87-92 (1979). - LETHAM, D.S., WETTENHALL, R.E.H.: Trans-
fer RNA and cytokinins, 129-193. In: The Ribonucleic Acids, 2nd Ed.,
eds. P.R. STEWART, D.S. LETHAM. Berlin-Heidelberg-New York: Springer
1977. - LETHAM, D.S., PARKER, C.W., DUKE, C.C., SUMMONS, R.E., MACLEOD,
J.K.: Ann. Bot. 41, 261-263 (1977). - LETHAM, D.S., GOODWIN, P.B.,
HIGGINS, T.J.V. (eds.): Phytohormones and Related Compounds: A Comprehensive Treatise. Vol. I: 641 p., Vol. II: 648 p. Amsterdam-Oxford-
New York: Elsevier/North-Holland, Biomedical Press 1978a. - LETHAM,
D.S., SUMMONS, R.E., ENTSCH, B., GOLLNOW, B.I., PARKER, C.W., MACLEOD,
J.K.: Phytochemistry 17, 2053-2057 (1978b). - LIBBENGA, K.R.: Hormone
receptors in plants, 325-333. In: Frontiers of Plant Tissue Culture,
ed. T.A. THORPE. Calgary: Int. Assoc. Plant Tissue Culture 1978. -
LIEBISCH, H.W.: Uptake translocation, and metabolism of labelled GA$_3$
glucosyl ester, 109-113. In: Biochemistry and Chemistry of Plant Growth
Regulators, Proc. Int. Symp. Cottbus, GDR, eds. K. SCHREIBER, H.R.
SCHÜTTE, G. SEMBDNER. Halle: Acad. Sci. GDR 1974. - LOCY, R., KENDE,
H.: Planta 143, 89-99 (1978). - LORENZI, R., SAUNDERS, P.F., HEALD,
J.K., HORGAN, R.: Plant Sci. Lett. 8, 179-182 (1977). - LUNNON, M.W.,
MACMILLAN, J., PHINNEY, B.D.: J. Chem. Soc. Perkin Trans. I, 2308-2316
(1977).

MAASS, H., KLÄMBT, D.: Planta 133, 117-120 (1977). - MACLEOD, J.K.,
SUMMONS, R.E., LETHAM, D.S.: J. Org. Chem. 41, 3959-3967 (1976). -
MACMILLAN, J.: Some aspects of gibberellin metabolism in higher
plants, 129-138, see PILET, P.E. (ed.) 1977. - MAKSYMOWYCH, R.,
ERICKSON, R.O.: Science 196, 1201-1203 (1977). - MAPELLI, S., RANIERI,
A.M.: Planta 142, 37-40 (1978). - MARRIOTT, K.M., NORTHCOTE, D.H.:
J. Exp. Bot. 28, 219-224 (1977). - MATSUBARA, S., SHIOJIRI, S., FUJII,
T., OGAWA, N., IMAMURA, K., YAMAGISHI, K., KOSHIMIZU, K.: Phyto-
chemistry 16, 933-937 (1977). - MATSUBARA, S., SUGIYAMA, T., HASHI-
ZUME, T.: Physiol. Plant. 42, 114-118 (1978). - MATTHYSSE, A.G.,
ABRAMS, M.: Biochim. Biophys. Acta 199, 511-518 (1970). - MAUK, C.S.,
LANGILLE, A.R.: Plant Physiol. 62, 438-442 (1978). - MCINNES, A.G.,
SMITH, D.G., DURLEY, R.C., PHARIS, R.P., ARSENAULT, G.P., MACMILLAN,
J., GASKIN, P., VINING, L.C.: Can. J. Biochem. 55, 728-735 (1977). -
MENNES, A.M., BOUMAN, H., VAN DER BURG, M.P.M., LIBBENGA, K.R.: Plant
Sci. Lett. 13, 329-339 (1978). - MIERNYK, J.A.: Physiol. Plant. 45,
63-66 (1979). - MIERNYK, J.A., BLAYDES, D.F.: Physiol. Plant. 39, 4-8
(1977). - MIKULOVICH, T.P., WOLLGIEHN, R., KHOKHLOVA, W.A., NEUMANN,

D., KULAEVA, O.N.: Biochem. Physiol. Pflanz. 172, 101-110 (1978). -
MILSTONE, D.S., VOLD, B.S., GLITZ, D.G., SHUTT, N.: Nucleic Acids
Res. 5, 3439-3455 (1978). - MONTAGUE, M.J., IKUMA, H.: Plant Physiol.
62, 391-396 (1978). - MORRIS, R.O.: Plant Physiol. 59, 1029-1033
(1977). - MORRIS, R.O., CHAPMAN, R.W.: Fed. Proc. 35, 587 (1976). -
MORRIS, R.O., ZAERR, J.B.: Anal. Lett. 11, 73-84 (1977). - MOUNLA,
M.A.Kh. Physiol. Plant. 44, 268-272 (1978). - MÜLLER, H., SCHUPHAN,
W.: Qual. Plant. 25, 297-309 (1976). - MÜLLER, P., KNÖFEL, H.D.,
LIEBISCH, H.W., MIERSCH, O., SEMBDNER, G.: Biochem. Physiol. Pflanz.
173, 396-409 (1978). - MURAI, N., TALLER, B.J., ARMSTRONG, D.J.,
SKOOG, F., MICKE, M.A., SCHNOES, H.K.: Plant Physiol. 60, 197-202
(1977). - MURAI, N., ARMSTRONG, D.J., TALLER, B.J., SKOOG, F.: Plant
Physiol. 61, 318-322 (1978). - MURAKOSHI, I., IKEGAMI, F., OOKAWA,
N., HAGINIWA, J., LETHAM, D.S.: Chem. Pharm. Bull. 520-522 (1977).
MURAKOSHI, I., IKEGAMI, F., OOKAWA, N., ARIKI, T., HAGINAWA, J.,
KUO, Y.-H., LAMBEIN, F.: Phytochemistry 17, 1571-1576 (1978). - MUREN,
R.C., FOSKET, D.E.: J. Exp. Bot. 28, 775-784 (1977). - MUROFUSHI, N.,
DURLEY, R.C., PHARIS, R.P.: Agric. Biol. Chem. 41, 1075-1079 (1977).

NADEAU, R., RAPPAPORT, L.: Plant Physiol. 54, 809-813 (1974). - NASH,
L.J., JONES, R.L., STODDART, J.L.: Planta 140, 143-150 (1978). -
NELLES, A.: Planta 137, 293-298 (1977). - NEUMANN, D., JANOSSY, A.G.S.:
Planta 134, 151-153 (1977a); - Planta 137, 25-28 (1977b). - NEWCOMB,
W., SYONO, K., TORREY, J.G.: Can. J. Bot. 55, 1891-1907 (1977). -
NOMURA, T., TANAKA, Y., ABE, H., UCHIYAMA, M.: Phytochemistry 16,
1819-1820 (1977).

OKAMOTO, T., SHUDO, K., TAKAHASHI, S., YATSUNAMI, T., ISOGAI, Y.:
YAMADA, K.: Chem. Pharm. Bull. 26, 3250-3252 (1978). - OKITA, T.W.,
DE CALEYA, R., RAPPAPORT, L.: Plant Physiol. 63, 195-200 (1976).

PACES, V., KAMINEK, M.: J. Chromatogr. 153, 291-294 (1978). - PARKER,
C.W., LETHAM, D.S., GOLLNOW, B.I., SUMMONS, R.E., DUKE, C.C., MACLEOD,
J.K.: Planta 142, 239-251 (1978). - PEGG, G.F.: Endogenous gibberellins
in healthy and diseased plants, 592-606. In: Encyclopedia of Plant
Physiology, New Series, Vol. IV, eds. R. HEITEFUSS, P.H. WILLIAMS.
Berlin-Heidelberg-New York: Springer 1976. - PENGELY, W., MEINS, F.:
Planta 136, 173-180 (1977). - PETERSON, J.B., MILLER, C.O.: Plant
Physiol. 59, 1026-1028 (1977). - PIETERSE, A.H., ARIS, J.J.A.M.,
BUTTER, M.E.: Nature (London) 260, 423-424 (1976). - PILET, P.E. (ed.):
Plant Growth Regulation. 305 pp. Proc. 9th Int. Conf. Plant Growth
Substances, Lausanne 1976. Berlin-Heidelberg-New York: Springer 1977.
- POLYA, G.M., DAVIS, A.W.: Planta 139, 139-147 (1978). - PROEBSTING,
W.M., DAVIES, P.J., MARX, G.A.: Planta 141, 231-238 (1978). - PUPPO,
A., RIGAUD, J.: Physiol. Plant. 42, 202-206 (1973).

RAILTON, I.D.: S. Afr. J. Sci. 73, 22 (1977a); - Z. Pflanzenphysiol.
81, 323-329 (1977b); - S. Afr. J. Sci. 74, 191-192 (1978). - RAILTON,
L.D., RECHAV, M.: Plant Sci. Lett. 14, 75-78 (1979). - RAPPAPORT, L.,
ADAMS, D.: Philos. Trans. R. Soc. Lond. B. 284, 521-539 (1978). -
RAPOPORT, E.N., HELLER, K.E., DAYANANDAN, P., HEBARD, F.V., KAUFMAN,
P.B.: Plant Physiol. 62, 807-811 (1978). - REEVE, D.R., CROZIER, A.:
Gibberellin bioassays, 35-64, see KRISHNAMOORTHY, H.N. (ed.) 1975;
- The analysis of gibberellins by high performance liquid chromato-
graphy, 41-77, see: HILLMAN, J.R. (ed.) 1978. - RODAWAY, S.J., KENDE,
H.: Plant Physiol. 61, 1-6 (1978). - ROPERS, H.J., GRAEBE, J.E.,
GASKIN, P., MACMILLAN, J.: Biochem. Biophys. Res. Commun. 80, 690-697
(1978).

SCHLEE, D.: Pharmazie 30, 345-349 (1975). - SCHNEIDER, G.: Z. Chem. 18,
217 (1978). - SCHNEIDER, G., SEMBDNER, G., SCHREIBER, K.: Tetrahedron

1391-1397 (1977a). - SCHNEIDER, G., MIERSCH, O., LIEBISCH, H.W.:
Tetrahedron Lett., 405-406 (1977b). - SCHRAUDOLF, H., FISCHER, A.:
Plant Sci. Lett. 14, 199-205 (1979). - SCHREIBER, K., SCHNEIDER, G.,
SEMBDNER, G., FOCKE, I.: Phytochemistry 5, 1221-1225 (1966). -
SCHROEDER, R.L., BURGER, W.C.: Plant Physiol. 62, 458-462 (1978). -
SCHÜTTE, H.R., GROSS, D. (eds.): Regulation of Developmental Processes
in Plants. 408 pp. Proc. Conf. Halle 1977. Jena: Fischer 1977. -
SHININGER, T.L., POLLEY, L.D.: Plant Physiol. 59, 831-835 (1977). -
SHIVAKUMAR, A.G., KHAN, S.A., JACOB, T.M., PADAYATTY, J.D.: Ind. J.
Exp. Biol. 14, 529 (1976). - SILK, W.K., JONES, R.L., STODDART, J.L.:
Plant Physiol. 59, 211-216 (1977). - SIMPSON, S.F., TORREY, J.G.:
Plant Physiol. 59, 4-9 (1977). - SONKA, J.: Experientia 32, 1010-1011
(1976). - SOOD, S., BRENNER, K., BOPP, M.: Planta 138, 299-301 (1978).
- SOSSINKA, J., FEIERABEND, J.: Biochem. Physiol. Pflanz. 178, 505-513
(1978). - SPONSEL, V.M., MACMILLAN, J.: Planta 135, 129-136 (1977);
- Planta 144, 69-78 (1978). - SPONSEL, V.M., HOAD, G.V., BEELEY, L.J.:
Planta 135, 143-147 (1977). - STEVENSON, R.M., LEJOHN, H.B.: Can. J.
Biochem. 56, 207-216 (1978). - STOBART, A.K., KINSMAN, L.T.: Phyto-
chemistry 16, 1137-1142 (1977). - STODDART, J.L., JONES, R.L.: Planta
136, 261-269 (1977). - STODDART, J.L., TAPSTER, S.M., JONES, T.W.A.:
Planta 141, 283-288 (1978). - STOLP, C.F., NADEAU, R., RAPPAPORT, L.:
Plant Cell Physiol. 18, 721-728 (1977). - STRUXNESS, L.A., ARMSTRONG,
D.J., GILLAM, I., TENER, G.M., BURROWS, W.J., SKOOG, F.: Plant Physiol.
63, 35-41 (1979). - STUART, D.A., JONES, R.L.: Plant Physiol. 59, 61-
68 (1977); - Planta 142, 135-145 (1978). - STUART, D.A., DARNAM, D.J.,
JONES, R.L.: Planta 135, 249-255 (1977). - SUGIYAMA, I., IMURA, M.,
HASHIZUME, T.: Agric. Biol. Chem. 42, 467-470 (1978). - SUMMONS, R.E.,
MACLEOD, J.K., PARKER, C.W., LETHAM, D.S.: FEBS Lett. 82, 211-214
(1977). - SUSSMAN, M.R., FIRN, R.D.: Phytochemistry 15, 153-155 (1976).
- SUSSMAN, M.R., KENDE, H.: Planta 137, 91-96 (1977); - Planta 140,
251-259 (1978). - SUTHERLAND, E.W.: Science 177, 401-408 (1972). -
SWAMINATHAN, S., BOCK, R.M.: Biochemistry 16, 1355-1360 (1977). -
SWAMINATHAN, S., BOCK, R.M., SKOOG, F.: Plant Physiol. 59, 558-563
(1977).

TAIZ, L., STARKS, J.E.: Plant Physiol. 60, 182-189 (1977). - TAKAHASHI,
S., SHUDO, K., OKAMOTO, T., YAMADA, K., ISOGAI, Y.: Phytochemistry 17,
1201-1207 (1978). - TAKEGAMI, T., YOSHIDA, K.: Plant Cell Physiol. 18,
337-346 (1977). - TANAKA, Y., ABE, H., UCHIYAMA, M., TAYA, Y.,
NISHIMURA, S.: Phytochemistry 17, 543-544 (1978). - TAYA, Y., TANAKA,
Y., NISHIMURA, S.: Nature (London) 271, 545-547 (1978a); - FEBS. Lett.
89, 326-328 (1978b). - TAYLOR, C.M., RAILTON, I.D.: Plant Sci. Lett.
9, 317-322 (1977). - TEPFER, D.A., FOSKET, D.E.: Dev. Biol. 62, 486-
497 (1978). - THEILER, J.B., LEONARD, N.J., SCHMITZ, R.Y., SKOOG, F.:
Plant Physiol. 58, 803-805 (1976). - THIMANN, K.V.: Hormone Action in
the Whole Life of Plants. 448 pp. Amherst: University of Massachusetts
Press 1977. - TREWAVAS, A.J.: Plant growth substances, 249-298. In:
Molecular Aspects of Gene Expression in Plants, ed. J.A. BRYANT.
New York: Academic Press 1976. - TURNER, J.V., MANDER, L.N., COOMBE,
B.G.: Aust. J. Plant Physiol. 5, 347-355 (1978).

VAN STADEN, J., DIMALLA, G.G.: Ann. Bot. 41, 741-746 (1977). - VAN
STADEN, J., PAPAPHILIPPOU, A.P.: Plant Physiol. 60, 649-650 (1977). -
VAN STADEN, J., SMITH, A.R.: Ann. Bot. 42, 751-753 (1978). - VENIS,
M.A.: Receptors for plant hormones, 53-88. In: Adv. Bot. Res., Vol. 5,
ed. H.W. WOOLHOUSE. New York: Academic Press 1977. - VOLFOVA, A.,
CHVOJKA, L., HANKOVSKA, J.: Biol. Plant. Acad. Sci. Bohemoslov. 19,
421-425 (1977). - VONK, C.R.: Physiol. Plant. 44, 161-166 (1978). -
VREMAN, H.J., SCHMITZ, R.Y., SKOOG, F., PLAYTIS, A.J., FRIHART, C.R.,
LEONARD, N.J.: Phytochemistry 13, 31-37 (1974). - VREMAN, H.J.,
THOMAS, R., CORSE, J., SWAMINATHAN, S., MURAI, N.: Plant Physiol. 61,
296-306 (1978).

WANG, T.L., HORGAN, R.: Planta 140, 151-153 (1978). - WANG, T.L.,
THOMPSON, A.G., HORGAN, R.: Planta 135, 285-288 (1977). - WAREING,
P.F., HORGAN, R., HENSON, I.E., DAVIS, W.: Cytokinin relations in the
whole plant, 147-153, see PILET, P.E. (ed.) 1977. - WEILER, E.W.:
Planta 144, 255-263 (1979). - WHEELER, C.T., HENSON, I.E.: New Phytol.
80, 557-565 (1978). - WOOD, A., PALEG, L.G.: Aust. J. Plant Physiol.
1, 31-40 (1974). - WRIGLEY, A., LORD, J.M.: J. Exp. Bot. 28, 345-353
(1977).

YAMANE, H., MUROFUSHI, N., OSADA, H., TAKAHASHI, N.: Phytochemistry 16,
831-835 (1977). - YOKOTA, T., REEVE, D.R., CROZIER, A.: Agric. Biol.
Chem. 40, 2091-2094 (1976). - YOKOTA, T., KOBAYASHI, S., YAMANE, H.,
TAKAHASHI, N.: Agric. Biol. Chem. 42, 1811-1812 (1978). - YOSHIDA, K.,
TAKEGAMI, T.: J. Biochem. 81, 791-799 (1977). - YOUNG, H.: Anal. Bio-
chem. 79, 226-233 (1977).

Professor Dr. NIKOLAUS AMRHEIN
Lehrstuhl für Pflanzenphysiologie
Ruhr-Universität
D 4630 Bochum

VII. Developmental Physiology

By Martin Bopp

When we survey the recent literature it becomes apparent that seed formation, germination, and development of the young growing seedling are major topics in the developmental physiology of higher plants. Besides the regulation of these processes, the biochemical events combined with these phenomena are in the foreground (BOPP, 1971).

Therefore this review is dedicated mainly to problems related to the formation and storage processes of seeds and seed germination.

1. Seed Formation

a) Carbohydrate Accumulation

It is well known that seed development in fruits is not readily changed or regulated by external factors (SINGH et al., 1978). Only the number of seeds in a fruit or grains in an ear can destine the grain volume, the number of cells in a seed (BROCKLEHURST, 1977) and the accumulation of starch, amino acids, and proteins (RADLEY, 1976, 1978). It is perhaps for this invariability of seed development that little is known about the causal analysis of this process. Our knowledge of physiology, biochemistry and molecular biology of seed development is inadequate in comparison with the amount of information available on special morphological and anatomical aspects of root, shoot, and cotyledon formation (cf. JONES, 1977; CLOWES, 1978a,b; TORREY and ZOBEL, 1977; BORNMAN et al., 1979) during seed development (DURE, 1975). Nevertheless some very important insights have been gained in the biochemical and physiological area in the last few years.

The physiology of seed development includes growth of endosperm (for growth curves see MONSELISE et al., 1978), development and organization of the embryo, accumulation of storage material in appropriate organs, hormones, formation of a transcription/translation apparatus activated very rapidly to supply the young embryo with protein synthesis, and finally, the process of desiccation (cf. MOUNLA, 1978).

The accumulation of substances in seeds of legume fruits has been investigated in some detail. Thereby the distribution of substances between pods and seeds plays the first role. The photosynthesis of the green pods of beans themselves contributes only marginally to the seed development. Applied ^{14}C sucrose is distributed either to the pod or to the seed, depending on the developmental stage. Only 30 days after anthesis the seeds contain a remarkable part of the total sucrose content (OLIKER et al., 1978b), indicating that the two parts do not act as alternative competing sinks, and also that only very few of the carbohydrates are transported from the pod to the seed.

On the other hand the pod and principally the inner epidermis of the
pod wall can be very active in the refixation of internal CO_2 produced
by the respiration of the developing seeds. Because malate is formed
during the CO_2 fixation by the seeds, this probably reflects the synthesis of oxaloacetate by PEP-carboxylase as the first CO_2 fixation
step, but also RuBP carboxylase is present in the seeds of *Lupinus*
(ATKINS and FLINN, 1978).

b) Nitrogen and Protein Storage

As for the accumulation of carbohydrates, the distribution of nitrogen
between seeds and pods is much more a temporal succession than a con-
currence between sinks; the accumulation of both groups of substances
(carbohydrates and proteins) in seeds begins only after this process
ceases in pods (OLIKER et al., 1978a). It is shown by STOREY and BEEVERS
(1978) that the leaf under fruits and pods furnishes the growing seeds
with the appropriate precursors. The activity of proteolytic enzymes
in the aging leaves makes it possible for these leaves to release amino
acids to the pods (STOREY and BEEVERS, 1978). Therefore the curves for
fresh weight of leaf, pod, and seed have their maxima one after the
other at different times during development. In the same order the activity of the enzymes glutamine synthetase and glutamate synthetase
changes in the three organs in corresponding curves (STOREY and BEEVERS,
1978). Evidence exists that reduced nitrogen in the plant is transported
to the seeds primarily as amide (PATE et al., 1977). Nitrate, found in
the leaves of legumes, is not transported to the pods, because almost
all nitrate is reduced in the subtending leaf (PATE et al., 1977).

The problem of protein storage in seeds has fascinated many research
groups in three different areas of interest, first human and animal
nutrition, second the specialization of cells to form only one or few
proteins at a certain time during development, and third the special
character of these proteins. Synthesis of such storage proteins is restricted to seeds and the formation of the proteins indicates a differential gene expression during seed formation, specifically in legumes,
during the development of cotyledons (MILLERD, 1975). Only the second
and third aspect of accumulation of storage proteins relates to development. The synthesis of pregermination proteins can be modified by
hormones. Gibberellic acid and cytokinin are stimulatory, whereas ABA
is inhibitory to protein synthesis (FOUNTAIN and BEWLEY, 1976).

> The main reserve proteins in the bean cotyledons are the globulins, vicilin,
> legumin, and phytohemagglutinin (BARKER et al., 1976; DERBYSHIRE et al., 1976).
> They are located in "protein bodies", which are typical organelles for protein
> storage found in haploid, diploid, and triploid tissue and bounded by a single
> membrane (PERNOLLET, 1978). Vicilin is, as in other legumes (MILLERD, 1975), a
> protein of $6,9\,S$ with 3 subunits (52, 49 and 46 kdalton), whereas the phyto-
> hemagglutinin has a size of $6,4\,S$ and 2 subunits (34 and 36 kdalton) (PUSZTAI
> et al., 1977; BOLLINI and CHRISPEELS, 1978). A third globulin has 11S. The
> ratio between the 7S and the 11S proteins can vary between 1:4 in *Vicia faba*
> and 9:1 in *Phaseolus vulgaris* (DERBYSHIRE and BOULTER, 1976). Furthermore in
> a number of legume species the seed storage proteins contain small amounts of
> bound oligosaccharide moities, suggesting the glycoproteid nature of these pro-
> teins (EATON-MORDES and MOORE, 1978).

All the proteins are catabolized during seedling growth after germination (ASHTON, 1976; BASHA and BEEVERS, 1976). In pumpkin seeds the degradation results in a decrease of the globulin to about 88% during
the first 4 days of germination (REILLY et al., 1978). The degradation
of reserve proteins of *Zea mays* was studied with an immunoassay proce-

dure, which shows a characteristic two-step pattern of mobilization
in the germinating caryopses (KHARKIN et al., 1978).

It is of interest that the genetic information regulating protein storage appears to be repressed in all tissues except for certain stages
during the development of cotyledons. But at different times in the
course of development, different species of globulins are formed.
This means that only some information about storage proteins is translated at the same stage of cotyledon development (SUN et al., 1978;
HALL et al., 1977). Therefore the proteins vary in relative concentration during development (CARASCO et al., 1978). The differential translation during seed formation may be confirmed by experiments on *Glycine
max*, in which it was shown that polyribosomes of developing seeds, translated in the wheat germ system, produce some, but not all, of the polypeptides formed as storage proteins in the cotyledons grown in in vitro
culture (BEACHY et al., 1978; THOMPSON et al., 1977).

In castor bean, where most of the storage material is deposited in
endosperm and not in the cotyledons as in legumes, the proteins, located in protein bodies too (NEUMANN and WEBER, 1978) consist of 40%
albumins (water soluble proteins), which are partly enzymes (ASHTON,
1976) and partly storage proteins (YOULE and HUANG, 1978); in addition
to globulins (insoluble in water), glutelins and prolamins are present,
which build the remaining 60%.

Seed globulin formation in cereals is also of particular interest.
(Extraction methods see SHEWRY et al., 1978.) The cereals contain,
with the exception of rice and oat, the alcohol-soluble prolamin as
the main storage protein (HARRIS and JULIANO, 1977). All the different
proteins (summarized by PERNOLLET, 1978) are deposited in protein bodies as in legumes and castor beans. The matrix protein of the protein
bodies is mainly glutelin. In the developing grain of rice the protein
bodies appear from 7 days after flowering onward. This appearance coincides with a drop in the content of the amino acid lysin and a rapid
synthesis of glutelin (VILLAREAL and JULIANO, 1978), as the main part
of the protein bodies.

The prolamin synthesis in rice also runs parallel to the start and formation of protein bodies in endosperm (MANDAC and JULIANO, 1978). In
Zea mais zein is the predominant protein composing the protein bodies.
It consists of a family of polypeptides differing in molecular weight
and amino acid composition (RIGHETTI et al., 1977; VIOTTI et al.,
1978). Polysomes are attached to the membrane of the protein bodies.
The in vitro translation products of these polysomes correspond to the
different polypeptides inside the protein bodies (VIOTTI et al., 1978),
which means that the storage proteins are synthesized at the surface
or in contact with the surface of the appropriate organelles.

Morphologically one finds at least three types of cells in respect to the protein bodies in *Zea*: Cells with protein bodies and numerous lipid bodies, cells
with peripheral lipid bodies and having or lacking protein bodies, and cells
always without protein bodies (BECHTEL and POMERANZ, 1978). All protein bodies
derive from ER and disappear during germination (YOULE and HUANG, 1976). Also
in such extreme plants as *Welwitschia* the stored reserves of protein exist as
protein bodies, described as a protein-carbohydrate complex (BUTLER et al.,
1979).

c) Lipid Accumulation

Lipid droplets or oleosomes (in *Sinapis alba*) are not bounded by the ER but by a separate lamellar structure or membrane containing nine major polypeptides separated in polyacrylamid gel electrophoresis as bands. They are formed during the build-up of these droplets near the plastids in very young cotyledons 12-14 days after pollination (BERGFELD et al., 1978). A hypothesis describing the degradation of these oleosomes during germination is presented by WANNER and THEIMER (1978). It is not possible here to go into more detail.

The relative amount of the different lipids in storage tissue changes very rapidly during seed development. It was demonstrated for the fat-accumulating cotyledons of soybean that the total fat content increases between 30 and 60 days after anthesis from 14% to 23% of the dry weight, and decreases only about 2% in the following 10 days of desiccation. The strong increase after 30 days indicates that the synthetic activity for lipids is highest at 30 days after flowering and decreases afterwards continuously up to 65 days (WILSON and RINNE, 1978b). In the last 10 days of development, when the total lipid content decreases, the monolinoleic triglycerides nearly double in content from 13% to 23%, whereas the trilinoleic triglycerides disappear completely. Similar changes were found for other triglycerides during the ripening process (WILSON and RINNE, 1978a; WILSON et al., 1976). The importance of lipids in the metabolism of developing seeds has been stressed by BEEVERS and MENZL, 1977).

d) Developmental Change During Embryo Formation

Not only lipids and storage proteins are accumulated during embryogenesis, also specific enzymes and their products are formed in different amounts in particular parts and stages of fruit and seed development (KATO and KUBOTA, 1978; MARBACH and MAYER, 1978). During barley grain development, e.g., many enzymes involved in amino acid synthesis and metabolisms are present in all stages of development and even in all tissues of the seeds. This is not surprising and only worth mention because certain exceptions exist: glutamat-dehydrogenase is not homogeneously distributed in the tissue but confined to the endosperm, other enzymes such as glutaminesynthetase are mainly located in the pericarp. One significance of such differential enzyme content is the character of the cells as storage tissue, growing embryo, seed coat, etc. (DUFFUS and ROSIE, 1978). In cotton most of the glyoxysomal enzymes showed a characteristic increase from 22 to 50 days after anthesis, but the synthesis of malate occurs just prior to seed desiccation after 45-50 days of development (CHOINSKI and TRELEASE, 1978). During the development of the seeds of *Sinapis alba* one finds an increase and afterwards a decrease of a small activity of phenylalanine-deaminase (PAL) (WELLMANN, personal communication). The main activity of the enzyme is found at the same time as the formation of sinapin (a derivative of sinapic acid as a part of the sinapin-p-hydroxybenzylglucosinolate) is highest in the seeds and sinapin is formed via the deamination of phenylalanine. The main period of sinapin accumulation is between 15 and 25 days after anthesis, which is about at the beginning of seed desiccation (LÜDICKE, 1976).

Very young, immature embryos can germinate when isolated from the seeds, but in such "germinating" embryos of *Sinapis* not all enzymatic activities can be induced during the developmental process by external factors as is the case in fully ripe seeds. For example, the activity of PAL cannot be stimulated by light in the germinating embryos unless they are isolated 20 days or later

after anthesis. One can assume that in the first stage either the external
factors do not find the reaction ability of the tissue or the genes necessary
for enzyme formation are not to be activated prior to a certain stage of cell
development. The transition from the first to the second stage when the enzyme
is inducible is accompanied by a rapid increase in fresh weight within 24 h
(MURACH, 1979). The behavior of younger and older embryos is also different in
other aspects, for example, when they are cultivated on a Murashige-Skoog medium. Fifteen-day-old embryos do not "germinate" on this medium but they grow
as a whole, forming callus. Older embryos, however, grow out in structures of
teratomas, which means that in this stage of ripeness of embryonic development
the "morphogenetic activity" is much more stable than in very young embryos
(PINNOW, 1978).

Sometimes isolated cotyledons of immature embryos can produce embryoids under cultural conditions (KRUL and WORLEY, 1977; HU et al.,
1978) whereas in other plant material 100% of cotyledon explants produce calli (GOSCH-WACKERLE et al., 1979). The formation of embryoids
may be dependent on the presence or absence of hormones such as 2,4 D
or other auxins (KAMADA and HARADA, 1979). In celery cultures 2,4 D
completely suppressed the formation of embryoids, whereas without that
substance many embryos are found in the culture (WILLIAMS and COLLIN,
1976; AL-ABTA and COLLIN, 1978). The formation of new embryoids on
isolated embryos as a source of genetically defined plant material
may have some importance for the cultivation of gymnosperms. Under
the influence of isopentenyladenin up to 40% of excised embryos of
Picea abies produce - if the other conditions are optimal - between 10
and 20 buds, which can grow into plants (v. ARNOLD and ERIKSSON,
1978). Also in other conifers, plantlet formation from embryonic tissue as a possible source of genetic homogeneous material was observed
(REILLY and WASHER, 1972). It would be worth studying the conditions
of embryo formation in embryonic material more in detail and from a
causal point of view to understand the embryonic character of a tissue.

It was mentioned earlier that the transition of a callus-forming embryo into
a teratoma-forming *Sinapsis* plant is accompanied by a rapid increase in fresh
weight of the embryo. During this phase of cell expansion in pea cotyledons the
nuclear DNA and also the RNA content increases greatly. Also the activity of
DNA-dependent RNA-polymerase and the template availability on the DNA (tested
with homologous RNA-polymerase) increase parallel to the developmental rate of
the cotyledons (CULLIS, 1978).

e) RNP-Particles

A very important aspect of seed formation is related to the deposition
of all elements essential for seed germination. In addition to the storage proteins, which supply the germinating seed with energy and metabolic material, are these the mRNA's. These are necessary to start new
protein synthesis immediately after watering. Therefore a feature of
early embryonic development is a rapid polysome formation without new
RNA-synthesis. This is only possible with stored mRNA's in the germinating seeds (BHAT and PADAYATTY, 1975). They exist as preformed long-
living mRNA-protein particles (= mRNP particle) (PAYNE, 1976), which
can be separated from cell material by fractionating methods as distinct particles, called informosomes (AJKTHOSHIN et al., 1976) and
which are very probably formed in the nuclei (TAKAHASHI et al., 1976).
Two classes of mRNA activity containing elements can be distinguished
in a cell-free extract and would include about 75% real "soluble" RNP-
particles, and 25% associated with membranous fractions (PEUMANS et
al., 1978). The particles are formed for example in rye during the
last five weeks of seed development and contain a population of het-

erogenous RNA-parts consisting of ribosomal as well as mRNA. The formation of RNP-particles in the seeds stops, as for other macromolecules, only upon desiccation of the seeds (PEUMANS et al., 1979) and
consequently one can find and analyze the RNP's in dry seeds (PEUMANS
et al., 1978). In an in vitro system isolated RNP-particles can be
translated even if the seeds are completely dormant (PEUMANS and
CARLIER, 1977).

It is assumed that such RNP's are found and function in most, if not
all, seeds (PAYNE, 1976; TAKAHASHI et al., 1977; AJKTHOSHIN et al.,
1976; TANEJA and SACHAR, 1976),even if one assumes that in different
types of plants the preformed messenger may not have an identical
character in size, protein content, etc. (PAYNE, 1976). These preformed RNP-particles allow a rapid increase in the rates of protein
synthesis independent of de novo synthesis of RNA in wheat embryos
(SPIEGEL and MARCUS, 1975; BROOKER et al., 1978), but it seems that
they are not involved in the temporal regulation of early seed development during germination. Rather their function is to start with a
quick resumption of growth upon exposure of the seeds to water (BROOKER
et al., 1978); on the other hand the messenger for certain special enzymes is conserved in this way (TANEJA and SACHAR, 1975). When wheat
embryos are studied 6 h after the start of germination, more than half
of all messenger RNA's present in the cells are actively involved in
translation, but only a part of them was preformed, the other part is
newly synthesized after soaking (CAERS et al., 1979). In rape seeds
the synthesis of newly formed RNA starts soon after the seeds are
placed in contact with water and well before physical imbibition is
complete. It is first detectable in peripheral cells of the embryo
closed to the testa (PAYNE et al., 1978). In the later stages of germination - which means later than 6 h - most of the mRNA's are newly
transcripted (CAERS et al., 1979). The stored polyadenylic acid and
mRNA form polysomes very rapidly and therefore disappear in the first
stages of germination, as was shown in the embryo axis of radish
(DELSENY et al., 1977; BHAT and PADAYATTY, 1975).

It should be mentioned that not only in seeds but also in plant cell cultures
the mRNP-particles can be identified on sucrose density gradients during the
logarithmically growing phase (PFISTERER, 1978). The meaning of such long-
lived and protected mRNA may be clarified by the finding of preserved mRNA in
the storage tissue of Jerusalem artichokes. When tissue slices are inoculated
in aerated water at room temperature, the amount of poly a-RNA is not changed
during the first few hours, but protein synthesis increases. From this it is
concluded that stored mRNA may exist in the quiescent cells of the organ (BYRNE
and SETTERFIELD, 1978). Therefore, it may be a general phenomenon that mRNA
is preformed and deposited as RNP-particles in tissue undergoing rapid development under certain conditions. This phenomenon has also been reported for egg
development in insects (WINTER, 1975; WIEMANN-WEISS, 1979).

f) Phytohormones

Another aspect of seed development is related to phytohormones and
their ability either to regulate embryo development itself or to supply the germinating seed with regulators necessary for coordinated
growth and activation in the different parts of a seed (cf. HALL and
BANDURSKI, 1978).

IAA was found in oat seeds in two protein compounds, one of the proteins belongs to the albumins, the other to the glutelins. It seems
that the two IAA fractions were distributed in different parts of the
seed. The albumin auxin is located in the embryo, the glutelin com-

pound in the endosperm (ZIMMERMANN, 1978). For gibberellins one can
find similar conditions. Different gibberellins are not distributed
in a homogeneous manner in the embryo and the endosperm. For example
in *Secale cereale* GA_8 is mainly found as a free gibberellin in the endo-
sperms during the milk ripeness, whereas GA_6 is located in the embryo
in a conjugated state (DATHE and SEMBDNER, 1978; DATHE et al., 1978).
During the ripening of the rye seed mainly in the course of dehydration, the content of GA decreases (SLOMINSKI et al., 1979), and no
gibberellin can be identified in a mature caryopse. These caryopses,
however, contain remarkable amounts of abscisic acid and 4'-dihydro-
phaseic acid (DATHE et al., 1978).

In barley, the time course of the content of free gibberellins followed more or less the fresh weight of the endosperm. The highest amount
of GA_3 was found when fresh weight was highest. Smaller peak of gibberellin-like activity appears in the embryo shortly before full ripeness of the caryopses (MOUNLA, 1978). This experiment demonstrates a
specific behavior of the different organs in respect to hormone formation or accumulation. Therefore it is of interest to look at the function of the suspensor in this regard: The suspensor of *Phaseolus coc-
cineus*, distinguished by endoreduplicated nuclei of different levels
in the proximal and distal parts (CREMONINI and CIONINI, 1977), plays
an important role in supplying the early embryo with hormones. In beans
it is possible that gibberellins are delivered from the suspensor, because embryos cultivated in vitro without suspensor, but on a medium
containing gibberellin between 10^{-6} and 10^{-8} mol, show the same development as intact suspensor containing embryos without GA supplement in
the medium (CIONINI et al., 1976). The auxin content of the suspensor
of *Tropaeolum* is higher than that in the embryo (PRZYBYLLOK and NAGL,
1977) and the source of cytokinins may also be located in the suspensor. At the heart-shaped stage of *Phaseolus* embryos the suspensor shows
cytokinin activity at the level of zeatin, 2iPA, and zeatinribosid,
whereas in the embryo more polar cytokinins, such as zeatinglucosid,
are present. In a late stage of development the embryo seems to become autonomous for cytokinin supply as well as for gibberellins
(LORENZI et al., 1978). THOMAS et al. (1978a,b) showed that the presence of an embryo is essential for the synthesis (or accumulation) of
GA and cytokinin in wheat grains, because normal embryo-containing
seeds possess a high hormone content in contrast to natural embryo-
less seeds with only very low amounts of the hormones.

> Content of abscisic acid is highest during milk- and wax-ripeness stages of
> grain development and starch accumulation (RADLEY, 1976). The amount of ABA
> decreases during seed dehydration (SLOMINSKI et al., 1979). A decrease of ABA
> was also found under artificial dehydration (KING, 1976).

2. Seed Germination and Seedling Growth

a) Phytohormones

One major aspect of this review should be directed to a discussion of
the internal regulatory and metabolic processes during the early stages
of germination and seedling growth. It is without doubt that all phytohormones are involved in the process of regulation during the first
stages of seedling development, first of all gibberellins and cytokinins (DIMALLA and van STADEN, 1977). Therefore one finds a change of
the hormone content during germination. Imbibited seeds of wheat lose
their cytokinin content completely from the embryonated halves and

only partly from the embryo-less halves (THOMAS et al., 1978a,b).
Soaking maize kernels results in an increase of the level of free
cytokinins as compared with the level in dry seed - a few hours later
it decreases anew. The ribotides also decrease slightly after 4 h
(JULIN-TEGELMAN, 1979). It seems that certain seeds lack a nucleoside
phosphorylase as well as a nucleoside kinase system for cytokinin-
nucleotide synthesis (CHEN and ECKERT, 1977). But in *Lactuca* exogenous
[14]C-kinetin is taken up and metabolized to the 5'nucleotide (MIERNYK,
1979). Other early metabolic products of applied kinetin are f[6]AMP and
f[6]-adenin, which appear as seed germination progresses (MIERNYK and
BLAYDES, 1977). During germination it appears that cytokinin glucosides, the storage form of cytokinin, in the endosperm of *Zea mays*,
are transported to the embryo axis; this is supported by the fact
that the level of β-glucosidase activity is highest in the embryo
three days after imbibition (SMITH and van STADEN, 1978).

The cytokinins play an important role in germination of light-requir-
ing seeds (THOMAS, 1977), in growth (MIKULOWICH and KULAEVA, 1977)
and chlorophyll formation of the cotyledons (PETERS et al., 1978;
FORD et al., 1977a). Potassium probably is a prerequisite for this
regulation because it enhances the expansion response of the cotyledons to cytokinin. With increasing age, the response of the cotyledons to cytokinin is reduced and this reduction is associated with
lower levels of internal potassium. Furthermore a high level of external KCl offsets the lower potassium content in the tissue and enhances the effect of cytokinin (GREEN and MUIR, 1978). The cytokinin
supply, however, has in general a much greater influence on plastid
biogenesis than on leaf growth (FEIERABEND and DE BOER, 1978). In
seedlings of *Agrostemma*, the inhibition of growth and pigment accumulation by ABA can be overcome by the cytokinin BAP (SCHMERDER et al.,
1978).

In excised cucumber cotyledons the formation of chlorophyll can be
promoted if the cotyledons are treated for 18 h in the dark with
bleeding sap and subsequently continuously illuminated (for the light
effect cf. GIRNTH et al., 1978). The active component of the bleeding
sap seems to be cytokinin, which moves from seedling roots into the
cotyledons (DEI, 1978) and is not produced in a high amount in detached leaves under the influence of light (UHEDA and KURAISKI, 1977).
From experiments of DEI and TSUJI (1978) it is concluded that light,
particularly short pulses of red light, does not interact with benzyladenin directly, even if the BA pretreatment does stimulate the red
light effect on chlorophyll formation like other hormones.

Further arguments for a transport of cytokinin from the roots to the
leaves in a germinating embryo are as follows: In derooted seedlings
of rye nearly all parameters such as dry weight, amino-N, protein,
DNA, and RNA levels are lowered, but after the addition of kinetin
these components increase by 70%-100%, suggesting that the accumulation of substrate in general is a cytokinin-controlled step which depends on hormonal products transported from the roots to the shoots
(DE BOER and FEIERABEND, 1978). Benzyladenin and its riboside are most
effective among the various plant hormones in increasing the arginine
decarboxylase activity in cotyledons of *Cucumis sativus* during germination (SURESH et al., 1978).

In respect to the greening of cotyledons it was assumed (HARDY et al., 1970)
that the light-dependent process independent of an effect of cytokinin is
mediated by the hypocotyl-hook. But FORD et al. (1977b) were able to show that
this may be an artifact caused by a submerged growth of the seedlings with a
low O_2 tension. Under aerobic conditions the greening of cotyledons was the

same with and without a hook. Light is one of the factors necessary for the formation of the large subunit of RUBP-carboxylase, coded in the plastides, because dark-grown seedlings contain an excess on free small subunits of this enzyme for which the genetic information is located in the nucleus (FEIERABEND and DE BOER, 1978).

A third effect of cytokinin is the participation of the hormone in regulation or breaking of seed dormancy (summary see TAYLORSON and HENDRICHS, 1977). In celery, cytokinin increases the response to dormancy-breaking red light treatment. On the other hand the red light changes the internal cytokinin content in the seeds qualitatively and quantitatively, probably because phytochrome controls the germination via the endogenous cytokinin level (THOMAS et al., 1978b) as was assumed also in some earlier papers of these authors.

The manner of action of gibberellin during the germination of barley seeds stimulating the hydrolytic enzymes in the aleuron layer has been well known for many years (cf. SCHRAUDOLF, 1973; DASHEK and CHRISPEELS, 1977; RODAWAY and KENDE, 1978; VARNER and HO, 1976). Interesting details give new examinations on the distribution of the GA_3-stimulated α-amylase. The data of these experiments support the hypothesis that the α-amylase, a glycoprotein with a heterogenous glycosylation (RODAWAY, 1978), is found in membrane vesicles which serve as an intermediate system in the secretion of the enzyme. Those isoenzymes of amylase were found with the particular membrane fraction which is secreted from the producing cells (LOCY and KENDE, 1978). The vesicles derived from the rough endoplasmic reticulum (COLBORNE et al., 1976).

Much less is known about the gibberellin effects on the germination process of other plants than barley. Under certain circumstances the light requirement can become compensated by GA_3, for example in lettuce seeds, because ethylene needs either light or gibberellic acid to preserve an osmotically induced dormancy (DUNLOP and MORGAN, 1977). But the action of gibberellin in such experiments is not yet understood at all (LESHEM and INBAR, 1978). A possible hypothesis that the GA effect is related to increased manganese interacting with some enzyme activities is discussed by BHARTI et al. (1978a,b).

Further effects of gibberellin in the course of germination are as follows: The caryopses of the grass *Phalaris* germinate after removing the husk, after surgical experiments on the seed coat, but the experimental gearing can be replaced by applying GA_3 to the seed (JUNTTILA et al., 1978).

In castor bean GA_3 applied to the endosperm affects the subcellular development of enzymes. First of all the enzymes of the glyoxylate cycle and of fatty acid β-oxydation are particularly responsive and increase in activity within 24 h after treatment (WRIGLEY and LORD, 1977). These observations are confirmed by GONZALEZ (1978) who extends them to the assembly of the enzymes and the functional segregation of the ER. At the same time as the enzymes the RNA associated with the ER increases. Also RNAse activity was stimulated by GA. These effects were antagonistically inhibited by abscisic acid (TAKAIWA and TANIFUJI, 1978).

The *tissue-specific effects* of hormones - the consequence of the competence - on the enzyme activity is well known. In a dwarf strain of water melon it was demonstrated that the development of catalase activity is stimulated by GA_3 in the whole embryo, whereas invertase activity was only stimulated in the embryonic axis and not in the cotyledons (EVENSON and LOY, 1978). Also *enzyme-specific effects* of GA_3 can be shown: In the after-ripened seeds or embryos of *Agrostemma githago*

the content of phosphatase, which is normally localized in the cotyledons, is enhanced by GA_3 treatment. This stimulation is specific
for phosphatase (BORRISS and SCHMERDER, 1978). A general influence
of GA on the protein pattern of wheat grains was not found (ANGUILLESI
et al., 1978).

HALL and BANDURSKI (1978) examined the question whether the auxin in
the embryo is produced by the endosperm (PISKORNIK, 1975) and transported from there to the embryo. Indeed they found that IAA can move
from endosperm to the young seedling and that tryptophan applied to
the endosperm appears as IAA in the seedling, but at rates that are
too slow to provide the shoot with sufficient IAA. Therefore the embryo has to produce its own auxin. During the transport from endosperm
to the seedling the IAA becomes esterified (PERCIVAL and BANDURSKI,
1976; SCHNEIDER and WIGHTMAN, 1974). One should, however, mention that
only 2% of the transported radioactivity was auxin or auxin conjugates.

A new aspect of controlling the ripening of bananas was found by DESAI and
DESHPANDE (1978). They dipped post-harvest bananas into solutions of IAA and
ABA. Both substances hasten the ripening, judging from increase in sugar, as-
corbic acid, and change in color. The results may be of theoretical interest -
stimulation of ethylene by the two hormones - for practical purpose the effects
are too small.

As was mentioned before (TAKAIWA and TANIFUJI, 1978) ABA can be considered as an antagonistic hormone in seed germination as in growth
processes etc. This observation is in fact not new, but it is supported by new experiments and results. In pine seeds (Pinus pinea) a germination inhibitor was identified with exactly the same properties as
ABA. This was, however, not the only one in the seed coat (MARTINEZ-
HONDUVILLA and SARTOS-RUIZ, 1978). Exogenous ABA was also found inhibitory, for example in the induction of isocitratlyase activity in
the endosperm of castor bean (MARRIOT and NORTHCOTE, 1977), in the
pigment accumulation of Agrostemma seeds (SCHMERDER et al., 1978), in
germination and coleoptile growth in rice seedlings (KARSAKAR and
BAJRACHARYA, 1978). In the last case the effect can be counteracted
by kinetin and gibberellin, but not by IAA. Unfortunately such observations cannot be generalized, because the peroxydase activity during
mung bean germination is inhibited by ABA to 87%, and this response
was substantially counteracted only by auxin, while gibberellic acid
and kinetin not only failed to alter the peroxydase activity per se,
they are also not able to counteract the inhibitory ABA-effect (DENDSAY
and SACHAR, 1978). Therefore in every system one has to look for the
special conditions, but to understand hormone action on germination
the antagonistic effect of a stimulating and an inhibiting hormone
may be of importance.

Ethylene finally works in part as a germination stimulator or as an
inhibitor: In lettuce seeds the removal of ethylene should prevent
germination, while addition of the hormone stimulates germination up
to 95%-100% depending on the ethylene concentration (NEGM and SMITH,
1978). In other more critical experiments it was shown that GA_3, cytokinin, and fusicoccin may substitute for ethylene action - namely
under stress conditions - but it is to be supposed that in a closed
system a gas like ethylene may have a more pronounced effect on germination (RUDNICKI et al., 1978).

A quite complicated situation is found in the different seeds of cocklebur:
There are large seeds in the lower part of the fruits, which have a high germination potential, and smaller ones in the upper part, which fail to germinate
under ordinary conditions, because they cannot produce a thrust sufficient to

overcome the block imposed by the seed coat. The stimulated germination of
such small seeds decreases with increasing periods of water imbibing, which
is called secondary dormancy. It is shown that ethylene can effectively prevent
the development of secondary dormancy, which is not possible with benzyladenin
and many other substances (ESASHI et al., 1978).

b) Germination-Regulating Substances

Beside the regular phytohormones of the higher plants many substances (of endogenous or exogenous origin) have an influence on seed germination: procaine,
which accelerates the tissue senescence in detached cotyledons, disturbs the
germination of lettuce seeds, perhaps by virtue of the interference with the
endogenous calciumion balance (MUMFORD, 1978). Many of the phenolic compounds,
often quoted as inhibitors of growth, are not particularly active against lettuce germination compared with abscisic acid or coumarine. Aromatic alcohols,
aldehydes and carboxylic acids, however, are strong inhibitors, whereby increased lipophily usually leads to increased inhibitory activity (REYNOLDS,
1978) in the same way as was demonstrated for aliphatic compounds (REYNOLDS,
1977).

Nonprotein amino acids, as well as protein amino acids, can act as germination
inhibitors. Imbibition of lettuce or legume seeds in solution of such amino
acids inhibits the germination more or less, the strongest inhibition was
found by glutamic acid (WILSON and BELL, 1978a). Eluates from *Glycine wightii*
containing free amino acids also strongly inhibit the germination or the growth
of lettuce seedlings (WILSON and BELL, 1978b). This inhibition may perhaps be
one of the roles of the secondary metabolites; to suppress the germination or
the growth of competing species (BELL, 1976; SEIGLER, 1977).

A completely different germination inhibitor, which works in physiological
concentrations, was found in seeds of parsley and described as heraclenol, a
substance also present in *Heracleum candicans* (KATO et al., 1978).

Barbiturates (amobarbital, barbital, secobarbital, and others) influenced the
normal development of germinating rice seedlings (KORDAN, 1978; KORDAN and
MUMFORD, 1978) stimulating the etiolation in light (KORDAN, 1977) and the formation of lateral roots (KORDAN and GLAUERT, 1978). Sterols in pine seedlings
are probably not involved in the germinating process because their content remains nearly unchanged throughout the experimental germination period (VU and
BIGGS, 1978).

c) Temperature Effects

Temperature (HAGESETH, 1978) and first of all change of temperature
has a remarkable influence on the regulation of seed germination. The
effect on the timing of seed germination by different constant temperatures was extensively studied in the seeds of *Dolichos bilorus* (LABOURIAU
and PACHECO, 1978). Of more interest, however, are the so-called thermoperiodic effects, with a regulating change of low and high temperatures, which have been known since 1923 (HARRINGTON, 1923). They cannot be explained up to now in a molecular or biochemical way. Therefore many more experiments are necessary. In cocklebur *(Xanthium)* nondormant seeds germinate in temperature of 33° 20 h after imbibition,
in 23° after one and a half days. At 28°, however, one finds a bio-
phasic germination: one part of the seeds germinates after 20, the
other after 36 h. From this and comparable results an endogenous temperature-sensitive rhythm was suggested, involved in the seed germination (SATOH and ESASHI, 1978). But nondormant and secondary dormant seeds (see above) have a different behavior, which make it difficult to explain the results in a general manner (ESASHI and TSUKADA,
1978). The nondormant seeds are incapable of germinating under con-

stant temperatures below 25° in air; but a maximal germination was
obtained in a thermoperiodic regime of 8 h at 23°C and 16 h at 8°C.
In the high temperature phase a process must occur which is aerobic
and had to precede the inductive dark period (ESASHI and TSUKADA),
both periods should show a diurnal fluctuation.

   Other effects of changing temperature are described for celery (and very similar
   for lettuce) seeds, in which a pretreatment with 32°C lowered the subsequent
   germination at 22°C but not at 17°C, an effect completely removable by gibberellin or cytokinin treatment (BIDDINGTON and THOMAS, 1978; HEYDECKER and JOSHUA,
   1977). In other cases a high temperature phase is a prerequisite for the germinating phase at lower temperature of about 12°C (THOMPSON and COX, 1978). The
   effect of low temperatures on embryos may be a consequence of a so-called chill-
   ing injury rather than the direct effect of the low temperature on metabolic
   steps (POWELL and MATTHEWS, 1978).

These few examples are sufficient to demonstrate that the field of
temperature effects is full of unsolved problems and it is necessary
either to look on the ecological background of the germination process (cf. BARON, 1978; BASKIN and BASKIN, 1977) or to go into more detail of the biochemical events changed by the environmental temperature. Under the last aspect it may be of interest that low temperatures
have a pronounced influence on the polyribosome content in embryos. One
finds the highest amount of polyribosomes during cold treatment of
wheat (TATEYAMA et al., 1978). Another aspect is the influence of low
temperatures on the unsaturated fatty acid ratio (BARTKOWSKI et al.,
1977). Therefore in some cases, demonstrated on a more chill-sensitive
and a less chill-sensitive cultivar of cotton, treatment with fatty
acids significantly increases the germination at low temperature in
the less sensitive seeds but not for the more sensitive cultivar
(BARTKOWSKI et al., 1978).

The free amino acid pool, which includes beside the proteinic L-amino
acids D-amino acids (OGAWA et al., 1978), is also drastically influenced by temperature during germination. At 10°C one finds a high concentration of glutamate and aspartate - the two major amino acids in
storage proteins of soybean - only in the cotyledons (DERBYSHIRE et
al., 1976; OCHIAI-YANAGI et al., 1977). Asparagine and glutamine are
differently distributed in cotyledons and axis. In contrast, at 23°C
the concentrations of the amides asparagine and glutamine are high in
cotyledons as well as in the embryo axis, and aspartate and glutamate
are low in both organs (DUKE et al., 1978). The temperature therefore
has a remarkable influence on the formation or distribution of reserve
material, but nothing is known about the manner of this distribution.
In germinating cotton seeds the amount of the amides is also high
(CAPDEVILLA and DURE, 1977). For some enzymes the activity is much
higher at 23°C than at 10°C. The low temperature effect appears to be
due in part to an enhanced activation energy for enzymes, involved in
energy transduction (DUKE et al., 1977).

d) Protein and Amino Acid Metabolism During Germination

The protein and amino acid metabolism during germination is an important
part of the whole process of the growth and coordination in the young
embryo or seedling. The tissues of the quiescent seeds contain free
amino acids besides quaternary nitrogen compounds. During germination
the amino acid level increases several-fold, whereby in wheat grain
glutamin is the predominant amino acid in the aleurone tissue and endosperm and asparagin in the embryo (CHITTENDEN et al., 1978). In both
types of tissue the increase in free amino acid levels occurs inde-

pendently for the first two days of germination, afterwards the further increase is dependent upon the presence of the embryo or of GA
(CHITTENDEN et al., 1978).

A major aspect of germination, however, is the interaction between
the different parts of the seed, and for this reason transport phenomena play an important role. In barley the peptide-transport system
from the endosperm to the embryo was studied extensively. One single
system was characterized which operates for di- and oligopeptides
(HIGGINS and PAYNE, 1977, 1978a,b). The very quick uptake of the dipeptide glycylglycin can be inhibited by other dipeptides, but not by
amino acids, which shows the specificity of the transport system or
transport site. These transport sites are the receptors for the substance, which should be transported by the particular system (SOPANEN
et al., 1978). For further investigations di- and oligopeptides were
taken, which are resistant to peptidase hydrolysis. These artificial
peptides are accumulated against a concentration gradient by the scutellum of germinating barley grains (HIGGINS and PAYNE, 1977b). The
transport in question is stereospecific, because the replacement of a
L-amino acid residue in a dipeptide by its D-stereoisomer decreases
the affinity to the transport site, leading to a drastic reduction in
transport. Therefore one can conclude that an active peptide transport
system exists in the scutellum functioning in the transfer to the embryo of products obtained by hydrolysis of the endosperm storage proteins (HIGGINS and PAYNE, 1978c; SOPANEN et al., 1978).

In other systems the interaction between different parts of the seeds
is also necessary to regulate the protein metabolism. In mung bean the
removal of the axis from the dry seed slows down the catabolism of reserve proteins to 25% of the rate in intact seedlings and the hydrolysis for the reserve protein vicilin, the major endopeptidase of mung
bean seedlings (BAUMGARTNER and CHRISPEELS, 1977) is reduced by 77%.
But many other activities, such as glutamine synthetase and asparagin
synthetase, are under the control of the cotyledons themselves. Removal of the axis has no effect in this case (KERN and CHRISPEELS,
1978). Because the starch synthesis and degradation in seedlings and
in isolated cotyledons proceeds similarly, it is concluded that in
some cases the results, which indicate the dependence of enzyme activity of the axis, can be an artifact (HOFFMANNOWA, 1978). The time course
of the vicilin degradation was followed in *Pisum*. Surprisingly, the protein was not metabolized in the cotyledons earlier than 72 h after
soaking (KONOPSKA, 1978), albeit the activity of enzymes for the degradation of proteins such as proteases and phosphatase is already high
after 1 h of inhibition (KONOPSKA and SAKOWSKI, 1978). But the timing
regulation of enzyme activities in seeds and seedlings cannot be generalized at all. Therefore it is nearly impossible to give a picture
universally valid for the seeds of all species of a certain storage
type. For example in pumpkin seeds the highest proteolytic activity
was found not earlier than 6 days after beginning of germination
(REILLY et al., 1978) and in castor bean two amino peptidases are described with the highest activity 4 days after germination, concurrent
with a rapid depletion of storage proteins (TULLY and BEEVERS, 1978).
Furthermore a comparison of the change in the proteins of chickpea during germination demonstrates a different degradation rate for three
polypeptides with the sedimentation coefficients of 2.2 S, 6.9 S and
10.3 S. During the early stages of germination the degradation rate of
6.9 S fraction is higher, while the 10.3 S fraction is broken down faster in the later stages more than 9 days after imbibition (KUMAR and
VENKATARAMAN, 1978).

Nevertheless in general the activity of endopeptidases located in the protein bodies is a prerequisite for the rapid metabolism of the reserve proteins which accompanies germination (PRESTON and KRUGER, 1977). In dry kernels of corn aminopeptidase was found in higher amounts in endosperm and scutellum than in the embryo. The enzyme decreases in all parts of the seedling as soon as they undergo rapid loss of nitrogen (FELLER et al., 1978).

During germination, some substances, including enzyme proteins, are leached out from the seeds. For proteins this leaching is independent of whether the testa is removed before imbibition or not (ABDEL SAMAD and PEARCE, 1978), therefore the assumption that early leachage during imbibition of seeds results from the death of peripheral cells of the cotyledons (POWELL and MATTHEWS, 1978) cannot be valid in general.

References

ABDEL SAMAD, I.M., PEARCE, R.S.: J. Exp. Bot. 29, 1471-1478 (1978). - AJKTHOSHIN, M.A., DOSHCHANOV, K.J., AKANOV, A.U.: FEBS Lett. 66, 124-126 (1976). - AL-ABTA, S., COLLIN, H.A.: Ann. Bot. 42, 773-782 (1978). - ANGUILLESI, M.C., FLORIS, C., GRILLI, I., MELETTI, P.: Biochem. Physiol. Pflanz. 173, 340-346 (1978). - ARNOLD, S. v., ERIKSSON, T.: Physiol. Plant. 44, 283-287 (1978). - ASHTON, F.M.: Annu. Rev. Plant Physiol. 27, 95-117 (1976). - ATKINS, C.A., FLINN, A.M.: Plant Physiol. 62, 486-490 (1978). - ATKINS, C.A., KUO, J., PATE, J.S., FLINN, A.M., STEELE, T.W.: Plant Physiol. 60, 779-786 (1977).

BARKER, R.D.J., DERBYSHIRE, E., YARWOOD, A., BOULTER, D.: Phytochemistry 15, 751-757 (1976). - BARON, F.J.: Am. J. Bot. 65, 804-810 (1978). - BARTKOWSKI, E.J., BUXTON, D.R., KATTERMANN, F.R.H., KIRCHER, H.W.: Agron. J. 69, 37-40 (1977). - BARTKOWSKI, E.J., KATTERMANN, F.R.H., BUXTON, D.: Physiol. Plant. 44, 153-156 (1978). - BASHA, S.M.M., BEEVERS, L.: Plant Physiol. 57, 93-97 (1976). - BASKIN, J.M., BASKIN, C.C.: Am. J. Bot. 64, 1242-1247 (1977). - BAUMGARTNER, B., CHRISPEELS, M.J.: Eur. J. Biochem. 77, 223-233 (1977). - BEACHY, R.N., THOMPSON, J.F., MADISON, J.T.: Plant Physiol. 61, 139-144 (1978). - BECHTEL, D.B., POMERANZ, Y.: Am. J. Bot. 65, 75-85 (1978). - BEEVERS, L., MENZL, R.M.: Plant Physiol. 60, 703-708 (1977). - BELL, E.A.: FEBS Lett. 64, 29-35 (1976). - BERGFELD, R., HONG, Y.N., KÜHNL, T., SCHOPFER, P.: Planta 143, 297-307 (1978). - BHARTI, S., SINGH, Y.D., LALORAYA, M.M.: J. Exp. Bot. 29, 1085-1090 (1978a); - J. Exp. Bot. 29, 1091-1098 (1978b). - BHAT, S.P., PADAYATTY, J.D.: Nature (London) 256, 227-228 (1975). - BIDDINGTON, N.L., THOMAS, T.H.: Physiol. Plant. 42, 401-405 (1978). - BOER, J. DE, FEIERABEND, J.: Planta 142, 67-73 (1978). - BOPP, M.: Fortschr. Bot. 33, 141-165 (1971). - BOLLINI, R., CHRISPEELS, M.J.: Planta 142, 291-298 (1978). - BORNMAN, C.H., BUTLER, V., JENSEN, W.A.: Z. Pflanzenphysiol. 91, 189-196 (1979). - BORRISS, H., SCHMERDER, B.: Biochem. Physiol. Pflanz. 172, 453-474 (1978). - BROCKLEHURST, P.A.: Nature (London) 266, 348-349 (1977). - BROOKER, J.D., TOMASZEWSKI, M., MARCUS, A.: Plant Physiol. 61, 145-149 (1978). - BUTLER, V., BORNMAN, C.H., JENSEN, W.A.: Z. Pflanzenphysiol. 91, 197-210 (1979). - BYRNE, H., SETTERFIELD, G.: Planta 143, 75-83 (1978).

CAERS, L.J., PEUMANS, W.J., CARLIER, A.R.: Planta 144, 491-496 (1979). - CAPDEVILLA, A.M., DURE, L.: Plant Physiol. 59, 268-273 (1977). - CARASCO, J.F., CROY, R., DERBYSHIRE, E., BOULTER, D.: J. Exp. Bot. 29, 309-323 (1978). - CHEN, C.M., ECKERT, R.L.: Plant Physiol. 59, 443-447 (1977). - CHITTENDEN, C.G., LAIDMAN, D.L., AHMAD, N., WYN JONES, R.G.:

Phytochemistry 17, 1209-1216 (1978). - CHOINSKI, J.S., Jr., TRELEASE,
R.N.: Plant Physiol. 62, 141-145 (1978). - CIONINI, P., BENNICI, A.,
ALPI, A., D'AMATO, F.: Planta 131, 115-117 (1976). - CLOWLES, F.A.L.:
Ann. Bot. 42, 1237-1239 (1978a); - New Phytol. 80, 409-419 (1978b). -
COLBORNE, A.J., MORRIS, G., LAIDMAN, D.L.: J. Exp. Bot. 27, 759-768
(1976). - CREMONINI, R., CIONINI, P.G.: Protoplasma 91, 303-313 (1977).
- CULLIS, C.A.: Planta 144, 55-62 (1978).

DASHEK, W.V., CHRISPEELS, M.J.: Planta 134, 251-256 (1977). - DATHE,
W., SEMBDNER, G.: Biochem. Physiol. Pflanz. 173, 440-447 (1978). -
DATHE, W., SCHNEIDER, G., SEMBDNER, G.: Phytochemistry 17, 963-966
(1978). - DEI, M.: Physiol. Plant. 43, 94-98 (1978). - DEI, M.,
TSUJI, H.: Plant Cell Physiol. 19, 1407-1414 (1978). - DELSENY, M.,
ASPANT, L., GUITTON, Y.: Planta 135, 125-128 (1977). - DENDSAY, J.P.S.,
SACHAR, R.C.: Phytochemistry 17, 1017-1019 (1978). - DERBYSHIRE, E.,
BOULTER, D.: Phytochemistry 15, 411-414 (1976). - DERBYSHIRE, E.,
WRIGHT, D.J., BOULTER, D.: Phytochemistry 15, 3-24 (1976). - DESAI,
B.B., DESHPANDE, P.B.: Physiol. Plant. 44, 238-240 (1978). - DEWDNEY,
S.J., McWHA, J.A.: J. Exp. Bot. 29, 1299-1308 (1978). - DIMALLA, G.G.,
van STADEN, J.: Z. Pflanzenphysiol. 82, 274-280 (1977). - DUFFUS, C.M.,
ROSIE, R.: Plant Physiol. 61, 570-574 (1978). - DUKE, S.H., SCHRADER,
L.E., MILLER, M.G.: Plant Physiol. 60, 716-722 (1977). - DUKE, S.H.,
SCHRADER, L.E., MILLER, M.G., NIECE, R.: Plant Physiol. 62, 642-647
(1978). - DURE, L.S.: Annu. Rev. Plant Physiol. 26, 259-278 (1975). -
DUNLOP, J.R., MORGAN, P.W.: Plant Physiol. 60, 222-224 (1977).

EATON-MORDES, C.A., MOORE, K.G.: Phytochemistry 17, 619-621 (1978). -
ESASHI, Y., TSUKADA, Y.: Plant Physiol. 61, 437-441 (1978). - ESASHI,
Y., OKAZAKI, M., YANAI, N., HISHINUMA, K.: Plant Cell Physiol. 19,
1497-1506 (1978). - EVENSON, K.B., LOY, J.B.: Plant Physiol. 62, 6-9
(1978).

FEIERABEND, J., DE BOER, J.: Planta 142, 75-82 (1978). - FLINN, A.A.,
ATKINS, C.A., PATE, J.S.: Plant Physiol. 60, 412-418 (1977). - FELLER,
U., SOONG, T.T., HAGEMAN, R.H.: Planta 140, 155-162 (1978). - FORD,
M.J., BLACK, M., CHAPMAN, J.M.: J. Exp. Bot. 28, 926-934 (1977a). -
FORD, M.J., SLACK, P., BLACK, M., CHAPMAN, J.M.: Planta 132, 205-208
(1977b). - FOUNTAIN, D.W., BEWLEY, J.D.: Plant Physiol. 58, 530-536
(1976).

GIRNTH, C., BERGFELD, R., KASEMIR, H.: Planta 141, 191-198 (1978). -
GOLDBACK, H., MICHAEL, G.: Naturwissenschaften 64, 488 (1977). -
GONZALES, E.: Plant Physiol. 62, 449-453 (1978). - GOSCH-WACKERLE, G.,
AVIVI, L., GALUN, E.: Z. Pflanzenphysiol. 91, 267-278 (1979). - GOTÔ,
N.: Plant Cell Physiol. 19, 1121-1127 (1978). - GREEN, J.F., MUIR,
R.M.: Physiol. Plant. 43, 213-218 (1978).

HAGESETH, G.T.: J. Exp. Bot. 29, 281-293 (1978). - HALL, P.L.,
BANDURSKI, R.S.: Plant Physiol. 61, 425-429 (1978). - HALL, T.C.,
McLEESTER, R.C., BLISS, F.A.: Plant Physiol. 59, 1122-1124 (1977).
- HARDY, S.J., CASTELFRANCO, P.A., REBEIZ, C.A.: Plant Physiol. 46,
705-707 (1970). - HARRINGTON, G.T.: J. Agric. Res. 23, 295-332 (1923).
- HARRIS, N., JULIANO, B.O.: Ann. Bot. 41, 1-5 (1977). - HARVEY, D.M.,
HEDLEY, C.L., KEELY, R.Y.: Ann. Bot. 40, 993-1001 (1976). - HEYDECKER,
W., JOSHUA, A.: J. Hortic. Sci. 52, 87-98 (1977). - HIGGINS, C.R.,
PAYNE, J.W.: Planta 136, 71-76 (1977a); - Planta 134, 205-206 (1977b);
- Planta 138, 211-215 (1978a); - Planta 138, 217-221 (1978b); - Planta
142, 299-305 (1978c). - HOFFMANNOWA, A.: Biochem. Physiol. Pflanz.
173, 181-185 (1978). - HU, C.Y., OCHS, J.D., MANEINI, F.M.: Z. Pflan-
zenphysiol. 89, 41-49 (1978).

JONES, P.A.: Planta 135, 233-240 (1977). - JULIN-TEGELMAN, Å.: Plant
Sci. Lett. 14, 259-262 (1979). - JUNTTILA, O., LANDGRAFF, A., NILSEN,
A.J.: Acta Hortic. 83, 163-166 (1978).

KAMADA, H., HARADA, H.: Z. Pflanzenphysiol. 91, 255-266 (1979). -
KARSAKAR, S., BAJRACHARYA, D.: Z. Pflanzenphysiol. 88, 189-199 (1978).
- KATO, T., KUBOTA, S.: Physiol. Plant. 42, 67-72 (1978). - KATO, T.,
KOBAYASHI, M., SASAKI, N., KITAHARA, Y., TAKAHASHI, N.: Phytochemistry
17, 158-159 (1978). - KERN, R., CHRISPEELS, M.J.: Plant Physiol. 62,
815-819 (1978). - KHARKIN, E.E., MISHARIN, S.J., MARKOV, Y.Y., PESHKOVA,
A.A.: Planta 143, 11-20 (1978). - KING, R.W.: Planta 132, 43-51 (1976).
- KONOPSKA, L.: Biochem. Physiol. Pflanz. 173, 322-326 (1978). -
KONOPSKA, L., SAKOWSKI, R.: Biochem. Physiol. Pflanz. 173, 536-540
(1978). - KORDAN, H.A.: Ann. Bot. 41, 257-259 (1977); - Ann. Bot. 42,
73-81 (1978). - KORDAN, H.A., GLAUERT, A.W.: Ann. Bot. 42, 1233-1235
(1978). - KORDAN, H.A., MUMFORD, P.M.: Ann. Bot. 42, 997-999 (1978).
- KRUL, W.R., WORLEY, J.F.: J. Am. Soc. Hortic. Sci. 102, 360-367
(1977). - KUMAR, K.G., VENKATARAMAN, L.V.: Phytochemistry 17, 605-609
(1978).

LABOURIAU, L.G., PACHECO, A.A.: Plant Cell Physiol. 19, 507-512 (1978).
LESHEM, Y.Y., INBAR, M.: J. Exp. Bot. 29, 671-675 (1978). - LOCY, R.,
KENDE, H.: Planta 143, 89-99 (1978). - LORENZI, R., BENNICI, A.,
CIONINI, P.G., ALPI, A., D'AMATO, F.: Planta 143, 59-62 (1978). -
LÜDICKE, W.: Dissertation Fakultät Biologie, Heidelberg (1976).

MANDAC, B.E., JULIANO, B.O.: Phytochemistry 17, 611-614 (1978). -
MARBACH, J., MAYER, A.M.: J. Exp. Bot. 29, 69-75 (1978). - MARTIN,
G., DENNIS, F.J., Jr., McMILLAN, J., GASKIN, P.: J. Am. Soc. Hortic.
Sci. 102, 16-19 (1977). - MARTINEZ-HONDUVILLA, C.J., SARTOS-RUIZ, A.:
Planta 141, 141-144 (1978). - MARRIOT, K.M., NORTHCOTE, D.H.: J. Exp.
Bot. 28, 219-224 (1977). - NIERNYK, J.A.: Physiol. Plant. 45, 63-66
(1979). - MIERNYK, J.A., BLAYDES, D.F.: Physiol. Plant. 39, 4-8 (1977).
- MIKULOWICH, T.P., KULAEVA, O.N.: Sov. Plant Physiol. 24, 418-423
(1977). - MILLERD, A.: Annu. Rev. Plant Physiol. 26, 52-72 (1975). -
MONSELISE, S.P., VARGA, A., BRUINSMA, J.: Ann. Bot. 42, 1245-1247
(1978). - MOUNLA, M.A.KH.: Physiol. Plant. 44, 268-272 (1978). -
MUMFORD, P.M.: Ann. Bot. 42, 491-492 (1978). - MURACH, K.F.: Staats-
examensarbeit Heidelberg (1979).

NEGM, F.B., SMITH, O.E.: Plant Physiol. 62, 473-476 (1978). -
NEUMANN, D., WEBER, E.: Biochem. Physiol. Pflanz. 173, 167-180 (1978).

OCHIAI-YANAGI, S., TAHAGI, T., KITAMURA, K., TAJIMA, M., WATANABE, T.:
Agric. Biol. Chem. 41, 647-653 (1977). - OGAWA, T., KAWASAKI, Y.,
SASAOKA, K.: Phytochemistry 17, 1275-1276 (1978). - OLIKER, M.,
POLJAKOFF-MAYBER, A., MAYER, A.M.: Am. J. Bot. 65, 366-371 (1978a). -
OLIKER, M., MAYER, A.M., POLJAKOFF-MAYBER, A.: Am. J. Bot. 65, 372-
374 (1978b).

PATE, J.S., SHARKEY, P.J., ATKINS, C.A.: Plant Physiol. 59, 506-510
(1977). - PAYNE, P.I.: Biol. Rev. 51, 339-363 (1976). - PAYNE, P.I.,
DOBRZANSKA, M., BARLOW, P.W., GORDON, M.E.: J. Exp. Bot. 29, 77-88
(1978). - PERCIVAL, F., BANDURSKI, R.S.: Plant Physiol. 58, 60-67
(1976). - PERNOLLET, J.C.: Phytochemistry 17, 1473-1480 (1978). -
PETERS, J.A., HELL, K.G., HANDRO, W.: Plant Cell Physiol. 19, 1483-
1487 (1978). - PEUMANS, W.J., CARLIER, A.R.: Planta 136, 195-201
(1977). - PNEUMANS, W.J., CARLIER, A.R., CAERS, L.J.: Planta 140,
171-176 (1978). - PNEUMANS, W.J., CAERS, L.J., CARLIER, A.R.: Planta
144, 485-490 (1979). - PFISTERER, J.: Z. Naturforsch. 33C, 359-362
(1978). - PINNOW, B.: Staatsexamenarbeit Heidelberg (1978). -

PISKORNIK, Z.: Acta Biol. Cracov. Ser. Bot. 17, 1-12 (1975). - POWELL, A.A., MATTHEWS, S.: J. Exp. Bot. 29, 1215-1229 (1978). - PRESTON, K.R., KRUGER, J.E.: Plant Physiol. Suppl. 59, 55 (1977). - BRZYBYLLOK, T., NAGL, W.: Z. Pflanzenphysiol. 84, 463-465 (1977). - PUSZIAI, A., CROY, R.R.D., GRANT, G., WATT, W.B.: New Phytol. 79, 61-71 (1977).

RADLEY, M.: J. Exp. Bot. 27, 1009-1012 (1976); - J. Exp. Bot. 29, 919-934 (1978). - REILLY, C.C., O'KENNEDY, D.T., TITUS, J.S., SPLITTSTOES-SER, W.E.: Plant Cell Physiol. 19, 1235-1246 (1978). - REILLY, K., WASHER, J.: N.Z.J. For. Sci. 7, 199-206 (1972). - REYNOLDS, T.: Ann. Bot. 41, 637-648 (1977); - Ann. Bot. 42, 419-427 (1978). - RIGHETTI, P.G., GIANNAZZA, E., VIOTTI, A., SOAVE, C.: Planta 136, 115-123 (1977). - RODAWAY, S.J.: Phytochemistry 17, 385-389 (1978). - RODAWAY, S.J., KENDE, H.: Plant Physiol. 61, 1-6 (1978). - RUDNICKI, R.M., BRAUN, J.W., KHAN, A.A.: Physiol. Plant. 43, 189-194 (1978).

SATOH, S., ESASHI, Y.: Physiol. Plant. 43, 271-273 (1978). - SCHMERDER, B., RABENSTEIN, F., BORRISS, H.: Biochem. Physiol. Pflanz. 173, 97-113 (1978). - SCHNEIDER, E.A., WIGHTMAN, F.: Annu. Rev. Plant Physiol. 25, 487-513 (1974). - SCHRAUDOLF, H.: Fortschr. Bot. 35, 121-145 (1973). - SEIGLER, D.S.: Biochem. Syst. Ecol. 5, 195-199 (1977). - SHEWRY, P.R., HILL, J.M., PRATT, H.M., LEGGATT, M.M., MIFLIN, B.J.: J. Exp. Bot. 29, 677-692 (1978). - SINGH, R., PEREZ, C.M., PASCUAL, C.G., JULIANO, B.O.: Phytochemistry 17, 1869-1874 (1978). - SLOMINSKI, B., REJOWSKI, A., NOWAK, J.: Physiol. Plant. 45, 167-169 (1979). - SMITH, A.R., van STADEN, J.: J. Exp. Bot. 29, 1067-1075 (1978). - SOPANEN, T., BURSTON, D., TAYLOR, E., MATTHEWS, D.M.: Plant Physiol. 61, 630-633 (1978). - SPIEGEL, S., MARCUS, A.: Nature (London) 256, 228-230 (1975). - STOREY, R., BEEVERS, L.: Planta 124, 77-87 (1977); - Plant Physiol. 61, 494-500 (1978). - SUN, S.M., MUTSCHLER, M.A., BLISS, F.A., HALL, T.C.: Plant Physiol. 61, 918-923 (1978). - SURESH, M.R., RAMAKRISHNA, S., ADIGA, P.R.: Phytochemistry 17, 57-63 (1978).

TAKAIWA, F., TANIFUJI, S.: Plant Cell Physiol. 19, 1507-1518 (1978). - TAKAHASHI, N., TAKAIWA, F., FUKUEI, F., SAHAMAKI, T., TANIFUJI, S.: Plant Cell Physiol. 17, 1175-1184 (1976); - Plant Cell Physiol. 18, 235-246 (1977). - TANEJA, S., SACHAR, R.: Experimentia 31, 1128-1130 (1975); - Phytochemistry 15, 1589-1594 (1976). - TATYAMA, M., ISHIKAWA, H.A., ISHIKAWA, K.: Plant Cell Physiol. 19, 411-418 (1978). - TAYLORSON, R.B., HENDRICHS, S.B.: Annu. Rev. Plant Physiol. 28, 331-354 (1977). - THOMAS, T.H., in: Biochemistry and Physiology of Seed Germination, ed. A.A. KHAN, 111-144. Amsterdam: North Holland-Elsevier 1977. - THOMAS, T.H., KHAN, A.A., O'TOOLE, D.F.: Physiol. Plant. 42, 61-66 (1978a). - THOMAS, T.H., BIDDINGTON, N.L., PALEVITCH, D.: Photochem. Photobiol. 27, 231-236 (1978b). - THOMPSON, J.F., MADISON, J.T., MEUNSTER, A.E.: Ann. Bot. 41, 29-39 (1977). - THOMPSON, P.A., COX, S.A.: Ann. Bot. 42, 51-62 (1978). - TORREY, J.G., ZOBEL, R., in: The Physiology of the Garden Pea, eds. J.F. SUTCLIFF, J.S. PATE. London, New York: Academic Press 1977. - TULLY, R.E., BEEVERS, H.: Plant Physiol. 62, 746-750 (1978).

UHEDA, E., KURAISKI, S.: Plant Cell Physiol. 18, 481-483 (1977).

VARNER, J.E., HO, D.T., in: The Molecular Biology of Hormone Action, ed. J. PAPACONSTANITNOU. London, New York: Academic Press, 173-194 (1976). - VILLAREAL, R.M., JULIANO, B.O.: Phytochemistry 17, 177-182 (1978). - VIOTTI, A., SALA, E., ALBERI, P., SOAVE, C.: Plant Sci. Lett. 13, 365-375 (1978). - VU, C.V., BIGGS, R.H.: Physiol. Plant. 42, 344-350 (1978).

WANNER, G., THEIMER, R.R.: Planta 140, 163-170 (1978). - WIEMANN-WEISS,
D.: Dissertation Fakultät Biologie, Heidelberg (1979). - WILLIAMS, L.,
COLLIN, H.A.: Ann. Bot. 40, 325-332 (1976). - WILSON, M.F., BELL, E.A.:
Phytochemistry 17, 403-406 (1978a); - J. Exp. Bot. 29, 1243-1247
(1978b). - WILSON, M.F., RINNE, R.W.: Plant Physiol. 57, 536-559
(1976); - Plant Physiol. 61, 830-833 (1978a); - Plant Physiol. 61,
1014-1016 (1978b). - WILSON, M.F., RINNE, R.W., BRIM, C.A.: J. Am.
Oil Chem. Soc. 53, 595-597 (1976). - WINTER, H.: Verh. Dtsch. Zool.
Ges. 1974 67, 201-204 (1975). - WRIGLEY, A., LORD, J.M.: J. Exp. Bot.
28, 345-353 (1977).

YOULE, R.J., HUANG, A.H.C.: Plant Physiol. 58, 703-709 (1976); -
Plant Physiol. 61, 13-16 (1978).

ZEEVAERT, J.A.D., MILBORROW, B.V.: Biochemistry 15, 493-500 (1976). -
ZIMMERMANN, H.: Z. Pflanzenphysiol. 89, 115-118 (1978).

Professor Dr. MARTIN BOPP
Botanisches Institut der
Universität
Hofmeisterweg 4
D 6900 Heidelberg

VIII. Bewegungen

Von WOLFGANG HAUPT

1. Wachstumsbewegungen

Im Berichtszeitraum ist in der neuen Encyclopedia of Plant Physiology der Band 7 *Movements* erschienen. Darin sind die Fortschritte der letzten 15 bis 20 Jahre in Beiträgen kompetenter Spezialisten dargestellt. Ein Bericht herkömmlicher Art an dieser Stelle - auch über einen kleinen Teilbereich - könnte nicht mit dem entsprechenden Kapitel dort konkurrieren und wäre insoweit überflüssig. Als Alternative bietet sich dagegen an, aus einigen Kapiteln die interessantesten Ergebnisse, die als wichtige Fortschritte bezeichnet werden können, kurz zusammenfassend hier darzustellen. Dabei wird auf die Angabe von Primärliteratur ganz verzichtet, da diese in den Beiträgen zu finden ist. Um einen einigermaßen geschlossenen Themenkreis behandeln zu können, wurden für diesen Bericht die Wachstumsbewegungen ausgewählt.

2. Polaritätsinduktion

Die Polaritätsinduktion in keimenden Sporen und Zygoten, in der Regel als typisches Gebiet der Entwicklungsphysiologie betrachtet, hat so nahe Beziehungen zu Wachstumsbewegungen, daß ihr ein Kapitel im Bewegungsband gewidmet ist (WEISENSEEL). So geht nicht nur das differentielle Wachstum, das zur polaren Keimung führt, kontinuierlich über in tropistisches Wachstum des Keimschlauches oder des Rhizoids, sondern auch Teilprozesse sind diesen Entwicklungs- und Bewegungsvorgängen gemeinsam, insbesondere Veränderungen der Zellmembran. Fortschritte in unseren Kenntnissen des Polaritätsgeschehens können demzufolge wichtige Stimulationen für die Interpretation von Wachstumsbewegungen geben.

Zu den klassischen Studienobjekten der Polaritätsinduktion (Fucaceenzygoten, *Equisetum*spore) kam als geeignetes Objekt der *Lilium*pollen hinzu. Die bisherigen Ergebnisse zeigen eine bemerkenswerte Einheitlichkeit in sehr verschiedenen Verwandtschaftsgruppen und Zelltypen.

Alle Faktoren, die die Polarität induzieren können, scheinen Membraneffektoren zu sein. Aus Versuchen mit den verschiedensten polarisierenden Faktoren, insbesondere mit Licht, elektrischen Feldern und mit Ionengradienten, ergibt sich ein erstes hypothetisches Bild vom Gesamtgeschehen bei der Polaritätsinduktion: Wichtigster Schritt im Transduktionsprozeß ist wahrscheinlich eine gesteigerte Aufnahme von Ca^{++} an bestimmten Orten; dort erfolgt das verstärkte Zellwachstum. Diese Ca^{++}-Aufnahme ist Teil eines durch die Zelle hindurchgehenden Ionenstromes: Am "nichtwachsenden" Pol werden Ionen aktiv nach außen gepumpt, und Ionen strömen am wachsenden (oder zukünftig wachsenden) Pol passiv ein. Die Regulierung kann sowohl durch Aktivierung von Pumpen auf der Efflux-Seite, als auch durch Erhöhung der Permeabilität

auf der Influx-Seite erfolgen. Daraus ergeben sich noch nicht voll
voraussagbare Wirkungsmöglichkeiten der polarisierenden Außenfaktoren;
ein und derselbe Ionengradient kann z.B. je nach Randbedingungen zum
Auswachsen in der einen oder anderen Richtung führen.

Der Ca^{++}-Strom scheint von einer ungleichen Verteilung der Ca^{++}-Pumpen
und Ca^{++}-Leckstellen in der Membran abzuhängen. Die polarisierenden
Faktoren sollen zu solchen Ungleichverteilungen führen, und der resul-
tierende Strom verstärkt und stabilisiert diese Ungleichverteilung
durch seinen Einfluß auf die Membraneigenschaften und das Cytoplasma.
Bei diesen Membranwirkungen ist einerseits an die Permeabilität, andererseits an das Membranpotential zu denken, da auch lokale Differenzen
im Membranpotential auftreten (insbesondere Pol-zu-Pol-Differenzen).

> Der Ionenstrom kann mit neuen empfindlichen Methoden gemessen werden, ohne die
> Zelle in ihrer Entwicklung zu stören. Dabei entspricht die Stromstärke derjeni-
> gen, die durch Anlegen äußerer Felder als polarisierend wirksam erkannt wurde.

Für differentielles Wachstum muß nicht nur die Differenz, sondern auch
das Wachstum selbst gewährleistet sein. Ein wichtiger erster Schritt
scheint bei Fucaceenzygoten und Lilienpollen eine autonome Hyperpola-
risierung zu sein, unter anderem verursacht durch aktive KCl-Aufnahme,
die zu erhöhtem osmotischem Wert und nachfolgender Turgorsteigerung
führt.

3. Phototropismus

Ein umfangreicher Bericht liegt von DENNISON vor. Aufgrund der For-
schungsergebnisse der letzten Jahre ist die Frage des Photorezeptorpigments zugunsten von Flavinen (und gegen Carotinoide) so gut wie
entschieden. Das gilt jedenfalls für die große Mehrzahl untersuchter
Systeme, von den Koleoptilen über *Phycomyces* bis zu *Vaucheria*. Ein wichtiger Anstoß hierzu kam von Untersuchungen an *Dictyostelium*, wo eine
lichtinduzierte Absorptionsänderung das gleiche Wirkungsspektrum hat
wie der Phototropismus von *Phycomyces*, und die gleiche Absorptionsänderung wurde dann auch bei diesem Pilz gefunden. Die Absorptionsänderung
beruht auf Reduktion eines b-Cytochroms, und diese kann sogar in zell-
freien Systemen nachgewiesen werden, falls diesen FAD oder FMN zuge-
setzt wird.

Darüberhinaus können für ein Flavin als Photorezeptor folgende Argu-
mente angeführt werden: Die dichroitische Anordnung der phototropischen Photorezeptormoleküle bei *Phycomyces* ist im blauen und im ultra-
violetten Bereich unterschiedlich, und dieser Unterschied entspricht
unterschiedlichen Richtungen der Übergangsmomente im Flavinmolekül.
Bei Verwendung von Laserlicht, das sich durch besondere spektrale
Reinheit und zugleich besonders hohe Energie auszeichnet, wird im Wirkungsspektrum im Bereich von 600 nm noch eine kleine Schulter gefunden, die theoretisch vorausgesagt werden konnte und mit dem Triplett-
zustand des Flavins erklärt wird. In Extrakten aus *Phycomyces* kann eine
Fluoreszenz beobachtet werden, deren Anregungs- und Emissionsspektrum
einem Flavin entspricht, und diese fehlt in "nachtblinden" Mutanten,
d.h. solchen, die keine Lichtreaktionen mehr durchführen können. Umgekehrt ergeben Messungen und Berechnungen an Albinomutanten von *Phycomyces*, die noch phototropisch reagieren, daß die maximal verbliebenen
Carotinoidmengen kaum mehr für eine Lichtrichtungsperzeption ausreichen dürften. Schließlich zeigen theoretische Überlegungen, daß aufgrund der photochemischen Reaktionsmechanismen Carotinoide als photo-

tropische Photorezeptoren kaum in Frage kommen. Wegen der Übereinstimmung der Wirkungsspektren darf bedenkenlos von *Phycomyces* mindestens
auf die Koleoptilen extrapoliert werden, und dort finden sich *in vitro*
Flavinsignale in membranhaltigen Fraktionen.

Trotzdem ist die Lokalisierung der Photorezeptoren noch weniger geklärt als ihre chemische Natur. Partialbelichtungen führen bei *Phycomyces* zu dem Schluß, daß das Photorezeptorpigment nicht auf den wand-
nahen Bereich beschränkt ist. Im Hinblick auf den Dichroismus einerseits, die Bestätigung der Rotationshypothese andererseits (siehe unten) muß allerdings damit gerechnet werden, daß diese Ergebnisse der
Partialbelichtungen durch Streulicht verfälscht sind.

Die Perzeption der Lichtrichtung erfolgt bekanntlich auf verschiedenem
Wege: Bei Koleoptilen nach dem Prinzip der Schattenwirkung, bei *Phycomyces* nach dem Prinzip der Linsenwirkung. Aber auch dies ist noch kom-
plizierter als es auf den ersten Blick erscheint. Nicht nur reagiert
Phycomyces im UV (< 300 nm) nach dem Prinzip der Schattenwirkung, sondern ebenso im sichtbaren Bereich, falls es sich um eine carotinreiche
Mutante handelt. Das ist einleuchtend und paßt mit den klassischen
Vorstellungen zusammen. Ungeklärt sind dagegen gewisse Beobachtungen
bei *Avena*koleoptilen. Die Reaktionsumkehr durch Paraffinöl, die gewöhnlich als Hinweis auf den Linseneffekt gilt, könnte noch als ein Effekt
interpretiert werden, der nichts mit der Veränderung des Strahlengan-
ges zu tun hat (z.B. Einfluß auf den Stoffwechsel); aber wie läßt sich
die Beobachtung erklären, daß bei Verwendung stark divergierenden Lich-
tes von einer kleinen, sehr nahen Lichtquelle der positive Phototropismus ebenfalls nach negativ invertiert wird, analog zu dem schon länger
bekannten Ergebnis bei *Phycomyces* (wo dieses Ergebnis die Linsentheorie
weiter gestützt hat)?

Am Ende der Reaktionskette steht das ungleiche Wachstum entgegengesetzter Flanken. Der Versuch, bei *Phycomyces* dieses auf differentielle
Lichtwachstumreaktion (LWR) gemäß der Blaauwschen Hypothese zurückzuführen, scheiterte bisher daran, daß die LWR infolge Adaptation schon
nach einigen Minuten ausklingt, die phototropische Krümmung in einseitiger Dauerstimulation jedoch stundenlang anhalten kann.

> Eine einleuchtende Hilfshypothese hätte diese Erklärung retten können: Wenn die
> Adaptation, die zum schnellen Aufhören der LWR führt, über den ganzen Querschnitt
> des Sporangienträgers "gemittelt" würde, befände sich die stärker belichtete
> Zone kontinuierlich im stimulierten Zustand, d.h. sie könnte sich nicht adaptie-
> ren. Neue Versuche zeigten jedoch eindeutig, daß die Adaptation nicht nur longi-
> tudinal, sondern auch azimutal autonom ist.

DENNISON hat daher aufgrund einer Diskussion mit dem Referenten seine
Rotationshypothese aufgestellt und experimentell bestätigt: Da das
Wachstum des Sporangienträgers mit Rotation der Wachstumszone verbunden ist ("Spiralwachstum"), werden in konstanter einseitiger Belich-
tung immer neue Bereiche der Wachstumszone an den hellen Brennstreifen
geführt, und zwar mit einer Geschwindigkeit von $10°/min$. Sie erfahren
dort einen "Licht-an-Stimulus" mit nachfolgender positiver LWR; und
diese Stimulation muß naturgemäß so lange anhalten, wie der Sporangien-
träger belichtet wird. Der Beweis ist verblüffend einfach: Rotation
des Sporangienträgers in entgegengesetzter Richtung und mit entspre-
chender Winkelgeschwindigkeit läßt den Phototropismus verschwinden.
Das ist zugleich aber auch ein Beweis, daß die Photorezeptoren mit
der gleichen Geschwindigkeit "rotieren" wie die Zellwand, und dies
spricht sehr stark für wandnahe Lokalisierung. Damit rückt aber die
Lichtrichtungsperzeption beim Phototropismus von *Phycomyces* als "dyna-
mischer" Meßvorgang in die Nähe der phototaktischen Lichtrichtungs-

perzeption von Euglena ("two-instant-mechanism", Beitrag HÄDER); in
beiden Fällen ist die Bewegung der Photorezeptoren eine Voraussetzung
für den Meßvorgang.

Wenig Kenntnisse haben wir noch von den Transduktionsmechanismen. Nach
den oben referierten Ergebnissen bei *Dictyostelium* bietet sich an, als
ersten Schritt in der phototropischen Kausalkette eine Cytochromreduk-
tion anzunehmen (das könnte prinzipiell bei allen phototropischen Systemen zutreffen). An einem weiteren Schritt könnte cAMP beteiligt
sein, dessen Konzentration bei *Phycomyces* innerhalb einer Minute durch
Licht stark reduziert wird und nach wenigen Minuten sein altes Niveau
wieder erreicht; ein cAMP-Analogon führt zu einer vorübergehenden
Wachstumshemmung. Da cAMP auch das Myzelwachstum beeinflußt, wird man
diese heute sehr "moderne" Substanz weiter im Auge behalten, zumal sie
offenbar auch bei *Vaucheria* in die phototropische Reaktion eingreifen
kann.

Mit Sicherheit sind dagegen schon lange beobachtete Änderungen des elektrischen
Potentials in Koleoptilen nicht echte Glieder der phototropischen Kausalkette;
es handelt sich vielmehr um Nebeneffekte der ungleichen Auxinverteilung, analog
zum geoelektrischen Effekt, der ja auch seinen Platz in der geotropischen Trans-
duktionskette verloren hat.

Für die weitere Aufklärung der Transduktionsschritte dürfte die Mutantenanalyse
erfolgversprechend sein, die bei *Phycomyces* schon recht weit fortgeschritten
ist; man kann hier - etwas vereinfacht - input- und output-Mutanten unterscheiden
(erstere sind in allen Lichtreaktionen blockiert, führen aber z.B. geotropische
Krümmungen aus; letztere können keinerlei *Orientierungskrümmungen* durchführen,
reagieren aber *morphogenetisch* weiterhin auf Licht); dazu gibt es Mutanten, die
man sinngemäß als "Transduktionsmutanten" bezeichnen könnte.

Die hormonale Steuerung des Phototropismus bei Keimlingen von Samen-
pflanzen ist immer noch nicht allgemein geklärt, obwohl die Querverschiebung von Auxin bei Koleoptilen als wichtigster Mechanismus kaum
mehr angezweifelt werden kann und auch nicht dadurch an Bedeutung verliert, daß möglicherweise ein durch Licht gehemmter Längstransport
noch mitwirkt. Es werden nämlich immer wieder Fälle bekannt, in denen
durch eine bestimmte Behandlung die Krümmung verstärkt wird, der laterale Auxinquotient aber unverändert bleibt. Die Schwierigkeiten ver-
größern sich, wenn dicotyle Keimlinge in die Betrachtung einbezogen
werden. *Pisum* scheint sich nicht wesentlich anders zu verhalten als
Avena oder *Zea*; aber bei *Helianthus* ist Auxin vermutlich ohne Bedeutung.
Stattdessen wurde hier eine ungleiche Verteilung von Gibberellin auf
Licht- und Schattenseite im Verhältnis 1:7 gefunden; es liegt daher
sehr nahe, bei *Helianthus* einem Gibberellin die steuernde Hormonfunk-
tion zuzuschreiben, zumal dies auch beim Geotropismus angenommen werden muß (siehe unten).

Der letzte Schritt der Reaktionskette ist mit Sicherheit bei den beiden am häufigsten analysierten Systemen verschieden (Koleoptilen
und *Phycomyces*), schon allein wegen der chemischen Verschiedenheit der
Zellwände, die dabei ja die ausschlaggebende Rolle spielen. Die Auxin-
wirkung auf das Wachstum der *Zellulosewand* gehört in den Bereich der
Wachstums- und Entwicklungsphysiologie. Bei *Phycomyces* wird das Wachstum der *Chitinwand* nicht durch Vermittlung eines Hormons gesteuert; man
denkt an eine "direkte" Wirkung des Lichts, z.B. über eine auch *in vitro*
nachweisbare Aktivitätserhöhung der Chitin-Synthetase. *In vivo* äußert
sich dies in einer durch Licht erhöhten plastischen Streckungsfähig-
keit der Zellwand.

Gegenüber diesen teilweise sehr beachtlichen Fortschritten entziehen sich gewisse Grundphänomene noch immer einer exakten Interpretation: Die klassische

Dosiseffektkurve der Koleoptilen mit erster, zweiter positiver, eventuell auch
negativer und dritter positiver Krümmung findet sich mit gewissen Modifikationen
auch bei dicotylen Keimlingen und sogar bei *Vaucheria*. Auch wenn jetzt versucht
wird, auf der Basis von Adaptationsvorgängen zu einem besseren Verständnis zu
kommen und damit verschiedenen Krümmungstypen ihre "Eigenständigkeit" abzuspre-
chen, ist doch noch kein echtes Verständnis erreicht.

Insgesamt beeindruckt der Phototropismus einerseits mit einer Einheitlichkeit im Bereich des Photorezeptorpigments, andererseits sind im
Bereich von Transduktion und Reaktion offenbar von verschiedenen Pflan-
zengruppen bzw. -systemen verschiedene Wege eingeschlagen worden. Dies
wird bei der Besprechung des Geotropismus noch deutlicher zutage treten.

4. Geotropismus

Seit vielen Jahrzehnten sind wir gewohnt, den ursprünglichen, sehr begrenzt ver-
wendbaren Begriff "Heliotropismus" durch den allgemeineren "Phototropismus" zu
ersetzen. Eine lange fällige Konsequenz ist der Ersatz des "Geotropismus" durch
den neu vorgeschlagenen Begriff "Gravitropismus" (SIEVERS und VOLKMANN).

Das *Chara*-Rhizoid gilt als Musterbeispiel für eine scheinbar lückenlos
aufgeklärte gravitropische Reaktionskette. Trotzdem bleiben auch hier
noch Fragen offen, auf die in dem kurzen Artikel von SIEVERS und VOLK-
MANN hingewiesen wird. Insbesondere ist nach wie vor unsicher, ob die
aus $BaSO_4$ bestehenden Statolithen wirklich nur den Weg für die Golgi-
Vesikel versperren; einige Beobachtungen deuten sehr darauf hin, daß
auch der Druck auf das Plasmalemma eine entscheidende Rolle spielt,
und damit würde dieses System wieder ganz in die Nähe der Graviperzeption Höherer Pflanzen rücken.

Entscheidende Fortschritte wurden bei der Analyse der Graviperzeption
der Wurzeln Höherer Pflanzen erzielt (VOLKMANN und SIEVERS). Die Statocyten in der Wurzelhaube sind durch ihren Bau, ihre Orientierung
und die Verteilung der Organelle und Kompartimente für ihre Funktion
optimal organisiert, und die Verteilung der Zellbestandteile wird durch
eine sehr stabile Polarität der Zelle bestimmt. Entscheidend für die
Graviperzeption scheint der Druck der Statolithen (d.h. der Amylopla-
sten) auf einen distalen Stapel von ER-Membranen zu sein, und unter
geeigneten Bedingungen kann dieser Druck durch eine Deformation des
"Kissens" sogar sichtbaren Ausdruck gewinnen. Die spezifische schräge
Orientierung der lateralen Statocyten führt dazu, daß in gravitropischer Reizlage der Druck auf die ER-Kissen in beiden Wurzelhälften
asymmetrisch verteilt ist.

Alternativ zur Druckwirkung der Statolithen auf den "Sensor" wurde
diskutiert, daß die Neuverteilung der Statolithen in der Zelle oder
das Entlanggleiten auf der Unterlage während der Verlagerung das we-
sentliche Ereignis für die Graviperzeption ist. Theoretische Überlegungen machen diese Alternativen jedoch höchst unwahrscheinlich; im
Schwellenbereich verlagert sich ein Statolith nur um 1/15 seines ei-
genen Durchmessers. Demgegenüber liegt der durch einen Statolithen
ausgeübte Druck in der Größenordnung thigmischer Schwellenreize.

Eine Stütze finden die Folgerungen durch die Ergebnisse grober Eingriffe:
Durch Anwendung elektrischer Felder oder durch geeignetes Zentrifugieren kann
die polare Struktur der Statocyten zerstört werden. Die gravitropische Reak-
tionsfähigkeit geht gleichzeitig verloren. Sie kehrt aber synchron mit der Wie-
derherstellung der Statocytenstruktur zurück.

Ungeklärt sind noch die Folgeprozesse, die durch den Druck der Statolithen auf das ER-Kissen in Gang gesetzt werden. Einerseits kann an
Änderungen von Enzymaktivitäten gedacht werden; hierfür gibt es pas-
sende Beispiele, und im Hinblick auf den Quertransport von Phytohor-
monen (siehe unten) ist diese Vorstellung verlockend. Andererseits
muß auch an elektrostatische Wirkungen zwischen den zusammengedrück-
ten Membranen des ER-Kissens gedacht werden, gerade auch im Hinblick
auf die Bedeutung bioelektrischer Phänomene bei der Polaritätsinduktion (siehe oben). Beide Vorstellungen schließen sich natürlich nicht
gegenseitig aus.

Sehr viel weniger klar sind die Verhältnisse in Statocyten negativ
gravitropischer oberirdischer Organe. Hier sind bis jetzt in keinem
Fall ER-Kissen als Sedimentationsunterlage für die Statolithen gefunden worden; man schreibt deshalb hier dem Plasmalemma die Funktion
des Gravisensors zu. Damit wäre über die Druckempfindlichkeit von Bio-
membranen doch noch eine gewisse Einheitlichkeit in der Funktion zwischen Wurzel- und Sproßstatocyten gewährleistet.

> Wichtig ist der Hinweis, daß es in diesen Zellen wegen der großen zentralen
> Vakuolen keine Zufallsverteilung der Statolithen über das gesamte Zellumen
> geben kann. Modelle, die von einer solchen Verteilung ausgehen und die Stato-
> lithen nicht auf den cytoplasmatischen Wandbelag beschränken, sind demzufolge
> unbrauchbar, und manche sonst gut durchdachten theoretischen Überlegungen in
> früheren Arbeiten müssen daraufhin in ihrer Aussagekraft relativiert werden.

Die Analyse der hormonellen Steuerung des Gravitropismus (WILKINS)
hat in den letzten Jahren zu überraschenden Ergebnissen geführt. Einerseits konnte die ausschlaggebende Bedeutung des Auxins für den
Gravitropismus der Koleoptilen weiter bestätigt werden - wobei ebenso
wie beim Phototropismus immer noch offen bleibt, ob die eindeutig
nachgewiesene Querverschiebung des Auxins die einzige Komponente für
die Auxinasymmetrie ist -, andererseits ist jetzt größte Vorsicht bei
Verallgemeinerungen geboten. Bei Hypo- oder Epikotylen von Dicotylen-
Keimlingen (jedenfalls bei den untersuchten Objekten *Helianthus* und
Phaseolus) spielt Auxin höchstens eine untergeordnete Rolle, während
eine Steuerung der Krümmungsbewegung durch Gibberellin an Wahrschein-
lichkeit gewinnt. Möglicherweise gibt es sogar qualitative Unterschiede zwischen etiolierten und grünen Keimlingen. Darüberhinaus ist bei
diesen Sproßorganen das Problem aktuell, ob eine hormonelle Steuerung
überhaupt sehr wichtig ist; denn Perzeptions- und Reaktionsort scheinen hier übereinzustimmen, die Epidermiszellen sind für das Reaktions-
geschehen von ausschlaggebender Bedeutung. Damit könnte die Notwendigkeit radialer, tangentialer und longitudinaler Informationsweitergabe
entfallen.

> Extrem ist dieses Fehlen eines Signaltransports beim Gravitropismus der Gras-
> knoten ausgebildet. Hier reagieren diejenigen Epidermiszellen mit Neuaufnahme
> des Wachstums, die in Horizontallage auf der Unterseite des Knotens sind, und
> in diesen Zellen hat die Schwerkraft offenbar unmittelbaren Einfluß auf Enzym-
> aktivitäten und damit auf den Stoffwechsel.

Der durchgreifendste Erfolg dürfte bei der gravitropischen Reaktionskette der Wurzel erreicht worden sein (als Ergänzung zu den großen
Fortschritten im Perzeptionsvorgang, siehe oben). Das steuernde Hormon
ist ein in der Wurzelhaube entstehender Inhibitor, der in gravitropischer Reizlage in der Wurzelhaube quer verschoben wird und der basal-
wärts zur Wachstumszone transportiert wird. Abscisinsäure (ABA) wird
in der Wurzelhaube gebildet, kann bei einseitiger Applikation zur
Krümmung führen und dürfte daher der genuine Inhibitor sein. Es fehlt
nur noch der letzte Beweis, daß ABA in der Wurzelhaube dem Quertrans-

port unterliegt und in der Wachstumszone einer "gereizten" Wurzel un-
symmetrisch verteilt ist.

Die hormonelle Steuerung des Gravitropismus verspricht somit zu einem
Musterbeispiel dafür zu werden, wie ein und dieselbe Reaktion in verschiedenen Objekten sich völlig verschiedener Mechanismen bedient. Auf
einer höheren Ebene kann später allerdings vielleicht doch wieder eine
gewisse Relativierung dieser Mannigfaltigkeit erreicht werden: In den
gibberellingesteuerten Dicotylen-Keimlingen ist eine gewisse zusätzliche Beteiligung des Auxins nicht auszuschließen, und bei den auxin-
gesteuerten Koleoptilen scheint auch dem Gibberellin eine ergänzende
Funktion zuzukommen.

Abschließend soll noch auf eine wichtige Voraussetzung für weitere
Fortschritte hingewiesen werden, die für alle hormongesteuerten Krüm-
mungsbewegungen gilt: Die früher fast ausschließlich und teilweise
mit beachtlichen Erfolgen verwendete Methode, die interessierenden
Substanzen in Agarblöcken abzufangen, kann immer nur eine sehr be-
grenzte Aussagekraft haben, und die Analyse der Hormone *in situ* ist
entscheidend an den neuen Ergebnissen beteiligt und für weitere Forschungen ein unverzichtbares Desiderat.

5. Circumnutation

In der Vergangenheit ist die Frage nach endogener oder exogener Steuerung der
Circumnutation allzu grundsätzlich gestellt worden, und die Detailanalyse hat
hierunter möglicherweise gelitten. JOHNSSON befaßt sich überwiegend mit dem
theoretischen Hintergrund dieser Alternative, und aus dem "entweder-oder" wird
wahrscheinlich ein "sowohl-als auch", wobei der Anteil endogener und exogener
Komponenten von Objekt zu Objekt schwankt. Wichtig erscheinen vor allem die
sorgfältigen Überlegungen, welche Aussagekraft Versuchen am Klinostaten und in
der Schwerelosigkeit (Biosatellit) zukommt.

Die theoretischen Betrachtungen, wie durch Koppelung oszillierender oder nicht-
oszillierender Systeme im Organ schließlich die geforderten und beobachteten
Rhythmen entstehen können, sind noch nicht als "Fortschritte" im Verständnis
der Circumnutation zu werten, sondern als wichtige Grundlage für die weitere
Analyse.

6. Epinastie

Die Epinastie (und ihre Umkehr, die Hyponastie) ist ein weiteres Beispiel für die Mannigfaltigkeit von Regulationsmechanismen bei Reaktio-
nen, die wir mit einem gemeinsamen Terminus belegen (KANG). Man kann
sich deshalb sogar fragen, ob dieser gemeinsame Terminus überhaupt
gerechtfertigt ist. Beispielhaft werden drei Fälle von Epinastie be-
handelt, und in jedem Fall kann nur die Annahme einer Interaktion
mehrerer Phytohormone zu einem Verständnis der Reaktion führen. Aber
in jedem dieser Fälle verliert auch die Frage "endogene oder exogene
Steuerung" an Bedeutung, da nur auf der Basis des Zusammenwirkens in-
nerer und äußerer Regulationsfaktoren die epinastischen (und hyponasti-
schen) Krümmungen verstanden werden können.

Die Epinastie der Blattstiele kann mit einem sehr komplizierten Zusam-
menwirken von Auxin und Äthylen erklärt werden, wobei die beiden Hormone teilweise synergistisch, teilweise antagonistisch wirken. Dieses

Zusammenspiel ist trotz intensiver Bemühungen (insbesondere des Autors selbst) immer noch nicht in allen Details verstanden. Insbesondere macht nach wie vor die differentielle Wirkung in Ober- und Unterseite des Blattstieles Schwierigkeiten - aber eine außerordentlich stimulierende Hypothese des Autors wird vorgestellt, die Schwerkraftwirkungen, gegensätzliche Effekte einer "Überstimulation", Adaptation und hormonelle Wirkungen auf die Adaptation kombiniert. Für jeden in der Hypothese geforderten Teilprozeß lassen sich Beispiele aus anderen Bereichen der Bewegungs- oder Entwicklungsphysiologie anführen. Eine Konkretisierung dieser Hypothese könnte das Verständnis des Regulationsgeschehens entscheidend weiterbringen.

Die Bildung (Schließen) oder Öffnung des Hypokotyl- oder Epikotylhakens von Dicotylenkeimlingen erfordert ebenfalls eine (überwiegend antagonistische) Wechselwirkung von Auxin und Äthylen, aber möglicherweise spielen hier Gibberellin und Cytokinin als zusätzliche Regulatoren ebenfalls eine Rolle. Die bei Hakenbildung und Hakenöffnung beteiligte Lichtwirkung kann über eine Änderung des Hormonverhältnisses verstanden werden. Daneben ist jedoch auch eine so schnelle Weitergabe eines im Phytochrom perzipierten Lichtsignals im Organ zu beobachten, daß man an einen nicht-chemischen Signaltransfer denken muß.

Völlig abweichende Regulation finden wir bei der Epinastie der Grasblätter (Entrollen), die auf einer Erhöhung der Wanddehnbarkeit in den Mesophyllzellen basiert. Hier ist Gibberellin der entscheidende Regulator, dessen Synthese oder Abgabe aus einem Reservoir durch Licht (über Phytochrom) gesteuert wird, und Abscisinsäure wirkt als Antagonist.

Man kann aus diesen Ergebnissen den Eindruck gewinnen, daß die Pflanze nur dort mit Regulation durch ein einzelnes Hormon arbeitet, wo ein Außenfaktor ganz überwiegend für die Steuerung einer Reaktion verantwortlich ist. Wo dies nicht der Fall ist, wird die Wechselwirkung mehrerer Phytohormone in Anspruch genommen, die naturgemäß eine besonders differenzierte Regulation erlaubt.

Literatur

DENNISON, D.S.: Phototropism, 506-566.
HÄDER, D.-P.: Photomovement, 268-309.
JOHNSSON, A.: Circumnutation, 627-646.
KANG, B.G.: Epinasty, 647-667.
SIEVERS, A., VOLKMANN, D.: Gravitropism in Single Cells, 567-572.
VOLKMANN, D., SIEVERS, A.: Graviperception in Multicellular Organs, 573-600.
WEISENSEEL, M.H.: Induction of Polarity, 485-505.
WILKINS, M.B.: Growth-Control Mechanisms in Gravitropism, 601-626.

Professor Dr. WOLFGANG HAUPT
Botanisches Institut
Schloßgarten 4
D 8520 Erlangen

C. Genetics

I. Replication

Replication of the Eukaryotic Chromosome

By Walter Nagl

During the last two years several informative reviews on DNA replication and chromatin biosynthesis have been published. DNA replication was reviewed by ALBERTS and STERNGLANZ (1977), KORNBERG (1977), HAND (1978), LAWRENCE (1978), REICHARD (1978), SHEININ et al. (1978), and WINTERSBERGER (1978); the mechanisms of action of inhibitors of DNA synthesis were discussed by COZZARELLI (1977), the incorporation of uracil into animal cell DNA by GRAFSTROM et al. (1978; see also WIST et al., 1978). Cell cycle aspects of DNA replication were reviewed by NAGL (1977b) and analyzed in a monograph edited by JETER et al. (1978). BUSCH (1978) edited new volumes of *The Cell Nucleus* dealing - among other topics - with chromatin reduplication. Chromatin organization and synthesis was also discussed in volume 42 of Cold Spring Harbor Symposia on Quantitative Biology (1977).

The *main topics* of this review will be the *initiation of DNA replication* at the origins, and the *unit of replication (the replicon) in plants*. Some selected aspects of the pattern of DNA replication, the regulation of replication, and differential DNA replication will be briefly discussed.

1. Origin and Initiation of DNA Replication

The present-day understanding of the initiation of DNA replication was mainly developed from studies of prokaryotic and viral systems. The detection of several similar proteins, however, which are required for replication in prokaryotic and eukaryotic systems, as well as the universality of DNA structure and function, allow the conclusion that eukaryotic DNA is replicated, in principle, as viral and bacterial DNA, although the system may be complicated by the packaging of the DNA into chromatin.

The DNA sequence organization in the vicinity of the origin of DNA replication was analyzed in several phages and in *Escherichia coli*. There is now overwhelming evidence that this region always contains one or more reverse repeats (= fold-back sequences = palindromes; Fig. 1), which in part are homologous in all species studied (e.g., BASTIA, 1977; RAVETCH et al., 1977; DHAR et al., 1978; GRAY et al., 1978; SOEDA et al., 1978; MESSER et al., 1979).

DNA is replicated by means of a multienzyme complex, sometimes called a "replication apparatus". Replication is more complicated than suggested a few years ago and depends on a relatively large number of proteins (in *E. coli* a minimum of 13 gene functions is required for DNA replication). Nucleoside triphosphate hydrolysis energy is used at several discrete steps. Much of that unexpected complexity of DNA replication may arise from the need for extreme copying fidelity (the mistakes are in the order of one per 10^6-10^9 base pairs replicated, although replication proceeds with the incredible speed of 10^3 nucleotides per s). The replication apparatus is held together by relatively

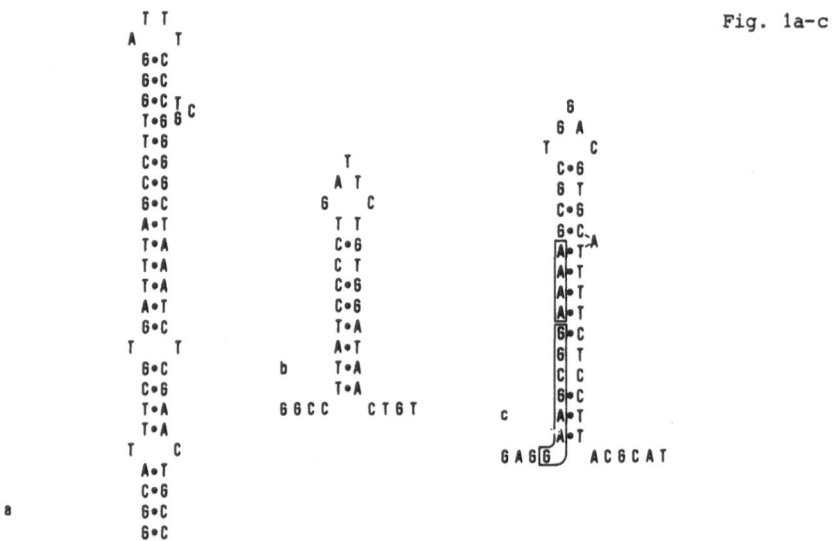

Fig. 1a-c

Fig. 1a-e. Potential secondary structures (hairpin models) of the palindromes present in the DNA sequences of the replication origin in (a) bacteriophage f1 (RAVETCH
et al., 1977), (b) and (c) the colicinogenic plasmid Col E₁ (BASTIA, 1977), (d)
human papovavirus BKV and (e) Simian virus SV 40 (DHAR et al., 1978). Note the similarities in all origins sequenced

weak protein-protein interactions and lacks a stable part like the
ribosome in the translational complex, so that only smaller individual protein components could be isolated so far. Moreover, different
phages, bacteria, and eukaryotes evidently replicate DNA by several
enzymatic pathways, which differ in the mechanism by which they initiate and elongate DNA chains. Thus, the following description is
still hypothetical and only shows the principal steps of replication.
Moreover, as proteins with comparable function have been described
under different terms, there exists a rather confusing terminology,
and several names must be indicated for each protein type mentioned
in this article (for reviews see JOVIN, 1976; CHAMPOUX, 1978a; WICKNER,
1978).

Naturally occurring DNA is in most cases supercoiled. The superhelical
turns (tertiary structure) in linear eukaryotic DNA may result entirely from the association of the DNA with histones to form nucleosomes,
while superhelical turns in intact closed DNA from bacteria is the
result of the action of DNA gyrase (see below). In viral DNA, the nu-
cleosome-induced turns are independent of the intrinsically superhelical state. Superhelical turns are, in addition, a result of the
unwinding of the double helix during replication: The high speed revo-
lution in front of the replication fork causes, particularly in closed
circular molecules and in locally fixed DNA molecules, a torque. Thus
one step in DNA replication is the removal of tertiary structures. Different enzymes that relax the rotational pressure by nicking one strand
ahead of the fork and sealing the nick after relaxing have been found
in many prokaryotic *and eukaryotic* systems. It is evident that they have
multiple roles and therefore, multiple names such as nicking-closing
enzyme, DNA unwinding protein (not to confuse with the DNA unwinding
enzyme), DNA binding protein, ω protein, swivelase, untwisting enzyme,
helix destabilizing protein (HD-protein), DNA relaxing enzyme, and

Fig. 1d,e

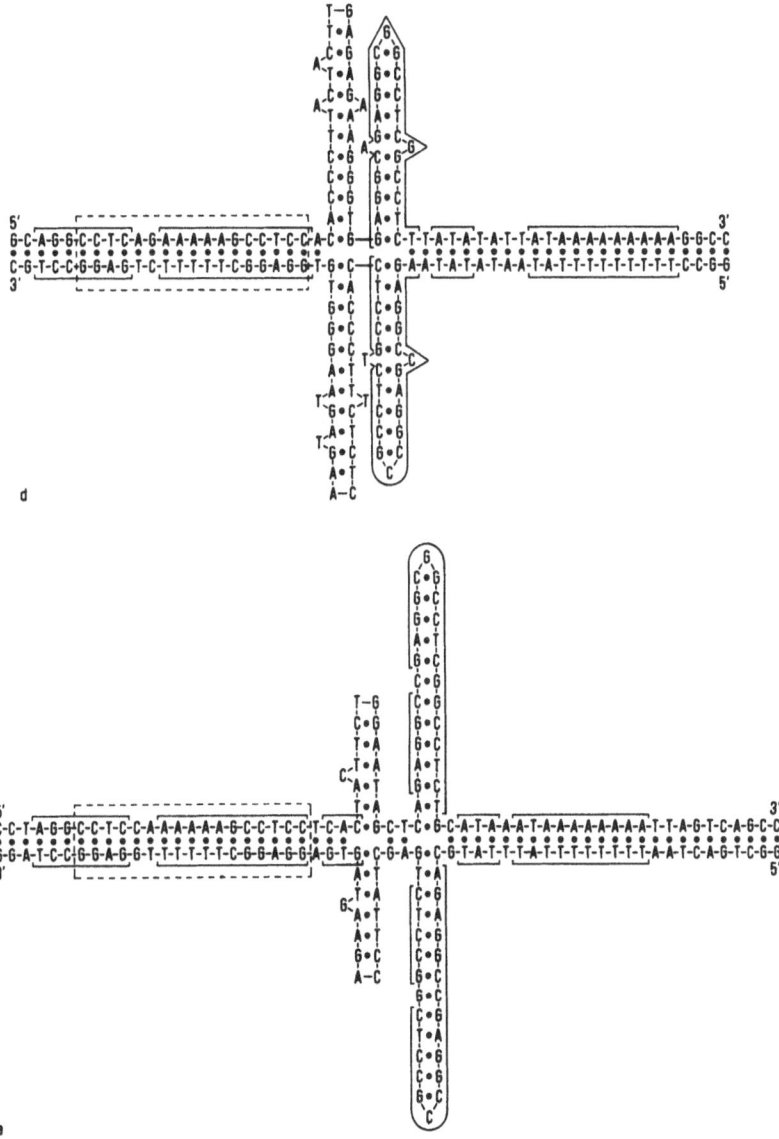

DNA topoisomerase. This type of protein introduces a transient single-strand break into duplex DNA and thereby provides a swivel for unwinding of superhelical coils. It becomes attached to the 3'-phosphate terminus of the nick (Fig. 2a); and this kind of linkage probably conserves the energy required for resealing the single-strand break, so that ATP is not necessary for the closing reaction. The protein further protects exposed single strands from nuclease attack, it binds (in vitro) particularly to single-stranded DNA at AT-rich regions (already two unpaired nucleotides are recognized by some of the DNA binding proteins).

Thus the protein contributes to the process of strand separation by modifying the conformation of DNA just by binding, so that the helix is destabilized, the melting temperature is lowered, and the palindromic sequences of the origin may be brought into the hair-pin form.

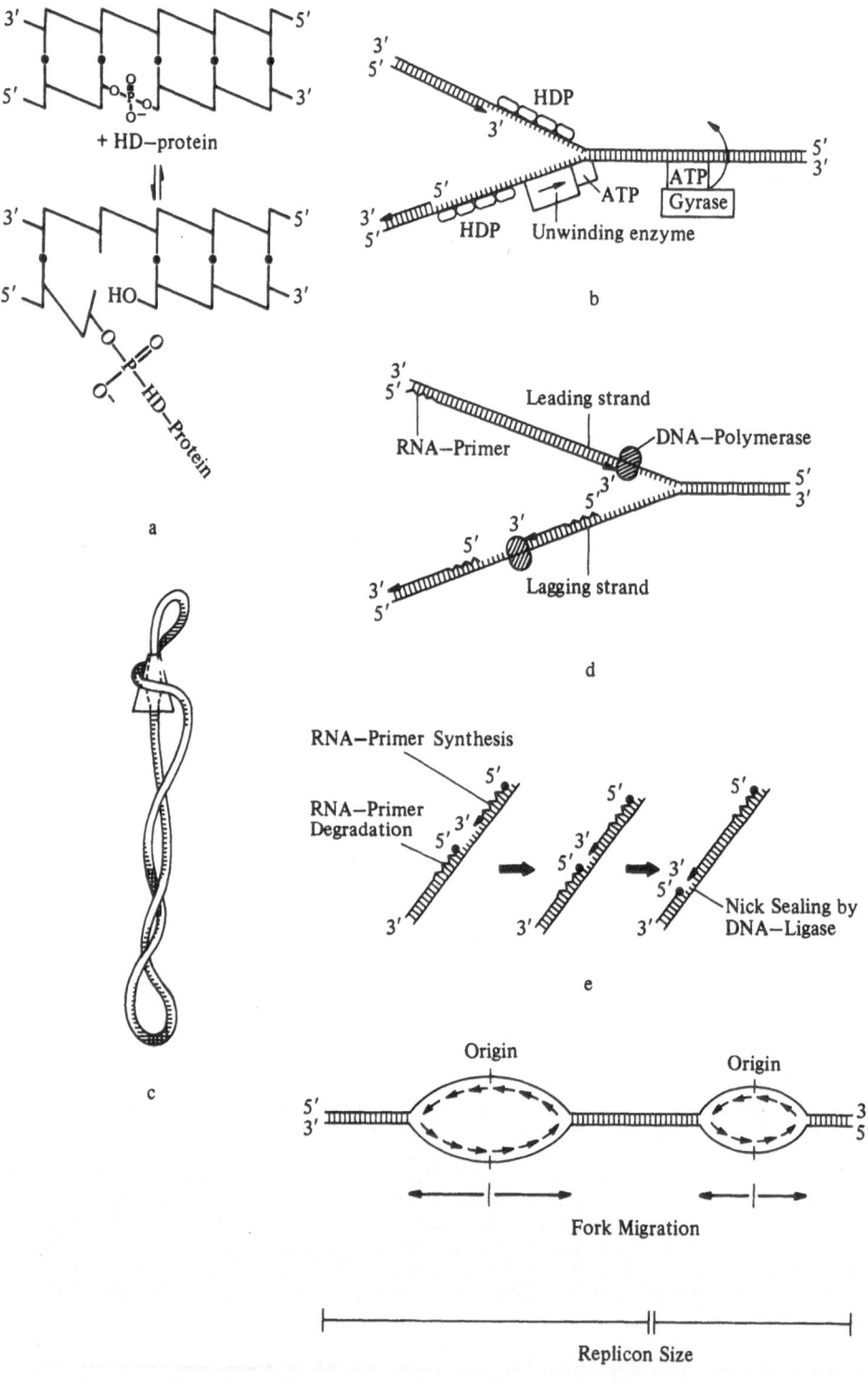

a

b

c

d

e

f

The destabilizing proteins, some of which also melt a short stretch
of the DNA, stimulate in addition DNA polymerase.

A further type of protein, the DNA unwinding or melting enzyme, can
be referred to as a single-stranded DNA-dependent ATPase; in meiotic
cells of *Lilium* this protein type was named U-protein (HOTTA and STERN,
1978). Evidently this protein alters the structure of DNA *enzymatically*.
Actually, this enzyme seems to perform the mechanical unwinding and
melting of the double helix by opening base pairs, and it may also
contribute to the removal of the resulting superhelical turns. It derives the energy for catalyzing strand separation from the hydrolysis
of ATP. As this enzyme unwinds the DNA in 5'-3'-direction, its leading
strand is the discontinuously replicated strand or lagging strand
(Fig. 2b).

In *E. coli*, an antagonistic enzyme was found and named DNA gyrase, which increases the superhelical density of covalently closed duplex DNA. Gyrase
couples the hydrolysis of ATP to the induction of negative superhelical turns
into a closed circular DNA molecule by transient nicking the DNA and promoting
unwinding of the helix during the life-time of the nick (in the absence of
ATP, gyrase acts like a swivelase). Gyrase could thus drive the replication
fork by unwinding the parental strands ahead of the fork and could reduce the
positive supercoil strain generated by unwinding of parental strands as they
are replicated (Fig. 2c). All the mechanisms of DNA priming and elongation
dependent on single-stranded DNA (given in many viruses) may resemble those
operating to replicate double-stranded DNA templates after the strands of the
double helix have been separated.

Details and references are given in the following reviews and papers:
JOVIN (1976), ABDEL-MONEM et al. (1977a,b,c), CHAMPOUX (1978a,b),
KOHWI-SHIGEMATSU et al. (1978), LE BON et al. (1978), LIU and WANG
(1978a,b), WICKNER (1978).

As all known DNA polymerases can only catalyze the extension of a nu-
cleotid strand at a free 3'-OH end, an important step of DNA replication is priming. For this, one of the palindrome hair-pins possesses

◀ Fig. 2a-f. Initiation and elongation of DNA chains. (a) Binding of a HD- (helix
destabilizing) protein (= nicking-closing enzyme) to the 3'-phosphorous terminus
of a DNA chain causing a nick and protecting the broken chain against nucleases;
partial melting may also occur (after CHAMPOUX, 1978b). (b) Possible energetic
contributions to helix unwinding: HDP (helix destabilizing protein) is firmly bound
to single-stranded DNA inhibiting renaturation; the arrows indicate the expected
directionality of the motion generated by DNA unwinding enzyme and by DNA gyrase,
respectively, due to conformational changes occurring when bound ATP is hydrolyzed
by these proteins (after ALBERTS and STERNGLANZ, 1977). (c) One of the models in
which negative supercoiling of the DNA is achieved by continued coiling of the
E. coli DNA, modulated by ATP hydrolysis, around a gyrase molecule (after LIU and
WANG, 1978a). (d) The replication fork during chain elongation; in this presentation
the synthesis of the leading (continuously replicated) strand and of the lagging
(discontinuously replicated) strand differs, but is carried out by the same type of
DNA polymerase (after ALBERTS and STERNGLANZ, 1977). (e) Okazaki piece priming and
joining: Each Okazaki piece is primed by RNA (probably at a palindromic sequence)
and elongated in 5'-3'-direction; after degradation of the primer RNA the gap is
filled by DNA polymerase and the nick sealed by DNA ligase (for filling the primer
gap at the 5'-end of linear DNA molecules see the palindromic models of CAVALIER-
SMITH, 1974, and DANCIS and HOLMQUIST, 1977). (f) Portion of a spread replicating
chromosomal DNA fiber displaying replication eyes (or bubbles). In this diagram
replication proceeds from the origin in each replicon into two directions, and both
strands are polymerized discontinuously. Note that fork migration is not identic
with the direction of replication in both strands

the recognition sequence for the primase, i.e., the RNA polymerase
which transcribes the primer RNA. The primer RNA and the one strand
of the hair-pin now form a stable DNA-RNA heteroduplex because the
hair-pin is not completely paired and displays single-stranded loops
(in *E. coli*, the primer is later removed by the exonuclease activity
of DNA polymerase I or, in other systems, by RNase H, which only di-
gests the RNA part in DNA/RNA hybrids). After initiation at the origin,
sequential DNA replication proceeds in one, or mostly two, directions.

The next step in DNA replication is the single-strand to double-strand
conversion, i.e., the replication or polymerization of the new strand
by DNA polymerase III. In eukaryotes, DNA polymerase-α seems to be the
replicase, but this is not yet uneqivocally proven (for details see
the reviews and original papers by WEISSBACH, 1977; WINTERSBERGER,
1977; BUTT et al., 1978; STEVENS and BRYANT, 1978; WAQUAR et al.,
1978). The extension of the new DNA strands occurs in 3'-5'-direction,
so that one strand has to be synthesized discontinuously in a direction that is contrary to fork migration (Okazaki pieces). There are
indications that also the other strand, which could be easily polymerized continuously, is actually synthesized dicsontinuously, Fig.
2f). MURAKAMI et al. (1977) suggested that any primer oligomere, which
surely is RNA also in eukaryotes (e.g., TSENG and GOULIAN, 1977), is
transcribed from a palindromic sequence. Thus, the high number of
palindromes in any DNA may be required in order to polymerize the dis-
continuous strand. It has also been found that the ends of Okazaki
pieces are nonrandomly distributed with respect to 6-methyl adenine
residues in the DNA (GÓMEZ-EICHELMANN and LARK, 1977) and that Okazaki
priming requires several proteins.

> The problem of telomere replication in linear DNA of eukaryotic chromosomes
> was again discussed. DANCIS and HOLMQUIST (1977) proposed a model basing on
> the ideas of CAVALIER-SMITH (1974), which requires transient telomere-telomere
> fusion along palindromic DNA sequences as a replication intermediate.

2. The Replicon

> After joining the Okazaki fragments, replicating eukaryotic DNA still appears
> in the form of fragments, the so-called intermediate fragments, which are the
> consequence of the multi-replicon nature of chromosomal DNA (Fig. 2f). Repli-
> cons could be directly visualized by electron microscopy (BALDARI et al.,
> 1978), but normally they are studied by DNA fiber autoradiography. Using the
> latter technique it was shown that DNA replication in the nucleus involves the
> temporally ordered replication of many subchromosomal DNA units (replicons),
> which often vary from 20 to 70 μm in length. When a particular replicon is
> activated, the synthesis of DNA usually begins in the middle of the unit and
> proceeds in both directions. The chain growth from this point in the overall
> 5'-3'-direction appears to proceed by the discontinuous formation of 5-6 S
> pieces (the above mentioned Okazaki pieces) that are ligated to form mature
> DNA. The opposing strand may proceed by the same mechanism or by direct ex-
> tension.

The molecular events which regulate the ordered initiation of specific segments, the chain growth rates, and ligation of adjacent replicons are as yet little understood (for references see PLANCK and
MUELLER, 1977). In soybean protoplasts and suspension-cultured cells,
about 5000 replicons of about 20 μm in size have been found to be active (CRESS et al., 1978). In animals and man, the size of the individual replicon varies between 30 and 400 μm, and the rate of repli-

cation is 0.2 to 1.2 µm/min/fork (e.g., YUROV, 1977; YUROV and LIAPU-
NOVA, 1977).

The portion of DNA chain replicated by an average replicon is often
completed within a fraction of the S period. Eukaryotic cells therefore may have several groups or families of replicons, the members
of which function more or less in unison at one time or another throughout S.

The distribution of replicon family members amongst the chromosomes is
unknown, but some observations show that adjacent replicons have a
high probability of initiating replication simultaneously. These and
other findings support the notion that replicon family members are
clustered along portions of the chromosome, and that they are activated and function in concert. PRESCOTT (1976) has suggested that mammalian cells may have an average of about 25 such families. VAN'T HOF
et al. (1978b) estimated that the angiosperm *Arabidopsis thaliana* may
have two replicon families that initiate DNA replication approximately 36 min apart. Individual replicons, on the average, function continuously throughout 74% of S in this species.

> The BrdU-Hoechst-Giemsa technique for examining the chronology and topology of
> the DNA replication in chromosomes of higher organisms allows comparison of DNA
> replication at the molecular and chromosome levels (PERRY and WOLFF, 1974; LATT,
> 1976, 1977). It was shown that the chromosome regions which are replicated
> simultaneously at the beginning, in the middle, and at the end of the S period
> correspond to R bands and positive G bands, respectively, of the metaphase
> chromosome. A human chromomere appears to contain 2 or 3 replicons, and 2 to
> 10 chomomeres may form a conglomeration observed as a prophase G band. Then
> the time necessary to complete G band replication will be determined by the
> replication time of the biggest replicon, i.e., 3-4 h (YUROV and LIAPUNOVA,
> 1977).

The total duration of the S period varies between different organisms
(it is positively correlated with the 2C DNA content), and even between different developmental stages and tissues. As the overall rate
of DNA replication on a chromosome is determined both by the number
of replicons active at any given time and by the actual rate of chain
elongation within individual replicons, it is not surprising that two
theories have been proposed in order to explain S time variation.

The one theory predicts that changes of S phase duration are caused
by changes in replicon *size*, because more or less initiation points
(origins) become activated. Evidence for this hypothesis was given
by CALLAN (e.g., 1976). TAYLOR (1977) studied CHO (Chinese hamster
ovary) cells under various conditions and estimated that there are
potential origins which are distributed in a regular repeating sequence along the DNA at about 12 kilobase intervals (= 4 µm), and
that the origins which are finally used under any experimental condition appear to be a random sample of the total potential origins.

The other hypothesis follows the observation that the *rate* of replication may change. VAN'T HOF et al. (1978a) found that the average
fork rate in root tip meristem cells of *Helianthus annuus* varies between
6 and 12 µm/h, dependent on the temperature (10° and 30°C, respectively). Similar results were found in cells of the fish *Pimphales promelas*
(fathead minnow), cultured at 14° and 34°C (GRABAR and FLICKINGER,
1977). Dependence of DNA synthesis period duration on the speed of
replication fork motion was also established in cell cultures of mammals and man (LIAPUNOVA and KHAITOVA, 1977). Variation in the rate of
DNA chain elongation was also found between SV40-transformed cells

exhibiting variable growth rates and residence in S phase, when cultured in the presence of different serum concentrations (VENKATESAN,
1977). In the case of Bloom's syndrome, the rate of replication fork
movement is retarded while replication length etc. are not altered
(HAND and GERMAN, 1977). However, the time and size parameters (9 μm/
h/fork, 18 μm replicon size) were similar in pea cells of various origin and under various culture conditions (VAN'T HOF and BJERKNESS,
1977).

As often, *both* theories may be true. ANANIEV et al. (1977) found that
in cultured cells of *Drosophila melanogaster* regulation of DNA synthesis
may be carried out by changing both DNA replication *rate* and replicon
size. GADDIPATI and SEN (1978) studied the genus *Vicia* by DNA fiber
autoradiography and detected that both the replicon size and the replication rate differ between species. In *V. faba* (2C = 22 pg) the mean
replicon size was 98 μm and the fork migration rate 30 μm/h, in *V. hirsuta*
(2C = 9 pg) the size was 79 μm and the rate 24 μm/h, and in *V. sativa*
(2C = 5 pg) the size was 53 μm and the rate 17 μm/h. From these data
it was calculated that each S phase nucleus of the three *Vicia* species
studied contains respectively about 16m, 6m, and 3m of DNA which is
divided into approximately 1,400,000, 72,000 and 59,000 replicons.

Equations for measuring the rate of DNA chain growth and replicon size by density labeling techniques were given by ROTI-ROTI and PAINTER (1977).

3. Pattern and Control of DNA Replication

These aspects of DNA replication will only briefly be discussed in
order to enable the reader to find the recent literature. The order-
ing of the initiation of DNA synthesis in specific replicons remains
unexplained. However, active replicons tend to be located in clusters
giving rise to distinctive labeling patterns in different chromosomes.
There have been several suggestions that the size of replication subunits (and thus the number of initiation points) may be regulated by
the organization of DNA into stoichiometric complexes with histones
to form nucleosomes and higher order structures of chromatin, and by
the interaction of initiation factors and DNA polymerases with these
complexes (BARLOW, 1972; KORNBERG and THOMAS, 1974; HEWISH, 1976;
BLUMENTHAL and CLARK, 1977). This aspect also involves the question
of chromatin biosynthesis, i.e., the structure of replicating chromatin
and the assembly of replicated DNA with histones. Chromatin biosynethis
will, however, not be discussed in detail, because no break-through
was made since the last review (Prog. Bot. 39, 132-152, 1977). It
seems that replicating DNA is less firmly wound around nucleosomes
with changed conformation, and that histone H1 is eliminated from the
replication fork; but it has not yet been discerned whether nucleosomes
are distributed conservatively, semi-conservatively or in a different
way (see, for instance, GRELLET et al., 1977; HEWISH, 1977; LEFRAK et
al., 1977; McKNIGHT and MILLER, 1977; AKIYOSHI and FUJII, 1978; HANCOCK,
1978; LEVY and JAKOB, 1978; RICHARDS et al., 1978; SEALE, 1978; WORCEL
et al., 1978). Also the question of whether heterochromatin is decon-
densed during DNA replication could not be decided unequivocally, but
it seems that DNA replication is possible in both condensed and decon-
densed heterochromatin (NAGL, 1977a; see also BARLOW, 1976, 1978; OONO
and HOTTA, 1976; DUTRILLAUX, 1977; HOLM, 1977; SPARVOLI et al., 1977;
SETTERFIELD et al., 1978).

SCHVARTZMAN and DIEZ (1977) studied late DNA replication in *Allium cepa* root
tips. Detailed maps of the replication patterns have been published for human
chromosomes (KONDRA and RAY, 1978), and for Chinese hamster chromosomes
(CAMPBELL and WORTON, 1977).

New evidence has been obtained that the sequence of replication is not irre-
versibly programmed into each chromosome. WILLARD (1977) found tissue-specific
heterogeneity in DNA replication patterns of human X chromosomes, FARBER and
DAVIDSON (1977) differences in the order of termination of DNA replication in
human chromosomes in peripheral blood lymphocytes and skin fibroblasts from
the same individual, and FARBER and DAVIDSON (1978) an altered pattern of
human chromosome replication in a human fibroblast-mouse cell hybrid.

An interesting enzyme that was recently detected is the Poly-(ADP-
ribosome)-polymerase of eukaryotic cells. This chromatin-bound enzyme
system transfers the adenosine diphosphate ribose moiety of NAD (nicotinamide adenine dinucleotide) to the nuclear proteins, adds succes-
sively ADP-ribose to form a homopolymer, and liberates nicotinamide.
There are several lines of evidence that ADP-ribosylation represents
an important step in histone cross-linking and thus chromatin condensation. In connection with DNA replication, contradictory findings
have been made: While some investigations led to the suggestion that
the formation of poly(ADP-ribose) suppresses the template activation
for DNA synthesis, an enhancement was observed in other cell systems
(for details see TANIGAWA et al., 1978). Further studies are necessary
to elucidate the relationship between ADP-ribosylation and DNA replication, but possibly some new aspects of regulation of DNA synthesis
can be found in this field.

4. Differential DNA Replication

In addition to the reports on differential DNA replication (i.e., DNA amplification and DNA underreplication), which have been reviewed already (Progr.
Bot. 37, 186-210, 1975; Progr. Bot. 39, 132-152, 1977; BUIATTI, 1977; NAGL,
1978a), many new findings have been published during recent years. In animals,
the somatic amplification of genes was found (ALT et al., 1978), as well as
that of repetitive DNA sequences (STROM et al., 1978). The DNA composition
of eukaryote nuclei is apparently variable, and may be correlated to differentiation from embryogenesis to senescence (e.g., KOVALEVA and RAIKOV, 1978;
KUENZLE et al., 1978; VIOLA-MAGNI et al., 1978).

In plants, numerous indications of differential DNA replication have
also been found (OLSZEWSKA, 1976; CREMONINI and CIONINI, 1977; DE
MARTINIS et al., 1977; DURANTE et al., 1977; KEOWN et al., 1977;
KESSLER and RECHES, 1977; NAGL, 1977c, 1978b,c; SHARMA, 1977; WINKLER,
1977; BROEKAERT and VAN PARIJS, 1978; SCHÄFER and NEUMANN, 1978;
SCHÄFER et al., 1978; BANERJEE and SHARMA, 1979; NAGL et al., 1979).
A model was proposed according to which differential DNA replication
and other somatic events which lead to nuclear DNA variation may be
predefined programs in order to substitute or improve phylogenetic
DNA changes (NAGL, 1977d,e, 1978a). However, many cases of differential DNA replication are not clearly characterized; particularly, they
must not be confused with the old - and probably incorrect - concept
of metabolic DNA (for a discussion see SCHEUERMANN, 1978).

The main problem in the understanding of differential DNA replication
lies in the fact that no method available so far can prove or disprove
the occurrence of the extra replication of part of the genome, or
underreplication during polyploidization. As a critical review on this

topic is to be published at a different place (NAGL, 1979), the reader
is referred to that article for further information.

References

ABDEL-MONEM, M., CHANAL, M.-C., HOFFMANN-BERLING, H.: Eur. J. Biochem.
79, 33-38 (1977a). - ABDEL-MONEM, M., DÜRWALD, H., HOFFMANN-BERLING,
H.: Eur. J. Biochem. 79, 39-45 (1977b). - ABDEL-MONEM, M., LAUPPE,
H.-F., KARTENBECK, J., DÜRWALD, H., HOFFMANN-BERLING, H.: J. Mol.
Biol. 110, 667-685 (1977c). - AKIYOSHI, H., FUJII, S.: J. Biochem. 84,
337-342 (1978). - ALBERTS, B., STERNGLANZ, R.: Nature (London) 269,
655-661 (1977). - ALT, F.W., KELLEMS, R.E., BERTINO, J.R., SCHIMKE,
R.T.: J. Biol. Chem. 253, 1357-1370 (1978). - ANANIEV, E.V., POLUKAROVA,
L.G., YUROV, Y.B.: Chromosoma 59, 259-272 (1977).

BALDARI, C.T., AMALDI, F., BUONGIORNO-NARDELLI, M.: Cell 15, 1095-1107
(1978). - BANERJEE, M., SHARMA, A.K.: Experientia 35, 42-43 (1979).
- BARLOW, P.: Cytobios 6, 55-80 (1972). - BARLOW, P.W.: Protoplasma 90,
381-392 (1976); - Nucleus 21, 1-11 (1978). - BASTIA, D.: Nucleic Acids
Res. 4, 3123-3142 (1977). - BLUMENTHAL, A.B., CLARK, E.J.: Exp. Cell
Res. 105, 15-26 (1977). - BROEKAERT, D., VAN PARIJS, R.: Z. Pflanzen-
physiol. 89, 169-184 (1978). - BUIATTI, M.: DNA amplification and
tissue cultures, 358-374. In: Plant, Cell and Tissue Cultures, eds.
J. REINERT, Y.P.S. BAJAJ. Berlin-Heidelberg-New York: Springer 1977.
- BUSCH, H.: The Cell Nucleus, Vols. IV (422 pp.), V (497 pp.).
New York: Academic Press 1978. - BUTT, T.R., WOOD, W.M., McKAY, E.L.,
ADAMS, R.L.P.: Biochem. J. 173, 309-314 (1978).

CALLAN, H.G.: Biol. Zentralbl. 95, 531-545 (1976). - CAMPBELL, C.E.,
WORTON, R.G.: Cytogen. Cell Genet. 19, 303-319 (1977). - CAVALIER-
SMITH, T.: Nature (London) 250, 467-470 (1974). - CHAMPOUX, J.J.:
Annu. Rev. Biochem. 47, 449-479 (1978a); - J. Mol. Biol. 118, 441-446
(1978b). - COZZARELLI, N.R.: Annu. Rev. Biochem. 46, 641-668 (1977).
- CREMONINI, R., CIONINI, P.G.: Protoplasma 91, 303-313 (1977). -
CRESS, D.E., JACKSON, P.J., KADOURI, A., CHU, Y.E., LARK, K.G.: Planta
143, 241-253 (1978).

DANCIS, B.M., HOLMQUIST, G.P.: Chromosomes Today 6, 95-104 (1977). -
De MARTINIS, P., BRUNORI, A., DEVREUX, M.: Z. Pflanzenphysiol. 84,
195-202 (1977). - DHAR, R., LAI, C.-J., KHOURY, G.: Cell 13, 345-358
(1978). - DURANTE, M., CREMONINI, R., BRUNORI, A., AVANZI, S.,
INNOCENTI, A.M.: Protoplasma 93, 289-303 (1977). - DUTRILLAUX, B.:
Humangenetik 35, 247-254 (1977).

FARBER, R.A., DAVIDSON, R.L.: Cytogen. Cell Genet. 18, 349-363 (1977);
- Proc. Natl. Acad. Sci. USA 75, 1470-1474 (1978).

GADDIPATI, J.P., SEN, S.K.: J. Cell Sci. 29, 85-91 (1978). - GÓMEZ-
EICHELMANN, M.C., LARK, K.G.: J. Mol. Biol. 117, 621-635 (1977). -
GRABAR, F., FLICKINGER, R.A.: Cell Tissue Res. 10, 505-507 (1977). -
GRAFSTROM, R.H., TSENG, B.Y., GOULIAN, M.: Cell 15, 131-140 (1978).
- GRAY, C.P., SOMMER, R., POLKE, C., BECK, E., SCHALLER, H.: Proc.
Natl. Acad. Sci. USA 75, 50-53 (1978). - GRELLET, F., DELSENY, M.,
GUITTON, Y.: Nature (London) 267, 724-726 (1977).

HANCOCK, R.: Proc. Natl. Acad. Sci. USA 75, 2130-2134 (1978). - HAND,
R.: Cell 15, 317-325 (1978). - HAND, R., GERMAN, J.: Humangenetik 38,
297-306 (1977). - HEWISH, D.R.: Nucleic Acids Res. 3, 69-78 (1976);

- Nucleic Acids Res. 4, 1881-1890 (1977). - HOLM, P.B.: Carlsberg Res.
Commun. 42, 249-281 (1977). - HOTTA, Y., STERN, H.: Biochemistry 17,
1872-1880 (1978).

JETER, J.R., Jr., CAMERON, I.L., PADILLA, G.M., ZIMMERMANN, A.M. (eds):
Cell Cycle Regulation. 266 pp. New York: Academic Press 1978. - JOVIN,
T.M.: Annu. Rev. Biochem. 45, 889-920 (1976).

KEOWN, A.C., TAIZ, L., JONES, R.L.: Am. J. Bot. 64, 1248-1253 (1977).-
KESSLER, B., RECHES, S.: Chromosomes Today 6, 237-246 (1977). - KOHWI-
SHIGEMATSU, T., ENOMOTO, T., YAMADA, M.-A., NAKANISHI, M., TSUBOI, M.:
Proc. Natl. Acad. Sci. USA 75, 4689-4693 (1978). - KONDRA, P.M., RAY,
M.: Humangenetik 43, 139-149 (1978). - KORNBERG, A.: Biochem. Soc.
Trans. 5, 359-374 (1977). - KORNBERG, R.D., THOMAS, J.O.: Science 184,
865-868 (1974). - KOVALEVA, V.G., RAIKOV, I.B.: Chromosoma 67, 177-
192 (1978). - KUENZLE, C.C., BREGNARD, A., HÜBSCHER, U.: Exp. Cell.
Res. 113, 151-160 (1978).

LATT, S.A.: Chromosomes Today 5, 367-394 (1976); - Can. J. Genet.
Cytol. 19, 603-623 (1977). - LAWRENCE, E.: Nature (London) 274, 210-
212 (1978). - LE BON, J.M., KADO, C.I., ROSENTHAL, L.J., CHIRKJIAN,
J.G.: Proc. Natl. Acad. Sci. USA 75, 4097-4101 (1978). - LEFFAK, I.M.,
GRAINGER, R., WEINTRAUB, H.: Cell 12, 837-845 (1977). - LEVY, A.,
JAKOB, K.M.: Cell 14, 259-267 (1978). - LIAPUNOVA, N.A., KHAITOVA,
N.M.: Dokl. Acad. Nauk SSR 236, 994-997 (1977). - LIU, L.F., WANG,
J.C.: Proc. Natl. Acad. Sci. USA 75, 2098-2102 (1978a); - Cell 15,
979-984 (1978b).

McKNIGHT, S.L., MILLER, O.L., Jr.: Cell 12, 795-804 (1977). - MESSER,
W., MEIJER, M., BERGMANS, H.E.N., BECK, E., SCHALLER, H., HANSEN,
F.G., von MEYENBURG, K.: Hoppe Seylers Z. Physiol. Chem. 360, 328
(1979). - MURAKAMI, H., TAIRA, S., MORI, H.: J. Theor. Biol. 68,
183-197 (1977).

NAGL, W.: Protoplasma 91, 389-407 (1977a); - Nuclear structures during
cell cycles, 147-193. In: Mechanisms and Control of Cell Division,
eds. T.L. ROST, E.M. GIFFORD, Jr. Stroudsburg, PA (USA): Dowden,
Hutchinson, Ross 1977b; - Experientia 33, 1040-1041 (1977c); - Nucleus
20, 10-27 (1977d); - Chromosomes Today 6, 151-152 (1977e); - Endopolyploidy and Polyteny in Differentiation and Evolution. Towards an
Understanding of Quantitative and Qualitative Variation of Nuclear DNA
in Ontogeny and Phylogeny. 283 pp. Amsterdam: North-Holland 1978a; -
Arch. Genet. 51, 20-21 (1978b); - Cell Chromosome Newslett. 1, 14-17
(1978c); - Z. Pflanzenphysiol., in press (1979). - NAGL, W., HEMLEBEN,
V., EHRENDORFER, F. (eds.): Genome and Chromatin. Organization, Function, Evolution. Vienna-New York: Springer 1979 (in press).

OLSZEWSKA, M.J.: Histochemistry 49, 157-175 (1976). - OONO, K., HOTTA,
Y.: Cell Struct. Funct. 1, 299-312 (1976).

PERRY, P., WOLFF, S.: Nature New Biol. 251, 156-158 (1974). - PLANCK,
S.R., MUELLER, G.C.: Biochemistry 16, 1808-1813 (1977). - PRESCOTT,
D.M.: Reproduction of Eukaryotic Cells. 301 pp. New York: Academic
Press 1976.

RAVETCH, J.V., HORIUCHI, K., ZINDER, N.D.: Proc. Natl. Acad. Sci. USA
74, 4219-4222 (1977). - REICHARD, P.: Fed. Proc. 37, 9-14 (1978). -
RICHARDS, B.M., PARDON, J.F., LILLEY, D.M.J., COTTER, R.I., WOOLEY,
J.C., WORCESTER, D.L.: Phil. Trans. R. Soc. London Sec. B 283, 287-289
(1978). - ROTI-ROTI, J.L., PAINTER, R.B.: J. Theor. Biol. 64, 681-696
(1977).

SCHÄFER, A., NEUMANN, K.-H.: Planta 143, 1-4 (1978). - SCHÄFER, A., BLASCHKE, J.R., NEUMANN, K.-H.: Planta 139, 97-101 (1978). - SCHEUER-MANN, W.: Cytobiologie 17, 232-245 (1978). - SCHVARTZMAN, J.B., DIEZ, J.L.: Cytobiologie 14, 310-318 (1977). - SEALE, R.L.: Proc. Natl. Acad. Sci. USA 75, 2717-2721 (1978). - SETTERFIELD, G., SHEININ, R., DARDICK, I., KISS, G., DUBSKY, M.: J. Cell Biol. 77, 246-263 (1978). - SHARMA, A.K.: Nucleus 20, 4-10 (1977). - SHEININ, R., HUMBERT, J., PEARLMAN, R.E.: Annu. Rev. Biochem. 47, 277-316 (1978). - SOEDA, E., KIMURA, G., MIURA, K.-I.: Proc. Natl. Acad. Sci. USA 75, 162-166 (1978). - SPARVOLI, E., GALLI, M.G., CHIATANTE, D., SGORBATI, S.: Exp. Cell Res. 110, 315-321 (1977). - STEVENS, C., BRYANT, J.A.: Planta 138, 127-132 (1978). - STROM, C.M., MOSCONA, M., DORFMAN, A.: Proc. Natl. Acad. Sci. USA 75, 4451-4454 (1978).

TANIGAWA, Y., KAWAMURA, M., KITAMURA, A.: Biochem. Biophys. Res. Commun. 81, 1278-1285 (1978). - TAYLOR, J.H.: Chromosoma 62, 291-300 (1977). - TSENG, B.Y., GOULIAN, M.: Cell 12, 483-489 (1977).

VAN'T HOF, J., BJERKNES, C.A.: Chromosoma 64, 287-294 (1977). - VAN'T HOF, J., BJERKNES, C.A., CLINTON, J.H.: Chromosoma 66, 161-171 (1978a). - VAN'T HOF, J., KUNIYUKI, A., BJERKNES, C.A.: Chromosoma 68, 269-285 (1978b). - VENKATESAN, N.: Biochim. Biophys. Acta 478, 454-460 (1977). - VIOLA-MAGNI, M.P., ROSSI, R., BIONDI, R., BENEDETTI, C.: Biochim. Biophys. Acta 520, 38-51 (1978).

WAQUAR, M.A., EVANS, M.J., HUBERMAN, J.A.: Nucleic Acids Res. 5, 1933-1946 (1978). - WEISSBACH, A.: Annu. Rev. Biochem. 46, 25-47 (1977). - WICKNER, S.H.: Annu. Rev. Biochem. 47, 1163-1191 (1978). - WILLARD, H.F.: Chromosoma 61, 61-73 (1977). - WINKLER, V.: Entwicklungsphysiologische Untersuchungen an der vielkernigen Grünalge Hydrodictyon reticulatum (L.) Lagerheim. Dissertation Universität Göttingen 1977. - WINTERSBERGER, E.: Trends Biochem. Sci. 2, 58-61 (1977); - Rev. Physiol. Biochem. Pharmacol. 84, 93-142 (1978). - WIST, E., UNHJEM, O., KROKAN, H.: Biochim. Biophys. Acta 520, 253-270 (1978). - WORCEL, A., HAN, S., WONG, M.L.: Cell 15, 969-977 (1978).

YUROV, YU.B.: Cell Differ. 6, 95-104 (1977). - YUROV, YU., LIAPUNOVA, N.A.: Chromosoma 60, 253-267 (1977).

Professor Dr. WALTER NAGL
Universität Kaiserslautern
Pfaffenbergstraße, Gebäude 13
D 6750 Kaiserslautern

II. Recombination

By Horst Binding and Reinhard Nehls

1. Introduction

The evolution of the sexual cycle has led to a sophisticated system
for the control of recombination which favors the interchange of genetic material in related and hence probably more harmonizing systems,
but prevents combinations of less compatible genomes. From the view-
point of a geneticist or plant breeder, this system has several dis-
advantages: (a) The range of sexual combinations is limited by homo-
genic and heterogenic incompatibility; (b) it is impossible to compile
the complete genetic information of heterozygous plants in the progeny
because gamete formation is preceded by meiosis in which reduction of
chromosome number and recombination takes place; (c) recombination is
restricted to chromosomal genes in most of the higher plants in which
cytoplasmic genophores of the male gamete are not transmitted to the
next generation. These are reasons why strong efforts are being made
to open up new ways to transfer genetic information in higher plants.

> Some aspects of gene transfer into higher plants by isolated DNA were briefly
> mentioned in Progress in Botany in 1977 (BINDING). In this issue (1979),
> HERZFELD discusses the most sophisticated intentions to mediate DNA incorpo-
> ration by plasmids and insertion factors. The integration of transplanted cell
> organelles has not unequivocally been achieved. Papers on this topic have been
> reviewed recently by GILES (1978).

Protoplast fusion is the most exploited tool to obtain asexual recombination in higher plants.

> BOPP has indicated this topic in this periodical already in 1975. A consider-
> able number of reviews have been written in the last few years on the regener-
> ation and fusion of plant protoplasts. The latest reviews have been presented
> at the Symposium on Plant Tissue Culture in Peking (COCKING, 1978a; GAMBORG et
> al., 1978a; MELCHERS, 1978b; KAO, 1978; McCOMB, 1978) and on the 4th Int. Cong.
> Plant Tissue and Cell Culture in Calgary (GLEBA, 1978b; CONSTABEL, 1978; COCKING,
> 1978b).

It will be attempted on the following pages to present some main features selected out of the wide-spread and partly prolifically written
information on protoplast fusion.

2. Regeneration of Plant Protoplasts

Preliminary experiments to cell fusion must be concerned with the elab-
oration of conditions for protoplast regeneration. The number of plant
species in which isolated protoplasts have been grown up to plants is
rather low. They belong mainly to the familiy of Solanaceae but also

to a few other taxa, summarized by SCHIEDER (1978c). In the economi-
cally important families of Fabaceae and Poaceae, for instance, positive results have not been achieved except for one report on brome
grass (KAO et al., 1973).

The regeneration of protoplasts depends on several factors: a) The
significance of the genetic constitution emerging from a number of
experiments.

   Isolated protoplasts of several dihaploid clones of potato, for example, responded very differently to the culture conditions (BINDING et al., 1978).

b) The physiological state of the plant material prior to protoplast
isolation is also rather critical. This is one reason why in vitro
cultures are often superior to greenhouse-grown plants.

   The importance of culture conditions on the yields of viable protoplasts has
   already been recognized in the course of investigations on the regeneration
   of isolated moss protoplasts (BINDING, 1966). Favorable conditions have later
   been evaluated for instance for greenhouse plants of tobacco (WATTS et al.,
   1974), for axenic shoot cultures of *Petunia* (BINDING, 1974d) and tobacco
   (BINDING, 1975) and for cell suspension cultures of soybean (KAO et al., 1970).

c) It is evident that the yields of viable protoplasts furthermore
depend on the conditions during the enzymic liberation of the protoplasts.

   Several enzyme preparations containing pectinase and cellulase activities have
   been used with varying success in different plant species (compiled by GAMBORG,
   1976). The significance of environmental factors such as light, temperature,
   plasmolyticum, and ions has been discussed by KAO (1978).

d) Obviously most critical are the culture conditions for the isolated
protoplasts. Plating densities, light regime, temperature, and especially the composition of the culture media are to be optimized for
each special protoplast system.

   Widely used culture media are NT medium (NAGATA and TAKEBE, 1970), medium B5
   (GAMBORG et al., 1968), DPD medium (DURAND et al., 1973), medium V-47 (BINDING,
   1974c) and medium KM 8p (KAO and MICHAYLUK, 1975), in several modifications.

Special arrangements must be made in fusion experiments in which the
surviving cells are left at low densities, namely if selective growth
conditions for the fusion products are applied.

   The regeneration of protoplasts at low densities has been obtained by using
   preconditioned media (KAO et al., 1970; BINDING, 1974d), feeder layers of inactivated protoplasts (RAVEH et al., 1973; RAVEH and GALUN, 1975), co-cultures
   (BINDING and NEHLS, 1978a,b) or highly qualified culture media (KAO and MICHAY-
   LUK, 1975).

Methods for the separate culture of single clones of fusion products
have been developed for cases in which the fusion products are easily
recognized only a few days after fusion and selective pressures cannot be applied.

   MENCZEL et al. (1978) cloned single fusion products which were able to form
   chloroplasts in suspensions of mutant white protoplasts as nurse cultures. KAO
   (1977) regenerated fusion products, each in a 25 µl droplet in KM 8p, a culture
   medium which is suited for the culture of protoplasts at low densities by the
   content of diverse organic supplements. GLEBA and HOFFMANN (1978) achieved a
   self-conditioning effect by using 1 µl droplets for each clone.

3. Subprotoplasts (Miniprotoplasts)

A subprotoplast is a protoplastlike structure but represents less than the complete content of a whole cell. Those of the subprotoplasts which lack at least one organelle type can be taken as carriers of selected organelles. Hence they are especially interesting for fusion experiments. They readily fuse to each other and to protoplasts under fusion-inducing conditions (BINDING, 1976; WALLIN et al., 1979).

> Subprotoplasts are formed (1) during plasmolysis, (2) during ripening of berries of some plant species, (3) during protoplast culture by buddying, (4) during incubation for protoplast fusion, etc. (cf. BINDING and KOLLMANN, 1976; BINDING, 1979). Recently, WALLIN et al. (1978) reported on the separation of subprotoplasts from protoplasts by the action of cytochalasin B and centrifugation. These subprotoplasts consisted of the nucleus surrounded by a small portion of cytoplasm and have been called miniprotoplasts.

> A reduced number of genetically active organelle types in the protoplasts has been obtained by ZELCER et al. (1978). The authors inactivated the nuclei of isolated protoplasts by X-rays and fused these protoplasts with untreated ones.

4. Protoplast Fusion

BOPP (1975) already mentioned the most effective fusion method which is based on the action of polyethylene glycol (PEG; KAO and MICHAYLUK, 1974; WALLIN et al., 1974), Ca^{2+} and high pH (KELLER and MELCHERS, 1973) in several modifications (BINDING, 1974b; SCHIEDER, 1974; and others). Using this method, plant protoplasts can be fused not only to one another but also to animal and human cells (JONES et al., 1976; WILLIS et al., 1977; DAVEY et al., 1978).

> Higher temperatures (BURGESS and FLEMING, 1974) and dimethylsulphoxide (HAYDU et al, 1977) were found to be suited to increase the effect of PEG. WEBER et al. (1976) focused attention on the fact that cell wall regeneration, which starts immediately after the removal of the hydrolytic enzymes, causes a fast drop of the frequencies of protoplast fusion.

> Agglutination of isolated protoplasts can be induced by a number of reagents besides PEG, as for instance concanavalin A (GLIMELIUS et al., 1974; CHIN and SCOTT, 1979), phytohemagglutinins (CHIN and SCOTT, 1979), antibodies (HARTMANN et al., 1973; BURGESS and FLEMING, 1974; LARKIN, 1977) and polyvinylalcohol (NAGATA, 1978).

The low tendency of isolated protoplasts to fuse under normal physiological conditions is due to the charges of the plasma membrane.

> This was already suggested by the observation that moss protoplasts readily fused in an electric field in sea water (BINDING, 1966). GROUT et al. (1972) demonstrated the negative surface charges of protoplasts by electrophoresis. The agents mentioned above which cause agglutination and fusion depolarize the membranes or even reverse the ζ-potentials (GROUT and COUTTS, 1974; NAGATA and MELCHERS, 1978). NAGATA and MELCHES could show that negative ζ-potentials are mainly due to phosphate groups which can be removed by acid phosphatase. NAGATA et al. (mentioned in MELCHERS, 1977) devised a method for the durable reversion of the protoplast charges by the addition of special positively charged phospholipids. This method makes possible the controlled induction of heteroplasmic fusion.

Electron microscopic investigations revealed some detailed information on the fusion process (recent review: FOWKE, 1978). The micrographs suggest that cell fusion is a gradual process beginning with a close contact of the plasmalemmas, followed by small cytoplasmic bridges which extend and fuse especially after the dilution of the PEG (WITHERS, 1973; BURGESS and FLEMING, 1974; WALLIN et al., 1974; FOWKE et al., 1975, 1976, 1977).

5. Markers and Techniques for Recognition, Isolation, Selection and Characterization of Fusion Products

Incubation of two different types of protoplast under fusion-inducing conditions usually results in protoplast populations which are hetero-karyotic at about 10%-20%. Only a few of the fusion bodies give rise to cell colonies. This demonstrates the advantage of markers to get hold of the fusion products. Furthermore, it is needed to analyze the cell clones or plants grown from the recovered fusion products with respect to their genetic constitution. As will be discussed later, the regenerated clones only rarely carry the complete karyotic and extra-karyotic genetic information of both partners. They may be chimeric resulting from the failure or delay of nuclear fusion, from organelle segregation or loss and rearrangements of chromosomes.

> Reviews on appropriate selection systems have recently been written by COCKING (1978b) and SCHIEDER (1978c).

Pairs of markers are easily available in combinations of unlikely dif-ferentiated protoplasts. They are useful for the recognition of recent fusion products and for the enrichment of fusion bodies by centrifuga-tion or manual selection. Unfortunately they disappear in the course of the early development of the fusion products.

> Differences in plastids of cells from suspension cultures and mesophyll were utilized by many authors (for example: KAO et al., 1974; KARTHA et al., 1974; CONSTABEL et al., 1975b, 1976; DUDITS et al., 1976; GOSCH and REINERT, 1976, 1978; KAO, 1976, 1977; REINERT and GOSCH, 1976; FOWKE et al., 1977; GLEBA and HOFFMANN, 1978; MELCHERS et al., 1978). In further experiments, mesophyll chloroplasts were paired to proplastids from shoot primordia (BINDING and NEHLS, 1978b) and to chromoplasts of fruit pericarp (BINDING, 1976). Pigmenta-tion of vacuoles of one partner has also been appropriate for the recognition of fusion products (POTRYKUS, 1971; REINERT and GOSCH, 1976). Density and gra-nulation of the cytoplasm and size of the vacuoles were additional markers in some of the experiments. Distinctive densities made possible the enrichment of fusion bodies by gradient centrifugation (HARMS et al., 1976; HARMS and POTRYKUS, 1978).

Preliminary results have been obtained with a selection system in which advantage is taken of the complementation of artificially induced meta-bolic blocks (NEHLS, 1978; NEHLS and BINDING, 1979).

> Selective survival of the fusion bodies has been observed. Their growth, how-ever, ceased at the latest after a few cell divisions. The usefulness of this method remains therefore to be proved.

Genetically stable markers proved to be powerful tools in somatic hy-bridization experiments, namely if they were expressed not only in early stages of regeneration.

Recognition and analysis of fusion products by different development of the
plastids was possible during a limited period of the formation of small cell
clusters (BINDING, 1976).

Frequently occurring nuclear and extrakaryotic mutations affect the
capability of chlorophyll formation. Chlorophyll-deficient mutants
have been utilized in a number of somatic hybridization experiments.

Either two protoplasts have been fused which both carried nuclear mutations
to obtain complementation in the hybrids (GILES, 1973, 1974; MELCHERS and
LABIB, 1973, 1974; BINDING, 1974a; COCKING et al., 1977; MALIGA et al., 1977;
MELCHERS and SACRISTAN, 1977; SCHIEDER, 1977, 1978b; WALLIN et al., 1979);
or a nuclear controlled chlorophyll deficiency of one partner was utilized
for the recognition of uniparental cell colonies. Another kind of marker was
used for the discrimination of the other parental type (MALIGA et al., 1978;
MELCHERS, 1978a; MELCHERS et al., 1978; SCHIEDER, 1978a); or genome and plas-
tome mutants were combined in order to recognize new genome/plastome combinations and mixed populations of plastids (GLEBA, 1974, 1979; GLEBA et al.,
1975). Both types of recombinant were called "cybrids" by the authors.

Most effective selection of somatic hybrids is possible if the uniparental protoplasts or their regenerants are eliminated from the
populations by selective conditions. A range of different systems
has become available in plants.

The classic systems of auxotrophic mutants which have been successful in
fungi and bryophytes since reports in 1974/1975 (FERENCZY et al., 1974; BINDING
and WEBER, 1974; SCHIEDER, 1974; ANNÉ and PEBERDY, 1975) could be established
recently also in higher plants (GLIMELIUS et al., 1978). A similar principle
of selection was verified in the interspecific hybridization system of *Nicotiana glauca* + *N. langsdorffii* which is independent of growth hormones. Somatic
hybridization of these plants has become a standard tool for basic studies on
protoplast fusion (see for example WETTER and KAO, 1976). Lack of proliferation of at least one parental protoplast type under the applied culture conditions (POWER et al., 1975, 1977; SCHIEDER, 1977, 1978a; MELCHERS, 1978a;
MELCHERS et al., 1978; ZELCER et al., 1978), different organogenic potentials
(MALIGA et al., 1977) and heterosis-like growth of the hybrid clones (SCHIEDER,
1978a) should also be mentioned in this connection.

An especially elegant selection procedure which is probably widely
applicable to plant systems has been introduced by MELCHERS and LABIB
(1973, 1974). Fusion hybrids of two different chlorophyll-deficient,
light-sensitive mutants of tobacco showed complementation to dark green
pigmentation and to normal light tolerance.

Positive selection pressures for hybrid clones can also be established
if one or both parental types differ in drug sensitivities which are
complemented in the hybrids.

MALIGA et al. (1978) used a kanamycin-resistant mutant strain, POWER et al.
(1977) and CONSTABEL et al. (1975b) naturally occurring differences in drug
sensitivities.

Karyological characteristics are appropriate for the identification
and especially useful for the analysis of fusion products.

The heterokaryotic nature of fusion bodies can easily be detected in most intergeneric systems by different staining properties (KELLER et al., 1973; KARTHA
et al., 1974; CONSTABEL et al., 1975, 1976; NEHLS, 1978; and others). Chromosomes have been counted to calculate the number of protoplasts which had con-
tributed to the original fusion body and to draw conclusions on karyotypic

mutations during the development of the fusion products (MELCHERS and LABIB, 1974; CONSTABEL et al., 1976; KAO, 1976, 1977; SMITH et al., 1976; DUDITS et al., 1977; MALIGA et al., 1977; MELCHERS and SACRISTAN, 1977; SCHIEDER, 1977; GLEBA and HOFFMANN, 1978; MELCHERS et al., 1978). More detailed analyses of karyotypic alterations can be performed if the chromosomes of the parents are clearly different in morphology. Heterokaryotic and hybrid cells have been investigated by these characteristics (KAO et al., 1974; CONSTABEL et al., 1975a, 1976, 1977; GOSCH and REINERT, 1976, 1978; KAO, 1976, 1977; BINDING and NEHLS, 1978b; KRUMBIEGEL and SCHIEDER, 1979).

Biochemical analyses have been carried out to prove the hybrid nature of regenerants, to obtain information on gene expression in the hybrids and to investigate the composition of the regenerants with respect to the cell organelles and nuclear genes.

Special emphasis has been put on investigations of the subunit composition of the ribulose-1,5-bisphosphate carboxylase, the so-called fraction I protein, of the plastids.

The enzyme consists of large subunits coded for by genes located on chloroplast DNA and small subunits which are specified by nuclear genes (cf. KOWALLIK and HERRMANN, 1977; v. WETTSTEIN et al., 1978); the subunits of different plant species differ in their electrophoretic mobilities. Hence, the fraction I protein can be used for the identification of the genomes and plastomes which are compiled in the regenerants of fusion bodies (KUNG et al., 1975; SMITH et al., 1976; CHEN et al., 1977; MELCHERS et al., 1978; ZELCER et al., 1978).

Several further enzymes such as alcohol dehydrogenases, glucose-6-phosphate dehydrogenases, glutamate dehydrogenases, lactate dehydrogenases, aspartate aminotransferases, aminopeptidases, peroxydases, and esterases have been used for the characterization of fusion products (WETTER and KAO, 1976; WETTER, 1977; GLEBA and HOFFMANN, 1978; MALIGA et al., 1978; SCHIEDER, 1978c).

An elegant method for the identification of the genetic material is to compare cleavage products of the DNA's.

BELLIARD et al. (1978) analyzed chloroplast DNA in cytoplasmic parasexual hybrids of tobacco by this method.

Analyses of meiotic segregants and sexual progenies of fusion products have been performed to obtain final confirmation of the hybrid nature and for investigations on the transmission of extrakaryotic genophores.

SCHIEDER (1978b) made use of anther cultures which exhibited segregation of the parental mutant types. Sexual progenies of somatic hybrids have been investigated by MELCHERS and LABIB (1974), SMITH et al. (1976), GLEBA (1979) and POWER et al. (1978).

6. Regeneration of Fusion Products

Intraspecific hybrid plants have been regenerated in the genera of *Nicotiana* (e.g. MELCHERS and LABIB, 1973) and *Datura* (SCHIEDER, 1977, 1978c), interspecific hybrids in *Nicotiana* (e.g. SMITH et al., 1976), *Petunia* (COCKING et al., 1977; POWER et al., 1977), *Datura* (SCHIEDER, 1978a,b,c), and *Daucus* (DUDITS et al., 1977), as well as intergeneric hybrids of potato + tomato (MELCHERS, 1978a; MELCHERS et al., 1978).

Shoot formation was initiated on hybrid callus of *Datura innoxia* +
Atropa belladonna (KRUMBIEGEL and SCHIEDER, 1979). Just root formation
was obtained on callus of *Arabidopsis* + *Brassica* (GLEBA et al., 1978).
All the other intergeneric fusion bodies which have been grown in
vitro formed at most cell clusters or calluses (KAO et al., 1974;
KAO, 1976, 1977; CONSTABEL et al., 1976, 1977; DUDITS et al., 1976,
1978; BINDING and NEHLS, 1978b; BRAR et al., 1978; EVANS et al.,
1978; GAMBORG et al., 1978b; GLEBA and HOFFMANN, 1978; GOSCH and
REINERT, 1978; ZENKTELER and MELCHERS, 1978).

7. The Fate of Extrakaryotic Genophores in Fusion Products

Aspects of extrakaryotic inheritance in experiments on protoplast fusion have been treated in recent review articles (BINDING, 1979; GLEBA,
1978). Preliminary information on the fate of plastids has been obtained from early developmental stages of fusion products. FOWKE (1978)
mentioned experiments of WEBER suggesting that the plastids of closely related plants both dedifferentiate regularly. A tendency of the
plastids to segregate has been observed in small cell clusters of interspecific fusion products (BINDING, 1976; BINDING and NEHLS, 1978b).
Ultrastructural studies of cell clusters grown from heterospecific
fusion bodies revealed evidence for the degeneration of one plastid
type (FOWKE et al., 1976, 1977).

The distribution of the plastids from either of the parents to the regenerated plants has been followed from analyses of the fraction I
protein. Most the regenerated plants grown from interspecific hybrids
contained only one kind of plastid.

> Only one chimeric plant of *Nicotinia langsdorffii* + *N. glauca* has been detected
> and one chimeric callus gave rise to plants with single but different plastid
> types (KUNG et al., 1975; CHEN et al., 1977). The somatic hybrids of tomato
> + potato contained - at least predominantly - one plastid type (MELCHERS et
> al., 1978). Both plastid types appeared to have had about the same chance to
> be retained in all these hybrids. Mixed populations of chloroplasts have been
> found to be carried on to regenerated plants and even to the sexual progeny in
> intraspecific hybrids of tobacco mutants and hybrids of tobacco + the closely
> related *Nicotiana debneyi* (GLEBA, 1974, 1979; GLEBA et al., 1975).

The existence of new combinations with respect to mitochondria was
suggested by some observations in somatic hybridization experiments
within the genera *Nicotiana* and *Petunia*.

> Regenerated plants from combinations of tobacco strains, one of which carried
> cytoplasmic male sterility from *Nicotiana debneyi*, exhibited hybrid nuclei and
> uniparental plastid DNA (proved by restrictase prints), but different degrees
> of male sterility and malformations of the flowers (BELLIARD et al., 1977,
> 1978). These variations were discussed to be probably due to mixed populations
> of mitochondria or even to recombinant DNA of these organelles. IZHAR and POWER
> (1979) combined cytoplasmic male sterility of *Petunia hybrida* with the genome
> of *P. axillaris* by protoplast fusion. Some of the plants which have been re-
> generated from fusion bodies of *Nicotiana sylvestris* + *N. suaveolens* (with the
> nuclei inactivated by X-rays) were also discussed to represent cybrids (ZELCER
> et al., 1978).

8. The Fate of Nuclei in Fusion Products

Fusion of parental interphase nuclei as well as the compilation of
the heterologous chromosomes during mitosis have been observed in
early developmental stages of fusion products. However, heterokaryotic cells and those with a uniparental nucleus were frequently found
in the cell progeny.

> The mitoses were more or less synchronized even if one of the parental proto-
> plast types was incapable of mitosis under the actual culture conditions.
> Occasionally, asynchrony of mitosis occurred resulting in chimeric cell clus-
> ters (KAO et al., 1974; BINDING and NEHLS, 1978b). Fusion of interphase nuclei
> was suggested by a few figures (CONSTABEL et al., 1975a; DUDITS et al., 1976;
> GOSCH and REINERT, 1976; KAO, 1977, 1978; BINDING and NEHLS, 1978b). KAO et
> al. (1974) were the first authors who found that the fusion of the parental
> genomes frequently occurred during the first cell division.

9. Chromosomal Behavior in Hybrid Cell Lines

In many fusion systems, the chromosomes of the parents are hardly
distinguishable, especially if these are closely related. In these
systems, it has been deduced from chromosome numbers that chromosomes
of both parents are retained in the hybrid plants (SMITH et al., 1976;
MALIGA et al., 1977; MELCHERS and SACRISTAN, 1977; CHUPEAU et al.,
1978; GLEBA et al., 1978; MELCHERS, 1978a; MELCHERS et al., 1978;
SCHIEDER, 1978a).

> A number of the plants were aneuploid. This might be attributed to the cul-
> ture conditions which cause karyotypic alterations in many plant cell systems.

Karyotypes with considerably different chromosome morphologies were
compiled in some hybridization experiments. The fates of the chromosomes could be followed up in a few cell clones over a longer period
(KAO, 1977; BINDING and NEHLS, 1978b; GLEBA and HOFFMANN, 1978). Already during early divisions of hybrid cell clones, anaphase bridges,
sticking and fragmentation of chromosomes, the formation of multi-
constrictional and ring chromosomes, and loss of chromosomes have
been observed.

> Incompatibilities of the genomes might have been responsible for these events
> as well as irregularities frequently occurring under in vitro conditions.

The decision which type of chromosome was going to be eliminated differed from system to system.

> In *Arabidopsis* + *Brassica* hybrids, chromosomes of both parents were retained
> (GLEBA and HOFFMANN, 1978); in hybrid callus of *Vicia* + *Petunia*, most of the
> chromosomes of either one or the other parent were lost (BINDING and NEHLS,
> 1978b); predominant elimination of *Nicotiana glauca* chromosomes happened in
> the soybean + *Nicotiana* hybrids (KAO, 1977).

Analyses of isoenzyme patterns revealed coincidence in the loss of
protein bands with proceeding chromosome elimination (WETTER, 1977).
Additionally, comparative investigations of isoenzyme patterns from
different cell lines indicated that the succession of chromosomes

getting lost was random (KAO, 1977). The numbers of retained chromosomes seemed to become stabilized (BINDING and NEHLS, 1978b).

References

ANNÉ, J., PEBERDY, J.F.: Arch. Microbiol. 105, 201-205 (1975).

BELLIARD, G., PELLETIER, G., FERAULT, M.: C. R. Acad. Sci. Paris
Ser. D 284, 749-752 (1977). - BELLIARD, G., PELLETIER, G., VEDEL, F.,
QUETIER, F.: Mol. Gen. Genet. 165, 231-237 (1978). - BINDING, H.:
Z. Pflanzenphysiol. 55, 305-321 (1966); - In: Haploids in Higher
Plants. Advances and Potential, ed. K.J. KASHA, 323-337. Guelph:
Huddleston & Barney Press 1974a; - Z. Pflanzenphysiol. 72, 422-426
(1974b); - Plant Sci. Lett. 2, 185-188 (1974c); - Z. Pflanzenphysiol.
74, 327-356 (1974d); - Physiol. Plant. 35, 225-227 (1975); - Mol. Gen.
Genet. 144, 171 (1976); - Prog. Bot. 38, 173-181 (1977); - In: Plant
Cell and Tissue Culture-Principles and Applications, eds. W.R. SHARP,
P.O. LARSEN, E.F. PADDOCK, V. RAGHAVAN, 789-805. Columbus: Ohio State
Press 1979. - BINDING, H., KOLLMANN, R., in: Cell Genetics in Higher
Plants, eds. D. DUDITS, G.L. FARKAS, P. MALIGA, 191-206. Budapest:
Akadémiai Kiadó 1976. - BINDING, H., NEHLS, R.: Z. Pflanzenphysiol.
88, 327-332 (1978a); - Mol. Gen. Genet. 164, 137-143 (1978b). -
BINDING, H., WEBER, H.J.: Mol. Gen. Genet. 135, 273-276 (1974). -
BINDING, H., NEHLS, R., SCHIEDER, O., SOPORY, S.K., WENZEL, G.:
Physiol. Plant. 43, 52-54 (1978). - BOPP, M.: Prog. Bot. 37, 155-176
(1975). - BRAR, D.S., RAMBOLD, S.K., CONSTABEL, F., GAMBORG, O.L.,
SHYLUK, J.P., KARTHA, K.K., in: Abstr. IAPTC Congress, 64, Calgary
1978. - BURGESS, J., FLEMING, E.N.: Planta 118, 183-193 (1974).

CHEN, K., WILDMAN, S.G., SMITH, H.: Proc. Natl. Acad. Sci. USA 74,
5109-5112 (1977). - CHIN, J.C., SCOTT, K.J.: Ann. Bot. 43, 33-44
(1979). - CHUPEAU, Y., MISSONIER, C., HOMMEL, M.-C., GOUJAUD, J.:
Mol. Gen. Genet. 165, 239-245 (1978). - COCKING, E.C., in: Proc. Symp.
Plant Tissue Culture, 255-264. Peking: Science Press 1978a; - Frontiers
of Plant Tissue Culture 1978, ed. T.A. THORPE, 151-158. Calgary: Univ.
Calgary 1978b. - COCKING, E.C., GEORGE, D., PRICE-JONES, M.J., POWER,
J.B.: Plant Sci. Lett. 10, 7-12 (1977). - CONSTABEL, F.: Frontiers of
Plant Tissue Culture 1978, ed. T.A. THORPE, 141-149. Calgary: Univ.
Calgary 1978. - CONSTABEL, F., DUDITS, D., GAMBORG, O.L., KAO, K.N.:
Can. J. Bot. 53, 2092-2095 (1975a). - CONSTABEL, F., KIRKPATRICK, J.W.,
KAO, K.N., KARTHA, K.K.: Biochem. Physiol. Pflanzen 168, 319-325
(1975b). - CONSTABEL, F., WEBER, G., KIRKPATRICK, J.W., PAHL, K.:
Z. Pflanzenphysiol. 79, 1-7 (1976); - C. R. Acad. Sci. Paris Ser. D
285, 319-322 (1977).

DAVEY, M.R., CLOTHUR, R.H., BALLS, M., COCKING, E.C.: Protoplasma 96,
157-172 (1978). - DUDITS, D., KAO, K.N., CONSTABEL, F., GAMBORG, O.L.:
Can. J. Genet. Cytol. 18, 263 (1976). - DUDITS, D., HADLACZKY, G.Y.,
LEVI, E., FEJER, O.L., HAYDU, Z., LAZAR, G.: Theor. Appl. Genet. 51,
127-132 (1977). - DUDITS, D., HADLACZKY, G.Y., LEVI, E., KONCZ, C.,
HAYDU, Z., PAAL, H., in: Abstr. IAPTC Congress, 67, Calgary 1978. -
DURAND, J., POTRYKUS, I., DONN, G.: Z. Pflanzenphysiol. 69, 26-34
(1973).

EVANS, D.A., GAMBORG, O.L., SHYLUK, J.P., WETTER, L.R., in: Abstr.
IAPTC Congress, 70, Calgary 1978.

FERENCZY, L., KEVEI, F., ZSOLT, J.: Nature (London) 248, 793-794 (1974). - FOWKE, L.C., in: Frontiers of Plant Tissue Culture 1978, ed. T.A. THORPE, 223-233. Calgary: Univ. Calgary 1978. - FOWKE, L.C., RENNIE, P.J., KIRKPATRICK, J.W., CONSTABEL, F.: Can. J. Bot. 53, 272-278 (1975); - Planta 130, 39-45 (1976). - FOWKE, L.C., CONSTABEL, F., GAMBORG, O.L.: Planta 135, 257-266 (1977).

GAMBORG, O.L., in: Cell Genetics in Higher Plants, eds. D. DUDITS, G.L. FARKAS, P. MALIGA, 107-127. Budapest: Akadémiai Kiadó 1976. - GAMBORG, O.L., MILLER, R.A., OJIMA, K.: Exp. Cell Res. 50, 151-158 (1968). - GAMBORG, O.L., KARTHA, K.K., OHYAMA, K., FOWKE, L.C., in: Proc. Symp. Plant Tissue Culture, 265-278. Peking: Science Press 1978a. - GAMBORG, O.L., SHYLUK, J.P., EVANS, D., WETTER, L.R., in: Abstr. IAPTC Congress 70, Calgary 1978b. - GILES, K.L., in: Proto-plastes et Fusion de Cellules Somatiques Végétales. Colloq. Int. C.N.R.S., 485-495. Paris 1973; - Plant Cell Physiol. 15, 281-285 (1974); - In: Frontiers of Plant Tissue Culture, ed. T.A. THORPE, 67-74. Calgary: Univ. Calgary 1978. - GLEBA, Y.Y.: Thesis, Kiev, USSR (1974); - In: Plant Cell and Tissue Culture - Principles and Applications, eds. W.R. SHARP, P.O. LARSEN, E.F. PADDOCK, V. RAGHAVAN, 775-788. Columbus: Ohio State Press 1979; - in: Frontiers of Plant Tissue Culture, ed. T.A. THORPE, 95-102. Calgary: Univ. Calgary 1978. - GLEBA, Y.Y., HOFFMANN, F.: Mol. Gen. Genet. 165, 257-264 (1978). - GLEBA, Y.Y., BUTENKO, R.G., SYTNIK, K.M.: Dokl. Akad. Nauk. USSR 221, 1196-1198 (1975). - GLEBA, Y.Y., KOHLENBACH, H.W., HOFFMANN, F.: Naturwissenschaften 65, 655 (1978). - GLIMELIUS, K., ERIKSSON, T., GRAFE, R., MÜLLER, A.J.: Physiol. Plant. 44, 273-277 (1978). - GLIMELIUS, K., WALLIN, A., ERIKSSON, T.: Physiol. Plant. 31, 225-230 (1974). - GOSCH, G., REINERT, J.: Naturwissenschaften 63, 534-535 (1976); - Protoplasma 96, 23-38 (1978). - GROUT, B.W.W., COUTTS, R.H.A.: Plant Sci. Lett. 2, 397-403 (1974). - GROUT, B.W.W., WILLISON, J.M., COCKING, E.C.: J. Bioenergetics 4, 311-328 (1972).

HARMS, C.T., POTRYKUS, I.: Theor. Appl. Genet. 53, 49-55 (1978). - HARMS, C.T., LÖRZ, H., POTRYKUS, I.: J. Cell Biol. 70, 208 (1976). - HARTMANN, J.X., KAO, K.N., GAMBORG, O.L., MILLER, R.A.: Planta 112, 45-56 (1973). - HAYDU, Z., LAZAR, G., DUDITS, D.: Plant Sci. Lett. 10, 357-360 (1977). - HERZFELD, F.: Prog. Bot., Vol. 41 (1979).

IZHAR, S., POWER, J.B.: Plant Sci. Lett. 14, 49-55 (1979).

JONES, C.W., MASTRANGELO, M.A., SMITH, H.H., LIU, H.Z., MECK, R.A.: Science 193, No. 4251 (1976).

KAO, K.N.: In: Cell Genetics in Higher Plants, eds. D. DUDITS, G.L. FARKAS, P. MALIGA, 149-152. Budapest: Akadémiai Kiadó 1976; - Mol. Gen. Genet. 150, 225-235 (1977); - In: Proc. Symp. Plant Tissue Culture, 331-339. Peking: Science Press 1978. - KAO, K.N., MICHAYLUK, M.R.: Planta 115, 355-367 (1974); - Planta 126, 105-110 (1975). - KAO, K.N., KELLER, W.A., MILLER, R.A.: Exp. Cell Res. 62, 338-340 (1970). - KAO, K.N., GAMBORG, O.L., MICHAYLUK, M.R., KELLER, W.A., MILLER, R.A.: Protoplastes et Fusion de Cellules Somatiques Végétales. Colloq. Int. C.N.R.S., 207-213. Paris 1973. - KAO, K.N., CONSTABEL, F., MICHAYLUK, M.R., GAMBORG, O.L.: Planta 120, 215-227 (1974). - KARTHA, K.K., GAMBORG, O.L., CONSTABEL, F., KAO, K.N.: Can. J. Bot. 52, 2435-2436 (1974). - KELLER, W.A., MELCHERS, G.: Z. Naturforsch. 28, 737-741 (1973). - KELLER, W.A., HARVEY, B.H., KAO, K.N., MILLER, R.A., GAMBORG, O.L.: Protoplastes et Fusion de Cellules Somatiques Végétales. C.N.R.S., 455-463. Paris 1973. - KOWALIK, K.V., HERMANN, R.G.: Prog. Bot. 39, 1-17 (1977). - KRUMBIEGEL, G., SCHIEDER, O.:

Planta 145, 371-375 (1979). - KUNG, S.D., GRAY, J.C., WILDMAN, S.G., CARLSON, P.S.: Science 187, 353-355 (1975).

LARKIN, P.J.: J. Cell Sci. 26, 31-46 (1977).

MALIGA, P., LÁZÁR, G., JOÓ, F., NAGY, A.H., MENCZEL, L.: Mol. Gen. Genet. 157, 291-296 (1977). - MALIGA, P., KISS, Z.R., NAGY, A.H., LAZAR, G.: Mol. Gen. Genet. 163, 145-151 (1978). - McCOMB, J.A., in: Proc. Symp. Plant Tissue Culture, 341-349. Peking: Science Press 1978. - MELCHERS, G.: In: International Cell Biology 1976-1977, eds. B. BRINKLEY, K. PORTER, 207-215. New York: Rockefeller Univ. Press 1977; - In: Production of Natural Compounds by Cell Culture Methods, eds. A.W. ALFERMANN, E. REINHARD, 306-311. München: Ges. Strahlen- und Umweltforschung 1978a; - In: Proc. Symp. Plant Tissue Culture, 279-283. Peking: Science Press 1978b. - MELCHERS, G., LABIB, G.: In: Protoplastes et Fusion de Cellules Végétales. C.N.R.S., 367-372. Paris 1973; - Mol. Gen. Genet. 135, 277-294 (1974). - MELCHERS, G., SACRISTAN, M.D.: La Culture des Tissus et des Cellules des Végétaux, 169-177. Paris: Masson 1977. - MELCHERS, G., SACRISTÁN, M.D., HOLDER, A.A.: Carlsberg Res. Commun. 43, 203-218 (1978). - MENCZEL, L., LÁZÁR, G., MALIGA, P.: Planta 143, 29-32 (1978).

NAGATA, T.: Naturwissenschaften 65, 263-264 (1978). - NAGATA, T., MELCHERS, G.: Planta 142, 235-238 (1978). - NAGATA, T., TAKEBE, I.: Planta 92, 301-308 (1970). - NEHLS, R.: Mol. Gen. Genet. 166, 117-118 (1978). - NEHLS, R., BINDING, H.: Hoppe Seylers Z. Physiol. Chem. 360, 332-333 (1979).

POTRYKUS, I.: Nature (London) 231, 57-58 (1971). - POWER, J.B., FREARSON, E.M., HAYWARD, C., COCKING, E.C.: Plant Sci. Lett. 5, 197-207 (1975). - POWER, J.B., BERRY, S.F., FREARSON, E.M., COCKING, E.C.: Plant Sci. Lett. 10, 1-6 (1977). - POWER, J.B., SINK, K.S., BERRY, S.F., BURNS, S.F., COCKING, E.C.: J. Hered. 69, 373-376 (1978).

RAVEH, D., GALUN, E.: Z. Pflanzenphysiol. 76, 76-79 (1975). - RAVEH, D., HABERMAN, E., GALUN, E.: In Vitro 9, 216-220 (1973). - REINERT, J., GOSCH, G.: Naturwissenschaften 63, 534 (1976).

SCHIEDER, O.: Z. Pflanzenphysiol. 74, 357-365 (1974); - Planta 137, 253-257 (1977); - Mol. Gen. Genet. 162, 113-119 (1978a); - Planta 141, 333-334 (1978b); - Habilitationsschrift, Bochum (1978c). - SMITH, A.H., KAO, K.N., COMBATTI, N.C.: J. Hered. 67, 123-128 (1976).

WALLIN, A., GLIMELIUS, K., ERIKSSON, T.: Z. Pflanzenphysiol. 74, 64-80 (1974); - Z. Pflanzenphysiol. 87, 333-340 (1978); - Z. Pflanzenphysiol. 91, 89-94 (1979). - WATTS, J.W., MOTOYOSHI, F., KING, J.M.: Ann. Bot. 38, 667-671 (1974). - WEBER, G., CONSTABEL, F., WILLIAMSON, F., FOWKE, L., GAMBORG, O.L.: Z. Pflanzenphysiol. 79, 459-464 (1976). - WETTER, L.R.: Mol. Gen. Genet. 150, 231-235 (1977). - WETTER, L.R., KAO, K.N.: Z. Pflanzenphysiol. 80, 455-462 (1976). - WETTSTEIN, v., D., POULSEN, C., HOLDER, A.A.: Theor. Appl. Genet. 53, 193-198 (1978). - WILLIS, G.E., HARTMANN, J.X., de LAMATER, E.D.: Protoplasma 91, 1-14 (1977). - WITHERS, L.A., in: Protoplastes et Fusion de Cellules Somatiques Végétales. Colloq. Int. C.N.R.S., 517-545. Paris 1973.

ZELCER, A., AVIV, D., GALUN, E.: Z. Pflanzenphysiol. 90, 397-407
(1978). - ZENKTELER, M., MELCHERS, G.: Theor. Appl. Genet. 52, 81-90
(1978).

Professor Dr. HORST BINDING
Dipl.-Biol. REINHARD NEHLS
Botanisches Institut
- Biologiezentrum -
der Universität
Olshausenstr. 40-60
D 2300 Kiel 1

III. Mutation: Higher Plants

By WERNER GOTTSCHALK

1. Methods for Inducing Gene and Chromosome Mutations

The number of chemicals which are found to be mutagenic is rapidly increasing. Some of these are closely associated with human life (fungicides, herbicides, insecticides, carcinogens, drugs, stimulants, factory effluents etc.). This is one of the main reasons why considerably more work is being done in chemical than in physical mutagenesis.

a) Physical Mutagens

The effectiveness of 230 ke V neutrons was tested on *Vicia faba* chromosomes. About 40% of protein recoils lead to observable cytological changes (GEARD, 1977). Long-wave UV radiation induces aberrations in chromosomes with 5-bromouracil-substituted DNA, whereas there is no mutagenic effect in chromosomes with unsubstituted DNA (KIHLMAN et al., 1978; *Vicia*). The intrachromosomal distribution pattern of chromatid aberrations was found to differ strikingly in *Vicia* after application of mutagens with non-delayed effects (X-rays, fast neutrons) as compared to those induced by mutagens with delayed effects (ethanol, maleic hydrazide; SCHUBERT and RIEGER, 1977). The mutagenic effectiveness of gamma rays was studied in combination with EMS* and DES (KAUL and BHAN, 1977) or with different metallic salts. A synergistic yield of M_2 chlorophyll mutants was obtained by combining the rays with Sr, Cd, Hg, Pb, or Cu (REDDY and VAIDYANATH, 1978; rice).

b) Chemical Mutagens

Review papers with many references were published on mutagenic effects on plant chromosomes or genes of the following substances: nitrous acid (ZIMMERMANN, 1977), acrolein (IZARD and LIBERMANN, 1978), maleic hydrazide (SWIETLIŃKSA and ŽUK, 1978), coumarins (GRIGG, 1977/78), *Vinca rosea* alkaloids (DEGRAEVE, 1978), beta-propiolactone (BRUSIK, 1977), caffeine (TIMSON, 1977), tetra-methyluric acid, a natural metabolite in some caffeine-producing plants (KIHLMAN, 1977), also of hycanthone and other drugs utilized against schistosomiasis in man (ONG, 1978).

Sodium azide (NaN_3) appears to be one of the most potent mutagens existing at present. In terms of number of mutated sectors, induced per unit of dose, NaN_3 was found to be more effective than EMS. EMS, however, is able to induce much greater damage per unit of seedling injury (CONGER and CARABIA, 1977; maize). Findings from embryonic barley shoots reveal that sodium azide may act only upon the replicating DNA

---

*Abbreviations used in the paper: EMS ethyl methanesulphonate, DES diethyl sulphate, EI ethyleneimine, MNU N-methyl-N-nitrosourea

(SANDER et al., 1978). In non-germinating barley embryos, single-strand breaks in DNA were found after NaN_3 treatment (VELEMÍNSKÝ et al., 1977a). There are certain indications that EMS induces different mutation spectra when applied to barley seeds or to a specific stage of later ontogenetic development, respectively (KISELEVA and KHVOSTOVA, 1978; M_2 chlorophyll mutants). DES, given to G_1 cells of barley seeds, induces predominantly chromatid changes. If, however, the G_1 phase is artificially prolonged, chromosome aberrations are mostly induced. Moreover, the storage of DES-treated seeds at 20% water content leads to an enhancement of the induced cytological and genetic injury (GICHNER and VELEMÍNSKÝ, 1977; GICHNER et al., 1977a,b). Isolated DNA, derived from mice, was found to induce chromatid aberrations in *Vicia faba* in a specific distribution pattern of the breakpoints (ŠLOTOVÁ et al., 1977).

Some insecticides (methyl-mercaptophos, thuricide) induce chromosome mutations and other mitotic irregularities (ISTAMOV, 1977, *Crepis capillaris*; SHARMA and SAHU, 1977, *Allium cepa*). This holds true for the fungicide dithane and the antibiotic tetracycline also (MANN, 1977, 1978; meiosis of *Allium cepa*). Earlier findings on cytogenetic effects of cigarette smoke and smoke constituents were confirmed and supplemented (PANDEY et al., 1978, *Allium cepa*; ARNOLD et al., 1978, *Allium sativum*). Furthermore, extracts of betle leaves, widely used in some Asiatic countries, proved to induce manifold mitotic anomalies when applied to onion root tip meristems (ABRAHAM and CHERIAN, 1978). The mutagenic effectiveness of lysergic acid diethylamide (LSD-25) is not yet clear. In vitro studies reveal not only the suppression of mitosis but also an enhanced chromosome damage. It is, however, very difficult to draw any conclusions from these findings for in vivo situation. According to COHEN and SHILOH (1977/78), LSD is at best a weak mutagen if mutagenic at all.

Clinically useful concentrations of some inhalational anaesthetics such as halothane or methoxyflurane are responsible for the occurrence of abnormal mitotic *Vicia* interphase nuclei (GRANT et al., 1977). Propane sultone - a potent carcinogen - seems to be a powerful mutagen inducing chromosome aberrations as well as gene mutations (SINGH and KAUL, 1978; barley). Gene mutations could also be induced by hydrazine and some of its derivatives which are utilized for synthesizing many pharmaceuticals and plant growth inhibitors (KIMBALL, 1977). Similar effects were observed after applying river water polluted by certain factory effluents (RAVINDRAN and RAVINDRAN, 1978; *Ornithogalum*). Also the non-edible oil from the seeds of *Argemone mexicana* is obviously a strong mutagen (REDDY and VAIDYANATH, 1977; M_2 chlorophyll mutants in rice). Ozone, preferably when given to early meiotic stages, leads to the formation of bridges, fragments, and micronuclei (JANAKIRAMAN and HARNEY, 1976). Chromatid aberrations were induced in *Crepis* and *Arabidopsis* through dichlorodiaminoplatinum treatment. Interestingly, the cis-isomer of the compound induced essentially more aberrations then the trans-isomer (SHEVCHENKO et al., 1977).

The interaction of some complexing substances and detergents with EI was studied in barley. In some cases, a wider spectrum of M_2 chlorophyll mutants was found (RANČELIS, 1977). Moreover, EI was combined with a non-toxic dose of caffeine, resulting in a strong increase of the number of chromosomal aberrations. Caffeine obviously influences the distribution pattern of the aberrations induced by EI specifically (GECHEFF, 1978).

It should be mentioned that it is now possible to perform scanning-electron microscopy of isolated and purified mitotic metaphase chromosomes having induced aberrations. In this way, aberrations become discernible which with conventional techniques remain undetected. This holds true for minute chromatid gaps and centromeric gaps among others (MACE et al., 1978 in Chinese hamster). It can be assumed that the method can also be used in plant material, thus opening new possibilities for analysing alterations of the chromosome and chromatid structure.

c) Antimutagenic Substances

The antimutagenic effect of cystine and cysteine, long known, was confirmed by CHERKASOV (1977; *Allium, Crepis*) and by REDDY and SMITH (1978; *Sorghum*). Gibberellic acid causes a decrease of the frequency of chromosome aberrations induced by X-rays. The highest radioprotective action of this substance is obtained at G_1 phase (ARARATIAN and VARDANIAN, 1978; seeds of *Crepis capillaris*). A 35%-75% decrease of the mutagenic effectiveness of EI was achieved by cadmium nitrate in root tips of *Crepis* (ROPUSHEV and GARINA, 1977).

Of particular interest are those cases in which repair syntheses are stimulated by mutagens. This effect is known for DNA single-strand breaks in barley seeds which could be repaired by MNU (VELEMÍNSKÝ et al., 1977b). An antimutagenic effect is also observed for sodium azide (GICHNER et al., 1978). Posttreatment of MNU-treated barley seeds with NaN_3 causes a reduction of the number of MNU-induced gene mutations and single-strand breaks in DNA (GICHNER et al., 1978). Also the storage of mutagen-treated seeds can have a protective effect. Storage of barley seeds for a period of 0-4 weeks at 30% water content leads to a recovery from EMS- or DES-induced injury. In seeds with 20% water content, however, an enhancement of the amount of induced genetic injury is observed (GICHNER et al., 1977a,b). A review paper on DNA repair in mutagen-injured higher plants was given by VELEMÍNSKÝ and GICHNER (1978).

2. Gene Mutations

The number of publications dealing with the characteristics and the behaviour of mutants, especially of experimentally obtained ones, has increased tremendously during the past years. Therefore, I can only give a very rough survey on the main problems studied in this field. Many details on various branches of applied mutagenesis are found in some symposium volumes of the International Atomic Energy Agency (1977a: methods; 1976b: vegetatively propagated crops; 1977b: grain legumes; 1976c, 1977c: disease resistance; 1976d, 1978: seed proteins; 1976a: cross-breeding). Moreover, a comprehensive review on the utilization of experimental mutagenesis for the improvement of vegetatively propagated crops and ornamentals was given by BROERTJES and van HARTEN (1978).

The problem of pleiotropic gene action has been studied in barley (GAUL and LIND, 1976). It was shown that individual characters of a pleiotropic complex vary independently from each other. Thus, it was possible to separate useful characters from undesirable ones, an optimistic result with regard to the application of mutagenesis in plant breeding. Some new "pleiotropic" spectra of partially similar pea mutants could be split by means of crosses. In this way, single components of the complex were attributed to the action of groups of independently functioning closely linked genes (GOTTSCHALK, 1976b).

The penetrance of two polymeric genes of *Pisum sativum* is highly influenced both by environmental factors and by other specific genes of the genome (GOTTSCHALK, 1978c). The problem of polymery was intensively studied in the group of barley *eceriferum* mutants differing in the wax coating from the initial lines. So far, 1310 *eceriferum* mutants isolated in Sweden referred to at least 65 different loci by diallelic crosses. For some loci, a significant mutagen specificity was found. Considering 1079 genotypes of this group, 1061 proved to be recessive and 18 dominant (LUNDQVIST, 1976, 1978). This is by far the best analysed

material existing at present as far as the phenomenon of polymeric
gene action is concerned.

Gene-ecological studies in peas revealed a strong dependence of the
realization of some mutant characters on climatic factors (chlorophyll
content, flowering and ripening time, stem fasciation and bifurca-
tion, internode length, seed production, composition of pleiotropic
spectra). The respective mutants were commonly grown with the mother
varieties at different localities of Russia (SIDOROVA and BOBODZHANOV,
1977) or in Germany, some African countries and several Indian locations (GOTTSCHALK, 1976a, 1978b). Four virescent maize mutants proved
to be temperature-sensitive with regard to chlorophyll formation. The
various genotypes of this group were found to have a specific threshold temperature below which greening does not occur (HOPKINS and
WALDEN, 1977).

a) Mutation Types

The problem of "directed mutations" is still completely open and no
findings are available in higher plants, indicating that some success
in this field can be expected in the near future. However, interesting
results have been obtained concerning the frequency of mutational
events in specific loci. Model experiments performed with sodium azide
in *barley* revealed distinct mutations in the following frequencies:

- 2.7 per 10,000 M_2 seedlings for *"waxy endosperm"*;
- 1.0 per 10,000 M_2 seedlings for *"vine (gigas)"*.

Similar results were obtained with regard to mutations causing distinct physiological anomalies in *barley* and *peas* (KLEINHOFS et al.,
1978). These encouraging results demonstrate that it is principally
possible to obtain a desired mutation in the frame of mutation treatments. Screening of 951,000 M_2 *barley* plants in Denmark resulted in
selecting 5 M_1 plants giving rise to powdery mildew-resistant M_2 seedlings. The respective genes are non-complementing alleles in the *ml-o*
locus (JÖRGENSEN, 1975). The rate of spontaneous somatic mutations in
the floral region of a *Canna* cultivar in India was found to be high
from January to April and low from July to September (GEORGE and NAYAR,
1976). Some chlorophyll mutants of *barley* were used for studying the
genetic regulation of chlorophyll synthesis, of membrane synthesis in
chloroplasts and related problems (v. WETTSTEIN, 1976; KAHN et al.,
1976; SMILLIE et al., 1976).

> The alteration of the photoperiodic behaviour is realized in *Gossypium hirsutum*
> ssp. *mexicanum*. This perennial wild species fructifies only under short-day
> conditions. An EI-induced mutant shows fructification under the conditions of
> normal day length (EGAMBERDIEV and PAYZIEV, 1977). A lowered sensitivity to
> photoperiod is also observed in an early flowering dwarf mutant of soybean
> (LEE CHOO KIANG and HALLORAN, 1977). - The expression of an unstable gene in-
> fluencing the chlorophyll production of *Glycine max* was found to depend on the
> temperature. There are pronounced differences in gene action at 19° and 29°C
> (SHERIDAN and PALMER, 1977).

b) Genes Controlling Meiosis

Problems related to the genetic control of meiosis have been reviewed
and discussed by BAKER et al. (1976). New *male sterile mutants* were selected in *Hordeum vulgare* (SETHI and BHATERIA, 1977), *Pennisetum americanum*
(RAO and KODURU, 1978) and *Corchorus capsularis* (MITRA, 1977). The expression of genetically conditioned male sterility in some *barley* mutants

is highly influenced by temperature, whereas light intensity remains
without any action in this respect (SHARMA and REINBERGS, 1976). In
Capsicum annuum, male sterile mutants were found to have a reduced number
of protein components in their anthers. Moreover, the activity of the
glucose-6-phosphate dehydrogenase is strongly reduced (MARKOVA and
DASKALOFF, 1976). The problem of genetic male sterility and its significance for plant breeding has been discussed by DRISCOLL and BARLOW
(1976), FOSTER (1976) and GOTTSCHALK (1976c).

An induced and some spontaneously arisen *desynaptic mutants* were isolated
in *Pennisetum typhoides* showing normal bivalent formation in pachytene and
a varying number of univalents in metaphase I (SINGH et al., 1977a).
Similar chemically induced or spontaneously arisen genotypes are available in *Pisum* (EZHOVA et al., 1977), *Glycine max* (WINGER et al., 1977)
and *Lolium perenne* (OMARA and HAYWARD, 1978). Pachytene has not been
studied in these genotypes. Therefore, it is not clear whether they
belong to the group of desynaptic or asynaptic mutants. A gamma-ray-
induced *tomato* mutant shows stickiness of chromosomes. At diakinesis
and metaphase I, the bivalents congregate into groups of various sizes
and lose their individuality. Regular anaphase distribution is not possible in the pollen mother cells. The meiotic anomalies cause a strong
reduction of pollen and seed fertility. They are obviously due to a
dominant mutation (RAO and RAO, 1977).

c) Induced Mutations in Plant Breeding

Since 1970, the number of officially released or approved varieties of many
crops, developed by means of induced mutations, has strongly increased. This
is especially valid for annual diploid and allopolyploid autogamous crops and
for vegetatively propagated species including ornamentals. At present, about
300 mutant varieties are available. The application of mutagenesis in plant
breeding is now going to become a well-functioning method. Many details and a
large number of references are found in the symposium volumes published by the
International Atomic Energy Agency (1976a,b,c,d, 1977a,b,c, 1978), moreover in
review papers given by SIGURBJÖRNSSON (1976), BROERTJES and van HARTEN (1978).
The Italian *durum wheat* variety "Augusto" has been developed by means of two
neutron-induced mutants (BAGNARA and PORRECA, 1977). Some mutant *barley* varieties were incorporated into cross-breeding programmes and were utilized for developing new improved varieties in Czechoslovakia and in the United Kingdom
(BOUMA, 1977; RUDDICK and MARSTERS, 1977). After application of ^{60}Co, *wheat*
mutants were obtained in India resembling phenotypically some existing commercial varieties arisen through conventional cross-breeding. These genotypes arose
obviously not exclusively by mutation, but possibly by alterations of the pattern of gene arrangement and by intragenic recombination (KULSHRESTHA and
MATHUR, 1978).

Plenty of work has been done in the field of selecting resistant mutants of many crops. I cannot mention the results in detail but I refer to two symposium volumes of the International Atomic Energy Agency
(1976c, 1977c) and to a paper given by FAVRET (1976), in which the
problems of disease resistance in connection with induced mutations
are discussed. Moreover, 10 *tobacco* mutants were isolated showing tolerance to two different herbicides causing strong damage on the leaves
of intact tobacco plants (RADIN and CARLSON, 1978). Some new cases of
monohybrid heterosis have been analysed in *Pisum sativum* and *P. arvense*
(JARANOWSKI, 1977; SHUMNY, 1978). Hybrids between fasciated and non-
fasciated peas in specific combinations surpass the capacity of the
better parental strain by 100%-250% in seed production (GOTTSCHALK,
1976d). The fasciated mutants just mentioned are homozygous for about
15 different genes. Therefore, a great diversity of different recom-

bination types was obtained after having crossed them with non-fasci-
ated genotpyes. Some of them are of agronomic interest because of a
favourable combination of some useful traits (GOTTSCHALK, 1977; GOTT-
SCHALK and BANDEL, 1978; BANDEL and GOTTSCHALK, 1978). High-yielding
Pisum recombinants, homozygous for different leaf genes, are available
in Poland (JARANOWSKI, 1977). The selection value of some *tomato* mutants
was improved by incorporating the respective genes into specific genetic backgrounds (BUTLER, 1977).

> The application of the methods of mutagenesis in *allogamous crops* is still in
> the very beginning. Some experiences are available in *Phleum* (BLIXT, 1976a)
> and *Trifolium pratense* (JARANOWSKI and BRODA, 1978). In *Phaseolus aureus*, an
> important pulse crop of India, a mutant with a protruding style arose, thus
> changing the breeding system of the species from self- to cross-pollination
> (RAGHUVANSHI et al., 1978).

Some progress in the methods of applied mutagenesis may facilitate mutation breeding to some extent. It is now possible in clonally propagated
plants to utilize the induced genetic variability through in vitro and in
vivo procedures via tissue cultures. In this way, many plants, heterozygous for mutant genes, originate from single cells and the chimerism
of the M_1 plants - a negative by-effect of mutation treatments in many
species - is avoided. A review of the problems associated with these
methods has been given by SKIRVIN (1978). The installation of gene
banks for genetically intensively studied crops is likewise a step forward with regard to the utilization of the genetic variability existing. For *Pisum sativum*, such a gene bank has long been available at
Weibullsholm/Sweden, but a computer system is now being used in order
to utilize the gene bank information for breeding and research (BLIXT,
1976b). In specific cases, applied mutagenesis is the only way to increase the genetic variability. This holds true for the widely used
turf and forage bermuda grasses. They represent sterile triploids between 4n *Cynodon dactylon* and 2n *C. transvaalensis* which cannot be improved
by conventional breeding methods. By treating dormant stolons with
gamma rays, some prospective mutants showing an improvement with regard to leaf size, internode length, spreading rate, nematode resistance and herbicide tolerance were obtained (BURTON, 1976).

3. Chromosome Mutations

Translocations were used for localizing genes in *barley* (HÄUSER and
FISCHBECK, 1976) and *maize* (BECKETT, 1978). By means of marker genes,
the translocation breakpoints were located in *barley* (PRASAD and DAS,
1976a,b). Moreover, it was possible to transfer the gene for mildew
resistance from *Avena barbata* to *A. sativa* by means of an induced translocation. The new line was found to be resistant to all the prevalent
mildew races (AUNG et al., 1977; AUNG and THOMAS, 1978).

a) Translocations in Natural Populations

> Besides genome mutations, chromosome mutations play an important role in the
> evolution of the plant kingdom. This holds particularly true for transloca-
> tions, certainly also for Robertsonian fusions which have been analysed so far
> only in a relatively small number of cases. General discussions and surveys on
> these problems were given by FLORY (1976, 1977; *Amaryllidaceae*), DEY (1977;
> *Compositae*) and JONES (1977, 1978; *Commelinaceae*).

By crossing different strains of *Triticum dicoccoides* of the Near East
area, 20 strains were found to belong to the same reciprocal translocation type, whereas two other strains have different chromosome
structures (KAWAHARA and TANAKA, 1978). The species *Secale cereale,
vavilovii*, and *africanum* have derived from *Secale montanum*. By means of
analysing the banding patterns of mitotic metaphase chromosomes,
SINGH and RÖBBELEN (1977) found that *S. montanum* is separated from *vavilovii* and *africanum* by one translocation each, whereas two translocations each separate *S. cereale* from the two species just mentioned.
Secale montanum and *africanum*, finally, are separated from *cereale* by
three translocations each. Also within the species *Secale cereale*, interchange heterozygosity is found (BAILEY et al., 1978). In Israel,
Vicia sativa forms an aneuploid series with 10, 12, and 14 chromosomes.
Different lines, having the same chromosome number (2n=10, 2n=12),
were found to differ from each other by two translocations, while
lines with 10 and 12 chromosomes differ by up to three translocations.
The 2n=10 types derived obviously from 2n=12 types via centric or tandem fusion (LADIZINSKY and TEMKIN, 1978).

In the populations of some species, a high degree of structural hy-
bridity - resulting in full sterility - is maintained through vegetative propagation. This holds true for plants of the Himalayan species
Allium consanguineum heterozygous for several translocations (GOHIL and
KOUL, 1978). Indian populations of *Allium cepa* contain even plants in
which all the chromosomes of the complement are structurally altered
by translocations. They form four rings or chains of four chromosomes
each and no bivalents in their pollen mother cells (KAUL, 1977). In a
specific clone of *Aloë barbadensis* an interchange mosaic was found. The
translocation between two long chromosomes of the complement appeared
only in some root tips and a few flowers, whereas all the other parts
of the plants were chromosomally normal (SAPRE, 1977).

A very interesting cytogenetic situation was analysed in some *Viscum*
species characterized by permanent translocation heterozygosity (BARLOW
et al., 1978). Male plants of *Viscum cruciatum* from Israel have 6 bivalents and 1 configuration of 8 chromosomes in metaphase I while female
plants have 10 bivalents. In European and Californian material of *Viscum
album*, configurations of 4, 6 and 8 chromosomes are formed in both male
and female plants. A similar situation is realized in Japanese material. The findings indicate that translocation heterozygosity in these
populations is maintained by a system of balanced lethals similar to
that known in *Oenothera*.

b) Experimentally Induced Translocations

The degree of fertility of plants heterozygous for reciprocal translocations depends upon the distribution of the chromosomes out of the
rings or chains in anaphase I. This distribution is influenced by
temperature in four different translocation heterozygotes in *barley*.
High temperatures (22⁰-27⁰C) strongly reduce the proportion of alter-
nate orientation of the chromosomes of the translocation quadrivalents
in metaphase I, thus drastically reducing the fertility of the plants
(AKAVIA and LADIZINSKY, 1978). A reciprocal translocation between an
A and a B chromosome was found in the progeny of X-irradiated *Lolium
perenne* (EVANS and MACEFIELD, 1977).

Lines homozygous for 2-5 translocations were selected in *barley* after
having treated the initial material several times with mutagenic agents.
By means of a translocation tester set, the chromosomes involved could
be identified (KÜNZEL, 1977). Ten different stocks of *Gossypium hirsutum*,

homozygous for translocations, were used for determining the break-points in 15 chromosome arms (MENZEL and BROWN, 1978). In *Pisum sativum*, 24 lines, homozygous for different radiation-induced translocations, are available. The fertility of this material is not reduced, indicating that no gene material was lost during the structural alterations of the chromosomes involved (GOTTSCHALK, 1978a). In *Secale cereale*, on the other hand, most of the spontaneously arisen interchange homozygotes proved to be inviable (BAILEY et al., 1978).

c) Inversions

In the progeny of the cross between *Aloë dawei* and *elgonica*, a tetraploid plant heterozygous for a pericentric inversion arose. In some of its gametes, duplicate/deletion chromosomes are present. They survive in the progeny of the tetraploid material, whereas they would not be able to be maintained in diploid material (BRANDHAM and JOHNSON, 1977). A tetraploid plant of *Gasteria nigricans*, carrying a paracentric inversion, was meiotically analysed by BRANDHAM (1977). Chromosomes with large terminal deletions arose due to dicentric bridges. They proved to be viable in the diploid gametes obviously because the second chromosome complement present has a buffering effect. Inversion heterozygosity is reported for the first time in the cluster bean (*Cyamopsis tetragonoloba*; SARBHOY, 1977).

4. Genome Mutations

The manifold problems of haploidy and polyploidy have been studied so intensively during the past years that I can only consider some of them in the present review. Allopolyploidy, trisomy, and monosomy will be reviewed in Volume 43 of Progress in Botany.

Principle questions of polyploidy research have been discussed by CAUDERON (1977), GUPTA (1978), and GOTTSCHALK (1978d). New examples of the action of genes influencing the pairing of homologous and homoeologous chromosomes in allopolyploids are available for *Avena* species (JAUHAR, 1977) and for *Lolium temulentum/perenne* amphiploids (TAYLOR and EVANS, 1977). In a hexaploid plant of an F_2 family of the cross 6n *Avena sativa* × 4n *A. murphyi*, diploid sectors arose. Obviously, they contain an entire genome of one of the diploid ancestors of the hexaploid oat (LADIZINSKY and FAINSTEIN, 1978). This is one of the very rare cases of genome partition in polyploid plants.

a) Haploids and Polyhaploids

The results in obtaining haploids from immature anthers on chemically defined agar media have been reviewed by COLLINS (1977) considering the methods applied and the possibilities for producing homozygous diploids. A review on the same problems specifically in *barley* with many references was given by JENSEN (1975). Very high frequencies of *rye* haploids up to 40% via anther culture were obtained by WENZEL et al. (1977). In *Gerbera*, the procedures for producing haploids were modified. Not anthers but in vitro culture of capitulum explants were successfully used (PREIL et al., 1977). *Barley* haploids at a frequency of less than 1% were selected after having pollinated *Hordeum vulgare* with *Secale cereale*. The haploid constitution is possibly due to the elimination of the rye genome (FEDAK, 1977). A similar mechanism was respon-

sible for the occurrence of *Aegilops* haploids after crossing 4n *Ae.
triuncialis* and 6n *Ae. crassa* with 4n *Hordeum bulbosum*. The *bulbosum* chromosomes were obviously eliminated in the hybrid embryo during the earliest stages of ontogenetic development (CHAPMAN and MILLER, 1977).
Haploids in high frequencies were spontaneously obtained in the progenies of two *maize* lines due to parthenogenetic development of the egg
cells (ZVERZHANSKAYA et al., 1977). In *flax*, haploids of maternal origin arise by apogamety (THOMPSON, 1977).

Doubling of chromosome number does not in all cases result in normal
diploid (= dihaploid) plants. In a large number of dihaploid *tobacco*
families, reduced growth and vigour were observed in comparison to
the original diploid cultivar. Interestingly, the concentration of reducing sugars and of alkaloids increased in this material (BURK and
MATZINGER, 1976). In the progeny of distinct *Antirrhinum* haploids, exclusively diploids arise through a mechanism not yet known in detail.
During the early meiotic stages of these haploids, bivalents are formed. The plants produce functionable haploid gametes giving rise to
pure diploid lines after selfing (KONVIČKA, 1977).

Polyhaploid (= diploid) plants, obtained from 4n *Sorghum vulgare*, were
found to have some altered morphological characters as compared to
the tetraploid initial material. Meiosis and seed set, however, were
normal (BHASKARA RAO and REDDI, 1977). Hexaploid plants of *Hordeum
parodii* and *procerum* gave rise to polyhaploids when pollinated with 2n
Hordeum bulbosum. All the *bulbosum* chromosomes were selectively eliminated
(SUBRAHMANYAM, 1977, 1978). Dihaploids (= diploids) of 4n *Solanum tuberosum* produced a large number of viable monohaploids after having pollinated with 2n *Solanum phureja*. They arose parthenogenetically from the
egg cells. Colchicine treatment led to homozygous diploid plants with
bad pollen quality but with good seed fertility (van BREUKELEN et al.,
1977). Moreover, monohaploid potatoes were produced from dihaploid
anthers (FOROUGHI-WEHR et al., 1977).

b) Experimentally Produced Autopolyploids

A modified technique of colchicine application to *cereals* proved to be very
effective. The seedlings float on the surface of the colchicine solution the
coleoptiles being immerged. This procedure has low lethality and leads reliably to polyploidization (KUMMER and MIKSCH, 1977). In different *Trifolium*
species, nitrous oxide was found to be essentially more effective for producing
polyploids than colchicine. For *Trifolium hirsutum* and *heldreichianum*, however,
the treatment was toxic (TAYLOR et al., 1976). Single triploid and tetraploid
plants of *Capsicum annuum* and *Pennisetum typhoides* were obtained after gamma
ray and EMS treatment (INDIRA and ABRAHAM, 1977; SINGH et al., 1977b).

The relations between meiotic behaviour and fertility were studied in
different species in order to clarify whether cytological or genetic
principles are responsible for the reduction in fertility. In an advanced population of tetraploid *rye*, an increase of the quadrivalent
frequency at the expense of univalents and trivalents was found after
20 years of chromosome doubling. There was, however, no correlation
between quadrivalent frequency and the degree of all the other meiotic
irregularities. The quadrivalent frequency in tetraploid *rye* clones
was found to be highly influenced by certain environmental factors
such as temperature and nitrogen supply (HOSSAIN, 1976, 1978). In autotetraploid plants of *Fragaria vesca*, the pollen fertility was not correlated with the meiotic anomalies observed. Therefore, the variation
of the strains in fertility is attributed to genetic factors (SEBASTIAM-
PILLAI and JONES, 1977). The fruits of *Solanum khasianum* are rich in the

alkaloid solasodine used as a precursor for the synthesis of corti-
sone and related pharmaceutical products. The seed fertility of autotetraploids was very low in C_1 generation, but there was a clear improvement in fertility, yield, and glycoalkaloid content of the fruits
from C_1 to C_3 generation. The total fruit weight per plant of the C_3
tetraploids is higher than that of the diploid initial material (BHATT,
1977). Another positive effect of polyploidy refers to nodulation in
Phaseolus aureus. The number of nodules on the tap root is higher in
tetraploid plants than in diploid ones. This does not, however, hold
true for the lateral roots. Moreover, the total number of nodules increases with increasing level of ploidy (2n-3n-4n) during the pre-
flowering, but not during flowering stage of the plants (KABI and
BHADURI, 1978).

The species *Apluda mutica* shows intraspecific polyploidy with diploid,
hexaploid, and heptaploid sibs. Colchicine-induced tetradecaploid
plants with 2n = 14x = 140 chromosomes are reduced in growth and
vigour. Their pollen fertility is only 5%, but they do not in principle have more meiotic disturbances than autotetraploid plants. This
is an unusual behaviour. The species is obviously relatively unsuscept-
ible to high levels of polyploidy (MURTY and SATYAVATHI, 1978).

References

ABRAHAM, S., CHERIAN, V.D.: Cytologia 43, 203-208 (1978). - AKAVIA,
N., LADIZINSKY, G.: Chromosoma 67, 145-150 (1978). - ARARATIAN, L.A.,
VARDANIAN, A.A.: Genetika USSR 14, 1571-1577 (1978). - ARNOLD, R.C.,
MANN, S.K., BHALLA, P.R., SABHARWAL, P.S.: Cytologia 43, 137-141
(1978). - AUNG, T., THOMAS, H.: Euphytica 27, 731-739 (1978). - AUNG,
T., THOMAS, H., JONES, I.T.: Euphytica 26, 623-632 (1977).

BAGNARA, D., PORRECA, G.: Mutat. Breed. Newslett. 10, 2-4 (1977). -
BAILEY, R.J., REES, H., ADENA, M.A.: Heredity 41, 1-12 (1978). - BAKER,
B.S., CARPENTER, A.P.C., ESPOSITO, M.S., ESPOSITO, R.E., SANDLER, L.:
Annu. Rev. Genet. 10, 53-134 (1976). - BANDEL, G., GOTTSCHALK, W.:
Z. Pflanzenzücht. 81, 60-76 (1978). - BARLOW, B.A., WIENS, D., WIENS,
C., BUSBY, W.H., BRIGHTON, C.: Heredity 40, 33-38 (1978). - BECKETT,
J.B.: J. Hered. 69, 27-36 (1978). - BHASKARA RAO, E.V.V., REDDI, V.R.:
Curr. Sci. 46, 58 (1977). - BHATT, B.: Environ. Exp. Bot. 17, 121-124
(1977). - BLIXT, S.: Agri Hort. Genet. 34, 59-82 (1976a); - Induced
Mutations in Cross-Breeding, 21-36. Vienna: IAEA 1976b. - BOUMA, J.:
Mutat. Breed. Newslett. 10, 6 (1977). - BRANDHAM, P.E.: Chromosoma 62,
69-84 (1977). - BRANDHAM, P.E., JOHNSON, M.A.T.: Chromosoma 62, 85-91
(1977). - BREUKELEN, E.W.M. VAN, RAMANNA, M.S., HERMSEN, J.G.T.:
Euphytica 26, 263-271 (1977). - BROERTJES, C., VAN HARTEN, A.M.:
Application of Mutation Breeding Methods in the Improvement of Vegetatively Propagated Crops. 316 pp. Amsterdam: Elsevier 1978. - BRUSICK,
D.J.: Mutat. Res. 39, 241-256 (1977). - BURK, L.G., MATZINGER, D.F.:
J. Hered. 67, 381-384 (1976). - BURTON, G.W.: Improvement of Vegetatively Propagated Plants and Tree Crops through Induced Mutations,
25-32. Vienna: IAEA: 1976. - BUTLER, L.: Can. J. Genet. Cytol. 19,
31-38 (1977).

CAUDERON, Y.: Proc. 8. Congr. Eucarpia Madrid, 131-143 (1977). -
CHAPMAN, V., MILLER, T.E.: Wheat Inf. Serv. 44, 21-22 (1977). -
CHERKASOV, O.A.: Genetika USSR 13, 609-613 (1977). - COHEN, M.M.,
SHILOH, Y.: Mutat. Res. 47, 183-209 (1977/78). - COLLINS, G.B.: Crop
Sci. 17, 583-586 (1977). - CONGER, B.V., CARABIA, J.V.: Mutat. Res.
46, 285-296 (1977).

DEGRAEVE, N.: Mutat. Res. 55, 31-42 (1978). - DEY, D.: Nucleus 20, 88-93 (1977). - DRISCOLL, C.J., BARLOW, K.K.: Induced Mutations in Cross-Breeding, 123-131. Vienna: IAEA 1976.

EGAMBERDIEV, A., PAYZIEV, P.: Genetika USSR 13, 1736-1741 (1977). - EVANS, G.M., MACEFIELD, A.J.: Chromosoma 61, 257-266 (1977). - EZHOVA, T.A., BALAKIREVA, M.D., GOSTIMSKY, S.A.: Genetika USSR 13, 424-429 (1977).

FAVRET, E.A.: Induced Mutations in Cross-Breeding, 95-111. Vienna: IAEA 1976. - FEDAK, G.: Can. J. Genet. Cytol. 19, 15-19 (1977). - FLORY, W.S.: Nucleus 19, 204-227 (1976); - Nucleus 20, 70-88 (1977). - FOROUGHI-WEHR, B., WILSON, H.M., MIX, G., GAUL, H.: Euphytica 26, 361-367 (1977). - FOSTER, C.A.: Barley Genet. III, 774-784 (1976).

GAUL, H., LIND, V.: Induced Mutations in Cross-Breeding, 55-69. Vienna: IAEA 1976. - GEARD, C.R.: Mutat. Res. 44, 345-358 (1977). - GECHEFF, K.I.: Mutat. Res. 50, 77-83 (1978). - GEORGE, K.P., NAYAR, G.G.: Environ. Exp. Bot. 16, 295-297 (1976). - GICHNER, T., VELEMÍNSKÝ, J.: Mutat. Res. 45, 205-211 (1977). - GICHNER, T., VELEMÍNSKÝ, J., POKORNÝ, V.: Environ. Exp. Bot. 17, 63-67 (1977a). - GICHNER, T., VELEMÍNSKÝ, J., ŠVACHULOVÁ, J., POKORNÝ, V.: Biol. Plant. 19, 284-291 (1977b). - GICHNER, T., VELEMÍNSKÝ, J., POKORNÝ, V.: Environ. Exp. Bot. 18, 27-31 (1978). - GOHIL, R.N., KOUL, A.K.: Cytologia 43, 243-247 (1978). - GOTTSCHALK, W.: Induced Mutations in Cross-Breeding, 37-44. Vienna: IAEA: 1976a; - Induced Mutations in Cross-Breeding, 71-78. Vienna: IAEA: 1976b; - Induced Mutations in Cross-Breeding, 133-140. Vienna: IAEA: 1976c; - Induced Mutations in Cross-Breeding, 189-197. Vienna: IAEA 1976d; - J. Nucl. Agric. Biol. 6, 27-33 (1977); - Nucleus 21, 29-34 (1978a); - Genetika (Beograd) 10, 43-61 (1978b); - Genetica 49, 21-29 (1978c); - Nucleus 21, 99-112 (1978d). - GOTTSCHALK, W., BANDEL, G.: Z. Pflanzenzücht. 80, 117-128 (1978). - GRANT, C.J., POWELL, J.N., RADFORD, S.G.: Mutat. Res. 46, 177-184 (1977). - GRIGG, G.W.: Mutat. Res. 47, 161-181 (1977/78). - GUPTA, P.K.: Nucleus 21, 117-124 (1978).

HÄUSER, J., FISCHBECK, G.: Z. Pflanzenzücht. 77, 269-280 (1976). - HOPKINS, W.G., WALDEN, D.B.: J. Hered. 68, 283-286 (1977). - HOSSAIN, M.G.: Can. J. Genet. Cytol. 18, 601-609 (1976); - Cytologia 43, 21-34 (1978).

INDIRA, C., ABRAHAM, S.: Cytologia 42, 371-375 (1977). - International Atomic Energy Agency: Induced Mutations in Cross-Breeding. Vienna: IAEA 1976a; - Improvement of Vegetatively Propagated Plants and Tree Crops through Induced Mutations. Vienna: IAEA 1976b; - Induced Mutations for Disease Resistance in Crop Plants (1975). Vienna: IAEA 1976c; - Evaluation of Seed Protein Alterations by Mutation. Vienna: IAEA 1976d; - Manual on Mutation Breeding, 2nd Ed. Vienna: IAEA 1977a; - Induced Mutations for the Improvement of Grain Legumes in South East Asia (1975). Vienna: IAEA 1977b; - Induced Mutations against Plant Diseases. Vienna: IAEA 1977c; - Seed Protein Improvement by Nuclear Techniques. Vienna: IAEA 1978. - ISTAMOV, K.I.: Genetika USSR 13, 621-626 (1977). - IZARD, C., LIBERMANN, C.: Mutat. Res. 47, 115-138 (1978).

JANAKIRAMAN, R., HARNEY, P.M.: Can. J. Genet. Cytol. 18, 727-730 (1976). - JARANOWSKI, J.K.: Genet. Pol. 18, 337-355 (1977). - JARANOWSKI, J.K., BRODA, Z.: Theor. Appl. Genet. 53, 97-103 (1978). - JAUHAR, P.P.: Theor. Appl. Genet. 49, 287-295 (1977). - JENSEN, C.J.: Barley Genet. III, 316-345 (1975). - JONES, K.: Chromosomes Today 6, 121-129. Amsterdam: Elsevier/North-Holland Biomedical Press 1977; -

Nucleus 21, 152-157 (1978). - JÖRGENSEN, J.H.: Barley Genet. III,
446-455 (1975).

KABI, M.C., BHADURI, P.N.: Cytologia 43, 467-475 (1978). - KAHN, A.,
AVIVI-BLEISER, N., WETTSTEIN, D.v.: Genetics and Biogenesis of Chloro-
plasts and Mitochondria, 119-131. Amsterdam: Elsevier/North-Holland
Biomedical Press 1976. - KAUL, M.L.H.: Cytologia 42, 681-689 (1977).
- KAUL, M.L.H., BHAN, A.K.: Theor. Appl. Genet. 50, 241-246 (1977).
- KAWAHARA, T., TANAKA, M.: Wheat Inf. Serv. 45/46, 29-31 (1978). -
KIHLMAN, B.A.: Mutat. Res. 39, 297-316 (1977). - KIHLMAN, B.A.,
NATARAJAN, A.T., ANDERSSON, H.C.: Mutat. Res. 52, 181-198 (1978). -
KIMBALL, R.F.: Mutat. Res. 39, 111-126 (1977). - KISELEVA, G.N.,
KHVOSTOVA, V.V.: Genetika USSR 14, 213-222 (1978). - KLEINHOFS, A.,
WARNER, R.L., MUEHLBAUER, F.J., NILAN, R.A.: Mutat. Res. 51, 29-35
(1978). - KONVIČKA, O.: Z. Pflanzenzücht. 78, 31-43 (1977). - KÜNZEL,
G.: Arch. Züchtungsforsch. 7, 51-59 (1977). - KULSHRESTHA, V.P.,
MATHUR, V.S.: Theor. Appl. Genet. 53, 125-128 (1978). - KUMMER, M.,
MIKSCH, G.: Arch. Züchtungsforsch. 7, 305-309 (1977).

LADIZINSKY, G., FAINSTEIN, R.: Theor. Appl. Genet. 51, 159-160 (1978).
- LADIZINSKY, G., TEMKIN, R.: Theor. Appl. Genet. 53, 33-42 (1978).
- LEE CHOO KIANG, HALLORAN, G.M.: Mutat. Res. 43, 223-230 (1977). -
LUNDQVIST, U.: Barley Genet. III, 162-163 (1976); - Experimental Muta-
genesis in Plants (Sofia), 263-265 (1978).

MACE, M.L., DASKAL, Y., WRAY, W.: Mutat. Res. 52, 199-206 (1978). -
MANN, S.K.: Environ. Exp. Bot. 17, 7-12 (1977); - Environ. Exp. Bot.
18, 201-205 (1978). - MARKOVA, M., DASKALOFF, S.: Z. Pflanzenzücht.
77, 296-303 (1976). - MENZEL, M.Y., BROWN, M.S.: Genetics 88, 541-558
(1978). - MITRA, G.C.: Genetica 47, 71-72 (1977). - MURTY, U.R.,
SATYAVATHI, D.: Nucleus 21, 34-38 (1978).

OMARA, M.K., HAYWARD, M.D.: Chromosoma 67, 87-96 (1978). - ONG, T.-M.:
Mutat. Res. 55, 43-70 (1978).

PANDEY, K.N., BENNER, J.F., SABHARWAL, P.S.: Environ. Exp. Bot. 18,
67-75 (1978). - PRASAD, G., DAS, K.: Indian J. Genet. Plant Breed.
36, 102-107 (1976a); - Indian J. Genet. Plant Breed. 36, 309-311
(1976b). - PREIL, W., HUHNKE, W., ENGELHARDT, M., HOFFMANN, M.:
Z. Pflanzenzücht. 79, 167-171 (1977).

RADIN, D.N., CARLSON, P.S.: Genet. Res. 32, 85-89 (1978). - RAGHUVANSHI,
S.S., PATHAK, C.S., SINGH, A.K.: Cytologia 43, 143-151 (1978). -
RANČELIS, V.: Genetika USSR 13, 1446-1454 (1977). - RAO, M.K., KODURU,
P.R.K.: Euphytica 27, 777-783 (1978). - RAO, P.N., RAO, R.N.: Theor.
Appl. Genet. 50, 247-252 (1977). - RAVINDRAN, P.N., RAVINDRAN, S.:
Cytologia 43, 565-568 (1978). - REDDY, C.S., SMITH, J.D.: Environ.
Exp. Bot. 18, 241-243 (1978). - REDDY, T.P., VAIDYANATH, K.: Mutat.
Res. 48, 191-194 (1977); - Mutat. Res. 52, 361-365 (1978). - ROPUSHEV,
A.R., GARINA, K.P.: Genetika USSR 13, 32-36 (1977). - RUDDICK, D.P.,
MARSTERS, M.: Mutat. Breed. Newslett. 10, 5 (1977).

SANDER, C., NILAN, R.A., KLEINHOFS, A., VIG, B.K.: Mutat. Res. 50,
67-75 (1978). - SAPRE, A.B.: Caryologia 30, 423-427 (1977). - SARBHOY,
R.K.: Cytologia 42, 147-156 (1977). - SCHUBERT, I., RIEGER, R.: Mutat.
Res. 44, 337-344 (1977). - SEBASTIAMPILLAI, A.R., JONES, J.K.: Cyto-
logia 42, 525-534 (1977). - SETHI, G.S., BHATERIA, S.D.: Indian J.
Genet. Plant Breed. 37, 73-79 (1977). - SHARMA, C.B.S.R., SAHU, R.K.:
Mutat. Res. 46, 19-26 (1977). - SHARMA, R.K., REINBERGS, R.: Indian
J. Genet. Plant Breed. 36, 59-63 (1976). - SHERIDAN, M.A., PALMER,
R.G.: J. Hered. 68, 17-22 (1977). - SHEVCHENKO, V.V., GRINIKH, L.I.,

IVANOV, V.B.: Genetika USSR 13, 1543-1551 (1977). - SHUMNY, V.K.:
Experimental Mutagenesis in Plants (Sofia), 376-382 (1978). -
SIDOROVA, K.K., BOBODZHANOV, V.A.: Genetika USSR 13, 583-592 (1977).
- SIGURBJÖRNSSON, B.: Barley Genet. III, 84-95 (1976). - SINGH, C.,
KAUL, B.L.: Mutat. Res. 56, 355-357 (1978). - SINGH, R.B., SINGH, B.D.,
LAXMI, V., SINGH, R.M.: Cytologia 42, 41-47 (1977a). - SINGH, R.B.,
SINGH, B.D., SINGH, R.M., LAXMI, V.: Cytologia 42, 633-637 (1977b).
- SINGH, R.J., RÖBBELEN, G.: Chromosoma 59, 217-225 (1977). - SKIRVIN,
R.M.: Euphytica 27, 241-266 (1978). - ŠLOTOVA, J., RIEGER, R.,
SCHUBERT, I., KARPFEL, Z., MICHAELIS, A.: Mutat. Res. 44, 247-256
(1977). - SMILLIE, R.M., NIELSEN, N.C., HENNINGSEN, K.W., WETTSTEIN,
D.v.: Proc. Aust. Biochem. Soc. 9, 90 (1976). - SUBRAHMANYAM, N.C.:
Theor. Appl. Genet. 49, 209-217 (1977); - Chromosoma 66, 185-192
(1978). - SWIETLIŃKSA, Z., ŽUK, J.: Mutat. Res. 55, 15-30 (1978).

TAYLOR, I.B., EVANS, G.M.: Chromosoma 62, 57-67 (1977). - TAYLOR,
N.L., ANDERSON, M.K., QUESENBERRY, K.H., WATSON, L.: Crop Sci. 16,
516-518 (1976). - THOMPSON, T.E.: Crop Sci. 17, 757-760 (1977). -
TIMSON, J.: Mutat. Res. 47, 1-52 (1977).

VELEMÍNSKÝ, J., GICHNER, T.: Mutat. Res. 55, 71-84 (1978). -
VELEMÍNSKÝ, J., GICHNER, T., POKORNÝ, V.: Mutat. Res. 42, 65-70
(1977a). - VELEMÍNSKÝ, J., ZADRAŽIL, S., POKORNÝ, V., GICHNER, T.:
Mutat. Res. 44, 43-51 (1977b).

WENZEL, G., HOFFMANN, F., THOMAS, E.: Theor. Appl. Genet. 51, 81-86
(1977). - WETTSTEIN, D.v.: Membranes and Disease, 123-130. New York:
Raven Press 1976. - WINGER, C.L., PALMER, R.G., GREEN, D.E.: Soybean
Genet. Newslett. 4, 36-40 (1977).

ZIMMERMANN, F.K.: Mutat. Res. 39, 127-148 (1977). - ZVERZHANSKAYA,
L.S., GRISHINA, E.V., KOMAROVA, P.I.: Genetika USSR 13, 969-972
(1977).

Professor Dr. WERNER GOTTSCHALK
Institut für Genetik
der Universität Bonn
Kirschallee 1
D 5300 Bonn 1

IV. Function of Genetic Material

Organization and Function of the Eukariotic Genome

By Frank Herzfeld and Manuel Kiper

1. Introduction

The new tool of recombinant DNA and cloning eukaryotic DNA fragments
in bacteria determines the direction of molecular biology. Characterization of cloned fragments by restriction mapping, hybridization and
sequencing on a molecular basis has become the equivalent of gene
mapping by complementation group analysis in classic genetics. The
finding, however, that genes were interrupted by noncoding inserts
was a striking result and raised questions concerning their function
and evolution which till now are beyond reasonable response. Molecular
plant genetics developed their own field due to the special features
of plant genomes and the occurrence of natural gene manipulation discovered recently.

2. Sequence Organization of DNA

a) Interspersion of Unique DNA

The sequence arrangement of interspersed short repetitive and longer
unique sequences was first established in *Xenopus* nuclear DNA (DAVIDSON
et al., 1973) and in the course revealed in most animals and plants
investigated to make up a substantial fraction of the whole genome
(DAVIDSON et al., 1975; WALBOT and DURE, 1976; ZIMMERMANN and GOLDBERG,
1977; KIPER and HERZFELD, 1978). However, it now turns out that *organisms
as different as some insects, birds, fungi, and some higher plants display sequence
organization not dominated by what was called short-period interspersion pattern.*

> Longer-period interspersion of the major unique fractions of the genome is pre-
> dominant in birds (EPPLEN et al., 1978; EDEN and HENDRICK, 1978); in fungi (HUDSPETH
> et al., 1978); in insects (MANNING et al., 1975; WELLS et al., 1976; CRAIN et
> al., 1976) which are all organisms with large single-copy DNA contents. Some
> higher plants investigated, on the contrary, show sequence interspersion of
> unique DNA at even shorter distances (FLAVELL and SMITH, 1976; SMITH and FLAVELL,
> 1977); this could be correlated to the high repetitive DNA content of plants
> (FLAVELL et al., 1974). MURRAY et al. (1978) concluded from their studies that
> in the pea genome most unique sequences are short and of similar size distribu-
> tion as repetitive sequences, i.e., most being about 300 N long. Though this
> analysis of the pea genome is controversial (PEARSON et al., 1978; WALBOT and
> GOLDBERG, 1978) probably because of the failure of other investigators to really
> drive renaturation to completion and measure single-copy DNA, its main conclu-
> sion is backed by other work on plant nuclear DNA. For wheat (FLAVELL and SMITH,
> 1976) and rye (SMITH and FLAVELL, 1977) and for parsley (KIPER and HERZFELD,
> 1978) was found that during renaturation of short nuclear DNA fragments the
> majority of all unique DNA still remains cryptic due to the excessive inter-
> spersion with repetitive sequences. This, too, argues in favor of most unique
> sequences being shorter than the vast majority of all mRNA's.

The *functional interpretation* of the short period interspersion pattern
as a means of coordinated regulation of gene transcription originally
proposed by DAVIDSON and BRITTEN (1973) even in a revised form (DAVIDSON
et al., 1977) will not easily hold for interpretation of plant genomes.

Convincing evidence accumulates that most, if not all, *plant mRNA's* (of
median length 1000-2000 nucleotides) *are transcribed from unique DNA sequences.*

> This was proven both for tobacco leaves (GOLDBERG and HOSCHEK, 1978) and parsley
> root callus (KIPER et al., 1979). The same result seems to be obtained for pea
> (MURRAY et al., 1978).

Though it cannot be excluded that these unique DNA transcripts might
be produced by a splicing mechanism now found in most animal mRNA's
investigated (this review, p. 204) and possibly by excising interspersed repetitive sequences, till now no such repetitive sequences were
detected in animal mRNA introns investigated (WOO et al., 1978; ROOP
et al., 1978; LEDER et al., 1978) nor in plant nuclear mRNA precursors
(BARTELS and KIPER, 1979). Hence it now seems to be more rational to
look for structural genes in long unique DNA sequences detected in
all animal genomes and most plant genomes in substantial amounts.

> Though the existence of such long stretches of unique DNA were not assured in
> pea (MURRAY et al., 1978) even in this case the accuracy of the data did not
> exclude that a fraction as large as the whole *Drosophila* genome would go unde-
> tected there.

b) Interspersion of Repetitive DNA

Though the role of repetitive DNA in regulation of gene expression
still is unclear, its existence as a main constituent of plant genomes
has attracted some thorough investigations to its sequential relationships, its interspersion patterns, and its evolutionary origins. Above
all cereal genomes were studied by intraspecies and interspecies hybri-
diziation (SMITH et al., 1976; FLAVELL et al., 1977). Using this approach almost ten different sequence organization units were detected
in the wheat and rye genome defined by their evolutionary origin
(i.e., repeated sequences common to wheat, oats, barley, and rye, or
only to some of the cereals) (RIMPAU et al., 1978). The most striking
finding which became evident in the organization maps produced was
that *short repeated sequences belonging to different evolutionary groups were highly
interspersed with each other* in every organization unit lacking nonrepeti-
tive sequences. These units occupy about half of the genomes investigated. In addition to the families of repeated sequences isolated from
many species as satellites with a basic repeated unit, in higher plants
the repeated sequences frequently have unrelated neighboring sequences
at different places in the chromosomes. This fact raises some questions
as to the origin and evolution of these sequences.

> This kind of repeated sequence interspersion pattern has also been found to oc-
> cur in parsley (KIPER and HERZFELD, 1978) and in much smaller amounts in animal
> species (CECH and HEARST, 1976; MARX et al., 1976; BRITTEN et al., 1976).

c) Origin and Evolution of Repetitive DNA

The evolution of repetitive DNA has been extensively studied by reas-
sociation kinetics of heteroduplexes formed between repeated sequences
of four cereals (oats, barley, wheat, and rye) (SMITH and FLAVELL, 1974;

RIMPAU, 1976). The number of common repetitive DNA sequence groups is the greater the more closely the species are related to each other.

As demonstrated for tandemly arranged satellite DNA (BERIDZE, 1975), for spacer sequences in functional repetitive genes like histone and rRNA genes (OVERTON and WEINBERG, 1978; ARNHEIM and SOUTHERN, 1977) and for intermediate repetitive DNA scattered throughout the genome (GILLESPIE, 1977) repetitive DNA sequences evolve at a higher rate than the overall genome (GALAU et al., 1976; MacGREGOR et al., 1976) and on the other hand remain rather homogeneous within a species (CHAMBERS et al., 1978; KLEIN et al., 1978).

Amplification of a certain sequence, diversification and reamplification processes were regarded to result in the formation of different repetitive sequence families (NAGL, 1978; RIMPAU et al., 1978). SMITH (1976, 1978) put forward the idea that illegitimate and unequal crossover are the initial steps in generating tandem duplications (Fig. 1). Once a DNA sequence has become repetitive, unequal crossover will then tend to maintain the homogeneity of the repeats in the tandem array without implying any natural selection or functional role for the repetitive sequence. This homogenization is due to the *crossover fixation*:

The deletion and duplication of blocks of repeats by unequal crossover lead to a sort of genetic drift of mutations occurring in a tandem repeat array. I.e. there exist some cell lines in which most mutations occurring in the array will

Illegitimate unequal crossover in unique DNA

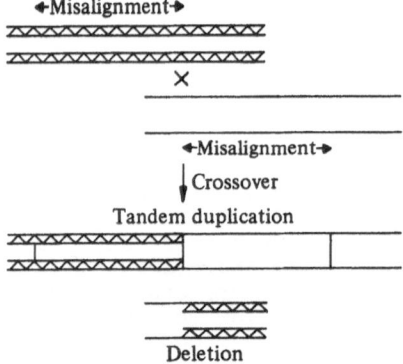

Legitimate unequal crossover in repetitive DNA

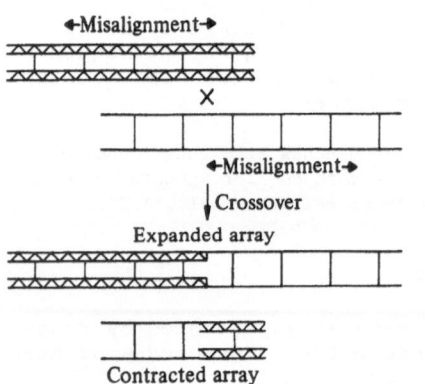

Fig. 1. In illegitimate unequal crossover the molecules are misaligned, resulting in a lengthened recombinant molecule carrying a tandem duplication of a segment equal in length to the misalignment distance, and a shortened recombinant molecule carrying a deletion of the same segment. X's marking the points of breakage and reunion. Vertical lines indicate the ends of the tandemly duplicated segments of the DNA sequence in the lengthened recombinant and the position of the deletion in the shortened recombinant. Legitimate unequal crossover occurs in repetitive DNA when the misalignment distance equals an integral number of repeats. Different mutations in the array will be excluded by contraction and subsequent expansion leading to a sort of genetic drift

be eliminated by the accumulation of deletions; thus an occasional mutation will spread by chance through the entire array when further duplications accumulate, thereby becoming fixed.

While crossover fixation leads to relative homogeneity of the repeats at any one moment, it would allow radical changes in the repeat pattern to occur gradually with time. Crossover fixation thus provides a simple explanation for the fact that repetitive sequences often differ markedly in different species yet are relatively homogeneous within a species.

MOORE et al. (1978) revealed in related sea urchins that nearly all the copies of all the families of repetitive sequences are different though these two animals are very similar in morphology, behavior and development. This says that *principal functions of genomic organization and the regulation of gene expression in development remain essentially unaltered while many individual repetitive sequence families have changed*. It thus seems that interspersed repetitive sequence families may be a mere product of chance displaying no special sequence dependent role or might retain their functional role even if their nucleotide sequence is altered but a family of related sequences established.

3. Fraction of the Genome Coding for Proteins

Higher animals and higher plants contain in their genomes a mass of DNA sufficient to code for more than 1.000.000 different proteins. Genetic considerations of mutation load (OHTA and KIMURA, 1971), complementation group analysis (YOUNG and JUDD, 1978) and UV irradiation studies (GOLDBERG et al., 1977; GIORNO and SAUERBIER, 1978) all indicated that only part of the whole nuclear genome may code for proteins. This can be tested directly by DNA-RNA hybridization, using saturation (GALAU et al., 1974) or kinetic analysis (BISHOP et al., 1974). Both methods have their shortcomings (AXEL et al., 1976).

The most serious problem is that due to internal limitations both methods would fail to detect a class of mRNA molecules being present only about once in 100 cells of the tissues examined (CAPLAN and ORDAHL, 1978). This could be somehow critical as in E. coli such a rare class of gene transcripts was found and as some measurements indicate that in eukaryotes genes not necessary in a specific cell type nevertheless are expressed at a very low rate (HUMPHRIES et al., 1976). Gene number measurements hence must be regarded to only provide minimum estimates. However, in Drosophila, rather consistent estimates of gene numbers were provided by genetic and biochemical methods used resulting in a minimum number of 5-10.000 structural genes (YOUNG and JUDD, 1978; LEVY and McCARTHY, 1975).

a) Gene Number Estimates in Plants

Gene number measurements for low and higher plants are given in Table 1. The large increase from a few thousand active genes in fungi to 10-20.000 in higher plants is obvious.

In tobacco leaves the 27.000 different mRNA's revealed by saturation analysis diverged largely from the 12.000 different species emerging from kinetic analysis. Based on similar divergent results with other plant tissues investigated KIPER (1979) argued that these high gene numbers found in plants by use of saturation hybridization methods result from some invalid assumptions underlying interpretation of data. According to his proposition the gene number would be reduced to some 15.000 in tobacco leaves.

No additional sequence complexity in poly(A)⁻mRNA was found, neither in parsley (KIPER et al., 1979a) nor in barley (KIPER et al., 1979b).

Though the division in different abundance classes is somewhat arbitrary the concept of different abundance classes in some cases at least seems valid (QUINLAN et al., 1978). In all cells and tissues examined so far, some genes are transcribed at a high rate and others at a rather low one. The differences between root callus and leaf given in the Table 1 for parsley really seem to reflect major differentiation processes as in leaf tissue some prominent proteins can be detected reflecting mRNA abundance (APEL and KLOPPSTECH, 1978) whereas in root callus suspension cells the protein spectrum is not ·dominated by some single species. Differential gene expression was directly proven for the gene of the large subunit of Ribulose Biphosphate Carboxylase in maize leaf cell types (LINK et al., 1978).

Table 1. Gene numbers and messenger abundance classes in plants

Organism	Gene number ScDNA saturation	Gene number RNA/cDNA kinetic analysis	Abundance	Reference
Yeast	3-4000	20	200	HEREFORD and ROSBASH (1977)
		450	20	
		2000	1-2	
Achlya	3300	29	1000	ROZEK et al. (1978)
		220	140	
		3000	11	
Tobacco	27.000	10	4500	GOLDBERG et al. (1978)
		770	340	
		11.300	17	
Parsley Root	13.700	960	60	KIPER et al. (1979a)
		10.500	4	
Leaf	10.000	28	1000	
		9.200	8	
Barley	14.000			KIPER et al. (1979b)

b) Gene Number Estimates in the Animal Kingdom

Whereas in the plant kingdom gene number estimates only recently were carried out the transcription of animal genomes was rather extensively surveyed. The emergent pattern from such studies on a variety of cultured cell types and tissues was that sequence complexities of mRNA populations cover the range of $7-15 \times 10^3$ different average sized mRNA sequences primarily transcribed from single-copy DNA, thus implying the use of only a very small proportion of the genome coding potential (GALAU et al., 1974; AXEL et al., 1976; HASTIE and BISHOP, 1976; RYFFEL and McCARTHY, 1975).

Some ·investigations reported gene number estimates exceeding the figure given above (BANTLE and HAHN, 1976; BISHOP et al., 1974; KAPLAN et al., 1978; SAVAGE et al., 1978). This might be due to rather complex tissues studied or cell lines which could be derepressed (BISHOP et al., 1974). Some results are controversial (BANTLE and HAHN, 1976; HASTIE and BISHOP, 1978).

Some thorough investigations were carried on the expression of different mRNA during differentiation. In sea urchin major differentiation
steps were studied and compared to individual functional tissues. It
emerged that all tissues shared a common set of mRNA (termed house-
keeping genes) of a few thousand species and contained another few
or many thousand active genes; early stages behaved the most active
(GALAU et al., 1976).

> Studies on structural gene expression in chicken (AXEL et al., 1976) and mouse
> (KAPLAN et al., 1978) showed that the vast majority of genes active in one
> tissue were active in other tissues studied, too. However, tissue-specific
> differences in mRNA populations were related to abundance as well as qualita-
> tive differences, i.e. mRNA most abundant in one tissue will sometimes be ab-
> sent in another tissue or more often will be expressed at a very low rate
> (YOUNG et al., 1976; HASTIE and BISHOP, 1976).

c) Fraction of the Genome Transcribed

The mRNA sequence complexity measurements revealed that in higher
eukaryotes only a minor fraction of the whole genome (below 5%) is
expressed in polysomal mRNA. However, in animals it was long realized,
that there exists a nucleus confined high complex and long RNA class
of heterodisperse length distribution (hnRNA) (LEWIN, 1975). Measurements indicated a four to ten times higher sequence complexity as compared to mRNA (HOUGH et al., 1975; GETZ et al., 1975; LEVY et al.,
1976); KLEIMAN et al., 1977). In *higher plants*, due to technical difficulties encountered, only recently *the existence of a high complex hnRNA
of size distribution longer than mRNA unambiguously was proven*. GOLDBERG et al.
(1978) found in tobacco leaves a hnRNA roughly four times as complex
as mRNA and BARTELS and KIPER (1979) in parsley demonstrated a hnRNA
more than three times as complex as mRNA and of mean length longer
than mRNA. This contrasts findings with lower plants where in yeast
(HEREFORD and ROSBASH, 1977), in Dictyostelium (FIRTEL and LODISH,
1973) and in achlya (TIMBERLAKE et al., 1977) total cell RNA complexity did not differ from that of mRNA.

In animals transcripts were investigated for their message content.
Though data obtained are somewhat conflicting it seems apparent that
in hnRNA message sequences often are separated from their 3'poly(A)
tail and their 5'cap structure by a non-message spacer (MACKEDOWSKI
and McCONKEY, 1978; HERMAN et al., 1976). In sea urchin it was shown
that mRNA sequences specific for one tissue nevertheless were exist-
ent in the hnRNA of other tissues examined. These individual hnRNA
molecules apparently did not reach the cytoplasm (WOLD et al., 1978).
The authors concluded that evidently a majority of hnRNA molecules
do not serve as mRNA precursors.

Obviously hence hnRNA could serve a further function, i.e. a regulatory one in gene expression. A model of hnRNA involved in gene regulation would have to take into account that all genes seem to be
transcribed into hnRNA whereas only tissues specific ones are transcribed and processed. This could mean that a DNA-hnRNA complex will
be the target or the means of turning off and on genes in differentiation.

4. Insertions in Genes

a) Genes Containing Intervening Sequences

The *new approach of cloning eukaryotic genes* in bacteria has confronted us with an *unexpected result*: In general, the coding sequences of genes on the DNA, those regions that will ultimately be translated in amino acid sequences are not contiguous but are interrupted by *"silent" DNA*.

An example is shown in Figure 2. The structural gene sequences coding for ovalbumin are separated by seven intervening sequences of various length. Base sequences expressed as amino acid sequences have been named *"extrons"* (GILBERT, 1978), sequences not expressed in this way after all that is known have been called *"introns"*. In the meantime, the number of genes is growing where such introns have been established (Table 2).

The limited knowledge gathered at present seems to indicate that eukaryotic genes which have not a split structure may be rare. No insertion was demonstrated in histone genes in general and in the Tyr-transfer RNA of Xenopus (CLARKSON et al., 1978) whereas in yeast Tyr-transfer RNA an insertion of fourteen nucleotides was shown near the anticodon coding region (GOODMAN et al., 1977).

β-globin genes as well as ovalbumin genes are encoded by single copy DNA as well as most other genes. Both these genes have been studied in very detail during the last two years. The *β-globin gene* of the mouse *shows two insertions,* one of 116 and another of 642 base pairs. In rabbit β-hemoglobin genes, the organization of the insertions, length and relative positions within the genes, corresponds to the β-globin gene of the mouse while the sequence of the insertions shows a strong evolutionary divergence (VAN DEN BERG et al., 1978). In hemoglobin and ovalbumin genes, the introns are comprised only of sequences which are present in the genome once or at highest, a few times. Because the protein coding as well as the intervening sequences belong to this sequence class, *unique sequences at least some thousand base pairs long constitute these structural genes* (this review, p. 199).

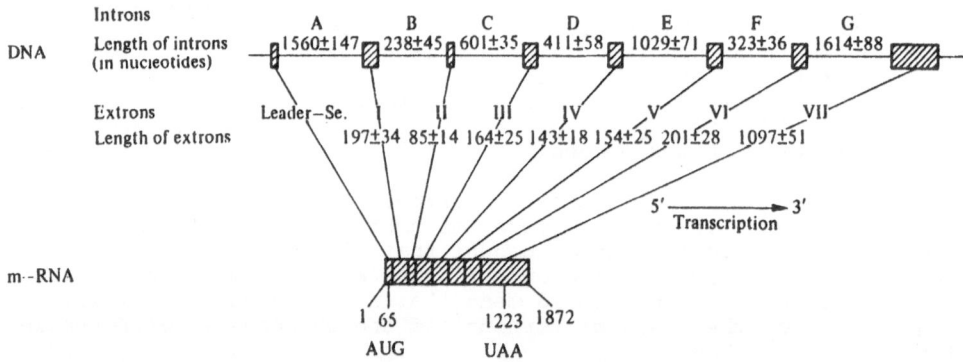

Fig. 2. Organization of the ovalbumin gene (GANNON et al., 1979). Introns A and E are only partially analyzed. The numbers given for these introns indicate only the portion of the introns known at present

Table 2. A survey of genes where intervening sequences have been shown

Gene	Reference
Transfer-RNA genes of yeast	GOODMAN et al. (1977)
Ribosomal RNA genes of Drosophila	GLOVER et al. (1975) GLOVER and HOGNESS (1977) WELLAUER and DAWID (1977)
β-Globin genes of rabbit and mouse	JEFFREYS and FLAVELL (1977a) JEFFREYS and FLAVELL (1977b) TILGHMAN et al. (1978)
Ovalbumin gene of chicken	BREATHNACH et al. (1977) DUGAICZYK et al. (1978)
Immunoglobulin genes	SAKANO et al. (1979)
Mitochondrial genes of yeast	BERNARDI (1978) HAID et al. (1979)

In the case of the hemoglobin genes, hybridization of the nucleus-confined precursor to the isolated gene sequences demonstrated that the *introns are transcribed* (TILGHMAN et al., 1978b). The overall length calculated for the precursor from m-RNA length plus total intron length is comparable to the length of one of several defined precursors isolated from the nucleus. However, this does not prove that the whole excess sequence complexity of the hnRNA compared to mRNA would be comprised of such introns.

Nothing is known about the enzymes which cut out the introns and splice the extrons within the nucleus. Addition or deletion of only one nucleotide would shift the frame of the triplett reading during translation. A hypothetic scheme of the splicing mechanism is presented in Figure 3.

b) Speculations Concerning the Functions of Inserted Sequences

The results which have been presented may be of importance in various areas of biology. *The splicing mechanism may be the basis for the evolution of new proteins.* Protein evolution was thought until now to proceed primarily by point mutations resulting from amino acid substitutions in already existing proteins (DICKERSON, 1971). It can be imagined that new protein functions have evolved by recombination of extrons coding for different parts of various established proteins. Protein evolution based on such a mechanism may be much faster as previously calculated on the basis of point mutations (GILBERT, 1978; DARNELL, 1979; REANNEY, 1979). It is interesting that TONEGAWA and coworkers (SAKANO et al., 1979) found that the hinge region and the domains of the immunoglobulin molecule are exactly represented by extrons of the respective immunoglobulin gene. Until now *no actual function of the introns was uncovered. In contrast, arguments have accumulated in favour of the described historical role* in protein evolution. This, however, will not explain the evolutionary constancy of introns like the β-globin intervening sequences showing only minimal divergence in length between mouse and rabbit.

The technique of cloning obviously will open new frontiers in determination of primary structure of defined DNA-fragments, i.e. nucleotide sequences. More difficult to foresee is the potential of the new methods to solve problems

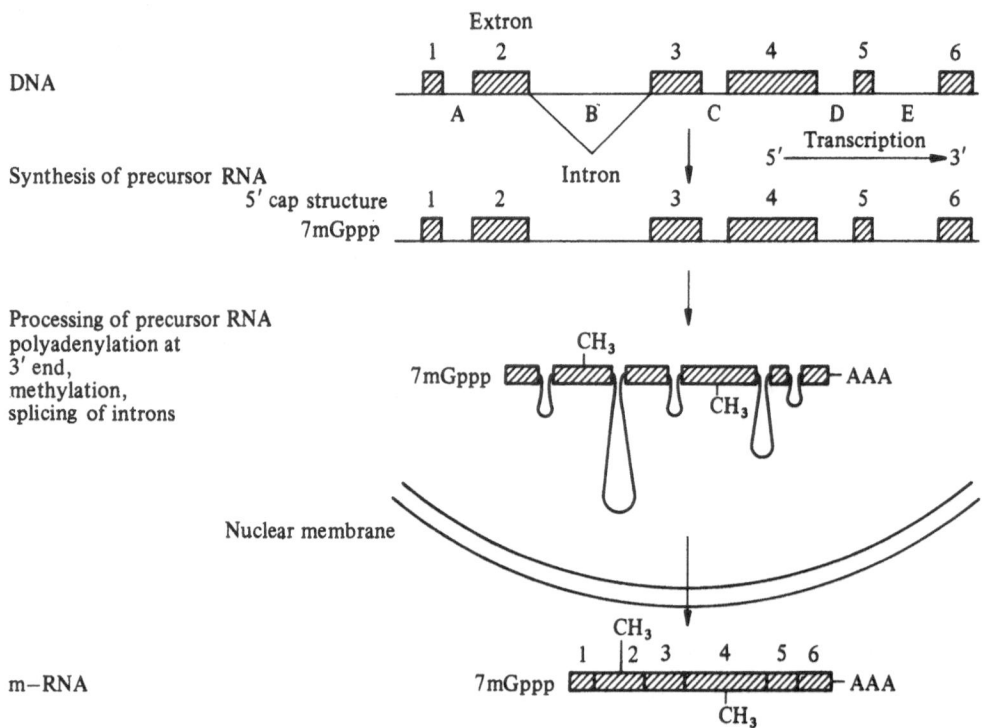

Fig. 3. Scheme of hnRNA processing. The basic but not strictly proved feature of
the model shown here is, that the regions of RNA which will become adjacent, are
brought together by looping out the intervening sequences and then covalently
joined via an intramolecular ligation reaction. The specificity of looping out
and subsequently ligation may be provided by nucleic acid-protein interactions

concerning the functions of DNA sequences in vivo. Provided the complete nucleotide sequence is known, how to identify promoters, regulatory sequences
or terminators? This difficulty is illustrated by our present ignorance about
the function of the introns described above, until now it was not possible to
develop an experiment which could test a special role in gene regulation or
any other function. One reason for this is that eukaryotic genes on plasmids
in bacteria are replicated but not expressed.

5. New Tools in the Analysis of Eukaryotic Gene Expression

a) Cloning in Eukaryotic Cells

The information encoded in eukaryotic genes can be properly recognized
and expressed only in eukaryotic cells. *Cloning in eukaryotic cells would
therefore be an important step in the elucidation of the in vivo function of this
genetic material.*

Systems have already been developed for cloning in yeast cells (BEGGS,
1978). *SV-40 Tumor Virus DNA* was modified to be used as a vector for animal genes in cell cultures (MULLIGAN et al., 1979). The genome of
SV-40 is a small, double-stranded, circular DNA molecule, in many aspects similar to the plasmids of bacteria (REDDY, 1978; FIERS et al.,
1978).

The double stranded Caulimo Virus DNA's may become a vector of higher
plant genes (HULL, 1978), yet the now *most promising candidate as a vector
of plant genes may be the Ti-Plasmid of Agrobacterium tumefaciens* (ZAENEN et al.,
1974; SCHELL, 1975). This plasmid, a constituent of the cells of Agrobacterium tumefaciens transforms plant cells. Part of the plasmid DNA
is integrated in the plant genome and replicated synchronously with
the host genome.

> For the maintenance and growth of the transformed cells, the presence of
> Agrobacterium is not further necessary (BROWN and WOOD, 1976). In contrast
> to untransformed cells, the growth of the transformed tissue on agar is not
> dependent on the addition of the hormones auxin and cytokinin. A second char-
> acteristic feature of the transformed cells is the production of unusual amino
> acids, the best studied of these are octopine and nopaline (PETIT et al.,
> 1970), derivatives of arginine which are not found in normal plant tissue.
> The enzymes which synthesize these compounds are encoded on the plasmid DNA
> (SCHELL and van MONTAGU, 1977; CHILTON et al., 1977). Agrobacterium is espe-
> cially adapted to utilize these amino acids for energy needs or synthetic
> pathways in its intermediary metabolism.

So, this is *a case of genetic colonisation of higher plant cells by bacteria*
evolved in evolution and not established due to genetic manipulation
by molecular biologists. Likewise, this system offers unique chances
for the genetic manipulation of higher plants and has been already
used to do this (SCHELL and van MONTAGU, 1978).

DNA-fragments to be introduced in plant cells were first ligated into
plasmid by standard methods developed in the field of gene technology
(HOLLENBERG, 1978). Subsequently Agrobacterium tumefaciens cultures
were transformed by the manipulated plasmid DNA and cells grown for a
number of generations so getting a sufficient multiplicity of the
plasmid necessary for transformation of plant cells. A portion of the
plasmid genome is integrated in a stable form in the plant genome
(CHILTON et al., 1978; DE PICKER et al. (1978), and may transfer additional foreign genetic material.

Besides strains of the plasmid which induce uncontrolled malignant
growth, other strains are known which integrate as well in the plant
genome but do not change the normal pattern of growth and morphogen-
esis. These strains may be more important for future experimentation
and applied research.

> The genetic manipulation of plants has a noticable advantage compared to equi-
> valent animal systems because of the until now unique system of the Ti-Plasmid.
> Yet, equivalent systems well suited for the cloning of animal genes, based pri-
> marily on SV-40 as a genetic vector, are rather promoted in their development.

b) "Reversed Genetics"

Eukaryotic cloning systems may serve as a major basic tool in the elucidation of signals encoded in the DNA and directing transcription,
processing of hn-RNA and translation but other more refined techniques
may become equally important. In advance of this, however, the nucleotide sequence of the genetic regions to investigate should be known
at least partially. *Due to the introduction of new sequencing techniques which
have greatly simplified* and fastened up *this task* (SANGER and COULSON, 1975;
MAXAM and GILBERT, 1977) *it is now possible to sequence molecules as large as
some thousand nucleotides within a limited time* (REDDY, 1978; FIERS et al.,
1978). The regions of the DNA fragment presumed to have an interesting
function may be sequenced after cloning in E. coli to provide DNA for
analysis and manipulation.

Subsequently other *new techniques permit to introduce discrete nucleotide substitutions in predetermined regions of a sequenced genome* (e.g. plasmid) in
vitro (FLAVELL and SMITH, 1976; SCHOTT and KÖSSEL, 1973). Using this
method *it should be possible to study structure-function relationships in structural as well as regulatory sequences.*

The classical genetic approach depends on mutations which arise at
random either spontaneously or as a consequence of mutagenic agents.
Mutants are usually selected or screened for by virtue of some desired
phenotype. Finally it is tried to analyze the genes at the molecular
level, at most without success. The advent of hybrid technology has
changed this situation. The gene of interest and its modified homologs
will be integrated in vitro, with its neighbouring regions, into a
so called vector, cloned and amplified in bacterial cells and rein-
troduced into a eukaryotic cell. This concept of reversed genetics
allows to study the effect of these manipulations in vivo and in
vitro (WEISSMANN, 1978).

References

APEL, K., KLOPPSTECH, K.: Eur. J. Biochem. 85, 581-588 (1978). -
ARNHEIM, N., SOUTHERN, E.M.: Cell 11, 363-370 (1977). - AXEL, R.,
FEIGELSON, P., SCHUTZ, G.: Cell 7, 247-254 (1976).

BANTLE, J.A., HAHN, W.E.: Cell 8, 139-150 (1976). - BARTELS, D.,
KIPER, M., in: Genome and Chromatin. Organization, Function, Evolution,
eds. W. NAGL, V. HEMLEBEN, F. EHRENDORFER. Wien, New York: Springer
1979 (in press). - BEGGS, J.D.: Nature (London) 275, 104-109 (1978).
- BERG, van den J., DOYEN, A. van, MARTEI, N., SCHAMBÖCK, A., GROSS-
VELD, G., FLAVELL, A.R., WEISSMANN, C.: Nature (London) 276, 37-44
(1978). - BERIDZE, T.: BBA 262, 393-396 (1975). - BERNARDI, G.:
Nature (London) 274, 328-333 (1978). - BISHOP, J.E., MORTON, J.G.,
ROSBASH, M., RICHARDSON, M.: Nature (London) 250, 199-204 (1974). -
BREATHNACH, R., MANDEL, J.L., CHAMBON, P.: Nature (London) 270, 314-
319 (1977). - BRITTEN, R.J., GRAHAM, D.E., EDEN, F.C., PAINCHAUD,
D.M., DAVIDSON, E.H.: J. Mol. Evol. 9, 1-23 (1976). - BROWN, A.C.,
WOOD, H.N.: Proc. Natl. Acad. Sci. USA 73, 496-500 (1976).

CAPLAN, A.J., ORDAHL, C.P.: Science 201, 120-130 (1978). - CECH, T.R.,
HEARST, J.E.: J. Mol. Biol. 100, 227-256 (1976). - CHAMBERS, C.A.,
SCHELL, M.P., SKINNER, D.M.: Cell 13, 97-110 (1978). - CHILTON, M.D.,
DRUMMOND, M.H., MERLO, D.J., SCIAKY, D., MONTOYA, A.L., GORDON, M.P.,
NESTER, E.W.: Cell 11, 263-271 (1977). - CHILTON, M.D., DRUMMOND,
M.H., MERLO, D.J., SCIAKY, D.: Nature (London) 275, 147-149 (1978). -
CLARKSON, S., MÜLLER, F., KUER, V.: Abstr. 12th FEBS-Meet. Dresden,
No. 3033 (1978). - CRAIN, W.R., DAVIDSON, E.H., BRITTEN, R.J.: Chromo-
soma 59, 1-12 (1976).

DARNELL, J.E.: Science 202, 1257-1260 (1979). - DAVIDSON, E.H., BRITTEN,
B.J.: Q. Rev. Biol. 48, 555-613 (1973). - DAVIDSON, E.H., HOUGH, B.R.,
AMENSON, C.S., BRITTEN, R.J.: J. Mol. Biol. 77, 1-23 (1973). - DAVIDSON,
E.H., GALAU, G.A., ANGERER, R.C., BRITTEN, R.J.: Chromosoma 51, 253-
259 (1975). - DAVIDSON, E.H., KLEIN, W.H., BRITTEN, R.J.: Dev. Biol.
55, 69-84 (1977). - DE PICKER, A., MONTAGU, M., SCHELL, J.: Nature
(London) 275, 150-152 (1978). - DICKERSON, R.: J. Mol. Evol. 1, 26-45
(1971). - DUGAICZYK, A., WOO, S.L., LAI, C.E., MYLES, L., MACE, J.,
McREYNOLDS, L., O'MALLEY, W.B.: Nature (London) 274, 328-333 (1978).

EDEN, F.C., HENDRICK, J.P.: Biochemistry 17, 5838-5844 (1978). -
EPPLEN, J.T., LEIPOLDT, M., ENGEL, W., SCHMIDTKE, J.: Chromosoma 69,
307-321 (1978).

FIERS, W., CONTRERAS, R., HAEGEMAN, G., ROGIERS, R., VOORDE, van den,
D., HEUVERSWYN, van, H., HERREWEGHE, van, J., VOLCHAERT, G., YSEBAERT,
M.: Nature (London) 273, 113-120 (1978). - FIRTEL, R.A., LODISH, H.F.:
J. Mol. Biol. 79, 295-314 (1973). - FLAVELL, R.A., SABO, D.L., BANDLE,
E.F., WEISSMANN, C.: J. Mol. Biol. 89, 255-272 (1974). - FLAVELL, R.A.,
WAALWIJK, C., JEFFREYS, A.J.: Biochem. Soc. Trans. 6, 742-746 (1978).
- FLAVELL, R.B., SMITH, D.B.: Hereditary 37, 231-252 (1976). - FLAVELL,
R.B., BENNETT, M.C., SMITH, J.B., SMITH, D.B.: Biochem. Genet. 12,
257-269 (1974). - FLAVELL, R.B., RIMPAU, J., SMITH, D.B.: Chromosoma
63, 205-222 (1977).

GALAU, G.A., BRITTEN, R.J., DAVIDSON, E.H.: Cell 2, 9-20 (1974). -
GALAU, G.A., KLEIN, W.H., DAVIS, M.M., WOLD, B.J., BRITTEN, R.J.,
DAVIDSON, E.H.: Cell 7, 487-505 (1976). - GANNON, F., O'HARE, K.,
PERRIN, F., LE PEMEC, L.P., BENOIST, C., COCHET, M., BREATHNACH, R.,
ROYAL, A., GARAPIN, A., CAMI, B., CHAMBON, P.: Nature (London) 278,
428-438 (1979). - GETZ, M.J., BIRNIE, G.D., YOUNG, B.D., Mac PHAIL,
E., PAUL, J.: Cell 4, 121-129 (1975). - GILBERT, W.: Nature (London)
271, 501 (1978). - GILLESPIE, D.: Science 196, 889-891 (1977). -
GIORNO, R., SAUERBIER, W.: Proc. Natl. Acad. Sci. USA 75, 4374-4378
(1978). - GLOVER, D.M., HOGNESS, D.S.: Cell 10, 167-176 (1977). -
GLOVER, D.M., WHITE, R.L., FINNEGAN, D.J., HOGNESS, D.S.: Cell 5,
149-157 (1975). - GOLDBERG, R.B.: Biochem. Genet. 16, 45-68 (1978). -
GOLDBERG, R.B., HOSCHEK, G., KAMALAY, J.C.: Cell 14, 123-131 (1978).
- GOLDBERG, S., WEBER, J., DARNELL, J. Jr.: Cell 10, 617-621 (1977).
- GOODMAN, H.M., OLSON, M.V., HALL, B.D.: Proc. Natl. Acad. Sci. USA
74, 5453-5457 (1977).

HAID, A., SCHNEYEN, R.J., GROSCH, G., BECHMANN, H., KAUDEWITZ, F.:
Hoppe Seylers Z. Physiol. Chem. 360, 275 (1979). - HASTIE, N.D.,
BISHOP, J.O.: Cell 9, 761-774 (1976). - HEREFORD, L.M., ROSBASH, M.:
Cell 10, 453-462 (1977). - HERMAN, R.C., WILLIAMS, J.G., PENMAN, S.:
Cell 7, 429-437 (1976). - HOLLENBERG, C.P.: Prog. Bot. 40, 211-230
(1978). - HOUGH, B.R., SMITH, M.J., BRITTEN, R.J., DAVIDSON, E.H.:
Cell 5, 291-299 (1975). - HUDSPETH, M.E.S., GOLDBERG, R.B., TIMBERLAKE,
W.E.: Proc. Natl. Acad. Sci. USA 74, 4332-4336 (1978). - HULL, R.:
TIBS 3, 254-256 (1978). - HUMPHRIES, S., WINDASS, J., WILLIAMS, R.:
Cell 7, 267-277 (1976).

JEFFREYS, A.J., FLAVELL, R.A.: Cell 12, 429-439 (1977a); - Cell 12,
1097-1108 (1977b).

KAPLAN, B.B., SCHACHTER, B.S., OSTERBURG, H.H., DE VELLIS, J.S.,
FINCH, C.E.: Biochemistry 17, 5516-5524 (1978). - KIPER, M.: Nature
278, 279-280 (1979). - KIPER, M., KÖCHEL, H., BARTELS, D., in: Genome
and Chromatin. Organization, Function, Evolution, eds. W. NAGL, V.
HEMLEBEN, F. EHRENDORFER. Wien, New York: Springer 1979b (in press).
- KIPER, M., HERZFELD, F.: Chromosoma 65, 335-351 (1978). - KIPER, M.,
BARTELS, D., HERZFELD, F., RICHTER, G.: Nucleic Acids Res. 6, 1961-
1978 (1979a). - KLEINMAN, L., BIRNIE, G.D., YOUNG, B.D., PAUL, J.:
Biochemistry 16, 1218-1223 (1977). - KLEIN, W.H., THOMAS, T.L., LAI,
C., SCHELLER, R.H., BRITTEN, R.J., DAVIDSON, E.H.: Cell 14, 889-900
(1978).

LEDER, P. et al., in: Cold Spring Harbor Symp. Quant. Biol. Vol. VLII,
Part 2, 915-920. Cold Spring Harbor (1978). - LEVY, B.W., McCARTHY,
B.J.: Biochemistry 14, 2440-2446 (1975). - LEVY, B.W., JOHNSON, C.B.,
McCARTHY, B.J.: Nucleic Acids Res. 3, 1777-1789 (1976). - LEWIN, B.:

Cell 4, 11-20 (1975). - LINK, G., COEN, D.M., BOGORAD, L.: Cell 15,
725-731 (1978). - MacGREGOR, H.C., MIZUNO, S., VLAD, M.: Chromosomes
Today 5, 331-339 (1976). - MACKEDOWSKI, V.V., McCONKEY, E.: Eur. J.
Biochem. 90, 397-404 (1978). - MANNING, J.E., SCHMID, C.W., DAVIDSON,
N.: Cell 4, 141-155 (1975). - MARX, K.A., ALLEN, J.R., HEARST, J.E.:
BBA 425, 129-147 (1976). - MAXAM, A.M., GILBERT, W.: Proc. Natl. Acad.
Sci. USA 74, 560-564 (1977). - MOORE, G.P., SCHELLER, R.H., DAVIDSON,
E.H., BRITTEN, R.J.: Cell 15, 649-660 (1978). - MULLIGAN, R.C., HOWARD,
B.H., BERG, P.: Nature (London) 277, 108-114 (1979). - MURRAY, M.G.,
CUELLAR, R.E., THOMPSON, W.F.: Biochemistry 17, 57-61 (1978).

NAGL, W.: Endopolyploidy and Polyteny in Differentiation and Evolution. Amsterdam: North-Holland 1978.

OHTA, T., KIMURA, M.: Nature (London) 233, 118-120 (1971). - OVERTON,
G.C., WEINBERG, E.S.: Cell 14, 247-257 (1978).

PEARSON, W.R., SMITH, S.L., WU, J.R., BONNER, J.: Plant Physiol. 62,
112-115 (1978). - PETIT, A., DELHAYE, S., TEMPÉ, J., NEVEL, G.:
Physiol. Vég. 8, 205-213 (1970).

QUINLAN, T.J., BEELER, G.W., COX, R.F., ELDER, P.K., MOSES, H.L.,
GETZ, M.J.: Nucleic Acids Res. 5, 1611-1623 (1978).

REANNEY, D.: Nature (London) 277, 598-600 (1979). - REDDY, V.B. et
al.: Science 200, 494-502 (1978). - RIMPAU, J.: Habilitationsschrift,
Universität Göttingen (1976). - RIMPAU, J., SMITH, D., FLAVELL, R.B.:
J. Mol. Biol. 123, 327-359 (1978). - ROOP, D.R., NORDSTROM, J.L.,
TSAI, S.Y., TSAI, M.Y., O'MALLEY, B.W.: Cell 15, 671-685 (1978). -
ROZEK, C.E., ORR, W.C., TIMBERLAKE, W.E.: Biochemistry 17, 716-722
(1978). - RYFFEL, G.U., McCARTHY, B.J.: Biochemistry 14, 1379-1385
(1975).

SAKANO, H., ROGERS, J.H., HÜPPI, K., BRACK, C., TRANNECKER, A., MAKI,
R., WALL, R., TONEGAWA, S.: Nature (London) 277, 627-633 (1979). -
SANGER, F., COULSON, A.R.: J. Mol. Biol. 94, 441-448 (1975). - SAVAGE,
M.J., SALA-TREPAT, J.M., BONNE, R.J.: Biochemistry 17, 462-467 (1978).
- SCHELL, J., in: Genetic Manipulation with Plant Material, Vol. 3,
ed. R. LEDOUX, NATO Advanced Study Institute Series, 163-181 (1975).
- SCHELL, J., van MONTAGU, M.: Brookhaven Symp. Biol. 29, 36-49 (1977);
- 12th FEBS-Meet. Dresden, Abstr. No. 3.533 (1978). - SCHOTT, H.,
KÖSSEL, H.: J. Am. Chem. Soc. 95, 3778-3785 (1973). - SMITH, D.B.,
FLAVELL, R.B.: Biochem. Genet. 12, 243-256 (1974); - BBA 474, 82-97
(1977). - SMITH, D.B., RIMPAU, J., FLAVELL, R.B.: Nucleic Acids Res.
3, 2811-2825 (1976). - SMITH, G.P.: Science 191, 528-535 (1976); -
TIBS 3, N 34-36 (1978).

TILGHMAN, S.M., TIEMEIER, D.C., SEIDMAN, J.G., PETERKI, B.M., SULLIVAN,
M., NAIZEL, J.V., LEDER, P.: Proc. Natl. Acad. Sci. USA 75, 725-729
(1978a). - TILGHMAN, S.M., CURTIS, P., TIEMEIER, D.C., LEDER, P.,
WEISSMANN, C.: Proc. Natl. Acad. Sci. USA 75, 1309-1313 (1978b). -
TIMBERLAKE, W.E., SHUMARD, D.S., GOLDBERG, R.B.: Cell 10, 623-632
(1977).

WALBOT, V., DURE, L.S.: J. Mol. Biol. 31, 349-370 (1976). - WALBOT,
V., GOLDBERG, R.B., in: Nucleic Acids in Plants, eds. J.W. DAVIES,
T. HALL. Cleveland, Ohio: CRC Press 1978. - WEISSMANN, C.: TIBS 3,
109-111 (1978). - WELLAUER, P.K., DAWID, J.B.: Cell 10, 193-212 (1977).
- WELLS, R., ROYER, H.D., HOLLENBERG, C.P.: Mol. Gen. Genet. 147,
503-509 (1976). - WOLD, B.J., KLEIN, W.H., HOUGH-EVANS, B.R., BRITTEN,
R.J., DAVIDSON, E.H.: Cell 14, 941-950 (1978). - WOO, S.L. et al.:

Proc. Natl. Acad. Sci. USA 75, 3688-3692 (1978).

YOUNG, B.D., BIRNIE, G.D., PAUL, J.: Biochemistry 15, 2823-2829
(1976). - YOUNG, M.W., JUDD, B.H.: Genetics 88, 723-742 (1978).

ZAENEN, I., van LAREBEKE, N., TEUCHY, H., van MONTAGU, M., SCHELL, J.:
J. Mol. Biol. 86, 109-127 (1974). - ZIMMERMANN, J.L., GOLDBERG, R.B.:
Chromosoma 59, 227-252 (1977).

Professor Dr. FRANK HERZFELD
Dr. MANUEL KIPER
Institut für Botanik
Universität Hannover
Herrenhäuser Str. 2
D 3000 Hannover 21

V. Extranucleäre Vererbung

Die Morphologie extranucleärer Erbträger im Verlauf des Lebenszyklus

Von CARL-GEROLD ARNOLD and KARL PETER GAFFAL

1. Einleitung

Im Jahre 1909 wurden von Correns an *Mirabilis jalapa* und von Baur an *Pelargonium zonale* erstmals nichtmendelnde Vererbungsphänomene beschrieben. Damit stellte sich die Frage nach Erbinformationsträgern außerhalb des Zellkerns. Heute wissen wir, daß für die extranucleäre Vererbung in erster Linie die DNS der Mitochondrien und Plastiden verantwortlich ist.

Durch den nichtmendelnden, oft einelterlichen Vererbungsmodus war das Auffinden extranucleärer Vererbungsvorgänge relativ einfach. Weitergehende Untersuchungen gestalteten sich dagegen schwierig, weil es lange Zeit nicht gelang, experimentell induzierte extranucleäre Mutanten sowie Rekombinanten aufzufinden. Dies wurde erst möglich, nachdem genetische Untersuchungen auch an eukaryotischen Mikroorganismen betrieben wurden, die es gestatteten, Mutanten des Mitochondrien- und Plastidengenoms mit Hilfe von Selektionsmedien aus einer großen Individuenzahl zu isolieren. Als man dann bei *Chlamydomonas* gefunden hatte, daß die Plastidengene nicht nur einelterlich, sondern in gewisser, wenn auch geringer Häufigkeit zweielterlich vererbt werden, konnte man mit diesen Selektionsmedien auch Rekombinationen des Plastidengenoms auffinden. Wenig später fand man bei *Saccharomyces* Rekombinanten der Mitochondrien-DNS. Die folgenden, zahlreichen Rekombinationsuntersuchungen ermöglichten Genkartierungen des Mitochondriengenoms der Hefe und des Chloroplastengenoms von *Chlamydomonas*. Die Ergebnisse dieser Forschungen wurden von FINCHAM et al. (1979), BIRKY (1978) und von GILLHAM (1978) ausführlich und übersichtlich dargestellt.

Nach der Entdeckung extranucleärer Rekombinationen stellte sich jedoch die Frage, wie die DNS dieser Zellorganellen in einen direkten Kontakt kommen kann und in welchem Moment des Lebenszyklus das geschieht. Eine Antwort ließ sich mit den früheren Vorstellungen kaum finden, wonach Mitochondrien und Plastiden mehr oder weniger unveränderliche morphologische Einheiten sind.

2. Variabilität der Mitochondrien- und Plastidengestalt

Neuere Untersuchungen ergaben, daß die *Ultrastruktur der Mitochondrien* nicht unveränderlich ist, sondern sogar *stark variieren kann*, zum Beispiel in Abhängigkeit vom ontogenetischen Entwicklungsstadium, von der Art des Gewebes, von Kulturbedingungen oder auch als Folge von Krankheiten (BINDER und SLUGA, 1977; DAMSKY, 1976; DUCKETT und TOTH, 1977; CAROTHERS et al., 1977; GRODUMS, 1977; HALLAM et al., 1972; HANZLIKOWA und SCHIAFFINO, 1977; HEGARTY et al., 1978; HERMAN und SWEENEY, 1977; HIGGINS et al., 1976; KALASHNIKOVA, 1978; KRAUSE et al., 1972; SALMONS et al., 1978; SENGER und SAACKE, 1970; SMITH et

al., 1972; SMITH und PAGE, 1977; VERTEL und FISCHMAN, 1977). Morphologische Veränderungen können aber auch induziert werden. So entstehen
nach Einwirkung von Cuprizon in Gewebekulturen der Maus kompakte Riesenmitochondrien (ASANO et al., 1978; KEYHANI, 1973; MALOFF et al.,
1978a,b; PUBLICOVER et al., 1977; SENDA, 1977; WAKABAYASHI et al.,
1977, 1978). Solche Riesenmitochondrien wurden nach Einwirkung von
Chloramphenicol und einigen anderen Antibiotika nicht nur in Gewebekulturen, sondern auch bei eukaryotischen Einzellern beobachtet
(ADOUTTE et al., 1972; ALBRING et al., 1973, 1975; BEN-SHAUL und MAR-
KUS, 1969; NEUMANN und PARTHIER, 1972; WABNER und RAFFAEL, 1975, 1977).

Besonders überraschend waren die Ergebnisse von Untersuchungen, bei
denen auf Grund elektronenmikroskopischer Serienschnitte die räumliche
Gestalt der Mitochondrien rekonstruiert wurde. So fand man in vegetativen Zellen von *Euglena* und *Chlamydomonas* große, *netzförmig verzweigte Mitochondrien* (ARNOLD et al., 1972; OSAFUNE, 1973; OSAFUNE et al., 1975a,
b,c; SCHÖTZ et al., 1972), in anderen Untersuchungen konnte bei *Chlorella, Euglena, Polytoma, Polytomella*, in Gameten von *Chlamydomonas*, bei *Try-
panosoma, Crithidia, Blastocrithidia, Cryptobia, Saccharomyces* und *Candida* nur
ein einziges, netzförmig verzweigtes Riesenmitochondrium festgestellt
werden; bei *Olpidium* waren es höchstens zwei Mitochondriennetze (ATKIN-
SON et al., 1974; BURTON und MOORE, 1974; CALVAYRAC et al., 1972;
CALVAYRAC et al., 1974; DAVISON und GARLAND, 1977; GAFFAL und KREUTZER,
1977; GROBE und ARNOLD, 1975; HOFFMANN und AVERS, 1973; LANGE und
OLSON, 1976; LEEDALE und BUETOW, 1970; PAULIN, 1975, 1977; PELLEGRINI,
1976; PELLEGRINI und PELLEGRINI, 1976; VICKERMAN, 1977). Netzförmig
verzweigte Mitochondrien sind aber nicht nur bei eukaryotischen Mikroorganismen gefunden worden, sondern auch in höheren Pflanzen sowie bei
Mäusen, Ratten und sogar Menschen, wenngleich hier nichts über die
Zahl solcher Netzmitochondrien pro Zelle ausgesagt wird (BAKEEVA et
al., 1978; KOUKL et al., 1977; RANCOURT et al., 1975; ROHFRITSCH, 1978;
ROHR, 1978; THIERY und BERGERON, 1976; WORTH und LUCAS, 1978).

> Wenn diese Ergebnisse auch besondere Aktualität besitzen, so darf nicht ver-
> schwiegen werden, daß netzartig verzweigte Mitochondrien bereits 1911 an Rat-
> tenpankreaszellen beobachtet wurden und auch bei Flagellaten, z.B. bei *Polytoma*,
> wurde das Chondriom auf Grund lichtmikroskopischer Untersuchungen als ein netz-
> förmiges Zellorgan beschrieben (MISLAVSKY, 1911; PRINGSHEIM, 1963 - dort weitere
> Literatur).

Bei neueren Untersuchungen an anderen Flagellaten, wie *Hemiselmis rufes-
cens* oder *Chromulina pusilla* wurde nur ein einziges, unverzweigtes, eher
wurmförmig aussehendes Mitochondrium entdeckt. Bei *Gonyostomum* und
Vacuolaria waren dagegen viele kleine Mitochondrien beobachtet worden,
die keinerlei Verzweigungen zeigten und auch bei *Paramecium* fand man
keine Hinweise auf verzweigte Mitochondrien, wenngleich hier keine
dreidimensionalen Rekonstruktionen durchgeführt wurden (HEYWOOD, 1977;
PERASSO und BEISSON, 1978; SANTORE und GREENWOOD, 1977).

Die Untersuchungen zur dreidimensionalen Gestalt der Mitochondrien
brachten also höchst unterschiedliche, zum Teil sogar widersprüchliche
Ergebnisse. Sie finden jedoch eine mögliche Erklärung durch die Fest-
stellung, daß die räumliche *Gestalt der Mitochondrien sich im Laufe des Lebenszyklus ändert* bzw. ändern kann. So beobachtete man bei der Hefe am Ende
der exponentiellen Wachstumsphase ein einziges großes, sehr stark verzweigtes Mitochondrium, in der stationären Phase dagegen 40 bis 50
kleine und unverzweigte Einzelmitochondrien. Bei *Chlamydomonas* wurden
in sich teilenden Zellen nur kleine, einfach gebaute Mitochondrien
gefunden und nicht die charakteristischen Netzmitochondrien adulter
Zellen (OSAFUNE et al., 1976; STEVENS, 1974, 1977).

Während die Mitochondrien-Morphologie auch innerhalb eines Organismus offenbar stark variieren kann, findet man nur wenige Angaben über morphologische Veränderungen von Plastiden, wenngleich die Umwandlung einer linsenförmigen Plastide in eine becherförmige bereits 1915 an Moosen beschrieben wurde (SAPEHIN, 1915). Bemerkenswert sind die Beobachtungen an *Euglena*, wonach in bestimmten Entwicklungsstadien auch die Plastiden netzartig verzweigt sind, während in anderen Stadien des Teilungszyklus nur Einzelplastiden vorliegen (CALVAYRAC und LEFORT-TRAN, 1976). Bei *Oenothera* konnte festgestellt werden, daß die Chloroplasten nicht nur glatte Außenmembranen besitzen, sondern auch tubuläre Ausstülpungen bilden können (SCHÖTZ und DIERS, 1975).

Bei Untersuchungen an höheren Pflanzen wurde gelegentlich die Vermutung geäußert, daß Plastiden fusionieren können (ESAU, 1972, dort weitere Literatur). Mit Sicherheit wurden solche Plastidenfusionen aber erst in der Zygote von *Chlamydomonas* nachgewiesen (BLANK et al., 1978; BROWN et al., 1968; CAVALIER-SMITH, 1970; GROBE und ARNOLD, 1975; LADYGIN et al., 1975). Diese Beobachtung ist jedoch für die Genetik von großer Bedeutung, denn sie liefert die morphologische Grundlage für die Rekombination von Plastidengenomen (siehe Seite 212).

Aus dieser Übersicht wird deutlich, daß *Mitochondrien und Plastiden keine unveränderlichen morphologischen Einheiten* darstellen. Beide Zellorganellen sind aber auch Träger genetischer Informationen und wir haben inzwischen tiefe Einblicke gewonnen über genetische Vorgänge in Mitochondrien und Plastiden. Umso erstaunlicher ist es, daß man heute, genau 70 Jahre nach Entdeckung der extranucleären Vererbung, nur oberflächlich über das morphologische Verhalten dieser Zellorganellen orientiert ist.

In jüngster Zeit wurden nunmehr systematische Untersuchungen über die Morphologie der Mitochondrien und Plastiden innerhalb des vegetativen und generativen Lebenszyklus durchgeführt (ARNOLD und GAFFAL, 1979; GAFFAL, 1978; GAFFAL und SCHNEIDER, 1978). Objekt dieser Untersuchungen ist in erster Linie der Phytoflagellat *Polytoma*, der trotz heterotropher Lebensweise enge phylogenetische Beziehungen zu *Chlamydomonas* hat. Erschwert werden diese Untersuchungen, weil es bisher noch nicht gelungen ist, die hier verwendete *Polytoma papillatum* in Synchronkulturen anzuziehen. Da aber andererseits die untersuchten Entwicklungsstadien aus einem gleichen Kulturansatz stammen, sind umweltbedingte Variationen, die besonders bei Mitochondrien zu befürchten wären, ausgeschlossen.

3. Die räumliche Gestalt der Mitochondrien im vegetativen Lebenszyklus

Wie bereits erwähnt (siehe Seite 213), ist bei *Polytoma* ein einziges, netzförmig verzweigtes Mitochondrium beobachtet worden, das vor allem den peripheren Raum der Zelle erfüllt und beispielsweise zwischen Cytoplasmamembran und Plastide liegt. Ein vollständig geschlossenes Netz ist aber nur in der jungen, eben aus der Mutterhülle entlassenen Interphasezelle vorhanden. Im Verlauf des nun folgenden Zellwachstums wächst auch das Mitochondriennetz mit, doch treten zusätzlich bis zu 40 kleine und unverzweigte Einzelmitochondrien auf, von denen der größere Teil in zentral gelegenen Bereichen der Zelle zu finden ist.

Die Mitose wird eingeleitet durch Verlagerung des Zellkerns von der Zellmitte zum apikalen Zellpol, wobei sich die eigentlich eiförmig gestaltete Polytomazelle abrundet. Im Zuge dieser Vorgänge werden Teile des Mitochondriennetzes zur Seite gedrängt, so daß apikal eine Öffnung entsteht. In der Prophase steigt die Zahl der kleinen, fast un-

verzweigten Einzelmitochondrien auf 100 bis 150 an. Sie sind jedoch
jetzt nicht auf das Zentrum der Zelle beschränkt, sondern liegen nunmehr auch peripher in den Maschen des Mitochondriumnetzes. Meta-, Ana-
phase- und Telophasestadien konnten noch nicht untersucht werden, weshalb ungeklärt bleibt, ob das Netz während der Mitose vollständig in
Einzelmitochondrien zerfällt. In der Cytokinese sind Einzelmitochondrien und ein Mitochondriumnetz vorhanden. Dieses teilt sich, so daß
in die Tochterzellen je ein etwa schalenförmig gestaltetes Netzmito-
chondrium gelangt. Die Einzelmitochondrien werden offenbar zufalls-
gemäß auf die Tochterzellen verteilt. Da in den jungen Tochterzellen
nur einziges Mitochondriennetz und fast keine Einzelmitochondrien vorhanden sind, muß angenommen werden, daß nach Abschluß der Zellteilung
die Einzelmitochondrien mit dem verbliebenen Mitochondriumnetz fusionieren. Es gibt zumindest keinen Hinweis für einen degenerativen Ab-
bau dieser Mitochondrien.

> Während des gesamten Lebenszyklus, besonders aber während der Mitose werden
> extracytoplasmatische Vesikel in den Raum zwischen Zellwand und Plasmalemma
> abgeschnürt. Diese Vesikel, die häufig auch Mitochondrienelemente enthalten,
> gehen spätestens nach der Entlassung der Tochterzelle aus der Mutterhülle als
> cytoplasmatische Restkörper zugrunde (GAFFAL und SCHNEIDER, 1979 zum Druck ein-
> gereicht).

Es ist bekannt, daß Mitochondrien polyploid sind, es ist deshalb anzunehmen, daß die DNS-Moleküle im Mitochondriennetz verteilt sind. Sowohl durch die Abschnürung von Einzelmitochondrien als auch durch Entmischungsvorgänge bei der Verteilung des Mitochondrieninventars auf
die Tochterzellen ist die Möglichkeit zur genetischen Segregation gegeben, die ein Charakteristikum der extranucleären Vererbung darstellt.
Wenn die Mitochondrien der nach der Zellteilung zugrundegehenden Rest-
körper ebenfalls Genomkopien enthalten, dann würde die Segregation be-
schleunigt, ein homozygoter Zustand möglicherweise früher erreicht.
Die Wiedervereinigung der Einzelmitochondrien mit dem Mitochondriennetz ist andererseits die morphologische Grundlage der Rekombination
von Mitochondriengenomen, die also sogar innerhalb eines jeden vegetativen Teilungszyklus möglich ist.

4. Die räumliche Gestalt der Plastiden im vegetativen Lebenszyklus

Polytoma besitzt einen Leukoplasten, der wie alle Plastiden als genetischer Informationsträger fungiert (SCHERBEL et al., 1974; SIU et al.,
1975, 1976). Er ist genau wie der *Chlamydomonas*-Chloroplast becherförmig
gestaltet, wobei die Basis dieses Bechers in Interphasezellen von
Polytoma kompakt und weniger perforiert erscheint, während die oberen
Seitenwände dünn und stark perforiert sind. Bei der Wanderung des
Zellkerns zum apikalen Zellpol zu Beginn der Mitose, verdrängt dieser
auch den oberen Rand des Bechers. Durch die gleichzeitig erfolgende
Abrundung der Zelle wird der Plastidenbecher gestaucht und schüsselförmig umgestaltet. Bereits in der Prophase wird die künftige Teilungs-
ebene durch eine Einschnürung deutlich. Während der Mitose wird die
Plastidenschüssel genau wie der gesamte Protoplast um 90° gedreht und
spätestens während der Cytokinese in zwei Hälften geteilt. In der eben
entstandenen Tochterzelle ist die Plastide noch schüsselförmig, kompakt und wenig perforiert. Im Verlauf der Rückverlagerung des Zellkerns
zur Zellmitte wird aus der Schüssel wieder ein langgezogener Plastidenbecher. Bei dieser Umgestaltung nimmt auch der Perforationsgrad wieder
zu. In den im Laufe der Mitose gebildeten extracytoplasmatischen Vesi-
keln sind auch schon Leukoplastenelemente beobachtet worden, doch wesentlich seltener als Mitochondrien.

Wenn man von seltenen Beobachtungen plastidärer Bruchstücke absieht,
dann kann gesagt werden, daß im Gegensatz zu den Mitochondrien im
vegetativen Teilungszyklus von *Polytoma* vorwiegend nur eine Plastide
vorhanden ist, zufällig segregierende Plastidenelemente sind nicht
beobachtet worden. Da aber auch die Plastiden polyploid sind, kann
eine genetische Segregation nur bei der Plastidenteilung, innerhalb
des Erbinformationsträgers erfolgen. Bei den meisten Pflanzen sind jedoch mehrere Plastiden in der Zelle vorhanden, weshalb die Segregation
in der Regel natürlich auch durch Entmischungsvorgänge bei der Verteilung der Zellorganellen auf die Tochterzellen erfolgt.

5. Die räumliche Gestalt der Mitochondrien im generativen Lebenszyklus

In Untersuchungen an *Chlamydomonas* konnte festgestellt werden, daß in
kopulationsbereiten Gameten ein netzförmiges Riesenmitochondrium vorhanden ist. In einigen Zellen sind noch bis zu vier kleine Einzelmitochondrien gefunden worden, die sicherlich als Bruchstücke des Netzes
zu betrachten sind. Bereits in einer 15 Minuten alten Zygote konnte
ein Mitochondriennetz nicht mehr gefunden werden, sondern etwa 20
lange, aber kaum mehr verzweigte Mitochondrien. Mit zunehmendem Zygotenalter nimmt die Zahl kleinerer, mehr oder weniger unverzweigter
Mitochondrien zu, bis nach sieben Stunden etwa 50 solcher Organellen
anzutreffen sind. Ältere Zygoten konnten nicht geprüft werden, da
dann eine bereits wieder vorhandene derbe Zygotenwand die Untersuchungen verhindert. Eine Fusion der Netzmitochondrien beider Gameten konnte nicht festgestellt werden, sie konnte jedoch auch nicht ausgeschlossen werden, da die ersten 15 Minuten des Zygotenlebens und mehr als
sieben Stunden alte Zygoten noch nicht untersucht sind (GROBE und
ARNOLD, 1975, 1977).

Dagegen konnte die *Fusion der Mitochondrien* in der Zygote von *Polytoma* ein-
wandfrei nachgewiesen werden. Bereits in einer so jungen Zygote, in
der die beiden Zellkerne die Karyogamie noch nicht begonnen hatten,
lag bereits ein einziges Mitochondriennetz (siehe Abb. 1) vor. Zusätzlich waren noch kleine, nicht oder nur wenig verzweigte Einzelmitochondrien vorhanden, deren Zahl mit zunehmendem Zygotenalter anwuchs. Eine
genaue Bestimmung des Zygotenalters ist hier nicht möglich, da bei
Polytoma die Gametenbildung nicht induziert werden kann, sondern spon-
tan erfolgt. Es ist jedoch offenbar, daß es genau wie bei *Chlamydomonas*
in der Zygote zu einem raschen Zerfall des Mitochondriennetzes kommt.
Wegen dieser Parallelität ist es denkbar, daß auch in der *Chlamydomonas*-
Zygote die Mitochondriennetze der Gameten vor ihrem Zerfall fusionieren. Ungeklärt bleibt die Frage, ob es bis zur Zygotenreife wieder zur
Bildung eines Netzes kommt oder ob dieses erst in der Zoospore neu entsteht.

Der bei *Polytoma* erstmalig gelungene einwandfreie Nachweis einer Mito-
chondrienfusion ist eine weitere morphologische Basis for Rekombinationen von Mitochondriengenomen. Es ist zu vermuten, daß Mitochondrien-
fusionen auch in Zygoten anderer Organismen stattfinden, z.B. bei
Saccharomyces, wo Rekombinationsuntersuchungen am mitochondrialen Genom
bereits zur Routine gehören.

Genetisches Interesse verdient auch der beobachtete Zerfall der Riesenmitochondrien in den Zygoten von *Chlamydomonas* und *Polytoma*. Es ist
unbekannt, wie sich die sicherlich in Mehrzahl vorhandenen Genomkopien
des Mitochondriennetzes auf die Einzelmitochondrien verteilen, ob je-
des dieser Einzelmitochondrien überhaupt ein Genom besitzt oder ob

Abb. 1. Ein aus 199 aufeinanderfolgenden
Serienschnitten rekonstruiertes Mitochon-
drieninventar (= Chondriom) einer Zygote
von Polytoma papillatum. Das Chondrium besteht aus 34 Einzelteilen, wobei besonders
das große korbähnliche Netz auffällt. Die
Existenz dieses Netzes kann keinen Zweifel
daran lassen, daß es durch die Fusion der
in den Gameten wahrscheinlich ebenfalls
vorwiegend als Netz vorliegenden Chondriome
entstanden ist. ca. 5000 ×

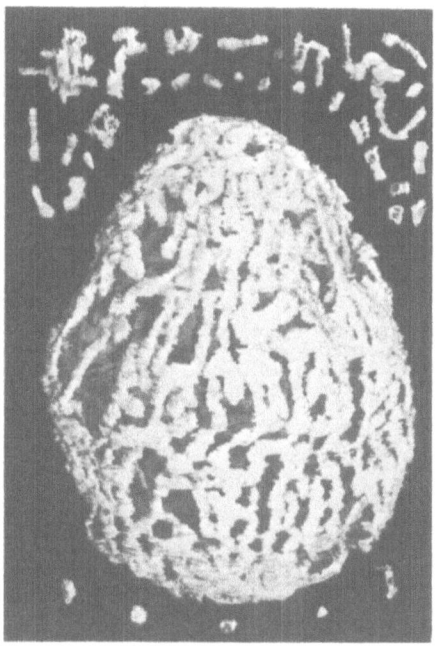

das Riesenmitochondrium in so viele Teile zerfällt, wie Einzelgenome
vorhanden waren. Eine Fülle von Problemen harren hier ihrer Klärung.
Zu ihnen gehört auch die Frage, ob das Phänomen der Mitochondrien-
fusion in der Zygote universell verbreitet oder nur auf einige Mikroorganismen beschränkt ist. Nach unseren bisherigen Erkenntnissen läßt
sich jedoch sagen, daß das morphologische Verhalten der Mitochondrien
mit dem bei *Chlamydomonas* festgestellten nichtmendelnden, aber in der
Regel zweielterlichen Vererbungsmodus der Mitochondriengene zu ver-
einbaren ist (ALEXANDER et al., 1974; WISEMAN et al., 1977).

6. Die räumliche Gestalt der Plastiden im generativen Lebenszyklus

Daß die beiden Plastiden der Gameten in der Zygote von *Chlamydomonas*-
Arten fusionieren, ist schon seit über zehn Jahren bekannt (BROWN et
al., 1968). Die Plastidenfusion erfolgt hier nach der Fusion der Zellkerne.

In der schon erwähnten jungen *Polytoma*-Zygote, bei der die Mitochondriennetze der Gameten schon vor der Kernkopulation fusioniert waren (siehe
Seite 216), lag ebenfalls nur ein einziger Leukoplast vor. Damit war
zum erstenmal eine *Plastidenfusion* außerhalb der Gattung *Chlamydomonas*
nachgewiesen worden. Bei *Polytoma* geschieht sie vor der Fusion der
Zellkerne. Der Fusionsmeridian ist nach dreidimensionaler Rekonstruktion der "Plastidenzygote" deutlich zu erkennen (siehe Abb. 2).

Die bei der isogamen *Chlamydomonas reinhardii* nachgewiesene einelterliche
Vererbung der Plastidengene läßt sich aus dem morphologischen Verhalten der Organellen nicht erklären. Die an *Ulva mutabilis* gemachte Beobachtung, wonach unmittelbar nach der Plasmogamie einer der beiden
Plastiden abgebaut werden soll, kann bei *Chlamydomonas* und *Polytoma* aus-

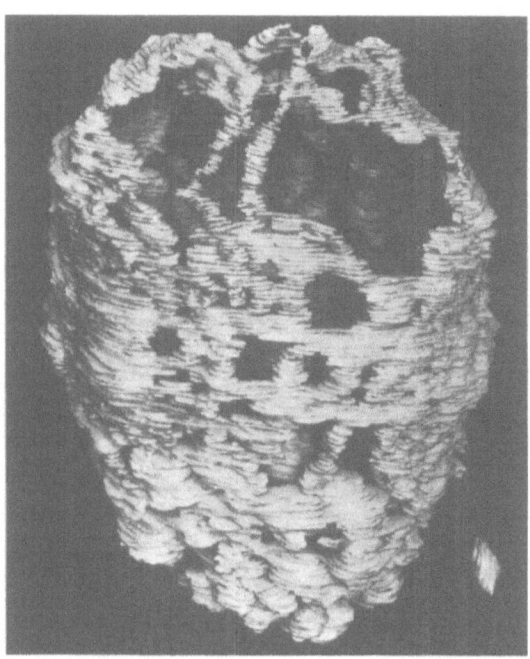

Abb. 2. Ein aus 186 aufeinanderfolgenden Serienschnitten rekonstruiertes Modell des Leukopla-
steninventars einer Zygote von
Polytoma papillatum. Neben einem
großen Plastidenbecher ist nur
noch ein wesentlich kleineres ei-
förmiges Element vorhanden. Die
Entstehung dieses großen Plastiden durch seitliche Fusion zweier
halb so großer Plastidenbecher in
den Gameten geht aus der Existenz
der beiden apikalen Öffnungen
hervor. Die Lage des Fusionsmeri-
dians wird noch durch die zwei
die apikalen Öffnungen trennenden Stege angezeigt. ca. 5700 x

geschlossen werden. Für die Eliminierung der Plastidengene des (-) Gameten müssen andere, vermutlich molekularbiologische Vorgänge verantwortlich sein (BRATEN, 1973; CHIANG, 1976; SAGER und LANE, 1972).

Genau wie bei den Mitochondrien, so stellt sich auch hier die Frage
nach der allgemeinen Verbreitung von Plastidenfusionen bei der Befruch-
tung. Eine Beantwortung wäre vor allem für solche Organismen wichtig,
bei denen die Übertragung von Plastiden aus ♂ Gameten in die Eizelle
bekannt ist, wie etwa bei *Oenothera* oder *Pelargonium*. Die Einbeziehung
eukaryotischer Mikroorganismen in die genetische Forschung hat unser
Wissen über die extranucleäre Vererbung wesentlich bereichert. Doch
nun ergibt sich die Notwendigkeit, die Erforschung höher organisierter
Lebewesen wieder zu aktivieren, um zu einem umfassenden Bild zu gelan-
gen.

Literatur

ADOUTTE, A., BALMEFRÉZOL, M., BEISSON, J., ANDRÉ, J.: J. Cell Biol.
54, 8-19 (1972). - ALBRING, M., RADSAK, K., RUMPELT, H.J., THOENES,
W.: Cytobiology 8, 168-174 (1973). - ALBRING, M., RADSAK, K., THOENES,
N.: Naturwissenschaften 62, 43 (1975). - ALEXANDER, N.J., GILLHAM,
N.W., BOYNTON, J.E.: Mol. Gen. Genet. 130, 275-290 (1974). - ARNOLD,
C.G., GAFFAL, K.P.: Biol. in unserer Zeit 9, 45-51 (1979). - ARNOLD, C.G.,
SCHIMMER, O., SCHÖTZ, F., BATHELT, H.: Arch. Mikrobiol. 81, 50-67
(1972). - ASANO, M., WAKABAYASHI, T., ISHIKAWA, K., KISLIMITO, H.:
Acta Pathol. Jpn. 28, 205-213 (1978). - ATKINSON, A.W., JOHN, P.C.L.,
GUNNING, B.E.S.: Protoplasma 81, 77-109 (1974).

BAKEEVA, L.E., CHENTSOV, Y.S., SKULACHEV, V.P.: Biochim. Biophys. Acta
501, 349-369 (1978). - BEN-SHAUL, Y., MARKUS, Y.: J. Cell Sci. 4,

627-644 (1969). - BINDER, S., SLUGA, E.: Klin. Monatsbl. Augenheilkd. 171, 786-791 (1977). - BIRKY, C.W.: Annu. Rev. Genet. 12, 471-512 (1978). - BLANK, R., GROBE, B., ARNOLD, C.G.: Planta 128, 63-64 (1978). - BRATEN, T.: J. Cell Sci. 12, 385-389 (1973). - BROWN, R.M., JOHNSON, C., BOLD, H.C.: J. Phycol. 4, 100-120 (1968). - BURTON, M.D., MOORE, J.: J. Ultrastruct. Res. 48, 404-413 (1974).

CALVAYRAC, R., BUTOW, R.A., LEFORT-TRAN, M.: Exp. Cell Res. 71, 422-432 (1972). - CALVAYRAC, R., LEFORT-TRAN, M.: Protoplasma 89, 353-358 (1976). - CALVAYRAC, R., BERTRAUX, O., LEFORT-TRAN, M., VALENCIA, R.: Protoplasma 80, 355-370 (1974). - CAROTHERS, Z.B., MOSER, J.W., DUCKETT, J.G.: Am. J. Bot. 64, 1107-1117 (1977). - CAVALIER-SMITH, T.: Nature (London) 228, 333-336 (1970). - CHIANG, K.S.: On the search for molecular mechanism of cytoplasmic inheritance: Past controversity, present progress and future outlook. In: Genetics and Biogenesis of Chloroplasts and Mitochondria, eds. T. BÜCHER, W. NEUPERT, W. SEBALD, S. WERNER. 896 pp. Amsterdam-London: North Holland 1976.

DAMSKY, C.H.: J. Cell Biol. 71, 123-135 (1976). - DAVISON, M.T., GARLAND, P.B.: J. Gen. Microbiol. 98, 147-154 (1977). - DUCKETT, J.G., TOTH, R.: Annals Bot. 41, 903-912 (1977).

ESAU, K.: Z. Pflanzenphysiol. 67, 244-254 (1972).

FINCHAM, J.R.S., DAY, P.R., RADFORD, A.: Fungal Genetics, 636 pp. Oxford-London-Edinburgh-Melbourne: Blackwell 1979.

GAFFAL, K.P.: Protoplasma 94, 175-191 (1978). - GAFFAL, K.P., KREUTZER, D.: Protoplasma 91, 167-177 (1977). - GAFFAL, K.P., SCHNEIDER, G.J.: Cytobiologie 18, 161-173 (1978); - Cytobios, in press. - GILLHAM, N.W.: Organelle Heredity. 602 pp. New York: Raven Press 1978. - GROBE, B., ARNOLD, C.G.: Protoplasma 86, 291-294 (1975); - Protoplasma 93, 357-361 (1977). - GRODUMS, E.I.: Cell Tissue Res. 185, 231-238 (1977).

HALLAM, N.D., ROBERTS, B.E., OSBORNE, D.J.: Planta 105, 293-309 (1972). - HANZLIKOVA, V., SCHIAFFINO, S.: J. Ultrastruct. 60, 121-133 (1977). - HEGARTY, P.V.J., DAHLIN, K.J., BENSON, E.E.: Experientia 34, 1070-1071 (1978). - HERMAN, E.M., SWEENEY, B.M.: J. Phycol. 13, Suppl. 160, 29 (1977). - HEYWOOD, P.: J. Cell Sci. 26, 1-8 (1977). - HIGGINS, E.S., ROGERS, K.S., MARKS, P.A.: Chem. Biol. Interact. 13, 295-305 (1976). - HOFFMAN, H.P., AVERS, C.J.: Science 181, 749-750 (1973).

KALASHNIKOVA, A.N.: Physiological variations of the mitochondria ultra-structure during ontogenesis, 691. In: Electron Microscopy 1978, eds. J.M. STURGESS, V.I. KALNINS, F.P. OTTENSMEYER, G.T. SIMON, Vol. 2, Biology; 9th Int. Congr. Electr. Microsc. Toronto, Ontario: Imperial Press 1978. - KEYHANI, E.: Exp. Cell Res. 81, 73-78 (1973). - KOUKL, J.F., VORBECK, M.L., MARTIN, A.P.: J. Ultrastruct. Res. 61, 158-165 (1977). - KRAUSE, W., DAVID, H., UERINGS, I., ROSENTHAL, S.: Acta Biol. Med. Germ. 28, 779-786 (1972).

LADYGIN, V.G., SEMENOVA, G.A., TAGEEVA, S.V.: Tsitologia 17, 115-121 (1975). - LANGE, L., OLSON, L.W.: Protoplasma 90, 33-45 (1967). - LEEDALE, G.F., BUETOW, D.E.: Cytobiology 1, 195-202 (1970).

MALOFF, B.L., SCORDILIS, S.P., REYNOLDS, C., TEDESCHI, H.: J. Cell Biol. 78, 199-213 (1978a). - MALOFF, B.L., SCORDILIS, S.P., TEDESCHI, H.: J. Cell Biol. 78, 214-226 (1978b). - MANTON, I.: J. Mar. Biol. Assoc. U.K. 38, 319-333 (1959). - MISLAVSKY, A.N.: Anat. Anz. 39, 497-505 (1911).

NEUMANN, D., PARTHIER, B.: Exp. Cell Res. 81, 255-268 (1972).

OSAFUNE, T.: J. Electr. Microsc. 22, 51-62 (1973). - OSAFUNE, T.,
MIHARA, S., HASE, E., OHKURO, I.: J. Electr. Microsc. 24, 33-40 (1975a);
- J. Electr. Microsc. 24, 283-286 (1975b); - Plant Cell Physiol. 16,
313-326 (1975c); - J. Electr. Microsc. 25, 261-269 (1976).

PAULIN, J.J.: J. Cell Biol. 66, 404-413 (1975); - Exp. Parasitol. 41,
283-289 (1977). - PELLEGRINI, M.: C.R. Acad. Sci. Paris 283, 911-913
(1976). - PELLEGRINI, M., PELLEGRINI, L.: C.R. Acad. Sci. Paris 282, 357-
360 (1976). - PERASSO, R., BEISSON, J.: Biol. Cellul. 32, 275-290
(1978). - PRINGSHEIM, E.G.: Farblose Algen. 471 S. Stuttgart: Fischer
1963. - PUBLICOVER, S.J., DUNCAN, G.J., SMITH, J.L.: Cell Tissue Res.
185, 373-386 (1977).

RANCOURT, M.W., McKEE, A.P., POLLACK, W.: J. Ultrastruct. Res. 51,
418-424 (1975). - ROHFRITSCH, O.: Protoplasma 95, 297-307 (1978). -
ROHR, R.: Biol. Cell. 33, 89-92 (1978).

SAGER, R., LANE, D.: Proc. Natl. Acad. Sci. USA 69, 2410-2413 (1972).
- SALMONS, S., GALE, D.R., SRÉTER, F.A.: J. Anat. 127, 17-32 (1978).
- SANTORE, U.J., GREENWOOD, A.D.: Arch. Microbiol. 112, 207-218 (1977).
- SAPEHIN, A.A.: Arch. Zellforsch. 13, 319-398 (1915). - SCHERBEL, G.,
BEHN, W., ARNOLD, C.G.: Arch. Mikrobiol. 96, 1-18 (1974). - SCHÖTZ, F.,
DIERS, L.: Planta 124, 277-285 (1975). - SCHÖTZ, F., BATHELT, H.,
ARNOLD, C.G., SCHIMMER, O.: Protoplasma 75, 229-254 (1972). - SENDA,
H.: Folia Pharmacol. Jpn. 73, 275-286 (1977). - SENGER, P.L., SAACKE,
R.G.: J. Cell Biol. 46, 405-408 (1970). - SIU, C.H., CHIANG, K.S.,
SWIFT, H.: J. Mol. Biol. 98, 369-392 (1975). - SIU, C.H., SWIFT, H.,
CHIANG, K.S.: J. Cell Biol. 69, 352-370 (1976). - SMITH, A.E., PAGE,
E.: Dev. Biol. 57, 109-117 (1977). - SMITH, D.G., WILKIE, D., SRIVA-
STAVA, K.C.: Microbios 6, 231-238 (1972). - STEVENS, B.J.: J. Microsc.
20, 90-91 (1974); - Biol. Cell. 28, 37-56 (1977).

THIERY, C., BERGERON, M.: Rev. Can. Biol. 35, 211-216 (1976).

VERTEL, B.M., FISCHMAN, D.A.: Dev. Biol. 58, 356-371 (1977). -
VICKERMAN, K.: J. Protozol. 24, 221-233 (1977). - VOSS, K., KEMMER,
C.: Exp. Pathol. 15, 311-318 (1978).

WABNER, T., RAFFAEL, J.: Biochim. Biophys. Acta 408, 284-296 (1975);
- Exp. Cell Res. 107, 1-13 (1977). - WAKABAYASHI, T., ASANO, M.,
ISHIKAWA, K., KISHIMOTO, H.: J. Electr. Microsc. 26, 137-140 (1977);
- Acta Pathol. Jpn. 28, 215-223 (1978). - WISEMAN, A., GILLHAM, N.W.,
BOYNTON, J.E.: Mol. Gen. Genet. 150, 109-118 (1977). - WORTH, E.R.,
LUCAS, F.V.: Am. J. Obstret. Gynecol. 130, 152-155 (1978).

Professor Dr. CARL-GEROLD ARNOLD
Dr. KARL PETER GAFFAL
Institut für Botanik und
Pharmazeutische Biologie der
Universität Erlangen-Nürnberg
Schloßgarten 4
D-8520 Erlangen

VI. Population Genetics

By ROBERT LICHTER

1. Introduction

Since the last survey on population genetics (Progr. Bot. $\underline{38}$, 230 (1976) no spectacular advances are to be mentioned. Actual progress is achieved by many small steps, such as the improvement of a model by omitting a restricting assumption or the extension of a single- (two-) locus model to a more realistic multilocus model. The contro- versy between the selectionists and the neutralists is still going on.

A modification of the species concept, based up to now mainly on mor- phological criteria, was proposed by HOFFMANN and ESSER (1978): "Popu- lations (races) belong to different species when the failure to inter- breed and to produce viable offspring is not caused by genetic para- meters operating in completion of the sexual cycle."

Some monographs may be mentioned: A reprint of 31 papers on the stochastic theory of population genetics edited by LI (1977a), starting with R.A. FISHER's 1922 paper up to recent work. A useful survey on the same topic, but readable only for biomathematicians or biologists with a solid mathematical background is given by MARUYAMA (1977). A review of the field of quantitative genetics and artificial selection is provided by the proceedings of a conference held at Ames, USA (POLLAK et al., 1977). Populations as biological communities were described by CHRISTIANSEN and FENCHEL (1977a) and an introduction into the par- ticular problems of plant populations was given by HARPER (1977). Molecular population genetics embraces mathematical theories and empirical results of molecular biology (NEI, 1975).

2. Gene Frequencies

The one- and two-locus systems (NAGYLAKI, 1977) or multilocus systems (EWENS and THOMSON, 1977) may be useful for the theoretical population geneticist to determine the frequencies of genes, genotypes, and mating types in successive generations and to detect the properties of equi- libria, but there is little attempt to elucidate the relevance of the mathematical model in face of the biological realities.

The joint evolution of gene frequency and population size was studied by ASMUSSEN and FELDMANN (1977). Homozygotes for recessive visible genes often appear in lines under artificial selection many generations from the start. In finite pop- ulations ROBERTSON (1978) found small values for the mean time until their detec- tion and a low dependence on population size. On the other hand, their standard deviation is high and, for example, in the order of 15 generations for a gene, not subject to selection, and occurring only once in the initial sample. This means that such a gene may appear after a few generations or only after many generations.

a) Mutation and Gene Substitution

The influence of mutation on population structure depends on the type
of mutation. Intragenic recombination can significantly affect the
number of neutral alleles and their distribution in a finite population (STROBECK and MORGAN, 1978). Intragenic recombination, when selectively neutral alleles are produced in a stepwise fashion (KIMURA
and OHTA, 1978), contradict EWENS' sampling theory which is based on
counting the actual number of alleles in the sample while assuming
every mutation to be unique and distinguishable. CHAKRABORTY and NEI
(1976) developed a formula for the bivariate probability of the number
of codon differences and the genetic variability in electromorphs in
finite populations, using the infinite site model and the stepwise
mutation model simultaneously. A mixture of the two models for the
study of electrophoretic variants in natural populations was proposed
by LI (1976a). It was shown that even if the proportion of nonstepwise
mutation is only 5%, the effective number of alleles given by the proposed model is considerably larger than that given by the model of
KIMURA and OHTA. Using a transient distribution of allele frequencies
in a finite population under the assumption that there are k possible
allelic states of a locus, and that mutation occurs in all directions
it is shown that when population size is suddenly increased the expected number of alleles increases more rapidly than the expected heterozygosity (NEI and LI, 1976).

> Despite the deleterious effect on the fertility of heterozygotes the chromo-
> some mutations can increase in a random mating population under the influence
> of random drift, segregation distortion, viability disadvantage, and recombi-
> nation modification (BENGTSON and BODMER, 1976).

> FISHER's model for the evolution of dominance, which indicates that the accu-
> mulation of dominance modifiers will be accelerated by (1) an increased fre-
> quency of the mutant heterozygote, and (2) increased selection for the pheno-
> type of the normal homozygote, has been criticized by HALDANE. In support of
> FISHER's model it was argued by MANNING (1977) that periods of intense selec-
> tion for the wild type are frequent in inbreeding populations. This situation
> should promote the accumulation of dominance modifiers.

Genes or sequences of DNA present in multiple copies per cell include
entire genomes of mitochondria and chloroplasts, nuclear ribosomal
RNA genes, and highly repetitive sequences in heterochromatin. All
copies in the cell are nearly identical, in spite of mutational pressure and weak selection. Evidence from yeast and *Chlamydomonas* suggests
that organelle gonophores undergo repeated recombination and that gene
conversion occurs when two molecules carrying different alleles at a
locus recombine. BIRKY and SKAVARIL (1976) argued that gene conversion
can result in the cell becoming pure for one allele. OHTA (1976), assuming homologous but unequal crossing over, proposed a model analogous
to that of fixation and loss of a mutant gene in finite populations.
This diffusion model is applicable also to the gene conversion model
of BIRKY and SKAVARIL (OHTA, 1977a). The possibility that some important quantitative characters of higher organisms may be subjected to
multigene families, such as that of immunoglobulin variable parts, requires a revision of a population genetics theory (OHTA, 1977b). Two
remarkable characteristics of multigene families are (1) the contraction or expansion of the gene number in a family, and (2) the coinci-
dental evolution. Studies on the probability of identity of two genes
within a family (clonality) have been carried out. Some statistical
analyses lead to the conclusion that somatic mutations cannot be the
major cause of the observed hypervariability and that the results can
readily be explained by the germ line theory (OHTA, 1978a,b).

A special problem in population genetics is that of fixation or loss of a gene. An improved solution of diffusion equations with one less approximation was developed by CASH (1977) to estimate the probability of ultimate fixation of a favourable allele in a monoecious population assuming different degrees of dominance (h = 0;1/2;1). The calculation of the fixation probability and the average time until its fixation was reexamined for a particular allele under random fluctuation of selection intensities using finite Markow chain methods. An approximate, but general expression for this probability was obtained and results of previous work were shown to be special cases of these methods (NARAIN and POLLAK, 1977). Considering the situations of substitution by an advantageous recessive allele, DARLINGTON (1977) pointed out that HALDANE had overestimated the cost of natural selection. The estimated cost, however, is still so great that it is to be supposed that most species are not fully adapted to their environments, but may be just a little better than their competitors.

b) Maintenance of Polymorphisms

The neutralist vs. selectionist controversy over the maintenance of polymorphisms in natural populations has now continued unabated for some ten years with undiminished vehemence.[1]

Therefore, there is a search for statistical tests which should allow in model or experimental populations either to prove or to reject one or the other hypothesis. Population homozygosity, for example, is a precise measure for testing departures from neutrality in the direction of heterozygote advantage or disadvantage (WATTERSON, 1977, 1978). If data from several generations are available, a further analysis may be used which is more sensitive to detect the action of directional selection on gene frequency at a certain locus (SCHAFFER et al., 1977). The differences between loci in the variance of gene frequency over subdivided populations can be used also to discern if natural selection is acting there or not (ROBERTSON, 1975). In a theoretical study the proportion of common neutral alleles in two related populations was found to be high even after 4N generations if the product 4Nv (N = effective population size, v = mutation rate) is small (LI and NEI, 1977).

The fact that most empirical correlations between the heterozygosity of common alleles and the number of rare alleles are significantly different from the mean correlations expected under the neutrality models does not necessarily deny the neutral hypothesis. They may be due to the inability of the current models to explain the observable patterns of allelic diversity (EANES and KOEHN, 1977).

In the case of two large populations of the same type with no genetic bridge, in the same environment, but of substantially different sizes, discrimination seems to be possible because the shape of the electrophoretic profile will depend on the population size for the mutation-drift model, but not for the selection-mutation model (MORAN, 1976). The standardized variance of gene frequency (NICHOLAS and ROBERTSON, 1976) and the variance of heterozygosity (LI, 1977b) may be a means to prove the operation of selection, though it is not efficient when selection effects are slight.

Several empirical results agree more or less with the neutral hypothesis. CHAKRABORTY et al. (1978), using data of different animal populations, examined

---

[1]The neutralist theory implies that the observed enzyme polymorphism can be explained by mutation of selectively neutral alleles and genetic drift, whereas the selectionists say that polymorphism is due to a balancing effect of mutation and selection

the relationship between genetic identity and the correlation of heterozygosities of different populations, and they found a good agreement with the theoretical expectations of the mutation-drift theory. The polymorphism for vernalization requirements in *Avena sativa* also showed neutral behavior (QUALSET and PETERSON, 1978). From a survey of the literature on associations between allozymes and gene arrangements (inversions) ISHII and CHARLESWORTH (1977) concluded that a neutralist explanation cannot be excluded.

Because it is difficult to deny any action of selection for the maintenance of polymorphisms, the neutralists MARUYAMA and KIMURA (1978) propose a perhaps more realistic model with simultaneous presence of selectively neutral and very slightly deleterious alleles. They assume that the selection coefficients of very slightly deleterious alleles follow a certain frequency distribution rather than having a fixed value. They emphasize, however, that although natural selection is involved, their point of view is fundamentally different from that of the selectionists. Nevertheless, one has the impression that this will be a first step to understanding.

In any case it seems not unreasonable to suppose that stabilizing selection could produce in natural populations, if not a stable polymorphism, then at least long periods of quasi-equilibrium, an effect which is not discernible from a mutation-drift situation (SEMEONOFF, 1977). Disruptive selection caused by heterogeneity of the environment is a relevant factor for the maintenance of polymorphisms (LOESCHKE, 1978).

Further situations where selection in combination with various environmental characters may lead to diverse polymorphisms are very common (BROWN, 1978). Some of them will be briefly mentioned: Genotype-environment interactions associated with temperature in *Lotus corniculatus* (ELLIS et al., 1977), an age-dependent genotype-environment interaction in radish (CRISP et al., 1977), the joint action of migration and selection (NAGYLAKI, 1976a), climatic gradients (clines) in different plant populations of western Europe (BERGMANN, 1978; HAMRICK, 1976; JONES, 1977; JONES and CRAWFORD, 1977), a multilocus model situation with spatially fluctuating environments (GILLESPIE and LANGLEY, 1976), certain multilocus or multiple allele systems in subdivided populations (GILLESPIE, 1977; KARLIN and CAMPBELL, 1978), frequency-dependent selection in random mating populations with partial inbreeding (HÜHN, 1978a), a frequency-dependent fitness of homozygotes for a reciprocal translocation (BAILEY et al., 1978), certain types of intergenotypic competition combined with frequency- and density-dependent selection (SMOUSE, 1976), a balance of forces between linkage and random assortment for loci having positive epistatic interactions (WILLS and MILLER, 1976) or having inequal gene effects (LINNEY, 1977), spatial and temporal variability in fitness (HOEKSTRA, 1978), and a ranking order of the fitness curve with respect to heterozygosity (GILLESPIE, 1976; WILLS, 1978).

There is some evidence from chemostat competition experiments with *bacteria* that single- and multi-locus heterosis may also be an important mechanism in maintaining gene-enzyme polymorphism in nature (BERGER, 1976). Different outcrossing rates and a heterozygosity higher than expected were found to be responsible for polymorphisms in *Lupinus succulentus* (HARDING and BARNES, 1977) and *Hordeum vulgare* (BROWN et al., 1978). Following LEWONTIN et al. (1978) numerical and analytical approaches demonstrated that heterosis alone does not sustain many alleles segregating at a locus. Stable equilibria for multiple alleles will probably be best explained by multiple niche selection.

Concerning the neutralist vs. selectionist controversy it has been pointed out already (Progr. Bot. $\underline{38}$, 234) that the question is not whether selection or mutation-drift is operating for the maintenance of polymorphisms. What is needed to resolve this controversy is a theory by which the statistical properties of a population under the

joint effect of mutation, random drift, and natural selection as a
factor of environmental conditions can be examined thoroughly and the
relative importance of these three factors can be evaluated (LI, 1978).

c) Gene Flow and Divergence Between Populations

Theoretical work on spatial and temporal variation, and the interaction of migration and variable selection have been reviewed by FELSEN-
STEIN (1976). The spatial pattern of rare alleles generated by new mutations in local populations was examined by SLATKIN and CHARLESWORTH
(1978) and they stated that gene flow is much more effective in dis-
persing alleles in a two-dimensional array than in one dimension. The
consequences of discontinuity in the migration rate and of a geographical barrier in the habitat were studied in a diffusion model by NAGY-
LAKI (1976b), and partial differential equations were deduced for the
expected gene frequency at each site and for the covariances between
gene frequencies at any two points of the habitat (NAGYLAKI, 1978).
Differences in fertilizer and liming can produce such a steep gradient
that the steepness of the cline in a population of *Anthoxanthum odoratum*
seemed to be determined more by the sudden change in environmental conditions than by the effects of gene flow (SNAYDON and DAVIES, 1976).

Two alternative methods for the colonization of new or vacant habitats
were examined, in the one model all populations contribute migrants to
a common pool, in the other, the so-called propagule pool model, each
propagule is derived from a single population. In the propagule model
group selection can be more effective and maintain genetic variability
between populations (BORMANN, 1978; SLATKIN and WADE, 1978).

> On the other hand it was demonstrated that when the variance of the migration
> in a certain direction is sufficiently large with respect to the number of sub-
> divisions, the population behaves as panmictic (MALECOT, 1975). However, ana-
> lyzing the variation of peroxidase enzymes in *Pinus ponderosa*, an almost out-
> crossing species, MITTON et al. (1977) found that selection specific for differ-
> ent environments is evidently strong enough to overcome the homogenizing force
> of migration and to produce population differentiation. CAISSE and ANTONOVICS
> (1978) showed by computer simulations that even in closely adjacent plant popu-
> lations which were connected by gene flow and subjected to linearly or stepwise
> varying selection, a monotonic cline was rapidly established in gene frequency
> at the selected locus.

NEI and ROYCHOUDHOURY's method of sampling variances of heterozygosity
for estimating the genetic distance between populations was improved
by MITRA (1976) and evaluated again by NEI (1978) for a small number
of individuals. The number of individuals to be used for estimating
the average heterozygosity can be very small if a large number of loci
are studied and the average heterozygosity is low. The diffusion equa-
tion method of OHTA and KIMURA as well as NEI's genetic distance were
revised, and corrected formulas in much simpler forms were presented
(LI, 1976b). Numerical computations indicated that the genetic distance based on electrophoretic data may give a serious underestimate
in comparisons between species. ADAMS (1977) used a principal component
technique to calculate a distance metric between any two cultivars of
Phaseolus vulgaris, and it was shown that these distances were highly in-
versely correlated with genetic relationship estimated from breeding
ancestry. It was suggested that this method would be of nearly uni-
versal applicability, being particularly useful for self-fertilizing
plant populations.

d) Linkage

Though the effect of linkage is of particular importance, many authors have investigated, mostly in a theoretical way, the significance of linkage disequilibrium. A model of gene flow and selection in two linked loci was analyzed by SLATKIN (1975). Gene flow can produce linkage disequilibrium between the loci, and, dependent on the spatial pattern of selection, linkage can either decrease or increase a population's response to local selection. However, if spatial variation in selective pressures is important as observed in nature, then one is forced to ask why more linkage disequilibrium is not found. BULMER (1976) studied the effect of stabilizing and disruptive selection on a polygenic character under random mating. Large and rapid changes in the genotypic variance result from the generation of linkage disequilibrium under selection. In artificial selection the effects of linkage disequilibrium will be more important than the slower and less dramatic effects due to changes in gene frequencies.

The effects on a neutral locus linked to a selected locus (the hitch-hiking effect) were considered again. HAIGH and SMITH (1976) argued that in aggregate hitch-hiking is more important than drift for reducing heterozygosity in large populations, whereas OHTA and KIMURA (1976) felt that hitch-hiking is not so important in this respect. It was, however, shown that significant disequilibrium can be generated between two neutral loci by the evolution of a linked selected locus (THOMPSON, 1977a). In case of multiple alleles the linkage disequilibrium between two loci has been evaluated by YASUDA (1978) and it was stated by WEIR and COCKERHAM (1978) that various tests of hypotheses about linkage disequilibrium are possible.

Models of two linked overdominant loci in moderately large, but finite, populations were examined by AVERY (1978). According to this, care must be taken before using deterministic results when a population is, or has recently been, finite. One has to be cautious about declaring that populations have different selection regimes because their linkage disequilibria are different, or on the contrary that populations have the same selection regime because their linkage disequilibria are equal. YAMAZAKI (1977) performed a multiplicative fitness model with overdominant multiple alleles at closely linked multiple loci.

Genetic linkage can influence the response to selection as reflected by probabilities of fixation of favorable alleles, for example to obtain genotypes or lines which are better than the best parent. Fixation probabilities of favorable alleles are decreased by repulsion but increased by coupling. There will be more coupling if one parent line is substantially better than the other (BAILEY and COMSTOCK, 1976).

The method of genotype assay (JINKS and TOWEY, 1976) for estimating the number of effective factors in the polygenic system was applied to the number of genes including the case of linked genes (HILL and AVERY, 1978). Though the genotype assay seems to be superior to other methods of estimating the number of genes, it is biased because the estimated number of genes increased steadily the more successive generations of selfing following an initial cross were involved. This might be explained by the fact that strong linkage disequilibrium generated by the cross would be resolved by recombination in subsequent generations.

3. Selection

Thirty two papers of a symposium in memory of the late Ove FRYDENBERG give a comprehensive oversight of the theory and the field of observation relevant to selection and of the relative importance of selection versus mutation and migration (CHRISTIANSEN and FENCHEL, 1977b).

a) Natural Selection and Fitness

Natural selection is a useful method for climatic adaptation of plant populations. That was shown with a bulked hybrid population of upland cotton which was grown without artificial selection for 10 generations in a nonirrigated, semi-arid region (QUISENBERRY et al., 1978). In oat 34 out of 58 bulk populations increased significantly in freezing resistance and changed similarly for winter survival in the field (MARSHALL, 1976). *Dactylis glomerata* was grown over five generations at different locations in Japan and USA. One Japanese increase differed significantly from the original population in five of six characters studied, indicating that environmental pressures caused shifts in the frequency distributions (RINKER et al., 1978).

It appears that there may be selection for a lower variance in a constant environment and selection for a larger variance when the optimal value of the character is changing in time, but both would apply only if the fluctuations in the optimum exceed a certain threshold amplitude (SLATKIN and LANDE, 1976). Spatial patterns in the mean and the variance of a quantitative character can result from interaction of spatially varying, optimizing selection and gene flow. It was argued that spatial variation in selection on a polygenic character can be much more effective in increasing heterozygosity than temporal variation because of the potentially greater increase in phenotypic variance (GILLESPIE, 1975). The spatial model of LEVENE as well is much more effective in retaining genetic variation in finite populations than analogous cyclic temporal models. This holds true especially for the haploid model (HEDRICK, 1978). Certain analogies to the theory of selection in temporal fluctuating environments were found by GILLESPIE (1975) in his diffusion model of selection for within-generation variances in offspring numbers, it means that the gametic pool is determined by the random number of offspring per individual.

Several models of selection for increased recombination were examined using computer simulations (CHARLESWORTH et al., 1977). A recessive allele conditioning recombination will be favored should there be favorable or deleterious mutants occurring at other loci (FELSENSTEIN and YOKOYAMA, 1976).

The selection effects of partially resistant multilines on the racial composition of a pathogen population depends on two factors: The degree of selection against unnecessary genes for virulence and the number of lines in the multiline. Results of simple theoretical models show that use of multiline varieties stabilizes the pathogen population, but complex races carrying two or more virulence genes will predominate. It was concluded that partially resistant multilines will provide stable disease control in crop plants only in limited and relatively rare circumstances (MARSHALL and PRYOR, 1978).

In a composite cross of barley it was stated that strong selection operates at all life cycle stages, although often in differing directions. Estimates of viability and fertility components were made on a three-locus level and they reveal large differences in viability and fertility among homozygous genotypes (CLEGG et al., 1978).

Fitness is the joint operation of phenotype and environment. Considering a nonsubdivided population, the fitness of a genotype is made up
of the fitness values of all individuals in the population having this
genotype, and the proportions to which they are exposed to the environmental situations of the population. Consequently, the fitness of the
genotype is equal to the mean of all these individuals. Since the fitness value may also be frequency- and densitiy-dependent, it should be
emphasized that, in general, functional and stochastic (statistical)
superiority in fitness should not be confused (GREGORIUS, 1977). The
mean fitness does not necessarily increase through selection for fecundity and it was shown that the genotypic frequencies that maximize the
mean fecundity of the population must not be the same as the stable
equilibrium frequencies (POLLAK, 1978).

The equilibrium conditions under density-dependent selection for n alleles at
one locus were investigated by GINZBURG (1977). Hermaphrodite populations with
partial female sterility show a complex form of frequency-dependent fitness
(ROSS, 1977). Frequency-dependent effects may reduce the efficiency of single
plant selection in wheat. Grain yield decreased as genotypic frequency increased, suggesting that high-yielding genotypes show a greater advantage at low
frequencies than lower-yielding ones (PHUNG and RATHJEN, 1976). In a finite
population model where the selection coefficient of the homozygotes varies
randomly with equal variance, while the fitness of the heterozygote is kept
fixed, a small average heterozygote advantage can cause an unexpectedly large
amount of heterozygosity (AVERY, 1977).

A new fitness function, a balance function, which results from a combination of
a turnover function defining the expression of favorable genetic factors and a
cost function describing the cost of metabolism was proposed by EGGERS-SCHUMACHER
et al. (1977). Based upon the double exponential model for fitness functions, a
new index for the intensity of natural selection was defined as a variance and
is such that a value of zero indicates no selection while a value of one indicates strong selection (MANLY, 1977).

Fitness estimates derived from genotypic frequencies determined at
the same developmental stage in two successive generations are the
true total fitnesses provided that selection is complete at the time
of observation. In case there is postobservational selection, even
though the true fitnesses are actually constant, the estimated fitnesses will give the false impression of frequency-dependent selection, namely in favor of the rare genotype (CHRISTIANSEN et al., 1977).

In natural populations of *Senecio vulgaris* the radiate morph produces more
capitula per plant and more seeds per capitulum than the nonradiate morph.
The nonradiate genotype shows faster growth rate and flowers sooner. Given that
these characters which might affect relative fitness vary between morphs, the
increase in frequency of the radiate morph is most probably the result of selection (OXFORD and ANDREWS, 1977) and not due to continued introgression from
S. qualidus, as was suggested by HULL (1976).

b) Artificial Selection and Response to It

The effect of t' generations of reverse selection after t generations
of forward selection was investigated in terms of the change in the
metric mean. If t' = t, then the ratio of the change in the metric
mean of the reverse and the previous forward selection equals 1 - F
where F is the inbreeding coefficient for a neutral locus at generation t (NICHOLAS, 1976). Joint selection for both extremes of mean
performance and of sensitivity to environmental conditions was done
in *Nicotiana* (BRUMPTON et al., 1977). A similar selection for growth
rate at the near ambient temperature of 20°C and at the higher tem-

perature of 30°C exhibited a considerable amount of genetical variation in *Schizophyllum commune* (WILLIAMS et al., 1976). The potential variation within these extreme isolates was greater than the variation observed in the whole sample. In *Nicotiana* the realized heritability for
sensitivity was only half that for performance.

> The superiority of single seed descent over conventional family selection was
> confirmed and it was indicated that mean performance and environmental sensi-
> tivity are largely under the control of different genes (JINKS et al., 1977).
> The heritability and the repeatability over diverse environments of a quadratic
> coefficient of the growth curve were both relatively high. Therefore, indirect
> selection for environmental sensitivity by selecting for the quadratic coeffi-
> cient is useful to overcome many problems of direct selection (BOUGHEY and
> JINKS, 1978; BOUGHEY et al., 1978).

Mass selection is still practised in certain situations, though it is
in general less effective than other methods where progeny tests are
involved (RINK and THOR, 1976). Recurrent mass selection appears to
be appropriate for recently introduced species which are becoming ad-
justed to a new environment and its hazards (KNOWLES, 1977). Better
results than by mass selection were obtained by a modified ear-to-row
selection in an open pollinated variety of corn. Grain yields have
continued to increase through the 10th cycle, i.e., more than 40% increase in ten years. There was some evidence of curvilinearity, but
COMPTON and BAHADUR (1977) feel that a plateauing for response to selection is unlikely, and that linear regression estimates of gain may
still be appropriate.

An individual's advantage often conflicts with the good of the group,
and therefore LEIGH (1977) examined how selection could reconcile individual advantage with the interest of the population. In a population individuals occur together in families and the performance of
family members for quantitative traits are correlated. This correlation affects the distribution of order statistics such as selection
differentials and breeding values. The regression of breeding value
on phenotype, however, is not equal to the heritability, even for
normally distributed observations, when extreme ranking individuals
are chosen. In this case selection response may be overestimated
(HILL, 1977).

Comparison was made between the results of individual and group selection. The groups were full sibs or chosen at random. In almost
all cases individual selection was found to be qualitatively and
quantitatively superior when used in conjunction with full sib groups.
This is due to the fact that change in gene frequency for full sib
groups is a function of both direct and associate effects, whereas
with random groups it is a function of direct effects only (GRIFFING,
1976a). Two extreme forms of nonrandom groups, that due to homozygosity and especially that due to homogeneity (e.g., clonal plants) can
further increase the efficiency of truncation selection (GRIFFING,
1976b).

> Recurrent selection experiments may also be used for estimation of gene num-
> ber, particularly when heritability is high. Effective ways to reduce the bias
> and the variability of the gene number estimate was the increase of the number
> of generations in one- and two-way selection programs. Different procedures
> were further proposed to minimize other biases due to dominance, linkage, in-
> equal gene effects, and epistatic effects (PARK, 1977a,b).

> Assuming an infinite population of autogamous diploid plants, selection by
> means of haploid lines should be more effective than selection among the ori-
> ginal diploid lines only where there is overdominance or epistatic interaction

between loci, and where the diploid population shows a high proportion of heterozygotes (FEYT and PELLETIER, 1976).

c) Environmental Interaction and Heritability

When lines or populations were grown in each of several locations the rankings seldom agree. The extent of disagreements depends on the heritability of the trait being measured. Variation in the relative performance of genotypes, formulated as genotype-environment inter- action, is normally studied in analysis of variance under the assump- tion of no correlation between residuals. In experiments with gerbera, tulips, and potatoes often positive correlation between residuals was found, but never a negative one. Positive correlation led to overesti- mation of genotypic effects (WEBER, 1976). The observed magnitude of additive genetic x environment interaction variance in *Pinus taeda*, derived by ANOVA components, was large enough to cause upward biases on heritability estimates and on genetic gain predictions of up to 100% for different selection methods (OWINO et al., 1977).

Half-sib estimates of heritability are almost free of bias from non- additive components of genetic variance and common environmental ef- fects, but in general full-sib estimates are not. A simple accurate approximation for this bias was derived, and it was found that the in- herent bias is usually negligible (PONZONI and JAMES, 1978). An alter- native concept was given by BURDON (1977): The relative performance of genotpyes for growth rate at two different environments can be ex- pressed as a genetic correlation between their growth rates and the two environments.

There might be an advantage of maximum likelihood methods over regression and sib covariance methods for estimating heritability when data are available either from one sex and for more than two generations of noninbred half sib progenies, or from both parents and offspring with hierarchical structure. For the latter case computational forms were developed that are useful when the full and half sib families are of unequal size (THOMPSON, 1977b,c). Linkage disequilibrium among small populations can cause a considerable amount of vari- ation of parameters such as the genetic variance, heritability and genetic correlation of quantitative traits. Estimates of these parameters made in dif- ferent generations from a single population will be unreliable predictors of these quantities in other replicate populations, or in the base population from which the replicate was drawn (AVERY and HILL, 1977).

An improvement of the method of SAKAI and MUKAIDE for estimating broad-sense heritability and other genetic parameters has been applied on empirical data of maize, sugar beet, and Norway spruce populations. The weakness of the method becomes evident if one looks at the very large variation of the parameter values in different years. There is also a question of unambiguity of the estimates, because for the determination of the best estimates the SQ-residuals should be at a minimum. Sometimes, however, these residuals do not vary at all or in a very small amount. Therefore, this method may be at the most helpful for the beginning of breeding work with long-living plants when no progeny tests are available (HÜHN, 1978b).

The estimator of the realized heritability in polygenic populations is biased by epistacy and linkage. BELLMANN and GÖSSEL (1977) found a mathematical term which determines the essential part of the difference between the realized heritability and its estimator. Two alternative ways of calculating heritabili- ty of an index were given by LIN (1978a). Since both gave the same results and the regression of the genetic index on the selection index is simpler to cal- culate, it is preferable to the analysis of covariance among relatives.

d) Selection Indexes

The basic theory and various modifications were reviewed in a paper
by LIN (1978b) and the limitations of selection indexes were discussed.
As with single trait selection, response to index selection would become impossible if the heritability of the index reduces to zero. Then
a negative correlation for favorable traits can arise after a period
of index selection for these traits which might have been uncorrelated
or even positively correlated. Some statistical properties of an index
of multiple traits including variance of the index and correlation between the index and an arbitrary linear combination of the genetic
effects were examined by LIN and ALLAIRE (1977); and by NORDSKOG
(1978).

A selection index for two traits was constructed which allows partial
restriction for one of the two traits. In practise the need for compensation by index selection for unwanted expected genetic changes
may arise whenever selection is applied at different stages of the
life cycle or in both sexes independently (ABPLANALP and EKLUND, 1978).
Secondary traits for which improvement is not important may also be
included in the selection index to assist selection with desired gains
for traits with economic importance (TAI, 1977). The selection index
procedure was extended for traits with direct, maternal, and grand-
maternal genetic components, each having different econonical value
(VAN VLECK, 1976).

> Selection to modify the entire annual growth curve may not be as effective as
> linear models may predict. The difficulties of using simple index selection
> to raise the entire growth curve have been shown for a population of *Nicotiana*
> *tabacum* (NAMKOONG and MATZINGER, 1975). Selection indices or similar criteria
> can be used also for indirect selection of complex characters, for instance,
> selection for metric tons per ha of sugarcane may be based on stalk length,
> stalk diameter, and stalk number (MILLER et al., 1978), or another example,
> selection for plant height, basal and mid-diameter, basal- and mid-sample
> fiber densities may give a useful nondestructive prediction criterion for the
> single plant fiber yield of jute (ARUNACHALAM and IYER, 1978).

e) Competition

Genetic parameters are often estimated from mixtures of genotypes,
but the models used rarely take into account neighbor effects, and
those that do are extremely complicated. A simple model was developed
that illustrates the effects of competition on genetic parameter estimation (HAMBLIN and ROSIELLE, 1978). Competition may either inflate or
reduce the apparent additive and dominance variance with consequent
effects on heritability estimates. Therefore, if early generation selection of individuals is to be effective, methods are needed that
assess the confounding effects of competition and which can be used
to correct for competitive effects.

> An extension of the Griffing theory might be a step toward redeveloping all
> breeding theory to account for competition effects. From the point of view of
> population genetics this extension reveals a new application for the coeffi-
> cient of kinship. This approach may be extended to groups of arbitrary size
> and admit also more complex situations (GALLAIS, 1976a). In a paper of HÜHN
> (1976) the conditions of positive mass selection with respect to competition
> were discussed. The expected response of a population of competing genotypes
> to selection of high-yielding individuals can be expressed in terms of a re-
> gression model of a type similar to that used to describe genotype-environment
> interactions (WRIGHT, 1977).

If competition influences the performance of the single plant then individual plant selection should be done at high population density comparable to normal population density (NASS, 1978). Competition can be seen also as a helpful force for selection in segregating populations to reduce the problem of numbers, provided the hazard of loss of valuable germ plasm during early generations of stress is acceptable (JENSEN, 1978).

4. Population Structure

a) Breeding Systems (Inbreeding and Heterosis)

In plants we find a wide range of systems ranging from complete, or virtually complete inbreeding to regular, enforced outcrossing.

Inbreeding is a word with several meanings. Therefore, JACQUARD (1975) recommended abandoning its use. He presented definitions of the various concepts to which it corresponds: (1) kinship, or the pedigree relation between individuals, measured by the coefficient of kinship ϕ, (2) the genetic drift of an isolated group, measured by its average coefficient of consanguinity α, (3) the deviation between the actual mating behavior and panmixis, measured by the coefficient δ, (4) the subdivision of a population into more or less isolated groups, characterized by the standardized variance $V/p(1-p)$, which may differ from one allele to another, (5) the deviation between the actual genotypic structure and that deriving from the Hardy-Weinberg law, a deviation which, for a locus with n alleles, can be measured by a set of $n(n-1)/2$ indices λ_{ij}. There are several relations of the above-mentioned coefficients and parameters to the coefficient of inbreeding (F), i.e., the probability that genes are identical by descent, either for a single plant or for a population.

> COCKERHAM and WEIR (1977) presented a system of digenic descent measures for finite populations. For recurrent selection procedures the effects of linkage and population size on levels of inbreeding were examined by using transition equations avoiding some intricacies and inaccuracies of earlier work (CHOY and WEIR, 1977). Another method allows exact calculation of inbreeding coefficients for both monoecious and dioecious populations with overlapping generations. Emphasis was laid on providing equations well suited for computer iterations (CHOY and WEIR, 1978).

> Analyzing isozymes in endosperm and embryo tissue is a suitable method to estimate rates of self-fertilization of individual trees in populations of *Picea abies* and Pinus sylvestris (MÜLLER, 1976, 1977). Population structure was studied in white spruce populations by comparing the effects of self-pollination and cross-pollination between neighboring and widely separated trees. Family relationships were indicated between near neighbors, but trees separated greater than 100 m were unrelated (COLES and FOWLER, 1976).

Though there is much variability in response to inbreeding among small populations, the average performance of a population decreases linearly with an increase in F. Therefore, estimation of inbreeding depression in a population requires careful sampling techniques and an adequate sample size. Different methods of inbreeding and comparison of populations with different inbreeding levels in maize imply that loss of heterozygosity per se is almost responsible for the depression. At least a minimum of three conditions is necessary for inbreeding to occur: (1) a population must have favorable and less favorable alleles at gene loci which control the character of interest, (2) there must be some degree of dominance among the alleles, and (3) upon inbreeding the loss of heterozygosity must occur in those loci (BURTON et al., 1978; GOOD and HALLAUER, 1977).

Population improvement by means of fixing a high number of favorable genes is important for breeding panmictic as well as hybrid varieties of rye. It was concluded that selection on the performance of S_1-lines could be advantageous especially in cases of higher degrees of dominance (WRICKE, 1976). Recurrent selection for intrapopulation improvement in maize using combined half sib and S_1 progeny tests showed that selection had reduced inbreeding depression in the population without decreasing heterosis for yield in half sib families (GOULAS and LONNQUIST, 1976).

> Estimates of phenotypic stability demonstrated that heterozygous populations display a considerably higher stability than the inbreds for each of the characters examined (WAHLE and GEIGER, 1978). As a result of recurrent selection experiments with maize, changes in heterosis were found to be compatible with a simple dominance genetic model (MOLL et al., 1978).

> To predict in the univariate case the distribution of the inbred lines that can be derived by single seed descent from the F_2 of a cross between two inbred lines, one needs only the genetical components of family means and variance, estimated in the first few generations. Beyond that one parameter, the additive genetic correlation between pairs of characters, is required to predict the properties of the inbred lines for two or more characters simultaneously (JINKS and POONI, 1976; POONI and JINKS, 1978).

Comparisons were made between the expected genetic means and variances of a quantitative trait in simulated populations of lines derived by diploidizing haploids on the one hand and by single seed descent on the other. In the absence of linkage no differences between populations were observed, but when linkage was present recombination was more frequent in populations of the single seed descent which leads to differences in means and variances. The direction and magnitude of the differences in means and variances depended upon linkage phase, recombination frequency, and whether nonallelic interactions are present. It was concluded that from theoretical considerations the single seed descent method is preferable (SNAPE, 1976; RIGGS and SNAPE, 1977).

Enforced complete homozygosity in tobacco may not be entirely responsible for the reduction in yield of dihaploid lines developed by anther culture. The inferiority of crosses of dihaploid lines compared to conventional entries suggests that this technique may not be suitable for a tobacco-breeding program where yield increase is of primary interest (ARCIA et al., 1978). There may be other reasons, however, to use the time-saving haploid technique. In two barley crosses the bulk population showed a higher mean grain yield, but less variation among lines, than the doubled haploid populations. The ten highest yielding lines of each method were selected and then there were no differences in average yield. In this case the doubled haploid method produced genotypes having the same yield potential as those developed by the bulk plot method (SONG et al., 1978).

> A model with altered male and female fertility was used to explore the possibility of the evolution of dioecy from monoecy, and the evolution of gynomonoecy and monoecy from the hermaphrodite state (CHARLESWORTH and CHARLESWORTH, 1978). In the gynodioecious species *Thymus vulgaris* the individual plants are either hermaphrodite or female. An appreciable rate of self-fertilization was found in four different populations, high enough to favor male sterility as a mechanism for restricting inbreeding (VALDEYRON et al., 1977).

b) Evolution of Synthetic Varieties

For the prediction of the evolution of synthetic varieties the knowledge of general combining ability of the parents is needed. A prediction formula using the number of parents, their general combining
ability and the values of their progenies from one generation of self-
fertilization was very efficient compared with predictions using only
the general combining ability (GALLAIS, 1976b).

Three generations of 21 synthetics of cocksfoot were grown at sward
density and spaced. In sward density the eight-parent synthetic showed
no specific effect from first to third generation, with four parents
there was a decrease in yield between first and second generation, and
the mean values of the two-parent synthetic were linearly decreasing
for successive generations. From the results it was concluded: as the
number of parents decreases, the number of parameters involved in the
description of evolution in advanced generations increases. For two-
parent synthetics, predictions based on the average value of the yield
in the first generation will often over- or underestimate the value
for the third generation (VINCOURT et al., 1977).

ROTILI (1977) concluded from his studies on synthetics of alfalfa,
derived from partly inbred lines, that it is possible to select the
best synthetics on the basis of the single cross performance. He further pointed out that the use of competition increases the efficacy
of phenotypic selection, and that the most promising approach seems
to be a synthetic variety based on four parents. A composite of maize
of all possible crosses of 12 local varieties was increased openpol-
linated for four generations. The yielding ability of the F_1 was equal
to the best local hybrid variety, but yield and most agronomic characters showed a gradual decline in advanced generations. On the other
hand one cycle of a modified mass selection in F_3 was effective in
raising the mean performance to the level of the F_1 (GALAL et al.,
1977). In some synthetics of sugar beet derived from a low sucrose
source there was an improvement of sucrose, but in no synthetic root
did yield and sucrose content increase simultaneously (HECKER, 1978).

Increasing the portions of heterozygotes in further generations could
be a means to preserve good performance. Assuming different ratios of
genic and cytoplasmic male steriles and different selfing rates, the
expected values of the portions of heterozygotes were higher only when
cytoplasmic male sterility was involved (KOBABE, 1978).

5. Populations of Polyploids, Especially Autotetraploids

In a review article GALLAIS (1976c) concluded that the general rules
for the use of heterosis in diploids apply for autotetraploids with
some modifications in the modalities, for example: double cross may
be better than single cross, an optimum level of inbreeding may exist
for single crosses, and an increase of the vigor of a synthetic is
possible in advanced generations with highly inbred parents.

In alfalfa for every level of inbreeding the general combining ability
was considerably greater than specific combining ability. Therefore,
the advantage of hybrids over synthetics did not appear great enough
to justify the higher cost in producing hybrid seed (ROTILI, 1976).
For characters affected by earlier selection the narrow-sense heritability has been rather low in tetraploid alfalfa. It seems that almost

all of the additive components have already been utilized and that
the variation found is due to heterozygosity and other interactions.
In hexaploids, however, an appreciable amount of genetic variation
was additive (SINGH, 1978).

Data from alfalfa progenies, derived from single clones, were analyzed
by multiple regression on the coefficient of inbreeding (GALLAIS' genetic model), and by a model for tetraploid populations developed by
HILL. Additive and digenic effects were responsible for more than half
of the total genetic variance among and within families. Trigenic or
quadrigenic effects were frequently significant, but mostly for individual families. Deviations from the genetic models were not significant, but better fits were obtained by HILL's model (HILL, Jr., 1976).

> An increase in bivalents and a decrease in quadrivalents indicated that in-
> breeding and selection for vigor and fertility on meiotic behavior have brought
> about a shift toward regular meiosis in the evolved tetraploid Job's tears
> (VENKATESWARLU and RAO, 1976).

> In spite of a low outcrossing rate (1%-3%) in the tetraploid *Hordeum jubatum*,
> a combination of inbreeding, polyploidy, and duplicated loci, which can be
> phenotypically heterozygous, has produced a genetic mechanism capable of main-
> taining high levels of genetic variability through fixed heterozygosity (BABBEL
> and WAIN, 1977).

> Based on certain defined assumptions a general model for tetraploid populations
> with heterosomes was presented by POHLMANN (1976). With respect to the hetero-
> somes an equilibrium is reached in a few generations. Thus it was concluded
> that natural dioecious populations ought to be in a balanced state.

References

ABPLANALP, H., EKLUND, J.: TAG 51, 277-280 (1978). - ADAMS, M.W.:
Euphytica 26, 665-679 (1977). - ARCIA, M.A., WERNSMAN, E.A., BURK,
L.G.: Crop Sci. 18, 413-418 (1978). - ARUNACHALAM, V., IYER, R.D.:
TAG 52, 129-134 (1978). - ASMUSSEN, M.A., FELDMANN, M.W.: J. Theor.
Biol. 64, 603-618 (1977). - AVERY, P.J.: Genet. Res. 29, 97-112
(1977); - Genet. Res. 31, 239-254 (1978). - AVERY, P.J., HILL, W.G.:
Genet. Res. 29, 193-213 (1977).

BABBEL, G.R., WAIN, R.P.: Can. J. Genet. Cytol. 19, 143-152 (1977). -
BAILEY, R.J., REES, H., ADENA, M.A.: Heredity 41, 1-12 (1978). -
BAILEY, Th.B., Jr., COMSTOCK, R.E.: Crop Sci. 16, 363-370 (1976). -
BELLMANN, K., GÖSSEL, M.: Biom. J. 19, 497-506 (1977). - BENGTSON,
B.O., BODMER, W.F.: Theor. Popul. Biol. 9, 260-281 (1976). - BERGER,
E.: Am. Nat. 110, 823-839 (1976). - BERGMANN, F.: TAG 52, 57-64 (1978).
- BIRKY, C.W., Jr., SKAVARIL, R.V.: Genet. Res. 27, 249-265 (1976).
BORMANN, S.A.: Proc. Natl. Acad. Sci. USA 75, 1909-1913 (1978). -
BOUGHEY, H., JINKS, J.L.: Heredity 40, 363-369 (1978). - BOUGHEY,
H., JINKS, J.L., COOMBS, D., SHUFFLEBOTHAM, W.: Heredity 41, 175-
183 (1978). - BROWN, A.H.D.: TAG 52, 145-157 (1978). - BROWN, A.H.D.,
ZOHARY, D., NEVO, E.: Heredity 41, 49-62 (1978). - BRUMPTON, R.J.,
BOUGHEY, H., JINKS, J.L.: Heredity 38, 219-226 (1977). - BULMER,
M.G.: Genet. Res. 28, 101-117 (1976). - BURDON, R.D.: Silvae Genet.
26, 168-175 (1977). - BURTON, J.W., STUBER, C.W., MOLL, R.H.: Crop
Sci. 18, 65-68 (1978).

CAISSE, M., ANTONOVICS, J.: Heredity 40, 371-384 (1978). - CASH,
W.W.: Biometrics 33, 528-532 (1977). - CHAKRABORTY, R., NEI, M.:

Genetics 84, 385-393 (1976). - CHAKRABORTY, R., FUERST, P.A., NEI, M.:
Genetics 88, 367-390 (1978). - CHARLESWORTH, D., CHARLESWORTH, B.:
Heredity 41, 137-153 (1978). - CHARLESWORTH, D., CHARLESWORTH, B.,
STROBECK, C.: Genetics 86, 213-226 (1977). - CHOY, S.C., WEIR, B.S.:
TAG 49, 63-77 (1977); - Genetics 89, 591-614 (1978). - CHRISTIANSEN,
F.B., FENCHEL, T.M.: Theories of Populations in Biological Communi-
ties. 144 pp. Berlin-Heidelberg-New York: Springer 1977a; - Measuring
Selection in Natural Populations. 564 pp. Berlin-Heidelberg-New York:
Springer 1977b. - CHRISTIANSEN, F.B., BUNDEGAARD, J., BARKER, J.S.F.:
Evolution 31, 843-853 (1977). - CLEGG, M.T., KAHLER, A.L., ALLARD,
R.W.: Genetics 89, 765-792 (1978). - COCKERHAM, C.C., WEIR, B.S.:
Genet. Res. 30, 121-147 (1977). - COLES, J.F., FOWLER, D.P.: Silvae
Genet. 25, 29-34 (1976). - COMPTON, W.A., BAHADUR, K.: Crop Sci. 17,
378-380 (1977). - CRISP, P., JOHNSON, A.G., ELLIS, P.R., HARDMANN,
J.A.: Heredity 38, 209-218 (1977).

DARLINGTON, P.J., Jr.: Proc. Natl. Acad. Sci. USA 74, 1647-1651 (1977).

EANES, W.F., KOEHN, R.K.: Genet. Res. 29, 223-230 (1977). - EGGERS-
SCHUMACHER, H.-A., FORMANN, G., WÖHRMANN, K.: TAG 49, 187-195 (1977).
- ELLIS, W.H., KEYMER, R.J., JONES, D.A.: Heredity 38, 339-347
(1977). - EWENS, W.J., THOMSON, G.: Genetics 87, 807-819 (1977).

FELSENSTEIN, J.: The Theoretical Genetics of Variable Selection and
Migration, 253-280. In: Annual Review of Genetics, Vol. 10, eds. H.L.
ROMAN, A. CAMPBELL, L.M. SANDLER, Palo Alto, CA, USA: Annual Reviews
Inc. 1976. - FELSENSTEIN, J., YOKOYAMA, S.: Genetics 83, 845-849
(1976). - FEYT, H., PELLETIER, G.: Ann.Amelior. Plant. 26, 365-386
(1976).

GALAL, S., Jr., HINDI, L.A., EL HATTAB, H.S., RADWAN, M.S., EL NAGGAR,
A.M.: Z. Pflanzenzücht. 78, 238-243 (1977). - GALLAIS, A.: TAG 47, 189-
195 (1976a); - Ann. Amelior. Plant. 26, 623-628 (1976b); - Ann. Amelior
Plant. 26, 639-646 (1976c). - GILLESPIE, J.H.: Genetics 81, 403-413
(1975); - Am. Nat. 110, 809-821 (1976); - Evolution 31, 85-90 (1977).
- GILLESPIE, J.H., LANGLEY, Ch.: Genetics 82, 123-137 (1976). -
GINZBURG, L.R.: J. Theor. Biol. 68, 545-550 (1977). - GOOD, R.L.,
HALLAUER, A.R.: Crop Sci. 17, 935-940 (1977). - GOULAS, Ch.K.,
LONNQUIST, J.H.: Crop Sci. 16, 461-464 (1976). - GREGORIUS, H.R.:
TAG 49, 165-176 (1977). - GRIFFING, B.: Genetics 82, 703-722 (1976a);
- Genetics 82, 723-731 (1976b).

HAIGH, J., SMITH, J.M.: Genet. Res. 27, 85-87 (1976). - HAMBLIN, J.,
ROSIELLE, A.A.: Crop Sci. 18, 51-54 (1978). - HAMRICK, J.L.: TAG 47,
27-34 (1976). - HARDING, J., BARNES, K.: Evolution 31, 247-255 (1977).
- HARPER, J.L.: Population Biology of Plants. 892 pp. New York:
Academic Press 1977. - HECKER, R.J.: Crop Sci. 18, 805-809 (1978). -
HEDRICK, P.W.: Genetics 89, 389-401 (1978). - HILL, R.R., Jr.: Crop
Sci. 16, 237-241 (1976). - HILL, W.G.: Biometrics 33, 703-712 (1977).
- HILL, W.G., AVERY, P.J.: Heredity 40, 397-403 (1978). - HOEKSTRA,
R.F.: Genet. Res. 31, 67-73 (1978). - HOFFMANN, P., ESSER, K.: TAG 53,
273-282 (1978). - HÜHN, M.: TAG 48, 105-118 (1976); - Z. Naturforsch.
33c, 755-768 (1978a); - Z. Pflanzenzücht. 81, 289-304 (1978b). -
HULL, P.: Heredity 36, 67-72 (1976).

ISHII, K., CHARLESWORTH, B.: Genet. Res. 30, 93-106 (1977).

JACQUARD, A.: Theor. Popul. Biol. 7, 338-363 (1975). - JENSEN, N.F.:
Crop Sci. 18, 622-626 (1978). - JINKS, J.L., POONI, H.S.: Heredity 36,
253-266 (1976). - JINKS, J.L., TOWEY, Ph.: Heredity 37, 69-81 (1976).
- JINKS, J.L., JAYASEKARA, N.E.M., BOUGHEY, H.: Heredity 39, 345-355

(1977). - JONES, D.A.: Heredity 39, 27-44 (1977). - JONES, D.A.,
CRAWFORD, T.J.: Heredity 39, 313-325 (1977).

KARLIN, S., CAMPBELL, R.B.: Genet. Res. 32, 151-169 (1978). - KIMURA,
M., OHTA, T.: Proc. Natl. Acad. Sci. USA 75, 2868-2872 (1978). -
KNOWLES, R.P.: Crop Sci. 17, 51-54 (1977). - KOBABE, G.: Z. Pflanzen-
zücht. 81, 149-158 (1978).

LEIGH, E.G., Jr.: Proc. Natl. Acad. Sci. USA 74, 4542-4546 (1977). -
LEWONTIN, R.C., GINZBURG, L.R., TULJAPURKAR, S.D.: Genetics 88, 149-
170 (1978). - LI, W.-H.: Genetics 83, 423-432 (1976a); - Genet. Res.
28, 119-127 (1976b); - (ed.) Stochastic Models in Population Genetics.
471 pp. Stroudsburg, PA: Dowden, Hutchinson, Ross 1977a; - Proc. Natl.
Acad. Sci. USA 74, 2509-2513 (1977b); - Genetics 90, 349-382 (1978);
- LI, W.-H., NEI, M.: Genetics 86, 901-914 (1977). - LIN, C.Y.: Can.
J. Genet. Cytol. 20, 485-487 (1978a); - TAG 52, 49-56 (1978b). - LIN,
C.Y., ALLAIRE, F.R.: TAG 51, 1-3 (1977). - LINNEY, R.: Heredity 38,
379-390 (1977). - LOESCHKE, V.: Die Erhaltung genetischer Variation
in heterogenen Umwelten: Theoretische Betrachtungen zum Beitrag der
disruptiven Selektion. 130 pp. Dissertation: FU Berlin 1978.

MALECOT, G.: Theor. Popul. Biol. 8, 212-241 (1975). - MANLY, B.F.J.:
Heredity 38, 321-328 (1977). - MANNING, J.T.: Heredity 38, 117-119
(1977). - MARSHALL, H.G.: Crop Sci. 16, 9-15 (1976). - MARSHALL, D.R.,
PRYOR, A.J.: TAG 51, 177-184 (1978). - MARUYAMA, T.: Stochastic Pro-
blems in Population Genetics. 245 pp. Berlin-Heidelberg-New York:
Springer 1977. - MARUYAMA, T., KIMURA, M.: Proc. Natl. Acad. Sci. USA
75, 919-922 (1978). - MILLER, J.D., JAMES, N.I., LYRENE, P.M.: Crop
Sci. 18, 369-372 (1978). - MITRA, S.: Genetics 82, 543-545 (1976). -
MITTON, J.B., LINHART, Y.B., HAMRICK, J.L., BECKMANN, J.S.: TAG 51,
5-13 (1977). - MOLL, R.H., COCKERHAM, C.C., STUBER, C.W., WILLIAMS,
W.P.: Crop Sci. 18, 641-645 (1978). - MORAN, P.A.P.: Genet. Res. 28,
47-53 (1976). - MÜLLER, G.: Silvae Genet. 25, 15-17 (1976); - Silvae
Genet. 26, 207-217 (1977).

NAGYLAKI, T.: Hereditary 37, 59-67 (1976a); - Genetics 83, 867-886
(1976b); - Selection in One- and Two-Locus Systems. 208 pp. Berlin-
Heidelberg-New York: Springer 1977; - Proc. Natl. Acad. Sci. USA 75,
423-426 (1978). - NAMKOONG, G., MATZINGER, D.F.: Genetics 81, 377-
386 (1975). - NARAIN, P., POLLAK, E.: Genet. Res. 29, 113-121 (1977).
- NASS, H.G.: Crop Sci. 18, 10-12 (1978). - NEI, M.: Molecular Population Genetics and Evolution. 286 pp. Amsterdam-Oxford: North Holland
1975; - Genetics 89, 583-590 (1978). - NEI, M., LI, W.-H.: Genet. Res.
28, 205-214 (1976). - NICHOLAS, F.W.: TAG 48, 101-104 (1976). -
NICHOLAS, F.W., ROBERTSON, A.: TAG 48, 263-268 (1976). - NORDSKOG,
A.W.: TAG 52, 91-94 (1978).

OHTA, T.: Nature (London) 263, 74-76 (1976); - Genet. Res. 30, 89-91
(1977a); - Nature (London) 267, 515-517 (1977b); - Proc. Natl. Acad.
Sci. USA 75, 5108-5118 (1978a); - Genet. Res. 31, 13-28 (1978b); -
OHTA, T., KIMURA, M.: Genet. Res. 28, 307-308 (1976). - OXFORD, G.S.,
ANDREWS, T.: Heredity 38, 367-371 (1977). - OWINO, F., KELLISON, R.C.,
ZOBEL, B.J.: Silvae Genet. 26, 131-134 (1977).

PARK, Y.C.: TAG 50, 153-161 (1977a); - TAG 50, 163-172 (1977b). -
PHUNG, T.K., RATHJEN, A.J.: TAG 48, 289-297 (1976). - POHLMANN, J.:
TAG 47, 257-261 (1976). - POLLAK, E.: Genetics 90, 383-389 (1978).
- POLLAK, E., KEMPTHORNE, O., BAILEY, T.B., Jr. (eds.): Proceedings
of the International Conference on Quantitative Genetics. 872 pp.
Ames, IA: Iowa State Univ. Press 1977. - PONZONI, R.W., JAMES, J.W.:
TAG 53, 25-27 (1978). - POONI, H.S., JINKS, J.L.: Heredity 40, 349-
361 (1978).

QUALSET, C.O., PETERSON, M.L.: Crop Sci. 18, 311-315 (1978). - QUISENBERRY, J.E., ROARK, B., BILBRO, J.D., RAY, L.L.: Crop Sci. 18, 799-801 (1978).

RIGGS, T.J., SNAPE, J.W.: TAG 49, 111-115 (1977). - RINK, G., THOR, E.: Silvae Genet. 25, 17-22 (1976). - RINKER, C.M., DEAN, J.G., MAY, R.G., GARRISON, C.S., CALHOUN, W.: Crop Sci. 18, 151-154 (1978). - ROBERTSON, A.: Genetics 81, 775-785 (1975). - ROBERTSON, A.: Genet. Res. 31, 255-264 (1978). - ROSS, M.D.: Heredity 38, 279-290 (1977). - ROTILI, P.: Crop Sci. 16, 247-251 (1976); - Crop Sci. 17, 245-248 (1977).

SCHAFFER, H.E., YARDLEY, D., ANDERSON, W.W.: Genetics 87, 371-379 (1977). - SEMEONOFF, R.: Heredity 39, 373-381 (1977). - SINGH, S.M.: TAG 52, 159-164 (1978). - SLATKIN, M.: Genetics 81, 787-802 (1975); - J. Theor. Biol. 70, 213-228 (1978). - SLATKIN, M., CHARLESWORTH, D.: Genetics 89, 793-810 (1978). - SLATKIN, M., LANDE, R.: Am. Nat. 110, 31-55 (1976). - SLATKIN, M., WADE, M.J.: Proc. Natl. Acad. Sci. USA 75, 3531-3534 (1978). - SMOUSE, P.E.: Am. Nat. 110, 849-860 (1976). - SNAPE, J.W.: Heredity 36, 275-277 (1976). - SNAYDON, R.W., DAVIES, M.S.: Heredity 37, 9-25 (1976). - SONG, L.S.P., PARK, S.J., REINBERGS, E., CHOO, T.M., KASHA, K.J.: Z. Pflanzenzücht. 81, 271-280 (1978). - STROBECK, C., MORGAN, K.: Genetics 88, 829-844 (1978).

TAI, G.C.C.: Crop Sci. 17, 182-183 (1977). - THOMPSON, G.: Genetics 85, 753-788 (1977a). - THOMPSON, R.: Biometrics 33, 485-495 (1977b); - Biometrics 33, 497-504 (1977c). - TOWEY, Ph., JINKS, J.L.: Heredity 39, 399-410 (1977).

VALDEYRON, G., DOMMÉE, B., VERNET, Ph.: Heredity 39, 243-249 (1977). - VAN VLECK, L.D.: Biometrics 32, 173-181 (1976). - VENKATESWARLU, J., RAO, P.N.: TAG 47, 165-169 (1976). - VINCOURT, P., GALLAIS, A., BERTHOLLEAU, J.-C.: Ann. Amelior. Plant. 27, 675-703 (1977).

WAHLE, G., GEIGER, H.H.: Z. Pflanzenzücht. 80, 211-222 (1978). - WATTERSON, G.A.: Genetics 85, 789-814 (1977); - Genetics 88, 405-417 (1978). - WEBER, W.E.: TAG 48, 85-93 (1976). - WEIR, B.S., COCKERHAM, C.C.: Genetics 88, 633-642 (1978). - WILLIAMS, S., VERNA, M.M., JINKS, J.L., BRASIER, C.M.: Heredity 37, 365-375 (1976). - WILLS, C.: Genetics 89, 403-417 (1978). - WILLS, C., MILLER, C.: Genetics 82, 377-399 (1976). - WRICKE, G.: TAG 47, 265-269 (1976). - WRIGHT, A.J.: TAG 49, 201-207 (1977).

YAMAZAKI, T.: Genetics 86, 227-236 (1977). - YASUDA, N.: Heredity 41, 155-163 (1978).

Dr. ROBERT LICHTER
Max-Planck-Institut
für Zellbiologie
D 6802 Ladenburg/Rosenhof

D. Taxonomy

I. Systematics and Evolution of Seed Plants

1. Introduction

The main developments in systematics in the past two years have taken
different directions. Firstly, the evolution of angiosperms documented by the combined fossil record of pollen and leaves led to a better
knowledge of early forms of angiosperms and of the time of their first
culmination. Secondly, the investigation of the evolution of repellent
chemical compounds forced by animals as predators is a new field in
chemical systematics. Still of interest and a field for unending discussion is the evolution of higher categories (symposium volume, ed.
KUBITZKI, 1977a). Such subjects certainly are much more attractive
to report and easier to present than results which appear mainly in
"small print"; these are often very briefly mentioned but quite frequently contain the material for ideas which are discussed more broad-
ly. So one could say that these subjects play the same role as the
"small print" in contracts: points only briefly formulated but of major importance. One should not hesitate to see especially in revisions
and, in general, in the increase of knowledge of the natural history
of species and genera, one of the main aspects of progress in systematics. This is also the reason for presenting two chapters dedicated
to special plant groups. The knowledge concerning the Compositae has
increased stimulated by a Reading symposium (symposium volume, ed.
HEYWOOD et al., 1977) and has led to many additional papers. But also
the Apiaceae seemed to deserve special treatment. Certainly other
groups too, e.g., the Leguminosae, with a good amount of special papers, and the *Erythrina*-symposium (ed. RAVEN, 1977) could have been
treated separately, but such a procedure would have led to another
concept in this report.

2. Evolutionary Processes

VAN STEENIS (1977, 1978) discussed the different methods of evolution
in plants and animals. He decided that within plants mutual competition is not an agent in their evolution. Animal evolution is, besides
the struggle against the environment, mainly influenced by competition.
Competition and survival of plants is more passive and mainly directed
against the inanimate environment. This tolerance led to a slower and
conservative evolution. In plants therefore the survival level lies
much lower and a greater free space ("patio ludens") for structural
development of nonadaptive character exists. BÖCHER (1977a) on the
other hand came to the conclusion that it is hardly possible to deny
that certain characters are neutral under present conditions, yet they
may have evolved a long time ago as adaptations to conditions which we
have no chance of estimating today: thus neutral characters may des-
cend from adaptive ones. In contrast VAN STEENIS interpreted the char-

acters of the "patio ludens" as a reservoir for new adaptations. In
this sense they cannot play a role in the explanation of some cases
of extinction. He also concluded that in plants the usual way of evolution was through sudden (saltatory) changes. On the other hand speciation by racial segregation in plants took place only in exceptional
cases. Co-evolution is considered as more independent. Normally changes in plant structure occurred first and adaptations by animals followed. Novelties arising in plants are exploited by animals for evolution and specialization. In plants, contrary to animals, reticulate
phylogeny has played an important role in evolution.

BÖCHER (1977a), dealing with life-form, discussed the importance of
convergence as an evolutionary process. In evolution convergence and
divergence represent opposite aspects of the same mechanism. It is
shown that parallel developments are not restricted geographically as
is sometimes thought (e.g., divaricateness, for this phenomenon see
also TOMLINSON, 1978). Convergent evolution may be eludicated as a
series of stages which show a decreasing number of common characters
and become more and more generalized, the characters defining the
stages finally being difficult to interpret as adaptive. While speciation may seem to have more or less clear-cut entities as end products,
the information of life-form and similar types may be characterized as
a process of lumping together by which the end products are abstrac-
tions that cannot be divided into well-separated units. Work with life-
forms is often neglected because the entities are not sharply circumscribed, but using several parallel systems of life-forms and biological types may help to express the sum of adaptive characters in a
species.

SACHS (1978) called phyletic diversity the existence of independent
phyletic lines and it is suggested that the coexistence of such groups,
e.g., families, is only possible by the existence of special traits in
these groups, e.g., chemical defence mechanisms, types of symbiosis or
special methods of CO_2 uptake. In any case it seems to be a little
artificial to cut evolution at a certain time and then to declare the
separated phyletic lines as independent. Therefore the expression
"phyletic diversity" loses importance.

3. Methods of Classification

a) Use of the Categories

CRONQUIST (1978b) took up again the discussion on the species concept.
In zoology the existence of reproductive isolation is more and more
the only criterion and leads to the recognition of sibling species.
The use of this criterion in botanical taxonomy is not practicable in
general. So CRONQUIST defined species as "the smallest groups that are
consistently and persistently distinct and distinguishable by ordinary means". However the decision about, for example, what are "ordinary means" etc. is individual. At the other side flexibility is necessary to do justice to the specific situation in different groups.
In a survey DAVIS (1978) discussed the principles and practice of
angiosperm classification. He points to the fact that taxonomists
need a practical species concept, because adequate experimental evidence is frequently lacking, which means that they have to deal mostly with "morphospecies". For the majority of plants, binomials must
be applied to representatives of populations that can be morphologically distinguished from another, allowing perhaps for some degree of

hybridization, especially in disturbed habitas. In any case phylogenetic speculation should be kept out of the classifactory process.

VALENTINE (1978) gave an overview of taxonomic possibilities and implications of ecologically different taxa. He argued that from an
evolutionary point of view habitat and environmental factors must always play a vital part. The use of the old expression "ecotype" is
recommended, but a formal recognition is needed for this to not get
lost. Since there are different cases for the application of infra-
specific categories, the rules for the use of subspecies and varieties
must not be too rigid and fixed. A pragmatic approach would be the
best. QUINN (1978) explained that the term "ecotype" is now being used
by ecologists to indicate almost any degree of genetic difference below the level of species. He argued that the evolutionary unit is the
population and the use of "ecotype" for the concept of genetically
based variation that is habitat-correlated can only contribute confusion, not additional information. STACE (1978) discussed more in detail the different application of ranks to different evolutionary
situations. He proposed the use of the expression "semispecies" or
the invention of a new two-tier binomial terminology.

> The application of taxonomical units in special cases is discussed by several
> authors. WELCH (1978) pointed out the difference between botanical and horti-
> cultural nomenclature. PARKER (1978) stressed the need of a good classification
> of crop plants, which is a neglected area. RICHARDSON (1978) exemplified the
> role of endemics in taxonomy. He came to the conclusion that some of the pro-
> blems caused by these groups are impossible to deal with the existing taxonomic
> framework.

> The amount of characters needed for keys has been investigated by SNEATH and
> CHATER (1978). Even though only few characters theoretically are sufficient
> to key out quite a good number of taxa, usually the number of characters has
> to be as large as the number of species concerned. The reasons and consequences
> of this fact have been discussed.

b) Numerical Taxonomy and Cladistics

Zoology and botany are contacting strongly in these two disciplines.
The elucidation of evolution by numerical techniques has been surveyed briefly from a zoological point of view by MOSS (1978). EDMONDS
(1978) presented a numerical study on *Solanum* sect. *Solanum*. JENSEN (1977)
investigated seven species of three series within the scarlet oak complex in the eastern U.S. The series borders do not appear in the numerical analysis and it is suggested that actual evolutionary relationships are better demonstrated in this way. Relatively minor and
taxonomically acceptable changes in the formation of characters have
been used in a taximetric study of the Genisteae (BISBY and NICHOLLS,
1977). The result reached is similar to former classification of the
tribe. The study should serve as an example.

BREMER and WANNTORP (1978) tried to call attention to cladistic taxonomy in the sense of Hennig. They reported principles, results, and
advantages of its application in botanical systematics. An increased
use of cladistics in plant systematics has been recommended by FUNK
and STUESSY (1978) and different examples are given. STUESSY and ESTA-
BROOK (1978) stressed the potential importance of cladistics for the
understanding of phylogeny and the development of phyletic diagrams.
This has been explained more in detail by ESTABROOK (1978). One of
the fundamental questions of cladistics, what is primitive, has been
discussed by ESTABROOK (1977). The importance of that question lies
in the fact that by detecting a character as primitive, the direction

of evolutionary trends can be traced out. One common way is to choose for the primitive status the one that occurs most frequently.

> ESTABROOK and ANDERSON (1978) provided an example of the application of compatibility analysis to the estimation of evolutionary relationships among the taxa of the genus *Crusea*. GARDNER and LA DUKE (1978) compared evolutionary trees obtained by character compatibility with an intuitive phylogram of *Lipochaeta*. Both types, the cladograms and the phylograms were in close agreement regarding relationships of the two sections to each other. It is concluded that character compatibility is quite useful in plant systematics, but it should be used only after the plant group is well understood, i.e., it should be used as a final step in a revisionary study.

4. Taxonomic Evidence

a) Morphology and Anatomy

α) *Microstructures*. The accumulation of protein and/or starch by sieve-element plastids and their systematic importance has been reviewed by BEHNKE (1977a). Sieve tube elements are very useful to characterize the monocotyledons. The dilated cisternae are a special feature of the Capparales (BEHNKE, 1977b) and the attachment of questionable taxa can, as in the case of *Gyrostemon* to the Capparales, be supported by this character. Succulent Centrospermae seem to be characterized by the possession of phytoferritin in phloem-parenchym cells. The whole Centrospermae are well circumscribed by the existence of the so-called P-III subtype of sieve-element plastids (BEHNKE, 1978). SPETA (1977) summarized the meaning of protein bodies for the classification of vascular plants (1977). The Antirrhineae can be rearranged in this way. Lactifers and idioblasts have been investigated in *Jatropha* (DEHGAN and CRAIG, 1978). The latter occur in the most advanced groups. Lactifers are significant in respect to the delimitation of sections and subsections and it has been suggested that these characters may be of taxonomical importance in the whole Euphorbiaceae. CZAJA (1978) gives a special idea of the systematics of vascular plants based on the types of starch grains. As often in such specialized investigations, important findings merge with hardly explainable data. Based on the types of starch grains the heterogeneity of the monocotyledons and the independent evolution of at least one of their groups from dicotyledons is being proposed.

β) *Morphology*. The genus *Hibbertia* shows an extreme diversity of leaf morphology and venation (RURY and DICKISON, 1977). Australian species exhibit the greatest diversity while New Caledonian Hibbertias possess remarkably similar venation patterns but are more variable in their floral structure. Within the whole Dilleniaceae, some species of *Hibbertia* in this respect are the most primitive ones. Normally evolutionary trends in the leaf morphology are correlated with trends in other features of the plant body.

> In the genus *Acer* a classification based on vein architecture is in accordance with a system of exomorphic characters (TAINAI, 1978). The ptyxis of leaves has been studied by CULLEN (1978) who found some special characters in some families. The presence of stipules may help to place correctly questionable genera as *Rhodoleia* (ENDRESS, 1978b) which belongs to the Hamamelidaceae. Hairs and stomata are successfully used for classification. WEBERLING (1977) reported that the Caprifoliaceae in respect of the trichomes show a great diversity and occupy a central position within the Dipsacales. EHLER (1977)

investigated structure, development, and function of trichomes within the Bromeliaceae. Hairs and stomata also can prove that the Urticales are a natural group with four distinct families (GANGADHARA and INAMDAR, 1977). The systematic relevance of associated stomata in the angiosperms at the moment is poor (EL-GAZZAR et al., 1977). The meaning of foliar stomata in connection with the subdivision of the Euphorbiaceae has been discussed by RAJU and RAO (1977). DEN HARTOG and BAAS (1978) showed that the overlap of stomatal types and crystalliferous epidermal cells supports to unite the Celastraceae and Hippocrateaceae into one family.

ERNET (1977a,b) presented a synthesis of all information concerning growth forms, inflorescences, comparative morphology, floral morphology and reproductive biology to obtain a more natural grouping of species in *Valerianella* and *Fedia*. Leaf and cymoid shape show the affinity between the two genera but serve for generic delimitation.

ENDRESS (1977c) summarized actual problems and results of floral morphology and their impact on systematics. For *Distylium*, *Distyliopsis*, and *Matudaea* he showed (1978a) that the hermaphrodite flowers of these genera are euanthia. They form a natural group within the Hamamelioideae.

A study of floral relationships within the Polygonaceae has been presented by GALLE (1977). The tepals are generally formed in a helical sequence. A division into groups with five and six tepals respectively is not supported. The androecia of the Capparales can become secondarily isomerous by a total reduction of a circular bulge (LEINS and SOBICK, 1977). With few exceptions all alismatalean flowers appear to be characterized by antipetalous stamen or staminodial pairs (SATTLER and SINGH, 1978).

An isolated view sometimes leads to not well-founded conclusions: The comparative morphology of the carpel and especially the vascularization of tepals and stamens in the Wurmbaeoideae (STERLING, 1977b) recalls the situation within the Rosaceae. This fact is used as an argument for a monophyletic origin of the angiosperms. Certainly more realistic is to join, basing on the carpel type, the genera *Saurauia*, *Actinidia*, and *Clemathoclethra* into one family Actinidiaceae of Theales relationship (GUÉDÈS and SCHMID, 1978). Within *Nigella* the various degrees of syncarpy and apparent apocarpy have originated from different growth patterns (LANG, 1977). The five different fruit types resulting from this may be used as taxonomic criteria.

The evolution of inflorescences has been studied by several authors. CORNER (1978c) found that within *Dillenia* branched, many-flowered inflorescences have developed from single terminal ones. The fruit evolution took place independently of these features. Within the Gesneriaceae, pair-flowered cymes stand at the beginning of the evolution and lead to normal ones (WEBER, 1978). It is suggested that multiple inflorescences in palms have several adaptive features, such as extending the flowering period and increasing protection against loss through herbivory (FISHER and MOORE, 1977).

γ) *Anatomy*. In many cases wood anatomy supports existing classifications, in other cases it gives new hints concerning the relationship of doubtful taxa. The information certainly is more useful, when it is connected with other data as from gross morphology, cytology etc.

WAGNER (1977) compared the distribution of vessel types within the monocotyledons with modern systems of classification and found only minor modifications necessary. Within the Rubiaceae the wood anatomy correlates with the arrangement of genera, tribes and even subfamilies

of modern authors (KOEK-NOORMAN, 1977). Nearly all exceptions to this
disposal occur in genera of uncertain taxonomical position.

> *Picrodendron*, a genus of uncertain position, is now placed, according to wood
> anatomy, into the Oldfieldioideae in the Euphorbiaceae (HAYDEN, 1977). The
> Penaeaceae now have found a place in the Myrtales (CARLQUIST and DEBUHR, 1977)
> which are better defined by wood anatomy. In the case of *Forsellesia*, a combi-
> nation of characters including the structure of the wood supports a transfer
> to Crossosomataceae (DEBUHR, 1978).

Subfamilial separation has been cleared up as in the case of Capparaceae, where the vessels of the Cleomeae and Cappareae do not justify
a distinction into two families (ALEYKUTTY and INAMDAR, 1978), or on
the generic level, where *Hydrangea* and *Cornidia* are clear-cut genera
(STERN, 1978).

The comparison between primitive and advanced vessel element types
gave interesting results. *Paeonia* sect. *Moutan* (KEEFE and MOSELEY, 1978)
has secondary wood with primitive and advanced features. Relationships
to the Dilleniaceae are supported in this way, but a separate status
of the family Paeoniaceae is justified. Within *Weinmannia* neotropical-
montane species show the most primitive vessel element type. Indo-
Malesian and Madagascan species on the other hand present the most
advanced vessel condition (DICKISON, 1977). The Magnoliales show only
partly a primitive wood structure. Six structural groups could be
established, each of them with a marked gradation from primitive to
advanced stages. Since the most primitive wood is not found in the
Magnoliales, an absolutely basic position of this order within the
Dicotyledons is not supported (GOTTWALD, 1977).

The presence or absence of transfer cells in leaves of the Leguminosae-
Papilionoideae is predictable for tribes and major series. The breaking up of Galegeae, Hedysareae, and Genisteae is emphasized (WATSON
et al., 1977). The Kranz syndrome seems to have arisen several times
independently in the Chenopodiales. It is based on different anatomical and structural combinations (CAROLIN et al., 1978). SEIBERT (1978)
showed for the Lithospermeae that there is no correlation between
fruit anatomy and fruit gross morphology. The position of the attachment area cannot be used as the only character for the circumscription
of the tribe. Fruit anatomy, calcareous cells, and absence of U-cells,
is useful to characterize the Lithospermeae, but the genera and species
are only rarely separable by this means.

b) Palynology

The importance of exine structure has been discussed by several authors
LEWIS (1977) gave an overview of recent and earlier findings concerning the morphology and adaptive significance of the exine. DOYLE (1977b
demonstrated again the possibility of separating the pollen of primitive angiosperms from otherwise similar pollen by exine structures.
Controversial fossil pollen types can be recognized in this way. HESSE
(1978a,b,c) summarized all that is known about "pollenkitt". It makes
pollen sticky in entomophilous species and is placed there on the surface of the exine. In anemophilous plants it is dry and mainly deposited in the cavities of the exine. SKVARLA et al. (1978) studied the viscin threads in Onagraceae pollen. There are different types characteristic for several groups. Ropy viscin threads characterize less specialized genera, while beaded ones are more common in moth-pollinated
genera and probably more derived. In the case of *Artemisia*, ROWLEY and
DAHL (1977) interpreted the exine as a dynamic surface zone.

Palynology very often helps to clear up or to support classification. Only in few cases is there little importance of pollen characters as in the sexine pattern grass pollen (PAGE, 1978). In *Pinus* the structure of the tectum surface is species-specific (KLAUS, 1978). The genus *Ptychopetalum* is divided in an African group with diploporate pollen and a S. American group with triporate pollen. West Africa is suggested as the center of aperture evolution (FEUER, 1978). Within neotropical Gesneriaceae there is no correlation between pollen structure and type of pollination (WILLIAMS, 1978). Pollen characters therefore can be used to trace evolutionary relationships. The Bignoniaceae on the other hand show a remarkable heterogeneity of pollen types often in one genus or even in one species (BUURMAN, 1977). A subdivision of the tribes by pollen characters therefore is not possible. *Collomia* (CHUANG et al., 1978) can be grouped by pollen types, but such a subdivision shows discrepancies with respect to present-day accepted classification. The pollen of Valerianaceae has been investigated twice. BLANKENHORN (1978) distinguished four basal pollen types according to sporoderm sculpture. Only *Valeriana* consists of more than one type. CLARKE (1978) differentiated two groups of genera within the family. The pollen of the *Valeriana* group has simple columellae, that of the *Valerianella* group elongate branched columellae. This example shows how contradictionary the interpretation of the same material may be.

c) Embryology

A symposium (ed. BINET, 1978) dealt with embryology in the broadest sense. Pollen genesis and cytogenesis, male and female gametogenesis, fertilization and development of embryo and endosperm were the main topics. HAMANN (1977) reported on convergence in embryological characters within the angiosperms. The helobial endosperm, for example, convergently arose in the Saxifragaceae as well. It evolved partly from the nuclear and cellular type. PHILIPSON (1977) gave a classification of the dicotyledons based on ovular morphology. He distinguished four types characterized by integument number and nucellar thickness. He stated that the unitegmic groups represent a modern, distinct evolutionary line within the dicotyledons called Unitegminae.

More special papers dealt with the embryology of Cyrillaceae, which are better placed in the Ericales than in the Celastrales (VIJAYRAGHAVAN and DHAR, 1978) and the meaning of proteins and glycoproteins in the ovule of primitive gymnosperms such as *Cycas* (PETTITT, 1977). These compounds may be involved in the discrimination between male gametes.

d) Karyology and Cytogenetics

The impact of chromosome studies on plant taxonomy has been discussed by MOORE (1978). He referred to the sources of chromosome data and gave examples of intensively overworked groups. Tropical floras and certain families are still very poorly investigated. For only 1% of the angiosperms is more detailed information (centromeric position, satellites etc.) available, and less information exists concerning the disposition of heterochromatic segments and similar traits. The use of chromosome data has been shown in various examples. The problem of incorrect data is also discussed. FAVARGER (1978a) wrote a "philosophy of chromosome counts". He discussed all the possible sources of wrong counts. These are mainly caused by: the influence of previous counts, an attempt to give an accurate chromosome number also in cases when the results are not absolutely clear, the wrong determination of the material, changing numbers in high polyploids, the role and nature of supernumerary chromosomes and the intraindividual variation of chromosome numbers. A real problem is to weed out evidently wrong counts from the literature, which are repeated now in every list and paper. FAVARGER proposed better documentation of the results and an obligate preservation of voucher specimens, as well as consultations between colleagues by counting difficult material and a better exchange of the material between investigators working on the same group.

Very often now chromosome numbers are one part of multivariate investigations. On the other hand special papers dealing with only chromosome numbers of certain regions (e.g., CARR, 1978 for Hawaii or FERNANDES et al., 1977 for Portugal) or certain groups (here the papers concerning the Compositae should be mentioned especially like KEIL and STUESSY, 1977; PINKAVA and KEIL, 1977; TOMB et al., 1978) give additional information.

The unique possibility of obtaining chromosome numbers from herbarium material has again been demonstrated for *Impatiens* (GILL and CHINNAPPA, 1977) while ENGELL (1978) showed that cytological differentiation as in *Polygonum* does not always have implications on the external morphology.

α) *Different Cytological Strategies.* *Brachycome lineariloba* with its cytotypes (2n = 2, 4, 5, 6, and 8) is becoming a more and more an important object of cytological investigations as e.g., *Haplopappus gracilis*. In *B. lineariloba* it has been suggested (KYHOS et al., 1977) that the basic karyotype comprises four pairs of chromosomes and dysploidy formated the other cytotypes. Connected with this phenomenon is the ability to enter more arid and harsh environments. This change is associated with increasing vigor. On the other hand race A (2n = 4) is the most vigorour and the most drought-tolerant of the five races. All this is seen in connection with the increasing aridity since Late Pleistocene. Quite different patterns are seen in the species of subgenus *Metabrachycome* (CARTER, 1978a) where all species cytologically investigated show the base number x = 9 (with some polyploids), sometimes living as apomicts. It seems that the more stable environmental conditions in W. Australia held the base number constant. A similar situation occurs in the also Australian genus *Calotis*. Arid zone annuals show a reduction from x = 8 to x = 4, while perennials remained at x = 7 (STACE, 1978). Connected with these phenomena are asexual reproduction and polyploidization.

Within N. American *Cirsium* (BLOOM, 1977) the species with high chromosome numbers show chromosome stability, while species with low numbers are characterized by substantial differences in the chromosomes with the construction of remarkable sterility barriers. In *Carduus* (GREMAUD, 1977) dysploidy is characteristic for the C. *defloratus* group while in the series *Crispi* polyploidy is more typical.

β) *B-chromosomes.* Several papers have dealt with the origin and effect of B-chromosomes. In *Gibasis* (BRANDHAM and BHATTARAI, 1977) the mean chiasma frequencies were significantly higher with increasing numbers of B's (0-6). This effect works rather at a cellular than at a whole anther level. Also in *Pennisetum typhoides* B-chromosomes influence chiasma frequency (RAO and PANTULU, 1978). More detailed investigations (PARKER et al., 1978) in *Hypochoeris maculata* showed that spindle behavior in presence of two or more B's is inefficient. EMC-meiosis is less susceptible to the presence of B's than PMC-meiosis. The increasing deleterious effect of B's limits the number of B's. Within *Cirsium* (BLOOM, 1977) supernumerary chromosomes seem to originate from the breakage of a metacentric chromosome. An interesting influence of B-chromosomes has been shown by CARTER (1978b). He found 1-3 B-chromosomes in *Brachycome lineariloba* in about 10% of all the plants investigated. In the presence of two B's there is pairing at meiosis. In the first pollen grain mitosis nondisjunction and preferential distribution occurs. In marginal populations the frequency of B's increased.

There are hints for a selective advantage of B-chromosomes under arid
conditions. This is one of the rare cases of an influence of B's on
plants outside of meiotic behavior. At the same time this shows, in
connection with other examples, the heterogeneity of the expression
"B-chromosome".

γ) *Cytology and Geography*. HEDBERG and HEDBERG (1977) gave an overview
of cytological investigations in Afroalpine plants. Chromosome counts
are known for 68% of all species. This showed an overall polyploidy
percentage of 49%. They suggest from these data a relatively undis-
turbed situation in recent geological time.

> More details in *Claytonia virginica* (LEWIS and SEMPLE, 1977) show that in
> N. America the diploids have a more northern, the triploids a more southern
> and the tetraploids a more eastern distribution. Within *Crocus cancellatus*
> from the mediterranean area (BRIGHTON, 1977a) a descending dysploid series
> from n = 8 to n = 4 occurs from west to east and may show the direction of
> development. In the *Lathyrus pratensis* complex autopolyploidy is probable. The
> tetraploids show a more eastern distribution while the diploids occur in Western
> and Central Europe (BRUNSBERG, 1977). In *Mentha* (HARLEY and BRIGHTON, 1977)
> the small chromosomes only give little additional information. The base numbers
> are x = 9, 10, 12 and 18. Increase in chromosome number may have been related
> to adaptation to mesic conditions.

SEAVEY and RAVEN (1977a,b) produced a differentiated analysis of the
genomes of *Epilobium* in relation to distribution. The changes are few
reciprocal translocations which form new typical genomes. An AA chromosome arrangement (separated by one reciprocal translocation from BB)
occurs in N. America, S. America and Europe. CC (different from AA by
two reciprocal translocations) is circumboreal. The BB chromosome arrangement is found in Australasia and is typical for this area (HAIR
et al., 1977). By this method several independent introductions of
Epilobium into S. America are demonstrable. Two separate introductions came from N. America, two from Australasia and one or two additional introductions from either source (SEAVEY and RAVEN, 1977c).

δ) *Fine Structure of Chromosomes*. C-banding is being used more and more
for detailed analysis in cytology. It seems at the moment that more
monocotyledons have been investigated in this way, certainly due to
the large chromosomes. BADR and ELKINGTON (1977) showed that with
this method in *Allium* subgen. *Molium* a grouping is possible. Good
agreement between external morphology and patterns of C-banding has
been shown in the continuing work on *Scilla* by GREILHUBER (1978) and
GREILHUBER and SPETA (1977, 1978). This genus may become one of the
first examples totally investigated by this method. In *Gibasis* C-band
polymorphism has been found within and between diploid and tetraploid
populations leading to an interpretation of the relationship between
the two species *G. consobrina* and *G. karwinskyana* (KENTON, 1978). *Anacyclus*
(EHRENDORFER et al., 1977) is a good example for the application of
C-banding. These characters show good but not complete agreement with
morphological and phytochemical data.

The role and distribution of heterochromatin has also been investigated (GREILHUBER, 1977). In *Lachenalia tricolor* centromeric heterochromatin
is uniform and a specific role is suggested (MOGFORD, 1978b). A comparison of the distribution of constitutive heterochromatin in mono-
centric and polycentric chromosomes is related to the mode of mitotic
division. In the holokinetic chromosomes of *Luzula*, besides the existence of kinetochors distributed all over the chromosome, the existence
of constitutive chromatin in a similar distribution also makes agmato-
ploidy possible (RAY and VENKETESWARAN, 1978).

ε) Chromosome Numbers and Classification of Higher Categories. GOLDBLATT and
ENDRESS (1977) can divide the Hamamelidaceae into two groups: The
Hamameloideae and Rhodoleioideae with n = 12 and the Liquidambaroi-
deae, Exbucklandoideae, and Disanthoideae with n = 8, 16, and 32. An
original base x = 7 is suggested leading to x = 6 and 12 on one hand
and x = 8 on the other. The Disanthoideae now hold an ancestral position. Relations to the Casuarinaceae (x = 8) are now imaginable. In
the Lecythidaceae the subfamilies show characteristic base numbers
(KOWAL et al., 1977): Planchonoideae x = 13, Napoleonoideae x = 16
and Lecythidoideae x = 17. No relations to other families can be drawn
from this information. Within the Cornales (GOLDBLATT, 1978), the
Nyssaceae (x = 22) and the Davidiaceae (x = 21) appear more closely
related to the Cornaceae (x = 11) while the exclusion of *Mastixia*
(n = 13, 11), *Curtisia* (n = 13) and *Helwingia* from the Cornaceae, is
indicated also on cytological grounds. In the Loasaceae, subfamily
Loasoideae (POSTON and THOMPSON, 1977), *Caiophora* is well separated
from the rest of the subfamily by its large chromosomes and chromosome numbers n = 8 and 7 (n = 14, 13, and 12 with small chromosomes
in the other genera investigated).

e) Biochemical Systematics

Techniques and instrumentation necessary for the isolation and structural characterization of the compounds have been refined. More and
more information on biosynthetic pathways and genetics of the secondary compounds has been accumulated. Structural and biosynthetic data
greatly enhance the utility of the compounds in phylogenetic interpretations (AVERETT, 1977). TURNER (1977d) stressed the importance of
chemosystematics and P.M. SMITH (1978) gave an optimistic valuation
of chemical evidence in plant taxonomy. GOTTLIEB (1977) gave an overview of possibilities and advantages of electrophoresis of plant enzymes for plant systematics. CAGNIN et al. (1977) have dealt with
quantitative methods and principles of biochemical systematics. Structural diversity of secondary compounds is considered as an independent,
primary criterion. Phyletic classifications based on chemical data
could be more useful than purely morphological ones.

M.C. KING (1977) summarized the interrelationships between molecular
evolution and systematics. WILSON et al. (1977) discussed evidence
and objections to the theory of biochemical evolution. In botany the
supposed rate of cytochrome c evolution indicates an age of 240 million years for the flowering plants. Since the oldest records of
flowering plants are not older than 130 million years, the discrepancy
can be solved in two ways. KING suggests a faster evolution of cytochrome c in plants, while WILSON et al. suppose an age of flowering
plants more ancient than thought before. KING stressed that the basic
assumption is that a fixed substitution in variable regions of structural genes has not been subject to selective pressures, since these
substitutions are largely irrelevant to the adaptive success of the
organisms.

α) Role of Secondary Metabolites. CRONQUIST (1977b) showed that evolution
could be interpreted as a sequence of inventions of repellents, which
one after another lost their effectiveness, as insects and other predators became resistent to them. Evolution has come to the point in
Asteraceae where iridoid compounds have lost their effectiveness and
been replaced by polyacetylenes and sesquiterpene lactones. This was
exemplified by R.O. GARDNER (1977) who showed the gradual replacement
of defence mechanisms based on tannins and crystals (primitive Rosidae)
by those based on a variety of toxic and repellent substances. Also

CATES and RHOADES (1977) stressed this continual chemical warfare between plants and their pests and predators by saying that all the toxic
substances seem to be evolved as a result of the sharp pressure of the
predators. The way in which insects become more and more adapted to
secondary plant substances has been investigated by BRATTSTEN and
WILKINSON (1977), who showed that secondary plant substances induce
the production of mixed-function oxidases in insect larvae. These oxidases play a major role in protecting herbivores against chemical
stress from secondary plant substances.

The defensive ecology of the Cruciferae is a good example for the way
plants and animals can interact (FEENY, 1977). Apart from mustard oil
glucosides, several other genera have evolved special typical defence
substances. In many cases in the Cruciferae secondary metabolites may
even attract insects. A qualitative change in these attractant products,
as shown for *Thlaspi arvense*, makes plants less detectible for crucifer-
adapted enemies. If it were possible to take advantage of the diversity
of compounds within the Cruciferae, a reduction of the application of
pesticides could result. In this connection it is interesting that
canavanine, a nitrogen-rich nonprotein amino acid, which almost exclusively occurs in Diocleinae and Kennediinae of the Papilionoideae
(LACKEY, 1977), has a role in food storage and defence for seeds and
seedlings (BELL et al., 1978).

The repellent system is also still working in Conifers (EDMUNDS and
ALSTAD, 1978). Rare toxins generally offer the greatest protection.
In trees, the variability of compounds (in Conifers tannins) gives
more protection. This system only works by sexual reproduction. Cloning
and selection programs in Conifers which are designed to obtain higher
progeny greatly narrow the variation in the pine gene pool and increase
the vulnerability of trees to insect damage. How intense the insecti-
cidal activity of foliage may be, is exemplified by the New Zealand
Conifers (SINGH et al., 1978). It is interesting to know that vegetation on low-nutrient soils contains a higher concentration of chemical
deterrent to herbivores and pathogens (MCKEY et al., 1978).

β) *Geographical Differentiation of Chemical Compounds*. The distribution of ter-
penoids has been investigated in several conifers.

> Within *Juniperus ashei* (ADAMS, 1977; KELLEY and ADAMS, 1978) the combined use
> of chemical and morphological characters shows that older relict populations
> are more variable, while the area reached by secondary migrations is populized
> by quite homogeneous individua. The recent distribution of monoterpenes in
> *Abies procera* and *A. magnifica* shows the influence of introgression, which is
> supported by paleobotanical evidence (ZAVARIN et al., 1978a). Mono- and ses-
> quiterpenoid constituents of E. Asian firs are similar to those from the New
> World ones. A chemosystematic dendrogram reflects quite good relationships
> based on morphological features (ZAVARIN et al., 1978b).

Luteolin 5-methyl ether only occurs in tropical but not in temperate
Cyperaceae (WILLIAMS and HARBORNE, 1977b). The existence of this
compound brings the family closer in chemical terms to the Juncaceae.
Five ecological groupings of *Parthenium* from throughout the Americas
show an ecogeographical gradation of secondary products as the probable result of adaptive responses to various physical and biotic
factors (RODRIGUEZ, 1977). Within *Phlox carolina* (LEVY and FUJII, 1978)
nearly every population of the totally investigated 64 expresses a
unique flavone pattern. This is interpreted as a consequence of dis-
section and contraction of a formerly more widespread distribution.

γ) Chemistry and Classification of Higher Categories. HARBORNE (1977c) summarized data on the distribution of flavonoids in combination with the evolution of angiosperms. In correlation with anatomical and morphological features different flavonoid types are regarded as primitive or advanced. DAHLGREN (1977b) separated within the "Sympetalae" two blocks following the distribution of iridoid compounds. The "noniridoid" groups (Araliiflorae, Asteriflorae, Campanuliflorae and Solaniflorae) which contain partly polyacetylenes or sesquiterpene lactones instead, are related but had been early differentiated. HEGNAUER and KOOIMAN (1978) discussed the importance of iridois within the Tubiflorae. These compounds are not only useful for classification of the higher taxa, but also at the level of genera and collective species. The ability to cyanogenesis is restricted mainly to the Magnoliopsida (HEGNAUER, 1977a). Pteridophyta, Gymnospermae and monocotyledons are characterized by different precursors in the synthesis of cyanogenic compounds. APARECIDA et al. (1978) demonstrated the importance of isoflavonoids as an auxiliary criterion to the conventional classification in the case of the Lotoideae. ROMEIKE (1978) demonstrated the independent origin of tropane alkaloids for several groups, which chiefly occur in the Solanaceae.

δ) Serology. Serology has been applied in several cases to eludicate relationships within families or between them. FAIRBROTHERS (1977a) gave an overview of serotaxonomy and showed its ability with some examples from the Cornales. In the field below the family level the tribe Genisteae has been investigated by CRISTOFOLINI and CHIAPELLA (1977). *Genista* is more complex than *Cytisus*. Two main groups are distinguished, *Cytisus, Lambotropis,* and *Calycotome* on one hand, *Genista, Teline, Pelteria, Spartium,* and less closely allied *Cytisanthus* and *Lygos* on the other. Some intermediate genera exist. Within the Primulaceae serological results are in accordance with the subdivision into subfamilies on a morphological base. *Androsace* proves to be more derived and isolated (JOHN, 1978). On the level above the family (FROHNE and JOHN, 1978) there are no similarities between Primulales and Plumbaginales, but between Primulales and Ericales and also with some species of Theales. LEE and FAIRBROTHERS (1978) have investigated the Rubiaceae according to serological affinities and found them related to Cornaceae, Caprifoliaceae, Apocynaceae, Asclepiadaceae, and Gentianaceae. Within the family, *Asperula* and *Galium* are close together but different from other Rubiaceae. The Moringaceae are excluded from the Capparales by serological results, while Capparaceae (as the central group) with Resedaceae, Brassicaceae, and Tovariaceae show clear relationships (KOLBE, 1978).

5. Reproductive Biology

Secondary sex characters in flowering plants are reviewed by LLOYD and WEBB (1977). There are differences in vigor, growth rate, vegetative reproduction, number of flowers produced, etc. Many of the observed differences between sexes are of direct selective value in relation to the distinct roles of males and females in sexual reproduction.

a) Flower and Pollination Ecology

Symposium volume: RICHARDS, The Pollination of Flowers by Insects. Linn. Soc. Symp. Ser. Nr. 6 (1978).

The symposium deals with insect pollination from different points of view.

FAEGRI (1978) gave an overview of the trends in research on pollen ecology. CORBET (1978) found marked fluctuation in the correlation between the nectar amount and sugar concentration of flowers and the behavior of visiting bees. STELLEMAN (1978) showed that in the case of *Plantago* syrphid flies also contribute to pollination. MEEUSE (1978a) pointed out that an insect visit in the case of *Salix* does not necessarily guarantee pollen transport to the stigmas. KEVAN (1978) found that not only ultraviolet flower color patterns are of importance for pollination. A colorimetric scheme has been developed to define floral color with respect to the daylight spectrum and insect color vision. VAN DER PIJL (1978) shows the problems caused by the different functions flowers may fulfill during their whole life-time and their answers. VOGEL (1978a) described ways of deceiving pollinators by simulating an abundance of pollen. MEEUSE (1978b) made an attempt to clear up the physiological and chemical background of flower opening in aroid sapromyophilous flowers. PROCTOR (1978) discussed the three trends in the evolution and speciation of magnoloid flowers. EISIKOWTICH (1978) noted a change in blooming time of coastal plants induced by changing wind velocity, while WOODELL (1978) investigated flight direction of bumblebees in dependence of wind direction. The meaning of flower color polymorphisms is discussed by KAY (1978) and also by MOGFORD (1978a).

a) Relation Flower-Pollinator. The coevolutionary relationships between plants and marsupials (in Australia) and lemures (in Madagascar) as pollinators seem to be relicts surviving from far back in evolutionary history. This type of interaction apparently is only present in regions where these animals have not been replaced in terms of this function by more advanced animals as bats in S. America for example (SUSSMAN and RAVEN, 1978). Pollination by bats is characteristic for many different species in C.-America (HOWELL, 1977). In spite of the great diversity of flowers visited by bats, there are some mechanisms to ensure conspecific pollen transfer. Pollen and nectar are available at different times in different plants and the deposition of pollen occurs at distinct regions of the body of the bats. Convergent floral evolution of Proteaceae in S. Africa (*Protea*) and (*Dryandra*) seems to have been influenced by the pollination of non-flying mammals (ROURKE and WIENS, 1977). Introduced (*Aloe* and *Echium*) and endemic species (*Muschia*) in Madeira are pollinated by Lizards; a probably old pollination system (ELVERS, 1977, 1978).

KENNEDY (1978) showed that closed-flowered *Calathea*-species which are more advanced have evolved from open-flowered conditions. It is supposed that bud-opening insects induced this situation by avoiding open flowers which they recognize as already visited. In *Chilopsis linearis* nectar is dispersed into pools and grooves. Bumblebees extract nectar from pools seven times faster than groove nectar. A coevolution is suggested since more flowers can be visited in the same time (WHITHAM, 1977). A special adaptation to fungus-gnats (VOGEL, 1978b,c) is shown in several plant groups such as *Asarum, Arisaema, Orchidaceae, Arachnites,* and *Aristolochia* which are all fungus mimetes. Pollination by beetles seem to have strongly influenced the evolution of angiosperm flowers (GOTTSBERGER, 1977), Cantharophily is present in plant groups of most diverse evolutionary levels. Even though cantharophily is very old it is also observed in advanced groups and therefore no longer a sign of primitivism. In the Magnoliidae specialized and unspecialized types of beetle pollination are distinguishable. Secondary polyandry is interpreted as response to beetle pollination and then accompanied by protogyny and the production of specially beetle-adapted odors. NILSSON (1978a,b) investigated the pollination ecology of different orchids. In *Platanthera chlorantha* the uniformity of pollinators causes little variation in flower morphology. In *Asclepias*, the large inflorescences normally produce only a limited number of mature follicles since mortality of developing follicles there is high. The evolution of large inflorescences therefore is interpreted only as a means of an increasing pollen amount (WILLSON and PRICE, 1977).

β) *Ways of Pollination.* Palms may be good examples for the transition from wind to insect pollination. The majority appear to be entomophilous. Structures relating to insects are present even in anemophilous genera (UHL and MOORE, 1977). Within the Amaranthaceae, floral characteristics are correlated with pollen characteristics in respect to anemogamy and entomogamy (ZANDONELLA and LECOCQ, 1977). In *Plantago* an evolution from entomophily to anemophily and finally a regression to autogamy is plausible (HAMMER, 1978). *Mazus* (KIMATA, 1978) is an example of the transition from cross-pollinated annuals to self-pollinated perennials. *Leavenworthia* shows a shift from self-incompatibility to self-compatibility depending on the loss of adapted pollinators (SOLBRIG and ROLLINS, 1977). The loss of anemophilous characters related to the transition of autogamy has been observed in *Hordeum vulgare* by HAMMER (1977). In *Oxalis alpina* (WELLER, 1978) distyly seems to have evolved out of tristyly within some isolated populations. Gynodioecy (GANDERS, 1978) probably became selectively advantageous when the outcrossing ancestors of *Nemophila* populations were subjected to a drastic increase in the rate of self-fertilization owing to changes in pollinator fauna.

γ) *Animal-Attracting Mechanisms.* VOGEL (1977) summarized the ecological meaning of nectaries and of the different ways nectaries are utilized by animals.

> In *Crescentia* and other Bignoniaceae, nectaries on the fruit attract ants which seem to function in an antiherbivore role (ELIAS and PRANCE, 1978). In *Krameria* free fatty oils attract bees which use these solely as part of the larval nest provision (SIMPSON et al., 1977). Anthocyanins are broadly correlated with pollination ecology, i.e., with special types of pollinators. Surprisingly, autogamous species retained the complexity in anthocyanins of related animal-pollinated species (HARBORNE and SMITH, 1978a). MEEUSE (1978c) discussed the presence or absence of nectarial excretion in reproductive regions of flowering plants in connection with the evolution of the flower, the phylogeny of the angiosperms and the paleoecology of the pollination syndromes.

b) Fruit and Seed Dispersal

Textbook: Verbreitungsbiologie (Diasporologie) der Blütenpflanzen: MÜLLER-SCHNEIDER (1977, 2nd ed.)

Within *Valerianella* and *Fedia* (ERNET, 1978) there is a great diversity of fruit types suitable for classification. Correlated with the morphology these types also show a great diversity in dispersal biology which contains nearly all possible features. Within the Epilobieae (SEAVEY et al., 1977) the evolution seems to be accompanied by increase in seed number and decrease in seed size. Amphicarpy in *Pisum fulvum* is basically genetically determined but a more recent phenomenon (MATTATIA, 1977); within the Cyperaceae it has been observed by WHEELER HAINES (1977) in *Trianoptiles*. SCHOLZ (1978b) reported synaptospermy in *Bromus*. The white sand campinas in Amazonia are regarded as "islands" surrounded by tropical rain forest (MACEDO and PRANCE, 1978). Of the 37 species there observed more than 75% are adapted for long-distance dispersal mainly by birds. NELSON (1978b) started a series concerning the drift of tropical seeds and fruits on coasts in the British Isles and western Europe.

c) Apomixis

Apomixis is investigated as well in the better-known older examples
as in some hitherto undetected cases. GRÖBER et al. (1978) studied
the effect of apomixis of fodder-grasses. Seed descendents of amphi-
mictic plants are stronger than those of apomicts in *Poa pratensis*.
The chromosome number has less influence on weakness. In *Ranunculus
auricomus*, probably the more widespread taxa in the Bothnian area are
of preglacial origin, the other of postglacial origin (JULIN, 1977).
A detailed investigation exists for a west mediterranean complex of
Limonium (ERBEN, 1978). The diploid species seem to be mainly amphi-
mictic, while triploids and higher polyploids are apomicts originated
by hybridization. The basis for this is the existence of restitution-
ary nuclei. The existence of a pollen and stigmatal polymorphism makes
it possible to prove apomixis directly in isolated population of one
type. Apomixis has been also suggested for several Dipterocarpaceae
of the tropical rainforest of Malaysia (KAUR et al., 1978). The existence of a series of allied taxa which include apomicts could be evidence for a continuing active diversification.

6. Ecology and Population Biology

The first textbook devoted to plant population biology was written
by HARPER (1977). The main aspect of this, in certain chapters more
than a textbook, is plant demography and in second line population
ecology.

COOK (1978) pointed out a special character complex which is probably
due to habitat factors, the "*Hippuris* syndrome". Several characters occur in unrelated groups influenced by aquatic habitat. The occurrence
of symmetrical tiers of leaves is still not clarified.

Introgression can be proved in different ways. The composition of mo-
noterpenoids shows introgression between *Abies grandis* and *A. concolor*,
a fact which can be explained by the past distribution of these species (ZAVARIN et al., 1978). On a genetical basis introgression has
been investigated in *Senecio squalidus* (diploid) and *S. viscosus* (tetraploid). Hybrids between the two species are triploid and secondarily
pentaploid. The pentaploids reduce chromosome number gradually to
the tetraploid level and may in such a way cross successfully with
S. viscosus (CRISP and JONES, 1978).

> The relationship between environment and hybrids is clear in *Cardamine*
> (URBANSKA-WORYTKIEWICZ and LANDOLT, 1978), where hybrids are supported in
> a habitat changed by man. In *Conostylis* (HOPPER, 1977) hybrid populations are
> confined to ruderal habitats. Within *Thymus vulgaris* (DOMMEE et al., 1978;
> ASSOUAD et al., 1978) natural selection adjusts the percentage of male sterili-
> ty; autogamic hermaphrodites are restricted to rocky habitats. The pollen fer-
> tility depends in Danish populations of *Armeria maritima* on environmental in-
> fluence (WOODELL et al., 1977). NEW (1978) compared populations of *Spergula
> arvensis* after 20 years and found a hairiness cline which probably can be at-
> tributed to climatic changes. Seed coat morphs for the same populations and
> the same periods remained ± unchanged. In *Capsella* there is a high degree of
> genetic variability and polymorphism between different populations (HURKA and
> WÖHRMANN, 1977). The raciation in *Epacris compressa* has been investigated in
> several papers (STACE and FRIPP, 1977a,b,c).

7. Plants and Man

BRÜCHER (1977) wrote a very intense treatment of tropical cultivated plants (except of timber and pasture plants). This work accentuates the genetic point of view, which means that the evolution and domestication of tropical plants is intensively treated. Six main chapters deal with starch plants, protein plants, technically useful plants, fruits and legumes, spices and drugs, and oil plants. The combination of an overall view of the literature with a rich practical experience produces many different aspects in the book. He discussed, for instance, little known tropical crops especially in the Solanaceae and the Palmae, which are certainly important resources for human nutrition. The important role that different, especially South American Leguminosae can play after the elimination of genes responsible for the production of toxic compounds, is also shown. These plants are an unused source for new crops. Domestication and cultivation is interpreted as man-induced evolution which is continuing.

PICKERSGILL (1977) wrote on the taxonomy, origin, and evolution of cultivated plants in the New World. Crops are often historically divided in many species which now are reduced to few taxa as, for example, in *Capsicum*. Speciation in crops may have preceded domestication or may have been accentuated by domestication. In many cases the relation to weedy ancestors now has been clarified. The occurrence of *Cucurbita*-specialized pollinators in S. America suggests an ancient presence of more than one species of *Cucurbita* there. This raises the possibility that S. America was the place of origin of some cultivated Cucurbitas. Especially in allopolploids a multiple origin may have occurred, as has been shown for New World cottons. Recent investigations give more information about the origin of cultivated plants. *Cocos nucifera* now seems to be of S.E. Asian origin, *Lagenaria* has its nearest relatives in Africa. Both genera comprise pantropical cultivated plants of hitherto unknown origin. The center of domestication may not coincide with the center of variability as has been the case for tomato and maize. Detailed analysis of archeologically preserved plant parts may produce unexpected results, as in *Daucus*, where fruits were found to belong to the indigenous species and not to the introduced species *Daucus carota*. It is now suggested that agriculture in the Americas developed independently in four different areas. The first two centers were Mesoamerica and the Andes, a third, probably later center was the humid lowland tropics and the last center was in eastern N. America. This area was, however, almost completely supplanted by the northward spread of the Mesoamerican domesticates.

8. Systematics in Special Biotopes - the Tropics

Symposium volume: Ann. Missouri Botan. Gard. 64, p. 657 et seq. (1977).

The humid tropics are an area of which still little is known, and therefore it is of interest as a reservoir of possible useful plants for the applied botanist, but also for the systematist. All disciplines there are still extremely neglected. TOMLINSON and RAVEN (1977) explained that in the tropics the greatest concentration of floristic and functional diversity occurs. There is a relative dearth of active research in the field of tropical botany and tropical ecosystems are being destroyed at a rapid rate. PRANCE (1977) gave a resumé of the floristic inventory of the tropics. FEDOROV (1977) thought that speci-

ation in the humid tropics follows at least in part the rule of homologous series postulated by VAVILOV. He explained this using the
Dipterocarpaceae as an example. There are only a few generic radicals
in this family, i.e., main characters which allow separation into different genera. Interspecific hybridization is a very rare event in
tropical rain forests. Convergence in tropical trees is adaptive, but
probably based on the law of homologous series. ASHTON (1977) believed
that in a tropical rain forest such as the Malaysian speciation is
still going on. But this vegetation type does not seem to be a reposi-
tory for botanical antiques. The influence of fire and frost on the
evolution of the geoxylic suffrutex has been investigated by WHITE
(1977). These life forms are adapted to fire, but evolved as a response
to unfavorable edaphic conditions. The tropical animal-plant interactions have been reviewed by JANZEN (1977). He strengthened the idea
that in many cases the adaptive value of a character is not known because of the lack of information on the natural history of the organisms. FAIRBROTHERS (1977b) marked the tropical rainforest as an untapped source of phytochemical data, both of theoretical and applied
importance. The information on tropical crops and their wild relatives
is very poor, and even simple morphological data are often lacking
(HAWKES, 1978). Therefore plant morphology and anatomy remain a central field of tropical inquiry (TOMLINSON, 1977).

9. Origin of Angiosperms

Symposium: Bot. Rev. 43 (1977).

What has been started in the last years is being continued with the
help of new fossil evidence concerning time and place of angiosperm
evolution. Paleobotany, especially the study of pollen and spores and
new methods of cretaceous leaf record instead of comparative morphology of modern plants now give the main information.

Several articles (DOYLE, 1977a, 1978; HICKEY and DOYLE, 1977) composed
an overview of angiosperm evolution which, even it it is still uncertain in some points, at least give a coherent, and plausible picture.
Unclear is still the exact time of origin of angiosperms and their ancestors as well. The earliest indubitable records of angiosperm mono-
sulcate pollen stems from the Cretaceous Barremian strata of England
(HUGHES, 1977) and Equatorial Africa (DOYLE et al., 1977). A hot dry
climate is suggested by the palynoflores for the latter area. Tricol-
pates there occur in Aptian horizons and predate their first occurrence
in Laurasia. Such records occur as well, completed by simple leaf
forms, in Siberia (KRASSILOV, 1977) and in the Potomac group in N.
America (HICKEY and DOYLE, 1977). The Barremian and Aptian pollen
grains show some morphological diversity including both monocot- and
dicot-like types, but also including "experimental" types with no
analogs in the living flora. Early-middle Albian is interpreted as
the phase of most rapid adaptive radiation. This event seems to have
occurred world-wide with a lag in invasion of higher altitudes. Large-
scale migrations could have played a major role in the dispersion.
The initial phase took place in areas with at least seasonal dry climate, a fact which is consistent with adaptations typical for angiosperms. These plants probably had been small-leaved shrubs. Later they
entered wetter areas as weeds in disturbed habitats, then radiated
into aquatic habitats, into the forest understory and finally came
in direct competition with trees such as conifers. Tricolpate prodi-

cotyledons originated in a time of increasing aridity and migrated polewards into Laurasia and Australasia.

The key biological innovations such as efficient reproduction, capacity for symbiotic interaction with the animal world, vegetative flexibility, etc. influenced certainly angiosperm radiation more than external physical or biotic environment changes. But without doubt climatic and geographic factors had an important influence on the pattern of angiosperm evolution.

Angiosperm evolution was not any more rapid or mysterious than that of other groups of organisms. The interval from first appearance to dominance, about 25-30 million years, was not shorter than, for example, the period of the first land plant evolution or of the radiation of mammals. KRASSILOV (1977) pointed to the remarkable connection in time between angiospermization and mammalization.

This coherent abstract of angiosperm evolution has been critisized in several points. One premise of this sight is the monophyletic origin which has been contradicted by several authors. HUGHES (1977) argued that all evidence for monophylesis is based on modern plants and on embryological features for which there is not and may never be any direct fossil evidence. In his opinion angiosperm phylogeny should now be approached only on the basis of purely paleontological data. He also doubted that it is correct to use names known from modern groups for paleobotanical record and he proposed to use a more neutral nomenclature.

MEEUSE (1977a) feels that it is at least possible that the forebears of recent flowering plants do not resemble even the primitive forms in our sense. On the other hand it is quite possible that the really primitive forms must have been highly advanced from the modern point of view. Therefore it is necessary to look at fossils at least from two different points of view. The pleiophyletic origin according to MEEUSE makes it impossible to look for one basic type.

Also KRASSILOV (1977) stresses a more polyphyletic origin. He argues that very different angiosperms simultaneously entered the fossil record. He believes that several angiosperm characters appeared in different groups and spread secondarily all over the whole group. HICKEY and DOYLE (1977) inferred that the relatively low diversity and generalized character of the fossils and the subsequent coherent pattern of morphological diversification are consistent with a monophyletic origin of the angiosperms.

Except for pollen, leaves, wood, and fruits, also sometimes flowers may contribute information as is shown by TIFFNEY (1977), who studied apparently angiospermous flower from the Upper Cretaceous of N. America.

Although the place of origin of angiosperms is suspected now more in general in low paleolatitudes with nearly no delay in reaching high paleolatitudes, THORNE (1977b) still holds S.E. Asia as the primary center of origin. Plate tectonics is still thought to be one of the major explanations for modern angiosperm disjunction (THORNE, 1978). But as HUGHES (1977) explained there is still a discrepancy between the age of modern groups proved by fossils and the time when, for example, Gondwanaland and Laurasia broke up.

10. Evolution and Classification of Higher Taxa of Seed Plants

Symposium: Evolution and Classification of Higher Categories. Plant Syst. Evol., Suppl. 1 (1977a), ed. KUBITZKI.

MERXMÜLLER (1977) summarized positive aspects but also some criticism on this item. He stressed again that within angiosperms real basic groups could hardly exist, and that rather mosaic evolution and its consequences on the distribution of characters is the main characteristic of living things. He deplored the inflation of categories and in connection with this the different content of often identically named higher taxa. Informal ranks could be the solution of the future. He warned against inserting each new fact into the system as fast as possible. The validity of this principle is well demonstrated by the case of the Gyrostemonaceae (see below) where the inclusion into the Capparales supported by phloem ultrastructure cannot at all be confirmed by palynology.

HEYWOOD (1977) pointed to the conflict of constructing a phylogenetic system which includes both the dynamic/historical phylogenetic component and a static horizontal present-day classification. KUBITZKI (1977) stressed that a fully phylogenetic classification of the flowering plants is impossible at the moment. MEEUSE (1977b) thought that higher taxa within the angiosperms can be delimitated and placed only when an agreement is reached concerning several fundamental morphological features within this group, as gymnospermy versus angiospermy, the sporophyll concept, the entomophilous syndromes etc. THORNE (1977a) explained some features of his system of classification of the angiosperms. DAHLGREN (1977a) commented and exemplified his well-known classificatory scheme. SPORNE (1977) tried to give objective criteria to fix the primitiveness of characters.

In Conifers, KENG (1977) supported by morphological investigations of the phylloclade the erection of a monogeneric family Phyllocladaceae.

Within the Liliatae CLIFFORD (1977) used an intensely clustering sorting strategy based on about fifty characters. By this method a subdivision is reached into four main groups. Only two of these (Zingiberales and Alismatidae) are traditional groupings. The two other groups do not represent any existing group, but are mainly characterized by wind pollination and insect pollination, respectively. HUBER (1977) explained that an increasing number of characters wipe out the gaps between the monocotyledons and the magnoliid dicotyledons, which in his sight appear to be two extreme wings of a single natural unit with Annonaceae, Aristolochiaceae, and Nymphaeaceae as links. A grouping into 12 units according to the number of dicotyledonous features and to the degree of their isolation has been presented. Groups with relatively more dicotyledonous characters (e.g., Arales and Helobiae) are more isolated, those with less or without such characters show a decreasing isolation. BURGER (1977) looked from the other side at monocotyledonous characters within the Piperales. The Piperales, Arales, and some Najadales seem to have so much conformity that they should represent very ancient lineages. The basic theory is the addition of single flowers consisting of a single bract, two stamens and a single adaxial pistil as in Chroranthaceae. The loss of internodes should have formed the three-part flower of the Piperales and many monocots (see Fig. 1).

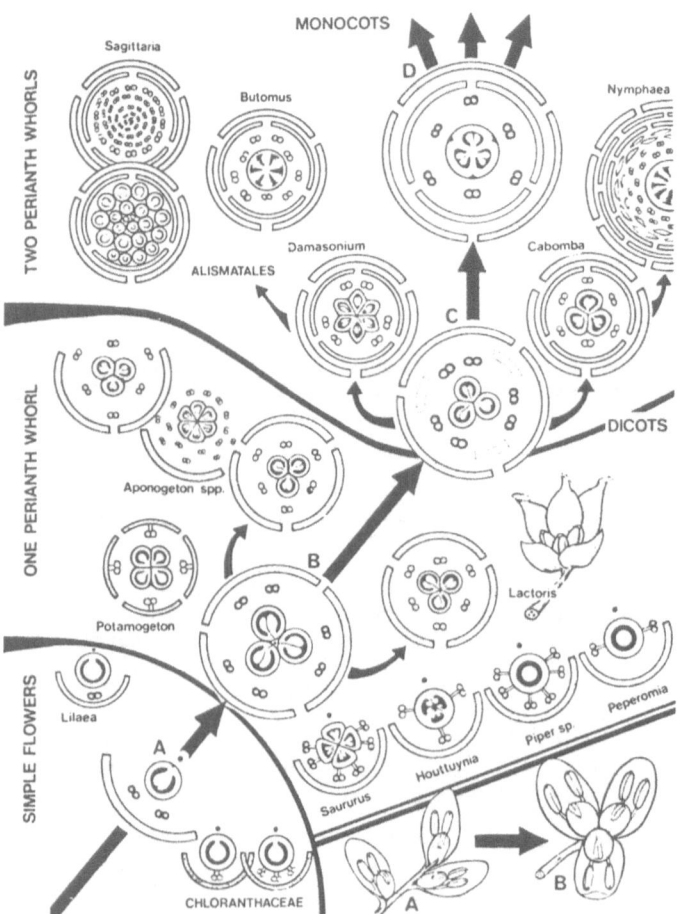

Fig. 1. Major stages in the evolution of flowers in Piperales (s.l.), Nymphaeales, Alismatales, and related monocotyledons. (After BURGER, 1977)

CRONQUIST (1978c) proposed a subclass Zingiberidae which includes the orders Zingiberales and Bromeliales. Within the Zingiberales only the Zingiberaceae and the Marantaceae are rich in leaf flavonoids (WILLIAMS and HARBORNE, 1977b). Different patterns of these compounds lead to the distinction of groups, genera etc. Pollen morphology supports separation of the Apostasiaceae as a separate family (SCHILL, 1978).

EHRENDORFER (1977) needed three premises to develop new ideas about the early differentiation of angiosperms: the monophyletic origin, the definition of stamens and carpels as appendicular sporangiophores and the adaptive nature of basic angiosperm characters. Based on these premises, EHRENDORFER said that it is no longer possible to believe that Magnoliidae link up directly to Rosidae-Dilleniidae. Rather the Hamamelididae show ambivalent affinities to both groups. ENDRESS (1977a) gave a perfect review of this coherent group with emphasis on the putative evolutionary trends. The Fagales may have originated from temperate deciduous, Hamamelidaceae-like ancestors. Adaptive radiation is strongly connected with wind pollination in mainly the latter group. The Hamamelididae have been interpreted by EHRENDORFER as a transitional field between Magnoliidae at one hand and Rosidae-Dilleniidae at the other. They correspond to an early phase of flower reduction connected with trends to anemophily, the Rosidae-Dilleniidae to a subsequent phase characterized by adaptation toward zoophily.

More in detail ENDRESS (1977b) showed that Eupomatiaceae, Himantandraceae, Magnoliaceae, and Annonaceae originated as a relatively coherent group. Floral anatomy and morphology support this. In this basal group there are still badly known and therefore misinterpreted genera. LEROY (1977) reported of a primitive angiosperm from Madagascar (described as *"Bubbia" perrieri*) with a bicarpellate, unilocular ovary. This unique species may serve the rank of at least a subfamily within the Winteraceae. The consequence would be closer relationship of the Winteraceae to the Canellaceae.

For the Eucryphiaceae a place near Cunoniaceae is supported by ana-tomical data (DICKISON, 1978) and also for *Bauera* a similar position is supported by embryology (PRAKASH and MC ALISTER, 1977). The Cunoni-aceae themselves have been discussed in respect to their chemistry by BATE-SMITH (1977). It is still uncertain where in the rosalean alli-ance this family should be placed. The presence of proanthocyanidins, flavonols, ellagitannins and the absence of iridois show similarities to Saxifragaceae and Rosaceae. But habit and distribution are very different from these families. The Cunoniaceae may be an anolog devel-opment to the Proteaceae. Basing on seed characters KRACH (1977) stressed the heterogeneity of Saxifragales. He acknowledged several more or less independent groups such as Crassulales, Escalloniales, a small group with *Ribes* and *Phyllonoma*, and the Cunoniales. In general a close affinity of the woody orders to the herbaceous families is very doubtful, but the information is still too poor to propose a new arrangement. CARLQUIST (1977c, 1978a,b) investigated the affinities of Grubbiacea and Bruniaceae by wood anatomy. The Bruniaceae are quite primitive and are linked by the Grubbiaceae with the Geissolomataceae. For all three families a generalized rosoid placement has been suggest-ed. The Crossosomataceae are chemically different from all putative associates. Some similarities exist with the Mimosoideae (TATSUNO and SCOGIN, 1978). A multivariate study of the Cneoraceae (LOBREAU-CALLEN et al., 1978) showed similarities to the Rutales, especially in paly-nology. Sieve-element plastids of the Vivianiaceae suggest an exclu-sion from the Centrospermae and a possible inclusion into the Gerani-ales (BEHNKE, 1977a). The Montiniaceae (DAHLGREN et al., 1977) occupy a peripheral position within the Cornales.

MABRY (1977) discussed all characters specific to the Centrospermae. The flavonols and the C-glycosylflavones of the Caryophyllales have been investigated by RICHARDSON (1978). The former constituents exist in all families, the latter are absent from the Aizoaceae and the Cactaceae but predominate in the Caryophyllaceae, the most advanced family in the order.

The sculpture of the pollen exine supports an inclusion of the doubtful genera *Gisekia, Limeum*, and *Hectorella* into the Centrospermae, but also excludes the Rhabdodendraceae, which are better placed near the Rutaceae (BEHNKE, 1977c). The genus *Stegnosperma* still has an uncertain position within the Centrospermae (HOFMANN, 1977).

The phloem ultrastructure supports a position of the Gyrostemonaceae close to Capparales, but this cannot be confirmed by palynology (BEHNKE, 1977c,d). The wood anatomy of the Sarraceniaceae shows pri-mitive features and similarities to the wood of Theales (DEBUHR, 1977). The reproductive anatomy of *Actinidia chinensis* suggests that the Actini-diaceae fit better into Theales than into Dilleniales or Ericales, while the Davidiaceae are different from the Actinidiaceae (SCHMID, 1978a,b). According to STRAKA and ALBERS (1978) the pollen morphology does not deny a position of the Diegodendraceae near Ochnaceae. The development of the androecium rather supports a relationship of the Datiscaceae to the Begoniaceae, while an inclusion into the Parietales or the Dilleniales is not possible (LEINS and BONNERY-BRACHTENDORF, 1977). The Urticales show a surprising diversity, and BERG (1977a) attributed them a central position within the angiosperms. An affinity to the Malvales is clear, while the relationship to the Hamamelidales is more vague. There are two complexes within the family. One centers round adaptation for wind pollination, the other round adaptation for insect pollination, the latter leading to several types of pseudanthous inflorescence. The flavonoid chemistry supports the separation of the Ulmaceae from the Celtidaceae at family level (GIANNASI, 1978).

WAGENITZ (1977) distinguished five main groups within the Asteridae: Gentianales-Rubiales; Polemoniales (including Solanaceae and Boraginaceae); Scrophulariales-Lamiales; Dipsacales, Campanulales-Asterales. For the Scrophulariales-Lamiales and the Dipsacales, affinities with the woody Saxifragales and Cornales are most probable, the phytochemical similarity of the Asterales with the Araliales is at least remarkable. There are not enough reasons for breaking up the Asteridae as proposed by several authors. The Fouquieriaceae (as well as Loasaceae and Columelliaceae) have only several characters in common with the Asteridae but are excluded by WAGENITZ. SCOGIN (1977) found in this family anthocyanins which support a close relationship to Solanales. *Pentaphragma* according to palynology (DUNBAR, 1978) is separated from the Campanulaceae as a separate family. Within the Gentianales, *Brunonia* has to be separated from the Goodeniaceae as a family or subfamily (CAROLIN, 1978). The relationship of Valerianaceae and Dipsacaceae are investigated by VERLAQUE (1977b) based on different characters. *Triplostegia* is inserted in the Valerianaceae and *Morina* a separate family with closer relationship to the Valerianaceae. BURTT (1977b) explained that the few linking genera between Scrophulariaceae and Gesneriaceae and between Dipsacaceae and Valerianaceae do not justify the union of the families. On the other hand the Selaginaceae are in general closer to the Scrophulariaceae, and should be reduced to tribal rank within the latter family.

11. Classification and Systematics of Seed Plants at the Family Level and Below

a) Biology and Chemistry of the Compositae

Symposium volumes: HEYWOOD et al. (1977)

It seems inadequate to cut off the two symposium volumes and to put the articles at different places in this report. Rather it appears more consequent to insert papers outside of the symposium into the following discussion. The large number of Compositae papers makes it possible to include only the main results. A much more complete list can be taken out of the Compositae Newsletter, appearing as one consequence of the symposium.

After this last inventory the family consists of about 1300 genera with about 21,900 species (TURNER, 1977c), a number of which, concerning the genera, in the meantime has certainly increased. The largest tribe, even after the exclusion of the Liabeae and without some parts of the Inuleae are the Senecioneae with ca. 2900 species, the smallest tribe are the Calenduleae with little more than 100 species. The number of genera is the highest within the Heliantheae with more than 200, while the Calenduleae again are the smallest tribe also in this respect.

Still unsettled is the question which other family is most closely related to the Compositae. Chemistry (HEGNAUER, 1977b, 1978) favors a relationship to Araliaceae and Apiaceae perhaps via the Boraginaceae. These two former families have in common with Compositae essential oils in schizogenous cavities, polyactylenes, and polyphenols (MABRY and BOHLMANN, 1977). Relations to the Calyceraceae are not supported by chemistry. From the point of polynology (SKVARLA et al., see below) different families are near to the Compositae, among others also the Calyceraceae, but not the Rubiaceae. CRONQUIST (1977a) still believed in a relation to Rubiaceae and Caprifoliaceae. The latter family and the Compositae are thought to have originated in the same stock from early members of the Rubiaceae. The Calyceraceae may have, more or less advanced, a parallel development. TURNER (1977a) named the Calyceraceae as "phyletic source" of the Compositae. It is questionable if there would be more clearness if we knew that this family is near to the Compositae! THORNER (1977a) finally believed that the closest relationships of this family have to be sought in the Cornales, Dipsacales, and Lamiales and ultimately in the rosalean Saxifragi neae. These discrepancies have been underlined by STEBBINS (1977) who concluded that "the Compositae cannot be regarded as descendent from or closely related to any other modern family".

Concerning the subdivision of the family, TURNER (1977c) discussed the different newer concepts, especially the classification into two groups such as the subfamily Cichorioideae with mainly anthemoid pollen and more primitive sesquiterpene lactones and the subfamily Asteroideae with mainly helianthoid or senecioid pollen and more advanced sesquiterpene lectones. As it seems from a palynological point of view, mainly the Anthemideae in this classification are questionable. But also a subdivision into three subfamilies, the monotribal Cichorioideae, the Vernonioideae, and the Asteroideae is still discussed.

MABRY and BOHLMANN (1977) gave an overview of the chemistry of the Compositae. Sesquiterpene lactones are of chemotaxonomic importance especially when biosynthetic relationships are known (HERZ, 1977a). The flavonoids (HARBORNE, 1977a) seem to have more taxonomic importance on the generic level or below than in higher categories. The importance of anthochlors and other flavonoids as honey guides is discussed by HARBORNE and SMITH (1978b). Polyacetylenes, reviewed by SØRENSEN (1977), are spread all over the family and it seems that this occurrence reflects an old ability of Compositae. ROBINS (1977b) reported on the alkaloids within Compositae. GREGER (1978) used data from flavonols and polyacetylenes to support morphological and cytological findings. Pharmaceutical and economic uses of the Compositae have been considered by WAGNER (1977a).

The fossil record, reported by TURNER (1977a), is very poor. This is underlined by the fact that there is no evidence that *Viguirea cronquistii* really was a composite (CREPET and STUESSY, 1978). TURNER defended the opinion that the Compositae form a very old phyletic line, whose origin stems back to at least the Cretaceous. Its distribution is mainly influenced by continental drift and the family arose in Western Gondwanaland. CRONQUIST (1977a) on the contrary said that the familiy with certainty is only to be traced back to the top of Oligocene. TURNER argued that the relative abundance of Compositae since Miocene is due to the increase of adaptation for disturbed habitats since that period.

STEBBINS (1977) summarized all results concerning growth habit. It is
generally accepted that the family is primitively woody (CRONQUIST,
1977a) with a frequent reduction to herbaceous conditions, but sometimes has reversed to woody habit. CRONQUIST stressed that secondarily woody habit may have occurred more often than thought since. The
existence of an active cambium in many composites facilitates such a
return. While concerning growth habit there is a more or less general
accordance, MABBERLY (1977) remains by the idea that at least in *Senecio*
even herbs derived from pachycaul ancestors.

BURTT (1977a) understands the differentiation of the capitula as consequences of selection by pollination mechanisms, breeding systems,
ovary protection, and fruit dispersal. LEPPICK (1977) thought of coevolution between capitula and pollinators. BURTT (1978a) argued that
when Compositae appeared probably the whole range of potential pollen
vectors was already in existence. The capitular forms therefore evolved
as an explosion of direct adaptive radiation toward the whole range of
pollinators. The deep ditches between the tribes with only few "intermediate" genera point to an early explosive diversification connected
with the evolution of the different capitular types. The fundamental
patterns in Compositae are heads of anthocyanic, relatively long,
tubular corollas (e.g., Vernonieae), liguliflorous heads (e.g., Lactuceae), radiate heads often reduced to discoid heads (Heliantheae)
and heads of bilabiate flowers (Mutisieae). JEFFREY (1977) formulated
the hypothesis that the ancestral corolla type was zygomorphic-bilabi-
ate (as in the Mutisieae), from which all other types in the family
are derivative. The ancestral inflorescence type should have been an
elongate raceme. To this hypothesis BURTT (1976a) opposed that it would
be more probable that Compositae originated from a corymbose raceme
with more regular and only peripherically differentiated flowers. Based
on different flower types JEFFREY presented a scheme of evolutionary
relationships. It remains the question if some of the relations do not
characterize only convergent developments. GARDNER (1977b) traced out
the occurrence of tretamerous flowers and found them independently recorded for most of the tribes. He thought of a connection between reduction of capitula size and the tendency of tetramerous flowers. In
Lasthenia tetramerous flowers are self-pollinated, while pentamerous
species are outbreeding. Microcharacters of the ligule have been used
by BAAGØE (1977a,b, 1978) for a tentative subdivision of the family.
Three main types are recognized; a helianthoid (with papillose cells),
a senecioid (with tabular cells) and a mutisioid type (with tabular
cells with crested outer walls). The different ligule type may have
developed several times independently in different groups, but it is
probable that these lines reappear in a quite systematic way. A grouping into two groups according to these characters coincides with a common subdivision of the Compositae.

As shown by SKVARLA et al. (1977), an apparently more primitive anthemoid or modified anthemoid pollen type is mainly characteristic for
the tribes Cardueae, Mutisieae, Liabeae, Vernonieae, Lactuceae, Anthemideae, Gorteriinae within Arctotideae and Tarchonanthinae within
Inuleae. This type is shared mainly by the Calyceraceae, but also with
the Valerianaceae, some Dipsacaceae, and Umbelliferae. The Rubiaceae and
Campanulaceae are found to have pollen characters unlike those of the
Compositae. The more advanced senecioid and helianthoid type is characteristic for the rest of the family. The presence or absence of
caveae in the exine, a character which is typical for the helianthoid
and the senecioid type, is explained by BOLICK (1978) as caused by
mechanical stress by volume changes. This fact makes this character
not as independent as thought.

SOLBRIG (1977) reported that for about one third of all species chromosomes counts are known, but more details are known only for a very low percentage of them. Typical, well-known features of Compositae cytology are the connection of the decrease of chromosome numbers with annual habit, while an increase is more frequent in perennial herbs. The well-known graph of chromosome number per number of species shows a gap at x = 7. The modal number of the family seems still to be x = 9. In the meantime several papers dealing with chromosome numbers mainly from N. America have been published. In addition to the above-mentioned papers there are those of MORTON (1977) for the Compositae of the British Isles and TOMB et al. (1978) for the Lactuceae. Some interesting features of Australian Compositae are reported above.

On a tribal base, the situation within the Compositae is as follows.

ROBINSON and KING (1977) reviewed the Eupatorieae, ca. 160 genera with 2000 species, and produced an almost complete revision of the tribe which now is subdivided into 19 groups. The chemistry here is only little known, but could give additional arguments for taxonomic conclusions (DOMINGUEZ, 1977). BOLICK (1977) found a similar style branch exposure in *Eupatorium havanense* as in the Astereae. The base number is probably ten. Additional chromosome counts have been reported by KING et al. (1977).

The Vernonieae consist of about 70 genera with ca. 1450 species (JONES, 1977). The subdivision of the tribe is not yet clear and also the most important genera have not yet been intensively examined. As in other Compositae, genera with Old and New World distribution seem to be separable in this respect for morphological reasons. Also the chemistry is not well known (HARBORNE and WILLIAMS, 1977). BURNETT et al. (1977a,b) investigated that sesquiterpene lactones which besides insects, also keep mammalian herbivores from feeding on Vernonias. Local revisions of *Vernonia* now are going on: KEELEY and JONES (1977a) for the Bahamas, CHAPMAN and JONES (1978) for Texas, KEELEY (1978) for the West Indies, the latter being based on pollen investigations (KEELEY and JONES, 1977b).

A review of Astereae has been made by GRAU (1977a). This tribe seems to be very coherent apart from a few doubtful exceptions. A subdivision in the traditional way does not make sense. At the moment there are ca. 135 genera with about 2500 species. Revisions appeared for *Poecilolepis* (GRAU, 1977b). *Aster* sect. *multiflori* (JONES, 1978), *Amellus* (ROMMEL, 1977), *Townsendia* for W.-Canada (SPEIRS, 1978). *Xylorhiza* separated from *Machaeranthera* (WATSON, 1977). Cytological data exist for *Bigelowia* (ANDERSON, 1978), concerning the cytogeography of *Aster pilosus* (SEMPLE, 1978a), for *Aster sonorae* which, based on genome homology, has to be included in *Machaeranthera* (STUCKY, 1978) of which the Sect. *Arida* has been investigated intensively (ARNOLD and JACKSON, 1978) and for *Chrysopsis* and *Heterotheca* (SEMPLE, 1977). The chemistry of the tribe has been hardly investigated (HERZ, 1977b). The polyacetylenes resemble those of the Anthemideae. In *Hazardia* relationships have been clarified by the distribution of flavonoids (CLARK and MABRY, 1978).

The Inuleae, ca. 180 genera with some 2100 species, are arranged provisonally into three subtribes (MERXMÜLLER et al., 1977). Initial investigations in fruit anatomy (*Inula* group) showed that the internal structure of the fruit wall offered promising characters for grouping of the tribe (MERXMÜLLER and GRAU, 1977). Recent revisions exist for *Pterocaulon* (CABRERA and RAGONESE, 1978), *Leysera* (BREMER, 1978), *Stoebe* for S.E. Africa and Angola (NOGUEIRA, 1977), *Pluchea* (GILLIS, 1977), *Helichrysum* in S.W. Russia (OVDIENKO et al., 1977). *Gnaphalium* in New Zealand has di- and polyploids with the base x = 7 (GROVES, 1977). Only little is known about the chemistry of the tribe (HARBORNE, 1977b).

STUESSY (1977a) presented a revised subtribal treatment for the 208 genera (with 2580 species) of the Heliantheae. Some parts of the Helenieae in the former sense are now included. Even though not very much is known about the chemistry of the Heliantheae (SWAIN and WILLIAMS, 1977) the results obtained give some hints to classification (e.g., the restriction of sesquiterpene lactones to certain subtribes). The flavonoids of *Plummera* show its proximity to *Hymenoxis* (BIERNER, 1978). Anthochlor pigments are responsible for the ultraviolet floral pattern of *Coreopsis* (SCOGIN, 1977). Revisionary work has been done for *Oparanthus* (STUESSY, 1977b), parts of *Perymenium* (FAY, 1978), *Chrysogonum* (STUESSY, 1977c), *Borrichia* (SEMPLE, 1978b), *Axiniphyllum* (TURNER, 1978), *Galinsoga* (CANNE, 1977, 1978), *Cymophora* (TURNER and POWELL, 1977b) and *Bebbia* (WHALEN, 1977). *Henricksonia* is for TURNER (1977b) an example for the fact that northern central Mexico must have been a primary center of diversification of the Asteraceae. The pollen wall and the development of the apertures in *Helianthus* have been investigated by HORNER and PEARSON (1978). The ancestral base numbers of the tribe are x = 8 and 9. Chromosome numbers have been determined for *Calycadenia* (CARR, 1977), *Lipochaeta* (GARDNER, 1977a) where the ploidy level is significant for the sections, and *Psilostrophe* with a revision of this genus (BROWN, 1977, 1978).

The Helenieae have been broken up by TURNER and POWELL (1977a). Some of the main splits are included in the Heliantheae. The displacement of other important parts into the Senecioneae is very problematic and a result has not been reached. *Flaveria* belonging to the latter group, was revised by POWELL (1978). The new tribe Coreopsideae is constituted and it only includes a few genera of the Jaumeinae. The Tagetinae are established as a separate tribe Tageteae (with 16-18 genera and ca. 140 species, STROTHER (1977a)). The uncertain situation of the Helenieae is certainly the weakest point of the current Synantherology. There is no chemical evidence for the inclusion of parts of the Helenieae into the Senecioneae, as, for example, the alkaloids are lacking although they show more helianthoid features (BOHM, 1977). The Tageteae (RODRIGUEZ and MABRY, 1977) also show chemically a more helianthoid pattern. In this group, revisions exist of *Pectis* (KEIL, 1977a, 1978) based on chromosome data (KEIL, 1977b), for *Chrysactinia, Harmackia* and *Lescaillea* (STROTHER, 1977b) as well as of *Nicolletia* (STROTHER, 1978).

The main and generally accepted tribal change is the separation of the Liabeae from the Senecioneae (NORDENSTAM, 1977a). The former tribe now consists of 15 genera with ca. 187 species in the New World and is situated near the Vernonieae. A review of the genus *Cacosmia* has been done by NORDENSTAM (1977b). The Senecioneae in the restricted sense without the allocations of the Helenieae are well circumscribed and also a chemically well-defined group (ROBINS, 1977a). The exclusion of few genera such as *Arnica* and its transfer to the old Helenieae is well established. The tribe with 100 genera (ca. 2900 species) is subdivided into two subtribes. The main subtribe consists of two complexes, the cacalioid and the senecioid. ROBINSON and CUATRECASAS (1978) gave a revision of central American *Pentacalia*. *Gynura* has been revised for Africa (DAVIES, 1978). Two noncoordinated attempts have been made to divide the large genus *Senecio*. NORDENSTAM (1978a) tried a subdivision according to a syndrome of morphological and anatomical characters which leads to smaller genera. JEFFREY et al. (1977) divided *Senecio* into 15 different genera based on a multifactorial analysis. *S. vulgaris* belonged there to *Senecio* s. str.

The Anthemideae contain 192 genera with about 1400 species (HEYWOOD and HUMPFRIES, 1977). The traditional subdivision into two subtribes is not justified. This is also supported by investigations of stomata types by NAPP-ZINN and EBLE (1978). The tribe now is under intense investigation so that the Anthemideae are comparatively well known with respect to their chemistry. The southern hemisphere genera can be divided according to the presence of anthemoid acetylenes and benzofurans. GREGER (1978) investigated the chemistry of *Anacyclus* with relation to the systematics. The position of *Heliocauta* was clarified by HUMPHRIES (1977). Further investigations on the cytology of *Achillea* were made by BISTE (1978) and ANDROSHCHUK et al. (1978). A revision of *Peyrousea* was done by BREMER (1977).

The Calenduleae (NORLINDH, 1977a) consist of 8 genera with 114 species. Little is known of further systematics and chemistry (VALADON, 1977). The same is to

be said of the Arctotideae (NORLINDH, 1977b), a tribe subdivided into 4 subtribes, with 15 genera and 194 species.

DITTRICH (1977) proposed to split the Cardueae into three distinct tribes,
which separates the groups, as chemistry shows, too far (WAGNER, 1977b). The
tribe in the old sense consists of 79 genera and about 2100 species. Other
papers deal with the cytology of *Carduus* (GREMAUD, 1977), the growth habit
(JÄGER, 1977) and embryology of *Cirsium* (ZABINSKA, 1977). Several papers deal
with *Centaurea* s.l. as the proposed subdivision of the Centaureinae for
Czeckoslovakia (DOSTAL, 1977), the investigation of hybrid populations of
Centaurea ornata and *C. cephalariifolia* and the taxonomical implications
(KUMMER, 1977) and the use of micromorphology within the Centaureinae (NORDEN-
STAM and EL-GHAZALY, 1977). For species from Egypt there is a strong correla-
tion between pollen morphological data and floral microcharacters.

The Mustisieae have 89 genera and about 970 species as treated by CABRERA
(1977). Chemical information is nonexistent. Four subtribes of various homogeneity are recognized. A polyphyletic origin of the tribe cannot be excluded.

The Lactuceae comprise 70 genera with about 2300 species (TOMB, 1977). Older
classifications are used while it has not been decided whether to use subtribes or only informal groups. The chemical knowledge of the Lactuceae is not
very advanced so that there is not much information (GONZALEZ, 1977). A revision of *Lactuca* for Europe now exists (FERAKOVA, 1977); the intense work of
ALDRIDGE treats macaronesian *Sonchus* subgen. *Dendrosochus* (ALDRIDGE, 1977a,b,
1978a,b). LACK (1978) included *Heywoodiella* into *Hypochoeris*. Pollen morphology
within the subtribe Microseridinae shows that it is essentially stenopalynous
(FEUER and TOMB, 1977). Suggested relationships are supported by these findings.
The comparative pollination ecology in *Pyrrhopappus* has been studied by BARBER
and ESTES (1978).

b) Systematics of the Apiaceae

Symposium: Actes 2. Symp. Int. Ombellifères (1978).

New World *Eryngia* are offshoots of Old World groups, with the base
number x = 8 and a tendency to have fewer chromosomes (CONSTANCE,
1978); hybrids are frequent, while in Europe interspecific hybrids
in the whole family are quite rare (HEYWOOD, 1978; GRACE and STEWART,
1978). The same base number together with the secondary number x = 11
gave rise in the Caucalideae to many aneuploids but few polyploids
(CAUWET-MARC and JURY, 1978). Similar findings for the Portugese
genera are presented by QUEIROS (1978) and SILVESTRE (1978) for Spanish material. Middle Asia is one of the most important centers of Old
World Apiaceae (PIMENOV, 1978), while S. India has quite isolated,
often monotypic genera with many endemics (BLASCO, 1978). A new record
for *Lilaeopsis* from Madagascar shows an interesting distribution of
this genus formerly known only from the Americas, Australia, and New
Zealand (RAYNAL, 1977).

New and old characters are used for classification as flower characters
(REDURON, 1978), fruit anatomy for separation of *Angelica* and *Archangelica*
(GUTIERREZ-BUSTILLO, 1978), phytodermology (GUYOT, 1978) and special features
in palynology as the internal structure of the endexine (CERCEAU-LARRIVAL and
ROLAND-HEYDACKER, 1978), which give good results when comparing *Seseli*-species
(PARDO, 1978).

JURY (1978) points out that fruit characters are sometimes overestimated as in *Ammiopsis*, while OKEKE (1978) showed the importance of such
characters for *Daucus* and SAENZ de RIVAS et al. (1978) for the whole
Caucalideae. The direction of evolution is suggested from the content
of DNA, which decreases from *Torilis* to *Daucus*, while within *Eryngium*,

derived species possess more DNA (LE COQ, 1978). Ontogenetic studies
in *Heracleum* stimulated new ideas about the direct of evolution (SATSY-
PETROVA, 1978), as do fossil records in the case of African *Steganotaenia*
(DECHAMPS, 1978). B-chromosomes in European *Heracleum* may influence the
recombination rate as shown by WEIMARCK (1978), and may originate from
Robertsonian translocations as in *Apium* (MARKS, 1978). Fossil records
of primitive umbelliferous pollen grains come from the Angloparisian
basin. The existence of Umbelliferes since early Eocene has therefore
been suggested by GRUAS-CAVAGNETTO and CERCEAU-LARRIVAL (1978). Several
papers dealt with the palynology of the Umbelliferae. The pollen of
Klotzschia shows a sexine stratification similar to many Araliaceae
(SHOUP and TSENG, 1977a); it may link the putative ancestral arali-
aceous type and the more specialized umbelliferous type. The ultrastructure of the tectum has been investigated by ROLAND-HEYDACKER and
CERCEAU-LARRIVAL (1978). A simple and a structural tectum can be distinguished, and the latter seems to be more derived and adapted to
xerophytic conditions and insolation. In *Eryngium maritimum* (VAN DER
PLUYM and HIDEUX, 1977) the variation in palynological parameters is
independent of the general geographical pattern. MAGIN (1977) showed
the accordance between the formation of the inferior syncarpic ovary
of Magnoliophytina and the ovary of Apiaceae. HEGNAUER (1978) gave an
overview of the recent investigations on the chemical characters of
Apiaceae and related families. From these data Apiaceae and Araliaceae
are closely related and the Umbelliflorae may have given rise to the
Asterales while the origin of the Umbelliflorae lies in groups similar
to recent Rutales. This is exemplified in more detail by PLOUVIER
(1978) who stressed the close chemical relationship of Apiaceae to
Pittosporaceae, while the Cornaceae and Saxifragaceae are more distinct. These connections are supported very well by investigations
on parasitic fungi, which demonstrate the same affinities. The investigation of DURRIEU (1978) also puts *Eryngium* in a quite separate place
and produces a good example of coevolution by comparing the ornamenta-
tion of teliospores in parasitic fungi with a system of Apiaceae based
on pollen morphology.

Detailed investigations on chemical compounds are in KUZNETSOVA (1978) for the
coumarins in Apiaceae of the URSS, CARBONNIER et al. (1978) for the coumarins
in the Peucedaneae. HILLER (1978) for the saponins in the Saniculoideae and
GONZALES and GALINDO (1978) for differences in sesquiterpene lactones of
Apiaceae and Asteraceae. Several genera have been investigated and their in-
dependence is proved, for example: *Tilingia*: GOROVOY (1978), *Vanasushava*:
BLASCO et al. (1978), a genus with a primitive pollen type: CERCEAU-LARRIVAL
et al. (1977), and *Tetrataenium*: MANDENOVA et al. (1978).

In the case of *Actinotus* MAGIN's (1978) work indicates that it probably
better belongs to the Cornales. As has been shown for Australian Apiaceae by DAWSON and WEBB (1978) the generic concept is, for historical
reasons, quite broad, when comparing the situation in the Northern Hemisphere. Embryology of the Apiaceae is homogenous, but *Hydrocotyle* is
quite isolated (KORDYUM, 1978). Problems of multidisciplinary synthesis
are discussed by HEYWOOD (1978b) for the Caucalideae; examples are
given for *Bupleurum* by ROUX et al. (1978) and CAUWET-MARC et al. (1978)
and for Bowlesieae by FROEBE (1978). The consequences for pollination
when the umbel is regarded as the reproductive unit are discussed by
BELL and LINDSEY (1978).

New revisions exist for *Prangos* (HERRNSTADT and HEYN, 1977) and the *Bupleurum
atlanticum* complex (CAUWET-MARC and CARBONNIER, 1977) a genus in which thorny
cushion-forming species are more derived (MEUSEL, 1978).

c) Other Families

Dicotyledons: Acanthaceae. *Oplonia* and *Elytraria*, rev. f. Cuba: BORHIDI and MUÑIZ (1977); *Justicia insularis-striata* compl. rev.: MORTON (1978). Amaranthaceae. *Chamissoa*, rev.: SOHMER (1977). Apocynaceae. Inclusion of some related taxa into *Tabernaemontana* supported by pollen characters: VAN CAMPO et al. (1979); *Ochrosia*, rev. for Hawaii: SAINT JOHN (1978a); class.: BOITEAU (1977) and differences between Plumeroideae and Tabernaemontoideae according to floral biology: BOITEAU and ALLORGE (1978). Araliaceae. *Schefflera*, pollen morphology, polymery in floral parts correlated with undifferentiated sexine, ancestral types in New Guinea: SHOUP and TSENG (1977b), TSENG and SHOUP (1978). Aristolochiaceae. *Asarum* in Europe: KUKKONEN and UOTILA (1977); *Pararistolochia* distinct genus: PONCY (1978). Asclepiadaceae. Palynology supports class. of the family in most cases, differences between subfam. Perioplocoideae and Cynanchoideae clear: SCHILL and JÄKEL (1978); *Tassadia*, rev.: PEREIRA (1977). Bonnetiaceae. *Caraipa*, *Mahurea*, rev., Bonnetiaceae related to Guttiferae but not to Theaceae: KUBITZKI (1978). Boraginaceae. Embryology of the family in wider sense heterogeneous, examples *Heliotropium*: KHALEEL (1978) and *Trichodesma*: KHALEEL (1977); *Tiquilia*, monogr., subgen. *Tiquilia* x = 8, subgen. *Eddya* x = 9: RICHARDSON (1977); *Pulmonaria*, detailed cytolog. data for Switzerland: BOLLIGER (1978); *Echium*, naturalized species of Australia: PIGGINN (1977); nodal anat. of Boraginaceae highly derived: NEUBAUER (1977). Brassicaceae. Basal body of the ovules perhaps useful for subdiv. of the family: PRASAD (1977a); Seed anatomy of *Thlaspi* and *Lepidium*: PRASAD (1977b); glabrous and subglabrous European taxa of *Arabis* polyphyletic: TITZ (1978); *Synthlipsis* with x = 10 included into *Nerisyrenia* with x = 9: BACON (1978); *Rhynchosinapis* included into *Hutera*, *H. leptocarpa* and *H. longirostra* schizoendemics, *H. rupestris* recent invader: GÓMEZ-CAMPO (1977a); *Hemicrambe townsendii*, geogr. disjunction between Tanger and Socotra: GÓMEZ-CAMPO (1977b); *Drabopsis*, monotypic: LEONARD (1977). Burseraceae. Pollen morphology and fruit characters correlated and useful for subdivision of the family: MITRA et al. (1977. Cactaceae. Chromosome nos. from 12 genera support base number x = 12: WEEDIN and POWELL (1978); *Trichocereus*, rev. for Argentina: KIESLING (1978), introgressive hybridization between *T. chilensis* and *T. litoralis*: RUNDEL (1977); areole of *Marnieria* a rudimentary lateral branchlet: PEUKERT (1977); Caryophyllaceae. Numerical study of *Silene* in N. America, generic delimitation of *Lychnis*, *Melandrium* and *Silene*: MC NEILL (1978), crosses between *Melandrium* and *Lychnis* possible: PRENTICE (1978); *Silene* rev. Balkan peninsula: MELZHEIMER (1977); chromosome counts in *Silene*: POHLMANN (1977); *Sagina*, rev. for N. America: CROW (1978); *Cerastium alpinum* and *C. arcticum* high polyploids, diploid ancestors extinct: BÖCHER (1977b); morphology of *Cucubalus* and *Drypis*: ROHWEDER and URMI (1978); pollen morphology of subfam. Paronychoideae in Spain: CANDAU (1978). Celastraceae. Rev. of Hippocrateae from Madagascar: HALLÉ (1978). Chenopodiaceae. Divided into subfamilies Chenopodioideae and Salsoloideae: BLACKWELL (1977); supraspecific conspectus of subtribe Chenopodiinae: SCOTT (1978a); *Beta* sect. *Corollinae*, rev.: BUTTLER (1977b); *Corispermum* in N. America: MAIHLE and BLACKWELL (1978); *Kochia* in N. America: BLACKWELL et al. (1978); *Suaeda* in N. America: HOPKINS and BLACKWELL (1977); *Suaeda* in Canada: BASSETT and CROMPTON (1978); *Esfandiaria* included into *Anabasis*: BOKHARI and WENDELBO (1978). Chloanthaceae. Tribe Physopsideae, rev.: ABID (1978a); *Cyanostegia*, rev.: ABID (1978b); *Hemiphora*, rev.: MUNIR (1978). Combretaceae. Vestured pits with restricted tax. significance: VAN VLIET (1978). Convolvulaceae. *Ipomoea batatas* compl., tax.: AUSTIN (1978). Crassulaceae. *Hylotelephium* separated from *Sedum*, ovary, flowering system and chromosome no. different: OHBA (1977); chromosome nos. of medit. *Sedum*: HÉBERT (1977); in *Sedum lanceolatum* group polyploidy important isolating mechanism: UHL (1977); *Crassula*, rev. for S. Africa: TÖLKEN (1977); *Kalanchoe* rev. for E. Africa: RAADTS (1977); Cucurbitaceae. Gen. rev. for Hawaii: SAINT JOHN (1977a, 1978b). Dipsacaceae. Fruit key to the family: VERLAQUE (1977a); *Tremastelma* monotypic: VERLAQUE (1977b). Dipterocarpaceae. New subfam. Pakaraimoideae: MAGUIRE et al. (1977). Elaeocarpaceae. Conspectus for Papuasia: COODE (1978). Ericaceae. Generic limits in the Cladothamneae: BOHM et al. (1978);

different alkane profiles assist in species recognition in *Arbutus*: SORENSEN et al. (1978); synopsis of *Rhododendron*: CULLEN and CHAMBERLAIN (1978); flavonoids in N. American *Rhododendron*: KING, B.L. (1977); *Diogenesia*, rev.: SLEUMER (1978). Flacourtiaceae. *Azara*, rev. SLEUMER (1977). Gentianaceae. *Gentiana*, rev. for Mexico and Central America: PRINGLE (1977); *Deianira*, rev.: GUIMARÃES (1977). Gesneriaceae. *Monophyllaea*, rev. BURTT (1978b); *Loxonia*, rev.: WEBER (1977); *Episcia, Alsobia, Nautilocalyx* and *Paradrymonia*, rev.: WIEHLER (1978). Grossulariaceae. *Phyllonoma*, rev.: MORI and KALLUNKI (1977). Grubbiaceae, rev.: CARLQUIST (1977a). Guttiferae. Infrageneric class. in *Hypericum*: ROBSON (1977); *Clusia* sect. *Cochlanthera*, rev.: MAGUIRE (1977). Haloragaceae. *Gunnera* in Argentina and Uruguay: MOLINA (1978); in Canadian *Myriophyllum* pollen is characteristic for every species: AIKEN (1978), MATHEWES (1978). Hamamelidaceae. Review of *Corylopsis*: MORLEY and CHAO (1977). Lamiaceae. *Leucas*, rev. of sect. *Lasiocorys*: SEBALD (1978); *Marrubium*, rev. for Iran: SEYBOLD (1978); *Coleus*, rev. for Sri Lanka: CRAMER (1978); *Teucrium*, subdivision into six sections according to characters of the inflorescences and flowers: KÄSTNER (1978); in *Thymus* the polymorphism of essential oils increases in populations of regions with more diversified climate. This corresponds with morphological traits: ADZET et al. (1977); chromosome nos. of Spanish taxa: FERNANDEZ CASAS et al. (1978). Lauraceae. *Aniba*, classification according to flavonoids: FERNANDES et al. (1978). Lecythidaceae. Embryo types and funicle-aril characters in Neotropical taxa tax. relevant, key: PRANCE and MORI (1978). Legumniosae. Chromosome nos. for all subfamilies: GOLDBLATT and DAVIDSE (1977). Mimosoideae. A multifactorial analysis suggests that the diversification and geographical distribution of *Acacia* took place in tropical regions of formerly western Gondwana, sect. *Filicium* probably most pimitive group: GUINET and VASSAL (1978); gum exudates are of tax. use in *Acacia*: ANDERSON, D.M.W. (1978); *Schleinitzia*, rev.: NEVLING and NIEZGODA (1978). Caesalpinioideae. Computer analysis gives two main series which fit with older class. and are consistent with data on seed amyloids and also with characters of leaf transfer cells: PETTIGREW and WATSON (1977); *Cassia* sect. *Abusus* and *Grimaldia*, rev.: IRWIN and BARNEBY (1978); *Cassia* ser. *Aphyllae*, rev. for Argentina: BRAVO (1978). Papilionoideae. *Erythrina* Symposium II, *Erythrina* with 108 species, different by porate pollen and x = 21 from other Erythrineae and therefore rather a monogeneric tribe, suited for biosyst. investigations: RAVEN (1977), pollen type advanced: GRAHAM and TOMB (1977), epidermal features: AYENSU (1977), seed characters do not correlate with current phylogenetic hypothesis: GUNN and BARNES (1977); delimitation of tribe Vicieae, *Cicer* as monogeneric tribe excluded: KUPICHA (1977); multifactorial class. of Phaseoleae into subtribe Glycininae with forest tendency and subtribes Phaseolinae and Cajaninae with savanna adaptations: BAUDET (1978); supported by the flavonoid chemistry: BAUDET and TORCK (1978); *Astragalus* rev. *Alopecuroidei*: BECHT (1978), sect. *Chronopus*: OTT (1978), new sect. *Laxiflori*: AGERER-KIRCHHOFF and AGERER (1977); nitrotoxins are more common in N. American than in S. American and Old World *Astragalus*: WILLIAMS and BARNEBY (1977a,b); very large rev. of tribe Amorpheae with *Errazurizia, Psorothamnus, Marina, Dalea*: BARNEBY (1977); *Dolichos* in the Himalaya: TATEISHI and OHASHI (1977); *Rhynchosia*, rev. for New World: GREAR (1978) and Nepal: OHASHI and TATEISHI (1977); *Psophocarpus*, rev.: VERDCOURT and HALLIDAY (1978); *Lupinus*, seed coat structure allows an arrangement of Old World species in four groups: HEYN and HERRNSTADT (1977), rev. of the *L. montanus* complex in Mexico and Central America: DUNN and HARMON (1977); multifactorial analysis of *Phaseolus* and *Vigna*: MARÉCHAL et al. (1978); seed protein profiles of African *Vigna* correspond with class.: CARASCO et al. (1977); chromosome nos. in *Ononis* ser. *Vulgares*: MORISSET (1978), chromosome nos. of taxa from Iraq: AL-MAYAH and AL-SHEHBAZ (1977). Lentibulariaceae. Embryology of *Utricularia*: SIDDIQUI (1978). Lythraceae. *Nesaea* distinct from *Heimia* and *Decodon*: GRAHAM (1977). Magnoliaceae. Integumentary studies, *Liriodendron* most advanced genus: BOUMAN (1977). Malpighiaceae. Byrsonimoideae new subfam.: ANDERSON (1977); Malvaceae. New genus *Allowissadula*: BATES (1978); dismembering of *Sida*, neotropical genera *Krapovickasia* and *Rhynchosida, Sida* more natural now: FRYXELL (1978). Menispermaceae. *Coscineae*, rev.: FORMAN (1978), leaf anat.: WILKINSON (1978) and

pollen morphology: FERGUSON (1978) show that *Coscinium* is more isolated.
Melastomataceae. Memecyleae, rev. Africa and Madagascar: JACQUES-FÉLIX (1978);
Centradenia, rev.: ALMEDA (1977). **Mesembryanthemaceae.** *Argyroderma*, monogr.:
HARTMANN (1977); *Oophytum*, morph., tax.: IHLENFELDT (1978); *Cephalophyllum*,
morph. and syst. position: HARTMANN (1978). **Monimiaceae.** *Hedycarya*, rev.:
JÉRÉMIE (1977). **Moraceae.** Rev. of some African genera: BERG (1977c); *Dorstenia*
sect. *Nathodorstenia*, rev., *Craterogyne* included in D.: BERG (1978); wood anat.
of Olmedieae, *Olmedia* excluded from the rest: MENNEGA and LANZING-VINKENBORG
(1977) and these renamed and redefined as Castilleae: BERG (1977b); *Ficus*,
new class.: RAMIREZ (1977), partly rev.: CORNER (1978a,b), origin of *F. syco-*
morus in the Middle East: GALIL et al. (1977). **Myoporaceae.** *Ranopisoa*, included
in Myoporaceae, supported by pollen characters: LOBREAU-CALLEN (1978). **Myristic-**
aceae. More primitive genera than *Myristica* are found in S. America, a New World
tropical origin of the family suggested: ARMSTRONG and WILSON (1978). **Myrtaceae.**
Octamyrtus, rev.: SCOTT (1978a); *Rhodomyrtus*, rev.: SCOTT (1978b). **Nyctaginaceae.**
Selinocarpus and *Ammocodon*, rev.: FOWLER and TURNER (1977); *Boerhavia*, class.:
FOSBERG (1978). **Nymphaeaceae.** *Nuphar lutea*, adaption for beetle pollination:
SCHNEIDER and MOORE (1977). **Onagraceae.** *Oenothera* sect. *Oenothera* subsect.
Munzia, rev. with cytological data, 58 species native in S. America, 20 homo-
zygous, rest heterozygous: DIETRICH (1978); *Oenothera* sect. *Kneiffia*, rev.:
STRALEY (1977); *Calylophus*, biosyst. rev.: TOWNER (1977); *Ludwigia*, fruit structures according with class.: EYDE (1978); flavonoids of *Epilobium* sect. *Epi-*
lobium only few: AVERETT et al. (1978). **Papaveraceae.** *Papaver somniferum* seems
to have originated in diploid strains of *P. setigerum*: HAMMER and FRITSCH
(1977). **Pittosporaceae.** *Pittosporum*, rev. for Hawaii: SAINT JOHN (1977b).
Plumbaginaceae. *Armeria maritima*, two not ecologically conditioned subspp. in
the Netherlands: LEFEBVRE and KAKES (1978). **Podostemonaceae.** *Macropodiella* divided into 5 spp.: CUSSET (1978); *Polypleurum*, embryology: NAGENDRAN et al.
(1977). **Polemoniaceae.** Flavonoid diversification parallel to gross morphology,
karyotypes, and pollen grains: SMITH et al. (1977). Divergence of separated
populations of *Collomia linearis* at the diploid level, floral and vegetative
characters independent: WILKEN (1978). **Polygalaceae.** Montabeae, comparative
anatomy and syst., one homogeneous tribe: STYER (1977). **Polygonaceae.** *Poly-*
gonum, chromosome nos. in Poland: WCISLO (1977). **Proteaceae.** *Adenanthes*, rev.:
NELSON (1978a). **Ranunculaceae.** *Clematopsis*, syst. discussion: RAYNAL (1978);
Adonis in Egypt, overview: IMAM et al. (1977); *Aconitum*, rev. for Pakistan and
Iran: RIEDL (1978), synopsis of subgen. *Paraconitum*: LAUENER and TAMURA (1978);
Ranunculus repens variable in Europe but no tax. recognition: COLES (1977);
karyology of *Ranunculus fluitans* in E. Germany, 3 cytotypes: TURALA-SZYBOWSKA
(1977). **Rosaceae.** *Crataegus*, chromosome nos. related to geography, N. America
x = 8, Europe to E. Asia x = 17: EL GAZZAR and BADAWI (1977); *Crataegus* in
Bavaria: LIPPERT (1978) and Poland: GOSTYŃSKA-JAKUSZEWSKA (1978); *Rosa* sect.
Pimpinellifolia, diploids and derived tetraploids: ROBERTS (1977). **Rubiaceae.**
Psychotria in Hawaii, sect. *Pelagomapouria* an older introduction with very
distinct species, sect. *Straussia* a younger introduction, species less differ-
entiated and less distinct when followed from the older to the younger islands:
SOHMER (1978); *Mastixiodendron*, corolla lobes free and ovaries partially supe-
rior, but morph. and anat. evidences for inclusion in Rubiaceae: DARWIN (1977);
Galium, rev. for Mexico and Central America: DEMPSTER (1978) and S. Africa with
chromosome nos.: PUFF (1978); *Perama*, rev.: STEYERMARK and KIRKBRIDE (1977);
Rutidea, rev.: BRIDSON (1978). **Rutaceae.** Acradenia, rev.: HARTLEY (1978);
Besistoa, rev.: HARTLEY (1977). **Salicaceae.** *Salix*, overview N.W. America:
DORN (1977) and S.E. Alps: PAIERO (1978). **Salicorniaceae.** Reinstatement:
SCOTT (1977). **Sapindaceae.** *Sinoradlkofera*, new genus: MEYER (1977). **Saxi-**
fragaceae. *Heuchera*, simplification of flavonoid chemistry associated with evolution within the genus: BOHM and WILKINS (1978a); chemically distinct *Tellina*
and *Elmera* possibly developed from ancestral stock with *Heuchera*-like flavonoids: BOHM and WILKINS (1978b). **Scrophulariaceae.** *Dermatobotrys*, embryology
supports inclusion: HAKKI (1977); *Linaria*, partly rev. for W. medit. area:
VIANO (1978a,b); *Calceolaria*, partly rev. for N.W.S. America: MOLAU (1978);
Castilleja, rev. for Costa Rica and Panama: HOLMGREN (1978) and chromosome nos.
from N. America, series with 2n, 4n, 8n, and 10n: HECKARD and CHANG (1977);

Rhabdotosperma, new genus separated from *Celsia* and *Verbascum:* HARTL (1977);
Wulfenia evolution in S.E. Europe: LAKUSIC (1978); *Lindernia*, class. supported
by flavonoid patterns in S. America (1977); *Parahebe*, chemosyst. overview:
GRAYER-BARKMEIJER (1978); host influences on style extension in *Orthocarpus*:
ATSATT and GULDBERG (1978); *Pedicularis*, key for Iran: WENDELBO (1977);
Odondites, chromosome nos.: SNOGERUP (1977); summary of *Euphrasia*: YEO (1978).
Solanaceae. *Solanum*, multifactorial analysis in sect. Solanum: EDMONDS (1977)
and serology may support class.: GRAY (1977); seed proteins; EDMONDS and
GLIDEWELL (1977). Development from allogamy to autogamy in *Lycopersicum*
pimpinellifolium: RICK et al. (1977). Sterculiaceae. *Maxwellia* included in
Sterculiaceae, tribe Lasiopetaleae: ROBYNS et al. (1977). Stilbaceae. *Xeroplana*,
very large rev.: ROURKE (1977). Tamaricaceae. *Tamarix*, very intense monogr.:
BAUM (1978a). Theophrastaceae. *Jacquinia*, rev. for Cuba: BORHIDI and MUÑIZ
(1978). Tiliaceae. *Triumfetta welwitschii* complex, tax.: BRUMMITT and SEYANI
(1978). Tremandraceae. Wood anat., in some features similarities to Pittosporaceae: CARLQUIST (1977b). Valerianaceae. *Nardostachys*, monogr.: WEBERLING
(1978); *Aretiastrum*, generical separated from woody *Valeriana* species: WEBERLING
and UHLARZ (1977); *Valeriana*, base number x = 11, dysploid series 6 - 7 - 8 and
6 - 12 - 11: FAVARGER (1978b). Violaceae. *Viola* sect. *Melanium*, chromosome nos.,
key for Italy: MERXMÜLLER and LIPPERT (1977). Vitaceae. Rev. for Argentina:
DE ROMERO (1978). Zygophyllaceae. Flavonoids and alkaloids: SALEH and EL HADIDI
(1977); chromosome nos. of different genera: POGGIO (1977). Monocotyledons:
Alliaceae. European *Allium* with 110 species: STEARN (1978); embryology shows
that *Brodiaea* is between *Allium* and *Muilla*: BERG, R.Y. (1978). Alstroemeriaceae.
Pardinae, new section of *Bomarea*: NEUENDORF (1977). Amaryllidaceae. *Ungernia*,
Lycorideae separated from Amaryllideae: MÜLLER-DOBLIES and MÜLLER-DOBLIES
(1978); *Crinum* in E. Africa: NORDAL et al. (1977). Araceae. *Cryptocoryne*, as-
cending and descending dysploidy, allopolyploidy, subdivision of the genus,
adaptive radiation in the way of island speciation: JACOBSEN (1977); *Anthurium*,
partly rev.: CROAT and BAKER (1978), MADISON (1978). Bromeliaceae. Ecology of
CAM-plants (1977). Commelinaceae. *Gibasis*, flavonoid patterns in relation to
chromosome changes: DEL PERO DE MARTINEZ and SWAIN (1977). Cyperaceae.
Eleocharis, chromosome nos. of N. American species: SCHUYLER (1977); *Cyperus*
subgen. *Pycreus*, two groups according to stomatal types in S. India:
GOVINDARAJALU (1978). Haemadoraceae. Differentiation between *Anigozanthos* and
Macropidia: HOPPER and CAMPBELL (1977). Hypoxidaceae. *Spiloxene* distinct from
Hypoxis, related to *Pauridia*, reduction from umbel to a single flower: THOMPSON
(1978). Iridaceae. *Crocus*: C. *minimus* example for Robertsonian change: BRIGHTON
(1978), wild relatives of C. *sativus* from Central Europe and S.W. Asia: MATHEW
(1977), C. *sativus* Autotriploid: BRIGHTON (1977b), chromosome nos. for 4 Iranian
species: MATHEW and BRIGHTON (1977), grouping of C. *vernus* agg. based on seed
coat structure: KUJAT and RAFIŃSKI (1978); *Iris* sect. *Oncocyclus*, cytology,
asymmetric and bimodal chromosome complement: AVISHAI and ZOHARY (1977); *Iris*,
karyotypes: CHAUDHARY et al. (1977); *Morea*, rev. for trop. Africa: GOLDBLATT
(1977). Juncaceae. *Juncus bufonius* agg. in W. Europe: COPE and STACE (1978).
Liliaceae. Uvularieae, anat.: STERLING (1977a); *Ornithogalum*, rev. for S. Africa:
OBERMEYER (1978); *Drimia* and allied genera, rev.: JESSOP (1977); *Tulbaghia*,
rev.: BURBIDGE (1978); *Wurmbea*, partly rev.: NORDENSTAM (1978b); *Chlorophytum*
and related genera, rev.: MARAIS and REILLY (1978), chromosome nos.: NAIK
(1977); *Urginea maritima* from Greece only tetraploid: DAMBOLDT and WULSCHE
(1977); *Lachenalia*, chromosome nos., extended dysploid series: ORNDUFF and
WATTERS (1978); chromosome nos. in Spanish genera: REJON (1978). Marantaceae.
Mass flowering in *Calathea gymnocarpa*: KENNEDY (1977); *Ischnosiphon*, rev.:
ANDERSSON (1977). Musaceae. Exine-less pollen in *Heliconia* as in some other
monocotyledons, sporophytic incompatibilities not known in this group because
there are no bacculate exines to store tapetal proteins: KRESS et al. (1978).
Orchidaceae. Palynology, separation in several subfamilies, surface structure
may characterize genera or even species: SCHILL and PFEIFFER (1977): epidermal
cell features characterize species in *Phragmipedium* and *Paphiopedilum*: ATWOOD
and WILLIAMS (1978); *Piperia*, biosystematics: ACKERMAN (1977); *Polystachya*,
partial rev.: CRIBB (1978a); *Solzia*, rev.: CRIBB (1978b); Maxillariinae from

Brasil, class.: BRIEGER (1977); *Epidendrum* in Ecuador, rev.: LØJTNANT (1977)
Orchidoideae, rev. for Thailand: SEIDENFADEN (1977); Neottioideae, rev. for
Thailand: SEIDENFADEN (1978); *Corymborkis*, rev.: RASMUSSEN (1977); *Orchis* sect.
Robustocalcare: HAUTZINGER (1977); WILLING and WILLING (1977) summarize orchid
literature for Europe and the medit. area. Palmae. *Tectiphiala*, new monotypic
genus: MOORE (1978a); *Hyophorbe*, rev.: MOORE (1978b); floral anatomy: UHL (1978a),
leaf anatomy: UHL (1978b); *Maxburretia, Liberbaileya* included: DRANSFIELD (1978),
floral anatomy: UHL (1978c). Pandanaceae. *Pandanus* ca. 500 species, separated
into 8 subgenera and 62 sections, micromorphological and anatomical data: STONE
(1977); in subgen. *Vinsonia* male flowers give characters to form natural groups:
HUYNH (1977a), subgen. *Kurzia*, monogr. based on leaf-micromorphology (1977b).
Poaceae. Triticeae, generic relationships: BAUM (1978b); *Oplismenus*, multiple
discriminant analysis: DAVEY and CLAYTON (1978); root anatomy of similar importance as leaf anatomy, Festucoideae even in this respect heterogeneous: GOLLER
(1977); anatomical characters of *Triticum* and *Aegilops*: FRITSCH et al. (1977);
Avena, very intense monogr.: BAUM (1977); *Pennisetum*, rev.: BRUNKEN (1977);
Chusquea and *Swallenochloa*, class.: SODERSTROM and CALDERÓN (1978a); *Chusquea*,
partly rev.: SODERSTROM and CALDERÓN (1978b); *Elionurus*, rev.: RENVOIZE (1978a);
Devenxia, rev. for Argentina and Chile: DE AGRASAR (1978); *Poa* and *Bellardiochloa*,
rev. for Spain: HERNÁNDEZ CARDONA (1978); *Chasmopodium*, rev. for W. Africa:
SCHOLZ (1977a); *Axonopus*, tax.: SCHOLZ (1977b); *Brachiaria* and *Megalachne*, tax.:
SCHOLZ (1978a); subtribes Aristareninae and Airinae, tax.: ALBERS and BUTZIN
(1977); *Panicum* group *Lorea*, rev.: RENVOIZE (1978b); *Festuca brachyphylla* group
in Greenland: FREDERIKSON (1977); *Vulpiella* and *Ctenopsis* distinct from *Vulpia*:
COTTON and STACE (1977); *Eleusine*, flavonoid patterns and systematics: HILU et
al. (1978); trop. American grasses, chromosome nos.: DAVIDSE and POHL (1978);
Trisetum from Japan, cytotax.: TATEOKA (1978); Western N. American grasses,
chromosome nos.: REEDER (1977); trop. African grasses, chromosome nos.: DUJARDIN
(1978). Zingiberaceae. *Alpinia* sect. *Pycnanthus* from New Guinea, rev.: SMITH
(1978). Gymnospermae: Araucariaceae. *Agathis*, rev. for Australia: HYLAND (1978).
Cupressaceae. *Juniperus* sect. *Sabina*, review for N. America: ZANONI (1978).
Phyllocladaceae. Rev.: KENG (1978). Pinaceae. *Pinus*, anatomical features of
the cone bract may be used as taxonomic character: MERKLE and NAPP-ZINN (1977,
1978); *Pinus* subgen. *Strobus* divided into three subgenera: LANDRY (1977);
SCHMIDT-VOGT gives an overall monograph of the spruce (1977).

References

ABID, M.A.: Brunonia 1, 407-692 (1978a); - Brunonia 1, 45-67 (1978b).
- ACKERMAN, J.D.: Bot. J. Linn. Soc. 75, 245-270 (1977). - ADAMS,
R.P.: Ann. M. Bot. Gard. 64, 184-209 (1977). - ADZET, T., GRANGER,
R., PASSET, J., SAN MARTINI, R.: Biochem. Syst. Ecol. 5, 268-272
(1977). - AGERER-KIRCHHOFF, C., AGERER, R.: Mitt. Bot. München 13,
203-233 (1977). - AIKEN, S.G.: Can. J. Bot. 56, 976-982 (1978). -
ALBERS, F., BUTZIN, F.: Willdenowia 8, 81-84 (1977). - ALBERS, F.,
HAAS, R.: Bot. Jahrb. 99, 462-467 (1978). - ALDRIDGE, A.E.: Bot.
Macaronésica 2, 25-58 (1977a); - Bot. Macaronésica 2, 81-93 (1977b);
Bot. J. Linn. Soc. 76, 249-285 (1978b); - Bot. Macaronésica 3, 41-59
(1978a). - ALEYKUTTY, K.M., INAMDAR, J.A.: Flora 167, 103-109 (1978).
- AL-MAYAH, A.-R.A., AL-SHEHBAZ, I.A.: Bot. Not. 130, 437-440 (1977).
- ALMEDA, F.: J. Arnold Arbor. Harv. Univ. 58, 73-108 (1977). -
ANDERSON, D.M.W.: Kew Bull. 32, 529-536 (1978). - ANDERSON, L.C.:
Syst. Bot. 2, 209-218 (1978). - ANDERSON, W.R.: Leandra VI-VII, 5-18
(1977). - ANDERSON, L.: Opera Bot. 43, 1-13 (1977). - ANDROSHCHUK,
A.F., KOSTINENKO, L.D., KHMEL', N.V.: Ukr. Bot. Zh. 35, 273-278
(1978). - APARECIDA, M., CAGNIN, H., GOTTLIEB, O.R.: Biochem. Syst.
Evol. 6, 225-228 (1978). - ARMSTRONG, J.E., WILSON, T.K.: Am. J. Bot.
65, 441-449 (1978). - ARNOLD, M.L., JACKSON, R.C.: Syst. Bot. 3,

208-217 (1978). - ASHTON, P.: Ann. M. Bot. Gard. 64, 694-705 (1977). - ASSOUAD, M.W., DOMMÉE, B., LUMARET, R., VALDEYRON, G.: Bot. J. Linn. Soc. 77, 29-39 (1978). - ATSATT, P.R., GULDBERG, L.D.: Plant Syst. Evol. 129, 167-176 (1978). - ATWOOD, J.T., WILLIAMS, N.H.: Selbyana 2, 356-366 (1978). - AUSTIN, D.F.: Bull. Torrey Bot. Club 105, 114-129 (1978). - AVERETT, J.E.: Ann. M. Bot. Gard. 64, 145-146 (1977). - AVERETT, J.E., KERR, B.J., RAVEN, P.H.: Am. J. Bot. 65, 567-570 (1978). - AVISHAL, M., ZOHARY, D.: Bot. Gaz. 138, 502-511 (1977). - AYENSU, E.S.: Lloydia 49, 436-453 (1977).

BAAGØE: J.: Microcharacters in the ligules of the Compositae, 119-139. In: The Biology and Chemistry of the Compositae, ed. V.H. HEYWOOD, J.B. HARBORNE, B.L. TURNER, Vol. I. London-New York: Academic Press 1977a; - Bot. Tidsskr. 71, 193-223 (1977b); - Bot. Tidsskr. 72, 125-147 (1978). - BACON, J.D.: Rhodora 80, 159-227 (1978). - BADR, A., ELKINGTON, T.T.: Plant Syst. Evol. 128, 23-35 (1977). - BARBER, S.C., ESTES, J.R.: Am. J. Bot. 65(5), 562-566 (1978). - BARNEBY, R.C.: Mem. N.Y. Bot. Gard. 27, 1-892 (1977). - BASSETT, J.J., CROMPTON, C.W.: Can. J. Bot. 56, 581-591 (1978). - BATES, D.M.: Gentes Herbarum 11, 329-354 (1978). - BATE-SMITH, E.C.: Biochem. Syst. Ecol. 5, 95-195 (1977). - BAUDET, J.C.: Bull. Jard. Bot. Natl. Belg. 48, 183-220 (1978). - BAUDET, J.C., TORCK, M.: Bull. Soc. Bot. Fr. 125, 479-483 (1978). - BAUM, B.R.: Oats: Wild and Cultivated. 380 S. Ottawa Biosystematics Research Institute, Monograph No. 14, 1977; - The Genus Tamarix. 209 S. Jerusalem: Israel Academy of Sciences and Humanities 1978a; - Can. J. Bot. 56, 2948-2954 (1978b). - BECHT, R.: Revision der Sektion Alopecuroides DC. der Gattung Astragalus L. 227 S. Vaduz: Cramer 1978. - BEHNKE, H.-D.: Plant Syst. Evol. Suppl. 1, 155-178 (1977a); - Ber. Dtsch. Bot. Ges. 90, 241-251 (1977b); - Plant Syst. Evol. 128, 227-235 (1977c); - Bot. Not. 130, 255-260 (1977d); - Bot. Jahrb. 99, 341-352 (1978). - BEHNKE, H.-D., MABRY, T.J.: Plant Syst. Evol. 126, 371-375 (1977). - BELL, C.R., LINDSEY, A.H.: The Umbel as a Reproductive Unit in the Apiaceae, 739-747. In: Actes 2. Symp. Int. Ombellifères. eds. A.-M. CAUWET-MARC, J. CARBONNIER. Perpignan 1978. - BELL, E.A., LACKEY, J.A., POLHILL, R.M.: Biochem. Syst. Ecol. 6, 201-212 (1978). - BERG, C.C.: Plant Syst. Evol. Suppl. 1, 349-374 (1977a); - Acta Bot. Neerl. 26, 73-82 (1977b); - Bull. Jard. Bot. Natl. Belg. 47, 267-407 (1977c); - Bot. Not. 131, 53-66 (1978); - Norw. J. Bot. 25, 1-7 (1978). - BIERNER, M.W.: Biochem. Syst. Evol. 6, 293-295 (1978). - BINET, P. (ed.): Cytobiologie de la reproduction sexuée des plantes ovulées. Bull. Soc. Bot. Fr. 125, 1-295 (1978). - BISBY, F.A., NICHOLLS, K.W.: Bot. J. Linn. Soc. 74, 97-121 (1977). - BISTE, C.: Feddes Repert. 88 (9-10), 533-613 (1978). - BLACKWELL, W.H.: Taxon 26, 395-397 (1977). - BLACKWELL, W.H., BAECHLE, M.D., WILLIAMSON, G.: Sida 7, 248-254 (1978). - BLANKENHORN, B.: Bot. Jahrb. 99, 108-138 (1978). - BLASCO, F.: Particularités biogéographiques des Ombellifères du sud de l'Inde, 71-83. In: Actes 2. Symp. Int. Ombellifères. eds. A.-M. CAUWET-MARC, J. CARBONNIER. Perpignan 1978. - BLASCO, F., CARBONNIER, J., CARBONNIER-JARREAU, N.-C., CAUWET-MARC, A.-M., CERCEAU-LARRIVAL, N.-T., GUYOT, M., MOLHO, D., REDURON, J.-P., ROLAND-HEYDACKER, F.: Etude multidisciplinaire de genre Vanasushava (Wight) Mukh. et Const. In: Actes 2. Symp. Int. Ombellifères, 663-674. eds. A.-M. CAUWET-MARC, J. CARBONNIER. Perpignan 1978. - BLOOM, W.L.: Syst. Bot. 2, 1-13 (1977). - BÖCHER, T.W.: Bot. J. Linn. Soc. 75, 1-19 (1977a); - Bot. Not. 130, 303-309 (1977b). - BOHM, B.A.: Heliantheae - chemical review, 739-767. In: The Biology and Chemistry of the Compositae, eds. V.H. HEYWOOD, J.B. HARBORNE, B.L. TURNER, Vol. II. London-New York: Academic Press 1977. - BOHM, B.A., WILKINS, C.K.: Can. J. Bot. 56, 1174-1176 (1978a); - Brittonia 30, 327-333 (1978b). - BOHM, B.A., BRIM, S.W., HEBDA, R.J., STEVENS, P.E.: J. Arnold Arbor. Harv. Univ. 59, 311-341 (1978). - BOITEAU, P.: Adansonia 17, 235-241 (1977). -

BOITEAU, P., ALLORGE, L.: Adansonia 17, 305-326 (1978). - BOKHARI, M.H., WENDELBO, P.: Bot. Not. 131, 279-292 (1978). - BOLICK, M.R.: Taxon 26, 239-240 (1977); - Plant Syst. Evol. 130, 209-218 (1978). - BOLLIGER, M.: Ber. Schweiz. Bot. Ges. 88, 29-62 (1978). - BORHIDI, A., MUÑIZ, A.: Acta Bot. Acad. Sci. Hung. 23, 303-317 (1977). - BORHIDI, A., MUÑIZ, O.: Plant Syst. Evol. 129, 1-11 (1978). - BOUMAN, F.: Acta Bot. Neerl. 26, 213-223 (1977). - BRANDHAM, P.E., BHATTARAI, S.: Chromosoma 64, 343-348 (1977). - BRATTSTEN, L.B., WILKINSON, C.F.: Science 196, 1349-1352 (1977). - BRAVO, L.D.: Darwiniana 21, 341-391 (1978). - BREMER, K.: Bot. Not. 130, 493-497 (1977); - Bot. Not. 131, 369-383 (1978). - BREMER, K., WANNTORP, H.-E.: Taxon 27, 317-329 (1978). - BRIDSON, D.M.: Kew Bull. 33, 243-278 (1978). - BRIEGER, F.G.: Bot. Jahrb. 97, 548-574 (1977). - BRIGHTON, C.A.: Kew Bull. 32, 33-45 (1977a); - Plant Syst. Evol. 128, 137-157 (1977b); - Plant Syst. Evol. 129, 299-314 (1978). - BROWN, R.C.: Rhodora 79, 169-189 (1977); - Madroño 25, 187-201 (1978). - BRÜCHER, H.: Tropische Nutzpflanzen. 529 S. Berlin-Heidelberg-New York: Springer 1977. - BRUMMITT, R.K., SEYANI, J.H.: Kew Bull. 32, 709-719 (1978). - BRUNKEN, J.N.: Am. J. Bot. 64, 161-176 (1977). - BRUNSBERG, K.: Opera Bot. 41, 1-78 (1977). - BURBIDGE, R.B.: Notes R. Bot. Gard. Edinburgh 36, 77-104 (1978). - BURGER, W.C.: Bot. Rev. 43, 345-393 (1977). - BURNETT, W.C., JONES, S.B., MABRY, T.J.: Taxon 26, 203-207 (1977a); - Plant Syst. Evol. 128, 277-286 (1977b). - BURTT, B.L.: Aspects of diversification in the capitulum, 41-59. In: The Biology and Chemistry of the Compositae, eds. V.H. HEYWOOD, J.B. HARBORNE, B.L. TURNER, Vol. I. London, New York: Academic Press 1977a; - Plant Syst. Evol. Suppl. 1, 97-109 (1977b); - Comp. Newsl. 7, 6-9 (1978a); - Notes R. Bot. Gard. Edinburgh 37, 1-60 (1978b). - BUTTLER, K.P.: Mitt. Bot. München 13, 255-336 (1977a); - Plant Syst. Evol. 128, 123-136 (1977b). - BUURMAN, J.: Pollen Spores 19, 447-519 (1977).

CABRERA, A.L.: Mutisieae - systematic review, 1039-1066. In: The Biology and Chemistry of the Compositae, eds. V.H. HEYWOOD, J.B. HARBORNE, B.L. TURNER, Vol. II. London-New York: Academic Press 1977. - CABRERA, A.L., RAGONESE, A.M.: Darwiniana 21, 185-261 (1978). - CAGNIN, M.A.H., GOTTLIEB, O.R.: Biochem. Syst. Ecol. 6, 225-238 (1978). - CAGNIN, M.A.H., GOMES, C.M.R., GOTTLIEB, O.R., MARX, M.C., DA ROCHA, A.L., DA SILVA, M.F. das G.F., TEMPERINI, J.A.: Plant Syst. Evol. Suppl. 1, 53-76 (1977). - CANDAU, P.: Lagascalia 7, 143-157 (1978); - CANNE, J.M.: Rhodora 79, 319-389 (1977); - Madroño 25, 81-93 (1978). - CARASCO, J.F., DERBYSHIRE, E., BOULTER, D.: Taxon 26, 513-516 (1977). - CARBONNIER, J., FATIANOFF, O., MOLHO, D.: Phytochimie comparée des taxons rattachés à la tribu des Peucedaneae (Umbelliferae: Apioideae) In: Actes 2. Symp. Int. Ombellifères, 387-513. eds. A.-M. CAUWET-MARC, J. CARBONNIER. Perpignan 1978. - CARLQUIST, S.: J.S. Afr. Bot. 43, 115-128 (1977a); - Am. J. Bot. 64, 704-713 (1977b); - J. S. Afr. Bot. 43, 129-144 (1977c); - Bot. Not. 131, 117-126 (1978a); - Aliso 9, 323-364 (1978b). - CARLQUIST, S., DEBUHR, L.: Bot. J. Linn. Soc. 75, 211-227 (1977). - CAROLIN, R.C.: Brunonia 1, 9-29 (1978). - CAROLIN, R.C., JACOBS, S.W.L., VESK, M.: Aust. J. Bot. 26, 683-698 (1978). - CARR, G.D.: Am. J. Bot. 64(6), 694-703 (1977); - Am. J. Bot. 65, 236-242 (1978). - CARTER, C.R.: Aust. J. Bot. 26, 699-706 (1978a); - Chromosoma 67, 109-121 (1978b). - CATES, R.G., RHOADES, D.F.: Biochem. Syst. Ecol. 5, 185-193 (1977). - CAUWET-MARC, A.-M., CARBONNIER, J.: Bull. Soc. Bot. Fr. 124, 167-176 (1977). - CAUWET-MARC, A.-M., JURY, S.: Données caryologiques sur le tribu des Caucalidées (Umbelliferae). In: Actes 2. Symp. Int. Ombellifères, 305-323. eds. A.-M. CAUWET-MARC, J. CARBONNIER. Perpignan 1978. - CAUWET-MARC, A.-M., CARBONNIER, J., CERCEAU-LARRIVAL, M.-T., DODIN, R., GUYOT, M.: Contribution pluridisciplinaire à la connaissance du genre Bupleurum L. In: Actes 2. Symp. Int. Ombellifères, 623-651. eds. A.-M. CAUWET-MARC, J. CARBONNIER.

Perpignan 1978. - CERCEAU-LARRIVAL, M.-T., ROLAND-HEYDACKER, F.:
Apport de la palynologie à la connaissance des Ombellifères actuelles
et fossiles. In: Actes 2. Symp. Int. Ombellifères, 213-229. eds.
A.-M. CAUWET-MARC, J. CARBONNIER. Perpignan 1978. - CERCEAU-LARRIVAL,
M.T., ROLAND-HEYDACKER, F., CARBONNIER-JARREAU, M.C.: Pollen Spores 19,
285-297 (1977). - CHAPMAN, G.C., JONES, S.B.: Sida 7, 264-281 (1978).
- CHAUDHARY, S.A., CHAUDHARY, G.A., AKRAM, M.: Bot. Not. 130, 263-267
(1977). - CHUANG, T.-J., HSIEH, W.C., WILKEN, D.H.: Am. J. Bot. 65,
450-458 (1978). - CLARK, W.D., MABRY, T.J.: Biochem. Syst. Ecol. 6,
19-20 (1978). - CLARKE, G.: Grana 17, 61-75 (1978). - CLIFFORD, H.T.:
Plant Syst. Evol. Suppl. 1, 77-95 (1977). - COLES, S.M.: Watsonia 11,
353-366 (1977). - CONSTANCE, L.: Some problems in New World Eryngium.
In: Actes 2. Symp. Int. Obellifères, 7-18. eds. A.-M. CAUWET-MARC,
J. CARBONNIER. Perpignan 1978. - COODE, M.J.E.: Brunonia 1, 131-299
(1978). - COOK, C.D.K.: The Hippuris syndrome, 163-176. In: Essays in
Plant Taxonomy, ed. H.E. STREET. London-New York: Academic Press 1978.
- COPE, T.A., STACE, C.A.: Watsonia 12, 113-128 (1978). - CORBET,
S.A.: Bees and the nectar of Echium vulgare, 21-30. In: The Pollination
of Flowers by Insects, ed. A.J. RICHARDS. Linn. Soc. Symp. Ser. 6
(1978). - CORNER, E.J.H.: Philos. Trans. R. Soc. London 281,347-371
(1978a); - Philos. Trans. R. Soc. London 281, 372-406 (1978b); -
Notes R. Bot. Gard. Edinburgh 36, 341-353 (1978c). - CORREA, S.,
PIPPEN, R.W.: Bot. Not. 131, 27-30 (1978). - COTTON, R., STACE, C.A.:
Bot. Not. 130, 173-187 (1977). - CRAMER, L.H.: Kew Bull. 32, 551-561
(1978). - CREPET, W.L., STUESSY, T.F.: Brittonia 30, 483-491 (1978).
- CRIBB, P.J.: Kew Bull. 32, 743-766 (1978a); - Kew Bull. 33, 79-89
(1978b). - CRISP, P., JONES, B.M.G.: Ann. Bot. 42, 937-944 (1978). -
CRISTOFOLINI, G., CHIAPELLA, L.F.: Taxon 26, 43-56 (1977). - CROAT,
T.B., BAKER, R.A.: Selbyana 2, 230-238 (1978). - CRONQUIST, A.:
Brittonia 29, 137-153 (1977a); - Plant Syst. Evol. Suppl. 1, 179-189
(1977b). - Brittonia 30, 505 (1978a); - Once again what is a species,
3-20. In: Beltsville Symposia in Agricultural Research. 2. Biosyste-
matics in Agriculture 1978b. - CROW, G.E.: Rhodora 80, 1-91 (1978). -
CULLEN, J.: Notes R. Bot. Gard. Edinburgh 37, 161-214 (1978). -
CULLEN, J., CHAMBERLAIN, D.F.: Notes R. Bot. Gard. Edinburgh 36,
105-126 (1978). - CUSSET, C.: Adansonia 17, 293-304 (1978). - CZAJA,
A.T.: Taxon 27, 463-470 (1978).

DAHLGREN, R.: Plant. Syst. Evol. Suppl. 1, 253-283 (1977a); - Publ.
Cairo Univ. Herbarium 7,8, 83-102 (1977b). - DAHLGREN, R., ROSENDAL
JENSEN, S., NIELSEN, B.J.: Bot. Not. 130, 329-332 (1977). - DAMBOLDT,
J., WULSCHE, A.: Mitt. Bot. München 13, 535-544 (1977). - DARWIN,
S.P.: J. Arnold Arbor. Harv. Univ. 58, 349-381 (1977). - DAVIDSE, G.,
POHL, R.W.: Ann. M. Bot. Gard. 65, 637-649 (1978). - DAVIES, F.G.:
Kew Bull. 33, 335-342 (1978). - DAVIS, P.H.: Notes R. Bot. Gard.
Edinburgh 36, 325-340 (1978). - DAVEY, J.C., CLAYTON, W.D.: Kew Bull.
33, 147-157 (1978). - DAWSON, J.W., WEBB, C.J.: Generic problems in
Australasian Apioideae (Umbelliferae). In: Actes 2. Symp. Int. Ombelli-
fères, 21-32. eds. A.-M. CAUWET-MARC, J. CARBONNIER. Perpignan 1978. -
DE AGRASAR, Z.E.R.: Darwiniana 21, 417-454 (1978). - DEBUHR, L.E.:
Plant Syst. Evol. 128, 159-169 (1977); - Aliso 9, 179-184 (1978). -
DECHAMPS, R.: Un cas intéressant d'évolution entre une espèce fossile
arborescente d'Afrique et son correspondant actuel Steganotaenia
araliacea (Ombellifères). In: Actes 2. Symp. Int. Ombellifères, 207-
212. eds. A.-M. CAUWET-MARC, J. CARBONNIER. Perpignan 1978. - DEHGAN,
B., GRAIG, M.E.:Am. J. Bot. 65, 345-352 (1978). - DEL PERO DE MARTINEZ,
M.A., SWAIN, T.: Biochem. Syst. Ecol. 5, 37-43 (1977). - DEN HARTOG,
R.M., BAAS, P.: Acta Bot. Neerl. 27, 355-388 (1978). - DEMPSTER, L.T.:
Univ. Calif. Publ. Bot. 73, 1-33 (1978). - DE ROMERO, M.E.M.: Darwiniana 21, 3-26 (1978). - DIAZ MIRANDA, D.: Bot. J. Linn. Soc. 75,
47-67 (1977). - DICKISON, W.C.: Bull. Torrey Bot. Club 104, 12-23

(1977); - Am. J. Bot. 65, 722-735 (1978). - DIETRICH, W.: Ann. Mo.
Bot. Gard. 64, 425-626 (1978). - DITTRICH, M.: Cynareae - systematic
review, 999-1015. In: The Biology and Chemistry of the Compositae,
eds. V.H. HEYWOOD, J.B. HARBORNE, B.L. TURNER, Vol. II. London-New
York: Academic Press 1977. - DOMINGUEZ, X.A.: Eupatorieae - chemical
review, 487-502. In: The Biology and Chemistry of the Compositae, eds.
V.H. HEYWOOD, J.B. HARBORNE, B.L. TURNER, Vol. I. London-New York:
Academic Press 1977. - DOMMEE, B., AASOUAD, M.W., VALDEYRON, G.:
Bot. J. Linn. Soc. 77, 17-28 (1978). - DORN, R.D.: Rhodora 79, 390-
429 (1977). - DOSTAL, J.: Zpr. Cesk. Bot. Spd. CSAV 12(2), 73-78
(1977). - DOYLE, J.A.: Patterns of evolution in early angiosperms,
501-546. In: Patterns of Evolution, ed. A. HALLAM. Amsterdam: Elsevier
Scientific Publishing Company 1977a; - Magnoliophyta, 289-292. In:
McGraw Hill Yearbook Science and Technology. McGraw-Hill Book Company
Inc. 1977b; - Annu. Rev. Ecol. Syst. 9, 365-392 (1978). - DOYLE, J.A.,
BIENS, P., DOERENKAMP, A., JARDINE, S.: Bull. Cent. Rech. Explor.
Prodr. Elf Aquitaine 1, 451-473 (1977). - DRANSFIELD, J.: Gentes
Herbarum. 11, 187-199 (1978). - DUJARDIN, M.: Can. J. Bot. 56, 2138-
2152 (1978). - DUNBAR, A.: Grana 17, 141-147 (1978). - DUNN, D.B.,
HARMON, W.E.: Ann. Mo. Bot. Gard. 64, 340-365 (1977). - DURRIEU, C.:
Les Champignons parasites et leur apport à la Systematique des Ombelli-
fères. In: Actes 2. Symp. Int. Ombellifères, 549-561. eds. A.-M.
CAUWET-MARC, J. CARBONNIER. Perpignan 1978.

EDMONDS, J.M.: Bot. J. Linn. Soc. 75, 141-178 (1977); - Bot. J. Linn.
Soc. 76, 27-51 (1978). - EDMONDS, J.M., GLIDEWELL, S.M.: Plant Syst.
Evol. 127, 277-291 (1977). - EDMUNDS, G.F., ALSTAD, D.N.: Science 199,
941-945 (1978). - EHLER, N.: Subtrop. Pflanzenw. 20, 1-40 (1977). -
EHRENDORFER, F.: Plant Syst. Evol. Suppl. 1, 227-234 (1977). - EHREN-
DORFER, F., SCHWEIZER, D., GREGER, H., HUMPHRIES, C.: Taxon 26, 387-
394 (1977). - EISIKOWITCH, D.: Insect visiting of two subspecies of
Nigella arvensis under adverse seaside conditions, 125-132. In: The
Pollination of Flowers by Insects, ed. A.J. RICHARDS. Linn. Soc. Symp.
Ser. 6 (1978). - EL GAZZAR, A., BADAWI, A.A.: Phytologia 35, 271-275
(1977). - EL GAZZAR, A., BADAWI, A.A., HAMZA, M.K.: Publ. Cairo Univ.
Herbarium 7,8, 47-55 (1977). - ELIAS, T.S., PRANCE, G.T., Brittonia 30,
175-181 (1978). - ELVERS, I.: Bot. Not. 130, 231-234 (1977); - Bot.
Not. 131, 159-160 (1978). - ENDRESS, P.K.: Plant Syst. Evol. Suppl. 1,
321-347 (1977a); - Ber. Dtsch. Bot. Ges. 90, 83-103 (1977b); - Ber.
Dtsch. Bot. Ges. 90, 1-13 (1977c); - Bot.Jahrb. 100, 249-317 (1978a);
- Plant Syst. Evol. 130, 157-160 (1978b). - ENGELL, K.: Bot. Tidsskr.
72, 113-118 (1978). - ERBEN, M.: Mitt. Bot. München 14, 361-631 (1978).
- ERNETT, D.: Plant Syst. Evol. 128, 1-22 (1977a); - Plant Syst. 127,
243-276 (1977b); - Plant Syst. Evol. 130, 85-126 (1978). - ESTABROOK,
G.F.: Syst. Bot. 2, 36-42 (1977); - Syst. Bot. 3, 146-158 (1978). -
ESTABROOK, G.F., ANDERSON, W.R.: Syst. Bot. 3, 179-196 (1978). -
EVANS, F.J., KINGHORN, A.D.: Bot. J. Linn. Soc. 74, 23-35 (1977). -
EYDE, R.H.: Ann. Mo. Bot. Gard. 65, 656-675 (1978).

FAEGRI, K.: Trends in research in pollination ecology, 5-15. In: The
Pollination of Flowers by Insects, ed. A.J. RICHARDS. Linn. Soc. Symp.
Ser. 6 (1978). - FAIRBROTHERS, D.E.: Ann. Mo. Bot. Gard. 64, 147-160
(1977a); - An untapped source of phytochemical data, 79-89. In: The
Role and Goals of Tropical Botanic Gardens, ed. B.C. STONE. Kuala
Lumpur, Malaysia: Rimba Ilmu Universiti Malaya, Penerbi University
Malaya 1977b. - FAVARGER, C.: Taxon 27, 441-448 (1978a); - Candollea
33, 11-21 (1978b). - FAY, J.J.: Allertonia 1(4), 235-296 (1978). -
FEDOROV, A.A.: Gard. Bull. Singapore 29, 127-136 (1977). - FEENY, P.:
Ann. Mo. Bot. Gard. 64, 221-234 (1977). - FERAKOVA, V.: The Genus
Lactuca L. in Europe. 122 S. Bratislava: Univerzita Komenského 1977.
- FERGUSON, J.K.: Kew Bull. 32, 339-346 (1978). - FERNANDES, A.,

QUEIROS, M., SANTOS, M.F.: Bol. Soc. Broteriana Sér. 2, 51, 37-90
(1977). - FERNANDES, J.B., GOTTLIEB, O.R., XAVIER, L.M.: Biochem.
Syst. Ecol. 6, 55-58 (1978). - FERNANDEZ CASAS, J., AGUILERA, J.G.,
REJON, M.R.: Anal. Inst. Bot. A. J. Cavanilles 34, 723-732 (1978). -
FEUER, S.M.: Am. J. Bot. 65, 759-763 (1978). - FEUER, S., TOMB, A.S.:
Am. J. Bot. 64, 230-245 (1977). - FISHER, J.B., MOORE, H.E.: Bot.
Jahrb. 98, 573-611 (1977). - FORMAN, L.L.: Kew Bull. 32, 323-338
(1978). - FOSBERG, F.R.: Smithson. Contrib. Bot. 39, 1-20 (1978). -
FOWLER, B.A., TURNER, B.L.: Phytologia 37, 177-208 (1977). - FREDERIK-
SON, S.: Bot. Not. 130, 269-277 (1977). - FRITSCH, R., KRUSE, J.,
OHLE, H., SCHÄFER, H.I.: Kulturpflanze 25, 155-265 (1977). - FROEBE,
H.A.: New aspects in the Systematics of the Bowlesiinae using flavo-
noid patterns, inflorescence and fruit structure. In: Actes 2. Symp.
Int. Ombellifères, 609-621. eds. A.-M. CAUWET-MARC, J. CARBONNIER.
Perpignan 1978. - FROHNE, D., JOHN, D.: Biochem. Syst. Ecol. 6, 315-
322 (1978). - FRYXELL, P.A.: Brittonia 30, 447-462 (1978). - FUNK,
V.A., STUESSY, T.F.: Syst. Bot. 3, 159-178 (1978).

GALIL, J., STEIN, M., HOROVITZ, A.: Gard. Bull. Singapore 29, 192-
205 (1977). - GALLE, P.: Bot. Jahrb. 98, 449-489 (1977). - GANDERS,
F.R.: Can. J. Bot. 56, 1400-1408 (1978). - GANGADHARA, M., INAMDAR,
J.A.: Plant. Syst. Evol. 127, 121-137 (1977). - GARDNER, R.C.: J. Bot.
64, 810-813 (1977a); - Rhodora 79, 139-146 (1977b). - GARDNER, R.C.,
LA DUKE, J.C.: Syst. Bot. 3, 197-207 (1978). - GARDNER, R.O.: Biochem.
Syst. Ecol. 5, 29-35 (1977). - GIANNASI, D.E.: Taxon 27, 331-344 (1978).
- GILL, L.S., CHINNAPPA, C.C.: Caryologia 30, 375-379 (1977). -
GILLIS, W.T.: Taxon 26, 387-591 (1977). - GOLDBLATT, P.: Ann. Mo. Bot.
Gard. 64, 243-295 (1977); - Ann. Mo. Bot. Gard. 65, 650-655 (1978).
- GOLDBLATT, P., DAVIDSE, G.: Ann. Mo. Bot. Gard. 64, 121-128 (1977).
- GOLDBLATT, P., ENDRESS, P.K.: J. Arnold Arbor. Harv. Univ. 58,
67-71 (1977). - GOLLER, H.: Beitr. Biol. Pflanz. 53, 217-307 (1977).
- GÓMEZ-CAMPO, C.: Bot. J. Linn. Soc. 75, 179-194 (1977a); - Anal.
Inst. Bot. A.J. Cavanilles 34, 151-155 (1977b). - GONZALEZ, A.G.:
Lactuceae - chemical review, 1081-1096. In: The Biology and Chemistry
of the Compositae, eds. V.H. HEYWOOD, J.H. HARBORNE, B.L. TURNER,
Vol. II. London-New York: Academic Press 1977. - GONZALEZ, A.G.,
GALINDO, A.: Lactonas sesquiterpénicas en Umbelíferas. In: Actes 2.
Symp. Int. Ombellifères, 365-377. eds. A.-M. CAUWET-MARC, J. CARBONNIER
Perpignan 1978. - GOROVOY, P.G.: The genus Tilingia Rgl. In: Actes 2.
Symp. Int. Ombellifères, 653-661. eds. A.-M. CAUWET-MARC, J. CARBONNIER
Perpignan 1978. - GOSTYNSKA-JAKUSZEWSKA, M.: Rocznik 31, 5-20 (1978).
- GOTTLIEB, L.D.: Ann. Mo. Bot. Gard. 64, 161-180 (1977). - GOTTSBERGER
G.: Plant Syst. Evol. Suppl. 1, 211-226 (1977). - GOTTWALD, H.: Plant
Syst. Evol. Suppl. 1, 111-121 (1977). - GOVINDARAJALU, E.: Adansonia
18, 95-128 (1978). - GRACE, J., STEWART, F.: Hybridization in the genus
Heracleum in the British Isles. In: Actes 2. Symp. Int. Ombellifères,
773-782. eds. A.-M. CAUWET-MARC, J. CARBONNIER. Perpignan 1978. -
GRAHAM, A., TOMB, A.S.: Lloydia 40, 413-435 (1977). - GRAHAM, S.A.:
Syst. Bot. 2, 61-71 (1977). - GRAU, J.: Astereae - systematic review,
539-565. In: The Biology and Chemistry of the Compositae, eds. V.H.
HEYWOOD, J.B. HARBORNE, B.L. TURNER, Vol. I. London-New York: Academic
Press 1977a. - Mitt. Bot. München 13, 243-254 (1977b). - GRAY, J.C.:
Plant Syst. Evol. 128, 53-69 (1977). - GRAYER-BARKMEIJER, R.J.: Bio-
chem. Syst. Ecol. 6, 131-137 (1978). - GREAR, J.W.: Mem. N.Y. Bot.
Gard. 31, 1-166 (1978). - GREGER, H.: Anthemideae - chemical review,
899-941. In: The Biology and Chemistry of the Compositae, eds. V.H.
HEYWOOD, J.B. HARBORNE, B.L. TURNER, Vol. II. London-New York: Academic
Press 1977. - GREGER, H.: Biochem. Syst. Ecol. 6, 11-17 (1978). -
GREILHUBER, J.: Plant Syst. Evol. 128, 243-257 (1977); - Plant Syst.
Evol. 130, 223-233 (1978). - GREILHUBER, J., SPETA, F.: Plant Syst.
Evol. 127, 171-190 (1977); - Plant Syst. Evol. 129, 63-109 (1978). -

GREMAUD, M.: Ber. Schweiz. Bot. Ges. 87, 173-181 (1977). - GRÖBER,
K., MATZK, F., ZACHARIAS, M.: Kulturpflanze 26, 303-327 (1978). -
GROVES, B.E.: N. Z. J. Bot. 15, 17-18 (1977). - GRUAS-CAVAGNETTO,
C., CERCEAU-LARRIVAL, M.-T.: Présence de pollens d'Ombellifères
fossiles dans les paléogène du bassin anglo-parisien: premiers résul-
tats. In: Actes 2. Symp. Int. Ombellifères, 255-267. eds.: A.-M.
CAUWET-MARC, J. CARBONNIER. Perpignan 1978. - GUÉDES, M., SCHMID, R.:
Flora 167, 525-543 (1978). - GUIMARAES, E.F.: Arq. Jard. Bot. Rio de
Janeiro 21, 45-124 (1977). - GUINET, Ph., VASSAL, J.: Kew Bull. 32,
509-527 (1978). - GUNN, C.R., BARNES, D.E.: Lloydia 49, 454-470
(1977). - GUTIERREZ-BUSTILLO, M.: Contribución al estudio anatómico
del fruto de los géneros Angelica L. y Archangelica Hoffm. In: Actes
2. Symp. Int. Ombellifères, 195-205. eds. A.-M. CAUWET-MARC, J.
CARBONNIER. Perpignan 1978. - GUYOT, M.: Intérêt des études de phyto-
dermologie dans la famille des Ombellifères. In: Actes 2. Symp. Int.
Ombellifères, 133-148. eds. A.-M. CAUWET-MARC, J. CARBONNIER.
Perpignan 1978.

HAIR, J.B., RAVEN, P.H., SEAVEY, S.R.: N. Z. J. Bot. 15, 1-4 (1977).
- HAKKI, M.J.: Bot. Jahrb. 98, 93-119 (1977). - HALLE,N.: Adansonia
17, 397-414 (1978). - HAMANN, U.: Ber. Dtsch. Bot. Ges. 90, 369-384
(1977). - HAMMER, K.: Kulturpflanze 25, 13-23 (1977); - Flora 167,
41-56 (1978). - HAMMER, K., FRITSCH, R.: Kulturpflanze 25, 113-114
(1977). - HARBORNE, J.B.: Flavonoid profiles in the Compositae, 359-
384. In: The Biology and Chemistry of the Compositae, eds. V.H.
HEYWOOD, J.B. HARBORNE, B.L. TURNER, Vol. I. London-New York: Academic
Press 1977a; - Inuleae - chemical review, 603-619. In: The Biology
and Chemistry of the Compositae, eds. V.H. HEYWOOD, J.B. HARBORNE,
B.L. TURNER, Vol. I. London-New York: Academic Press 1977b; - Biochem.
Syst. Evol. Ecol. 5, 7-22 (1977c). - HARBORNE, J.B. SMITH, D.M.:
Biochem. Syst. Ecol. 6, 127-130 (1978a); - Biochem. Syst. Ecol. 6,
287-291 (1978b). - HARBORNE, J.B., WILLIAMS, C.A.: Vernonieae - chemi-
cal review, 523-537. In: The Biology and Chemistry of the Compositae,
eds. V.H. HEYWOOD, J.B. HARBORNE, B.L. TURNER, Vol. I, London-New York:
Academic Press 1977. - HARLEY, R.M., BRIGHTON, C.A.: Bot. J. Linn. Soc.
74, 71-96 (1977). - HARPER, J.L.: Population Biology of Plants. London-
New York: Academic Press, 1-892 (1977). - HARTL, D.: Beitr. Biol.
Pflanz. 53, 55-60 (1977). - HARTLEY, T.G.: J. Arnold Arbor. Harv. Univ.
58, 416-436 (1977); - J. Arnold Arbor. Harv. Univ. 58, 171-181 (1978).
- HARTMANN, H.: Mitt. Inst. Allg. Bot. Hamburg 15, 121-235 (1977); -
Bot. Jahrb. 99, 264-302 (1978). - HASSALL, D.C.:Aust. J. Bot. 25,
429-453 (1977). - HAUTZINGER, L.: Ann. Naturhist. Mus. Wien 81, 31-73
(1977). - HAWKES, J.G.: The taxonomist's role in the conservation of
genetic diversity, 125-142. In: Essays in Plant Taxonomy, ed. H.E.
STREET. London-New York: Academic Press 1978. - HAYDEN, W.J.: J.
Arnold Arbor. Harv. Univ. 48, 257-279 (1977). - HÉBERT, L.-P.: Bull.
Soc. Neuchâtel. Sci. Nat. 100, 113-120 (1977). - HECKARD, L.R.,
CHUANG, T.-J.: Brittonia 29, 159-172 (1977). - HEDBERG, J., HEDBERG,
O.: Bot. Not. 130, 1-24 (1977). - HEGNAUER, R.: Plant Syst. Evol.
Suppl. 1, 191-209 (1977a); - The chemistry of the Compositae, 283-335.
In: The Biology and Chemistry of the Compositae, eds. V.H. HEYWOOD,
J.B. HARBORNE, B.L. TURNER, Vol. I. London-New York: Academic Press
1977b; - Phytochemie und Klassifikation der Umbelliferen, eine Neube-
wertung im Lichte der seit 1972 bekannt gewordenen phytochemischen Tat-
sachen. In: Actes 2. Symp. Int. Ombellifères, 335-363. eds. A.-M.
CAUWET-MARC, J. CARBORNNIER. Perpignan 1978. - HEGNAUER, R., KOOIMAN,
P.: Planta Med. 33, 11-33 (1978). - HERNÁNDEZ CARDONA, A.M.: Estudio
monográfico los géneros Poa y Bellardiochloa en la Peninsula Ibérica
e islas Baleares, Diss. Bot. 46, 365 (1978). - HERRNSTADT, J., HEYN,
C.C.: Boissiera 26 (1977). - HERZ, W.: Sesquiterpene lactones in the

Compositae, 337-357. In: The Biology and Chemistry of the Compositae, eds. V.H. HEYWOOD, J.B. HARBORNE, B.L. TURNER, Vol. I. London-New York: Academic Press 1977a; - Astereae - chemical review, 567-576. In: The Biology and Chemistry of the Compositae, ed. V.H. HEYWOOD, J.B. HARBORNE, B.L. TURNER, Vol. I. London-New York: Academic Press 1977b. - HESSE, M.: Plant Syst. Evol. 130, 13-42 (1978a); - Linzer Biol. Beitr. 9, 181-201 (1978b); - Linzer Biol. Beitr. 9, 237-258 (1978c). - HEYN, C.C., HERRNSTADT, J.: Bot. Not. 130, 427-435 (1977). - HEYWOOD, V.H.: Plant Syst. Evol. Suppl. 1, 1-12 (1977); - Introduction to the Taxonomy of the Umbelliferae. In: Actes 2. Symp. Int. Ombellifères, 107-112. eds. A.-M. CAUWET-MARC, J. CARBONNIER. Perpignan 1978a; - Multivariate taxonomic synthesis of the tribe Caucalideae. In: Actes 2. Symp. Int. Ombellifères, 727-736. eds. A.-M. CAUWET-MARC, J. CARBONNIER. Perpignan 1978b. - HEYWOOD, V.H., HUMPHRIES, C.J.: Anthemideae - systematic review, 851-898. In: The Biology and Chemistry of the Compositae, eds. V.H. HEYWOOD, J.B. HARBORNE, B.L. TURNER, Vol. II. London-New York: Academic Press 1977. - HEYWOOD, V.H., HARBORNE, J.B., TURNER, B.L. (eds.): The Biology and Chemistry of the Compositae, Vols. I, II. London-New York: Academic Press 1977. - HICKEY, L.J., DOYLE, J.A.: Bot. Rev. 43, 3-104 (1977). - HILLER, K.: Zur Phytochemie der Saniculoideae. In: Actes 2. Symp. Int. Ombellifères, 379-386. eds. A.-M. CAUWET-MARC, J. CARBONNIER. Perpignan 1978. - HILU, K.W., DE WET, J.M.J., SEIGLER, D.: Biochem. Syst. Ecol. 6, 247-249 (1978). - HOFMANN, U.: Ber. Dtsch. Bot. Ges. 90, 39-52 (1977). - HOLMGREN, N.H.: Brittonia 30, 182-184 (1978). - HORNER, H.T., Jr., PEARSON, C.B.: Am. J. Bot. 65(3), 293-309 (1978). - HOPKINS, C.O., BLACKWELL, W.H.: Sida 7, 147-173 (1977). - HOPPER, S.D.: Aust. J. Bot. 25, 395-411 (1977). - HOPPER, S.D., CAMPBELL, N.A.: Aust. J. Bot. 25, 523-544 (1977). - HOWELL, D.J.: Nature (London) 270, 509-510 (1977). - HUBER, H.: Plant Syst. Evol. Suppl. 1, 285-298 (1977). - HUGHES, N.F.: Bot. Rev. 43, 105-127 (1977). - HUMPHRIES, C.J.: Bot. Not. 130, 155-161 (1977). - HURKA, H., WÖHRMANN, K.: Bot. Jahrb. 98, 120-132 (1977). - HUYNH, K.-L.: Beitr. Biol. Pflanz. 53, 447-471 (1977a); - Bot. Jahrb. 98, 199-249 (1977b). - HYLAND, B.P.M.: Brunonia 1, 103-115 (1978).

IHLENFELDT, H.-H.: Bot. Jahrb. 99, 303-328 (1978). - IMAM, M., CHRTEK, J., SLAVI-KOVA, Z.: Publ. Cairo Univ. Herbarum 7,8, 261-272 (1977). - IRWIN, H.S., BARNEBY, R.C.: Mem. N.Y. Bot. Gard. 30, 1-297 (1978).

JACOBSEN, N.: Bot. Not. 130, 71-87 (1977). - JACQUES-FELIX, H.: Adansonia 18, 221-236 (1978). - JÄGER, E.J.: Flora 166(1), 75-92 (1977). - JANZEN, D.H.: Ann. M. Bot. Gard. 64, 706-736 (1977). - JEFFREY, C.: Corolla forms in Compositae - some evolutionary and taxonomic speculations, 111-118. In: The Biology and Chemistry of the Compositae, eds. V.H. HEYWOOD, J.B. HARBORNE, B.L. TURNER, Vol. I. London-New York: Academic Press 1977. - JEFFREY, C., HALLIDAY, P., WILMOT-DEAR, M., JONES, S.W.: Kew Bull. 32, 47-67 (1977). - JENSEN, R.J.: Syst. Bot. 2, 122-133 (1977). - JÉRÉMIE, J.: Adansonia 18, 25-54 (1978). - JESSOP, J.P.: J. S. Afr. Bot. 43, 265-319 (1977). - JOHN, J.: Biochem. Syst. Ecol. 6, 323-327 (1978). - JONES, A.G.: Rhodora 80, 319-357 (1978). - JONES, S.B.: Vernonieae - systematic review, 503-521. In: The Biology and Chemistry of the Compositae, eds. V.H. HEYWOOD, J.B. HARBORNE, B.L. TURNER, Vol. I. London-New York: Academic Press 1977. - JULIN, E.: Bot. Not. 130, 287-302 (1977). - JURY, S.L.: Tuberculate fruits in the Umbelliferae (tribe Caucalideae). In: Actes 2. Symp. Int. Ombellifères, 149-159. eds. A.-M. CAUWET-MARC, J. CARBONNIER. Perpignan 1978.

KÄSTNER, A.: Flora 167, 485-514 (1978). - KAUR, A., HA, C.O., JONG, K., SANDS, V.E., CHAN, H.T., SOEPADMO, E., ASHTON, P.S.: Nature (London

271, 440-442 (1978). - KAY, Q.O.N.: The role of preferential and as-
sortative pollination in the maintenance of flower colour polymor-
phisms, 175-190. In: The Pollination of Flowers by Insects, ed. A.J.
RICHARDS. Linn. Soc. Symp. Ser. 6 (1978). - KEEFE, J.M., MOSELEY, M.F.:
J. Arnold Arbor. Harv. Univ. 59, 274-297 (1978). - KELLEY, S.C.: J.
Arnold Arbor. Harv. Univ. 59, 360-413 (1978). - KEELEY, S.C., JONES,
S.B.: Rhodora 79, 147-159 (1977a); - Am. J. Bot. 64, 576-584 (1977b).
- KEIL, D.J.: Rhodora 79, 32-79 (1977a); - Rhodora 79, 79-94 (1977b);
- Rhodora 80, 135-146 (1978). - KEIL, D.J., STUESSY, T.F.: Am. J. Bot.
64, 791-798 (1977). - KELLEY, W.A., ADAMS, R.P.: Rhodora 80, 107-134
(1978). - KENG, H.: Plant Syst. Evol. Suppl. 1, 235-251 (1977); -
J. Arnold Arbor. Harv. Univ. 59, 249-273 (1978). - KENNEDY, H.: Bot.
Not. 130, 333-339 (1977); - Univ. Calif. Publ. Bot. 71, 1-90 (1978).
- KENTON, A.: Chromosoma 65, 309-324 (1978). - KEVAN, P.G.: Floral
coloration, its colorimetric analysis and significance in anthecology,
51-78. In: The Pollination of Flowers by Insects, ed. A.J. RICHARDS.
Linn. Soc. Symp. Ser. 6 (1978). - KHALEEL, T.F.: Bot. Not. 130, 441-
452 (1977); - Plant Syst. Evol. 129, 45-62 (1978). - KIESLING, R.:
Darwiniana 21, 263-330 (1978). - KIMATA, M.: Plant Syst. Evol. 129,
243-253 (1978). - KING, B.L.: Am. J. Bot. 64, 350-360 (1977). - KING,
M.C.: Ann. Mo. Bot. Gard. 64, 181-183 (1977). - KING, R.M., KYHOS,
D.W., POWELL, M.A., RAVEN, P.H., ROBINSON, H.: Ann. Mo. Bot. Gard.
63(4), 863-888 (1977). - KLAUS, W.: Grana 17, 161-166 (1978). -
KOEK-NOORMAN, J.: Ber. Dtsch. Bot. Ges. 90, 183-190 (1977). - KOLBE,
K.-P.: Bot. Jahrb. 99, 468-489 (1978). - KORDYUM, E.L.: La cytoembryo-
logie des espèces d'Ombellifères en rapport avec leur phylogenie et
leur evolution. In: Actes 2. Symp. Int. Ombellifères, 269-279. eds.
A.-M. CAUWET-MARC, J. Carbonnier. Perpignan 1978. - KOWAL, R.R., MORI,
S.A., KALLUNKI, J.A.: Brittonia 29, 399-410 (1977). - KRACH, J.E.:
Plant Syst. Evol. Suppl. 1, 141-153 (1977). - KRASSILOV, V.A.: Bot.
Rev. 43, 143-176 (1977). - KRESS, W.J., STONE, D.E., SELLERS, S.C.:
Am. J. Bot. 65, 1064-1076 (1978). - KUBITZKI, K. (ed.): Evolution on
classification of higher categories. Plant. Syst. Evol. Suppl. 1,
1-416 (1977a); - Plant Syst. Evol. Suppl. 1, 21-31 (1977b); - Mem.
N.Y. Bot. Gard. 29, 81-138 (1978). - KUJAT, R., RAFIŃSKI, J.N.: Plant
Syst. Evol. 129, 255-260 (1978). - KUKKONEN, J., UOTILA, P.: Ann. Bot.
Fenn. 14, 131-142 (1977). - KUMMER, C.: Mitt. Bot. München 13, 129-203
(1977). - KUPICHA, F.K.: Bot. J. Linn. Soc. 74, 131-162 (1977). -
KUZNETSOVA, G.A.: Coumarines d'Ombellifères de la flore d'U.R.S.S. In:
Actes 2. Symp. Int. Ombellifères, 515-524. eds. A.-M. CAUWET-MARC, J.
CARBONNIER. Perpignan 1978. - KYHOS, D.W., CARTER, C.R., SMITH-WHITE,
S.: Chromosoma 65, 81-101 (1977).

LACK, W.: Willdenowia 8, 329-339 (1978). - LACKEY, J.A.: Bot. J. Linn.
Soc. 74, 163-178 (1977). - LAKUSIC, R.: Bot. Jahrb. 99, 443-461 (1978).
- LANDRY, P.: Bull. Soc. Bot. Fr. 124, 469-474 (1977). - LANG, B.:
Bot. Jahrb. 98, 289-335 (1977). - LAUENER, L.A., TAMURA, M.: Notes
R. Bot. Gard. Edinburgh 37, 113-124 (1978). - LE COQ, C., GUERVIN,
C., HAMEL, J.-L., JOLINON, D.: La quantité d'A.D.N. nucléaire et la
garniture chromosomique chez quelques Umbellifères: application à
l'étude de leur évolution. In: Actes 2. Symp. Int. Ombellifères, 281-
291. eds. A.-M. CAUWET-MARC, J. CARBONNIER. Perpignan 1978. - LEE, Y.S.,
FAIRBROTHERS, D.E.: Taxon 27, 159-185 (1978). - LEFEBVRE, C., KAKES,
P.: Acta Bot. Neerl. 27, 17-26 (1978). - LEINS, P., BONNERY-BRACHTEN-
DORF, R.: Beitr. Biol. Pflanz. 53, 143-155 (1977). - LEINS, P., SOBICK,
U.: Bot. Jahrb. 98, 133-149 (1977). - LEONARD, J.: Publ. Cairo Univ.
Herbarium 7,8, 295-298 (1977). - LEPPICK, E.E.: The evolution of capi-
tulum types of the Compositae in the light of insect - flower interac-
tion, 61-89. In: The Biology and Chemistry of the Compositae, Vol. I.
London-New York: Academic Press 1977. - LEROY, J.-F.: Science 196,
977-978 (1977). - LEVY, M., FUJII, K.: Biochem. Syst. Ecol. 6, 117-125

(1978). - LEWIS, W.H.: Sida 7, 95-102 (1977). - LEWIS, W.H., SEMPLE, J.C.: Am. J. Bot. 64, 1078-1082 (1977). - LIPPERT, W.: Ber. Bayer. Bot. Ges. 49, 165-198 (1978). - LLOYD, D.G., WEBB, C.J.: Bot. Rev. 43, 177-216 (1977). - LOBREAU-CALLEN, D.: Adansonia 18, 237-242 (1978). - LOBREAU-CALLEN, D., NILSSON, S., ALBERS, F., STRAKA, H.: Grana 17, 125-139 (1978). - LØJTNANT, B.: Bot. Not. 130, 321-328 (1977).

MABBERLY, D.J.: Gard. Bull. Singapore 29, 41-55 (1977). - MABRY, T.J.: Ann. Mo. Bot. Gard. 64, 210-220 (1977). - MABRY, T.J., BOHLMANN, F.: Summary of the chemistry of the Compositae, 1097-1104. In: The Biology and Chemistry of the Compositae, eds. V.H. HEYWOOD, J.B. HARBORNE, B.L. TURNER, Vol. II. London-New York: Academic Press 1977. - MACEDO, M., PRANCE, G.T.: Brittonia 30, 203-215 (1978). - MADISON, M.: Selbyana 2, 239-282 (1978). - MAGIN, N.: Ber. Dtsch. Ges. 90, 53-66 (1977); - Blütenmorphologische Untersuchungen an Actinotus Lab. (Hydrocotyloideae) unter besonderer Berücksichtigung des Gynoeceums. In: Actes 2. Symp. Int. Ombellifères, 749-764. eds. A.-M. CAUWET-MARC, J. CARBONNIER. Perpignan 1978. - MAGUIRE, B.: Caldasia 11, 129-146 (1977). - MAGUIRE, B., ASHTON, P.S., DE ZEEUW, C., GIANNASI, D.E., NIKLAS, K.J.: Taxon 26, 341-385 (1977). - MAIHLE, N.J., BLACKWELL, W.: Sida 7, 382-391 (1978). - MANDENOVA, I.P., CARBONNIER, J., CARBONNIER-JARREAU, M.-C., CAUWET-MARC, A.-M., CERCEAU-LARRIVAL, M.-T., GUYOT, M., MOLHO, D., REOURON, J.-P.: Contribution à l'étude du genre Tetrataenium (DC.) Manden. (Pastinaceae K.-Pol. emend. Manden. Apioideae). In: Actes 2. Symp. Int. Ombellifères, 675-725. eds. A.-M. CAUWET-MARC, J. CARBON-NIER Perpignan 1978. - MARAIS, W., REILLY, J.: Kew Bull. 32, 653-663 (1978). - MARÉCHAL, R., MASCHERPA, J.M., STAINIER, F.: Boissiera 28 (1978). - MARKS, G.E.: Chromosoma 69, 211-218 (1978). - MATHEW, B.: Plant Syst. Evol. 128, 89-103 (1977). - MATHEW, B., BRIGHTON, C.A.: Iran. J. Bot. 1, 123-135 (1977). - MATHEWES, R.W.: Can. J. Bot. 56, 1372-1380 (1978). - MATTATIA, J.: Bot. Not. 130, 27-34 (1977). - MCKEY, D., WATERMAN, P.G., MBJ, C.N., CARTLAN, J.S., STRUHSAKER, T.T.: Science 202, 61-64 (1978). - MCNEIL, J.: Can. J. Bot. 56, 297-308 (1978). - MEEUSE, A.D.J.: Phytomorphology 27, 314-322 (1977a); - Plant Syst. Evol. Suppl. 1, 13-19 (1977b); - Entomophily in Salix: theoretical considerations, 47-50. In: The Pollination of Flowers by Insects, ed. A.J. RICHARDS. Linn. Soc. Symp. Ser. 6 (1978a); - The physiology of some sapromyophilous flowers, 97-104. In: The Polli-nation of Flowers by Insects, ed. A.J. RICHARDS. Linn. Soc. Symp. Ser. 6 (1978b); - Proc. K. Ned. Akad. Wet. 81, 300-326 (1978c). - MELZHEIMER, V.: Bot. Jahrb. 98, 1-92 (1977). - MENNEGA, A.M.W., LANZIG-VINKENBORG, M.: Acta Bot. Neerl. 26, 1-27 (1977). - MERKLE, M., NAPP-ZINN, K.: Bot. Jahrb. 97, 475-502 (1977); - Bot. Jahrb. 98, 549-572 (1978). - MERXMÜLLER, H.: Plant Syst. Evol. Suppl. 1, 397-405 (1977). - MERXMÜLLER, H., GRAU, J.: Publ. Cairo Univ. Herbarium 7,8, 9-20 (1977). - MERXMÜLLER, H., LIPPERT, W.: Mitt. Bot. München 13, 503-534 (1977). - MERXMÜLLER, H., LEINS, P., ROESSLER, H.: Inuleae - systematic review, 577-602. In: The Biology and Chemistry of the Com-positae, eds. V.H. HEYWOOD, J.B. HARBORNE, B.L. TURNER, Vol. I. London-New York: Academic Press 1977. - MEUSEL, H.: Bot. Jahrb. 99, 222-248 (1978). - MEYER, G.F.: J. Arnold Arbor. Harv. Univ. 58, 182-188 (1977). - MITRA, K., MONDAL, M., SAHA, S.: Grana 16, 75-79 (1977). - MOGFORD, D.J.: Pollination and flower colour polymorphism, with special refer-ence to Cirsium palustre, 191-199. In: The Pollination of Flowers by Insects, ed. A.J. RICHARDS. Linn. Soc. Symp. Ser. 6 (1978a). - J. S. Afr. Bot. 44, 111-117 (1978b). - MOLAU, U.: Bot. Not. 131, 293-316 (1978). - MOLINA, A.M.: Darwiniana 21, 473-490 (1978). - MOORE, D.M.: The chromosomes and plant taxonomy, 39-56. In: Essays in Plant Taxo-nomy, ed. H.E. STREET. London-New York: Academic Press 1978. - MOORE, H.E.: Gentes Herbarum 11, 284-290 (1978a); - Gentes Herbarum 11, 212-245 (1978b). - MORI, S.A., KALLUNKI, J.A.: Brittonia 29, 69-84 (1977).

MORISSET, P.: Watsonia 12, 145-153 (1978). - MORLEY, B., CHAO, J.-M.:
J. Arnold Arbor. Harv. Univ. 58, 382-415 (1977). - MORTON, J.K.:
Watsonia 11, 211-223 (1977); - Kew Bull. 32, 433-448 (1978). - MOSS,
W.W.: Proc. Acad. Natl. Sci. Philadelphia 129, 87-98 (1978). - MÜLLER-
DOBLIES, D., MÜLLER-DOBLIES, U.: Bot. Jahrb. 99, 249-263 (1978). -
MÜLLER-SCHNEIDER, P.: Veroeff. Geobot. Inst. Eidg. Tech. Hochsch.
Stift. Ruebel Zürich 61, 1-226 (1977). - MUNIR, A.A.: J. Adelaide Bot.
Gard. 1, 161-166 (1978).

NAGENDRAN, C.R., AREKAL, G.D., SUBRAMANYAM, K.: Plant. Syst. Evol.
128, 215-226 (1977). - NAIK, V.N.: Bot. J. Linn. Soc. 74, 297-308
(1977). - NAPP-ZINN, K., EBLE, M.: Plant Syst. Evol. 130, 167-190
(1978). - NELSON, E.C.: Brunonia 1, 303-406 (1978a); - Watsonia 12,
103-112 (1978b). - NEUBAUER, H.F.: Bot. Jahrbuch 98, 362-371 (1977).
- NEUENDORF, M.: Bot. Not. 130, 55-60 (1977). - NEVLING, L.J.,
NIEZGODA, Ch.J.: Adansonia 18, 345-364 (1978). - NEW, J.K.: Watsonia
12, 137-143 (1978). - NILSSON, L.A.: Bot. Not. 131, 35-51 (1978a); -
Bot. Not. 131, 355-368 (1978b). - NOGUEIRA, I.: Bol. Soc. Broteriana
51, 127-133 (1977). - NORDAL, J., RØRSLETT, B., LAANE, M.M.: Norw.
J. Bot. 24, 195-212 (1977). - NORDENSTAM, B.: Senecioneae and Liabeae -
systematic review, 799-830. In: The Biology and Chemistry of the Com-
positae, eds. V.H. HEYWOOD, J.B. HARBORNE, B.L. TURNER, Vol. II.
London-New York: Academic Press 1977a; - Bot. Not. 130, 279-286
(1977b); - Opera Bot. 44, (1978a); - Notes R. Bot. Gard. Edinburgh 36,
211-233 (1978b). - NORDENSTAM, B., EL-GHAZALY, G.: Publ. Cairo Univ.
Herbarium 7,8, 143-156 (1977). - NORLINDH, T.: Arctoteae - systematic
review, 943-959. In: The Biology and Chemistry of the Compositae. eds.
V.H. HEYWOOD, J.B. HARBORNE, B.L. TURNER, Vol. II. London-New York:
Academic Press 1977a. - NORLINDH, T.: Calenduleae - systematic review,
961-987. In: The Biology and Chemistry of the Compositae. eds. V.H.
HEYWOOD, J.B. HARBORNE, B.L. TURNER, Vol. II. London-New York: Academic
Press 1977b.

OBERMEYER, A.A.: Bothalia 12, 323-376 (1978). - OHASHI, H., TATEISHI,
Y.: Bot. Mag. Tokyo 90, 219-233 (1977). - OHBA, H.: Bot. Mag. Tokyo
90, 41-56 (1977). - OKEKE, S.: Morphological variation of bracts,
bracteoles and fruits in Daucus. In: Actes 2. Symp. Int. Ombellifères,
161-174. eds. A.M. CAUWET-MARC, J. CARBONNIER. Perpignan 1978. -
ORNDUFF, R., WATTERS, P.J.: J. S. Afr. Bot. 44, 387-390 (1978). -
ORLIEB, U., WINKLER, S.: Bot. Jahrb. 97, 586-602 (1977). - OTT, E.:
Revision der Sektion Chronopus Bge. der Gattung Astragalus L. 142 S.
Vaduz: Cramer 1978. - OVDIENKO, O.A., LITVINENKO, V.I., SHRETER, A.I.:
Byull. Mosk. Ova. Ispyt. Prir. Biol. 82(6), 74-87 (1977).

PAGE, J.S.: Kew Bull. 32, 313-319 (1978). - PAIERO, P.: Webbia 32, 271-
339 (1978). - PARDO, C.: Contribution polynologique à la taxonomie des
espèces espagnoles du genre Seseli L. (Umbelliferae). In: Actes 2. Symp.
Int. Ombellifères, 243-253. eds. A.-M. CAUWET-MARC, J. CARBONNIER.
Perpignan 1978. - PARKER, J.S., AINSWORTH, C.C., TAYLOR, S.: Chromosoma
67, 123-143 (1978). - PARKER, P.F.: The classification of crop plants,
97-124. In: Essays in Plant Taxonomy, ed. H.E. STREET. London-New York:
Academic Press 1978. - PEREIRA, J.F.: Arq. Jard. Bot. Rio de Janeiro
21, 235-392 (1977). - PETTIGREW, C.J., WATSON, L.: Taxon 26, 57-64
(1977). - PETTITT, J.M.: Nature (London) 266, 530-532 (1977). - PEUKERT,
D.: Bot. Jahrb. 97, 459-474 (1977). - PHILIPSON, W.R.: Plant Syst.
Evol. Suppl. 1, 123-140 (1977). - PICKERSGILL, B.: Nature (London) 268,
591-595 (1977). - PIGGINN, C.M.: Muelleria 3, 215-244 (1977). - PIMENOV,
M.G.: Les Ombellifères d'Asie Moyenne. In: Actes 2. Symp. Int. Ombelli-
fères, 33-45. eds. A.-M. CAUWET-MARC, J. CARBONNIER. Perpignan 1978. -
PINKAVA, D.J., KEIL, D.J.: Am. J. Bot. 64, 680-686 (1977). - PLOUVIER,
V.: Ombellifères et familles voisines: leurs analogies et leurs distinc-

tions biochimiques. In: Actes 2. Symp. Int. Ombellifères, 535-548.
eds. A.-M. CAUWET-MARC, J. CARBONNIER. Perpignan 1978. - POGGIO, L.:
Darwiniana 21, 139-151 (1977). - POHLMANN, J.: Mitt. Inst. Allg. Bot.
Hamburg 15, 69-73 (1977). - PONCY, O.: Adansonia 17, 465-494 (1978).
- POSTON, M.E., THOMPSON, H.J.: Syst. Bot. 2, 14-27 (1977). - POWELL,
A.M.: Ann. Bot. Gard. 65, 590-636 (1978). - POWELL, A.M., POWELL, S.A.:
Madroño 25, 160-169 (1978). - PRAKASH, N., MCALISTER, E.J.: Aust. J.
Bot. 25, 615-622 (1977). - PRANCE, G.T.: Ann. M. Bot. Gard. 64, 659-
684 (1977). - PRANCE, G.T., MORI, S.A.: Brittonia 30, 21-33 (1978). -
PRASARD, K.: Bot. Jahrb. 98, 266-272 (1977a); - Bot. Jahrb. 97, 508-
514 (1977b). - PRENTICE, H.C.: Bot. J. Linn. Soc. 77, 203-216 (1978).
- PRINGLE, J.S.: Sida 7, 174-217 (1977). - PROCTOR, M.C.F.: Insect
pollination syndromes in an evolutionary and ecosystemic context,
105-116. In: The Pollination of Flowers by Insects, ed. A.J. RICHARDS.
Linn. Soc. Symp. Ser. 6 (1978). - PUFF, C.: J. S. Afr. Bot. 44, 203-
279 (1978).

QUEIROS, M.: Sur la caryologie des Ombellifères du Portugal. In: Actes
2. Symp. Int. Ombellifères, 325-334. eds. A.-M. CAUWET-MARC, J.
CARBONNIER. Perpignan 1978. - QUINN, J.A.: Bull. Torrey Bot. Club 105,
58-64 (1978).

RAADTS, E.: Willdenowia 8, 101-157 (1977). - RAMIREZ, W.B.: Ann. Mo.
Bot. Gard. 64, 296-310 (1977). - RAJU, V.S., RAO, P.N.: Bot. J. Linn.
Soc. 75, 69-97 (1977). - RAO, M.V.S., PANTULU, J.V.: Chromosoma 69,
121-130 (1978). - RASMUSSEN, F.N.: Bot. Tidsskr. 71, 161-192 (1977).
- RAVEN, P.H.: Lloydia 40, 401-412 (1977). - RAY, J.H., VENKETESWARAN,
S.: Chromosoma 66, 341-350 (1978). - RAYNAL, J.: Adansonia 17, 151-154
(1977); - Adansonia 18, 3-18 (1978). - REDURON, J.-P.: Contribution à
l'étude morphologique du pétale chez les Ombellifères. In: Actes 2.
Symp. Int. Ombellifères, 121-131. eds. A.-M. CAUWET-MARC, J. CARBONNIER.
Perpignan 1978. - REEDER, J.R.: Am. J. Bot. 64, 102-110 (1977). -
REJON, R.: Anal. Inst. Bot. A.J. Cavanilles 34, 733-759 (1978). -
RENVOIZE, S.A.: Kew Bull. 32, 665-672 (1978a); - Kew Bull. 32, 419-
428 (1978b). - RICK, C.M., FOBES, J.F., HOLLE, M.: Plant Syst. Evol.
127, 139-170 (1977). - RICHARDSON, A.T.: Rhodora 79, 467-572 (1977).
- RICHARDSON, J.B.K.: Endemic taxa and the taxonomist, 245-260.
In: Essays in Plant Taxonomy, ed. H.E. STREET. London-New York: Aca-
demic Press 1978. - RICHARDSON, M.: Biochem. Syst. Ecol. 6, 283-286
(1978). - RIEDL, H.: Willdenowia 8, 317-328 (1978). - ROBERTS, A.V.:
Bot. J. Linn. Soc. 74, 309-328 (1977). - ROBINS, D.J.: Senecioneae -
chemical review, 831-850. In: The Biology and Chemistry of the Com-
positae, eds. V.H. HEYWOOD, J.B. HARBORNE, B.L. TURNER, Vol. II.
London-New York: Academic Press 1977a; - Comp. Newsl. 5, 1-11 (1977b).
ROBINSON, H., CUATRECASAS, J.: Phytologia 40, 37-50 (1978). - ROBINSON,
H., KING, R.M.: Eupatorieae - systematic review, 437-485. In: The Bio-
logy and Chemistry of the Compositae, eds. V.H. HEYWOOD, J.B. HARBORNE,
B.L. TURNER, Vol. I. London-New York: Academic Press 1977. - ROBSON,
N.K.B.: Bull. Br. Mus. Bot. 5, 291-355 (1977). - ROBYNS, A., NILSSON,
S., DECHAMPS, R.: Bull. Jard. Bot. Natl. Belg. 47, 145-153 (1977). -
RODRIGUEZ, E.: Biochem. Syst. Ecol. 5, 207-218 (1977). - RODRIGUEZ,
E., MABRY, T.J.: Tageteae - chemical review, 785-797. In: The Biology
and Chemistry of the Compositae, eds. V.H. HEYWOOD, J.B. HARBORNE,
B.L. TURNER, Vol. II. London-New York: Academic Press 1977. - ROHWEDER,
D., URMI, E.: Bot. Jahrb. 100, 1-25 (1978). - ROLAND-HEYDACKER, F.,
CERCEAU-LARRIVAL, M.-T.: Grana 17, 81-89 (1978). - ROMEIKE, A.: Bot.
Not. 131, 85-96 (1978). - ROMMEL, A.: Mitt. Bot. München 13, 579-728
(1977). - ROURKE, J.P.: J. S. Afr. Bot. 43, 1-8 (1977). - ROURKE, J.,
WIENS, D.: Ann. Mo. Bot. Gard. 64, 1-17 (1977). - ROUX, M., CARBONNIER,
J., CAUWET-MARC, A.-M.: Un Exemple d'analyse cladistique: le genre
Bupleurum L. (Umbelliferae). In: Actes 2. Symp. Int. Ombellifères,

575-592. eds. A.-M. CAUWET-MARC, J. CARBONNIER. Perpignan 1978. -
ROWLEY, J.R., DAHL, A.O.: Pollen Spores 19, 169-284 (1977). - RUNDEL,
P.W.: Plant Syst. Evol. 127, 1-9 (1977). - RURY, P.M., DICKISON, W.C.:
J. Arnold Arbor. Harv. Univ. 58, 209-256 (1977).

SACHS, T.: Plant Syst. Evol. 130, 1-11 (1978). - SAENZ DE RIVAS, C.,
HEYWOOD, V.H., JURY, S., AL ATTAR, A.: Etude micromorphologique et
anatomique du fruit des Caucalideae Bentham (Umbelliferae). In: Actes
2. Symp. Int. Ombellifères, 175-193. eds. A.-M. CAUWET-MARC, J.
CARBONNIER. Perpignan 1978. - SAINT JOHN, H.: Bot. Jahrb. 99, 490-497
(1977a); - Phytologia 38, 75-98 (1977b); - Adansonia 18, 199-220
(1978a); - Bot. Jahrb. 100, 246-248 (1978b). - SALEH, N.A.M., EL-HADIDI,
M.N.: Biochem. Syst. Ecol. 5, 121-128 (1977). - SATSYPEROVA, I.F.:
Ontogenèse et formes vitales des espèces d'Heracleum L. de la flore
de l'U.R.S.S. In: Actes 2. Symp. Int. Ombellifères, 765-772. eds.
A.-M. CAUWET-MARC, J. CARBONNIER. Perpignan 1978. - SATTLER, R.,
SINGH, V.: Bot. J. Linn. Soc. 77, 141-156 (1978). - SCHILL, R.: Bot.
Jahrb. 99, 353-362 (1978). - SCHILL, R., JÄKEL, U.: Trop. Subtrop.
Pflanzenw. 22 (1978). - SCHILL, R., PFEIFFER, W.: Pollen Spores 19,
5-118 (1977). - SCHMID, R.: Bot. Jahrb. 100, 149-195 (1978a); - Bot.
Jahrb. 100, 196-204 (1978b). - SCHMIDT-VOGT, H.: Die Fichte, Bd. I,
647 S. Hamburg, Berlin: Parey 1977. - SCHNEIDER, L.E., MOORE, L.A.:
Brittonia 29, 88-99 (1977). - SCHOLZ, H.: Willdenowia 8, 5-16 (1977a);
- Willdenowia 8, 93-99 (1977b); - Willdenowia 8, 383-387 (1978a); -
Willdenowia 8, 341-350 (1978b). - SCHUYLER, A.E.: Brittonia 29, 129-
133 (1977). - SCOGIN, R.: Biochem. Syst. Ecol. 5, 265-267 (1977). -
SCOGIN, R., YOUNG, D.A., JONES, C.E.: Bull. Torrey Bot. Club 104,
155-159 (1977). - SCOTT, A.J.: Bot. J. Linn. Soc. 75, 357-374 (1977);
- Kew Bull. 33, 303-309 (1978a); - Kew Bull. 33, 311-329 (1978b);
- Bot. Jahrb. 100, 205-220 (1978c). - SEAVEY, S.R., RAVEN, P.H.:
Plant Syst. Evol. 127, 107-119 (1977a); - Plant Syst. Evol. 128, 195-
200 (1977b); - J. Biogeogr. 4, 55-59 (1977c). - SEAVEY, S.R., MAGILL,
R.E., RAVEN, P.H.: Ann. Mo. Bot. Gard. 64, 18-47 (1977). - SEBALD, O.:
Stuttg. Beitr. Naturkd. Ser. A 308 (1978). - SEIBERT, J.: Diss. Bot.
44 (1978). - SEIDENFADEN, G.: Dan. Bot. Ark. 31 (1977); - Dan. Bot.
Ark. 32 (1978). - SEMPLE, J.C.: J. Can. Bot. 55, 2503-2513 (1977); -
Can. J. Bot. 56 (10), 1274-1279 (1978a); - Ann. Mo. Bot. Gard. 65,
681-693 (1978b). - SEYBOLD, S.: Stuttg. Beitr. Naturkd. Ser. A 310,
1-31 (1978). - SHOUP, J.R., TSENG, C.C.: Am. J. Bot. 64, 461-463
(1977a); - Grana 16, 81-84 (1977b). - SIDDIQUI, S.A.: Bot. Jahrb. 100,
237-245 (1978). - SILVESTRE, S.: Lagascalia 7, 163-172 (1978). -
SIMPSON, B.B., NEFF, J.L., SEIGLER, D.: Nature (London) 267, 150-151
(1977). - SINGH, P., FENEMORE, P.G., DUGDALE, J.S., RUSSELL, G.B.:
Biochem. Syst. Ecol. 6, 103-106 (1978). - SKVARLA, J.J., TURNER, B.L.,
PATEL, V.C., TOMB, S.A.: Pollen morphology in the Compositae and in
morphologically related families, 141-248. In: The Biology and Chemistry
of the Compositae, eds. V.H. HEYWOOD, J.B. HARBORNE, B.L. TURNER, Vol. I.
London-New York: Academic Press 1977. - SKVARLA, J.J., RAVEN, P.H.,
CHISSOE, W.F., SHARP, M.: Pollen Spores 20, 5-143 (1978). - SLEUMER,
H.: Bot. Jahrb. 98, 151-175 (1977); - Notes R. Bot. Gard. Edinburgh
36, 251-258 (1978). - SMITH, D.M., GLENNIE, C.W., HARBORNE, J.B.,
WILLIAMS, C.A.: Biochem. Syst. Ecol. 5, 107-115 (1977). - SMITH, P.M.:
Chemical evidence in plant taxonomy, 19-38. In: Essays in Plant Taxo-
nomy, ed. H.E. STREET. London-New York: Academic Press 1978. - SMITH,
R.M.: Notes R. Bot. Gard. Edinburgh 36, 273-292 (1978). - SNEATH,
P.H.A., CHATER, A.O.: Information content of keys for identification,
79-96. In: Essays in Plant Taxonomy, ed. H.E. STREET. London-New York:
Academic Press 1978. - SNOGERUP, B.: Bot. Not. 130, 121-124 (1977). -
SODERSTROM, T.R., CALDERÓN, C.E.: Brittonia 30, 297-312 (1978a); -
Brittonia 30, 154-154 (1978b). - SOHMER, S.H.: Bull. Torrey Bot. Club
104, 111-126 (1977); - Brittonia 30, 256-264 (1978). - SOLBRIG, O.T.:

Chromosomal cytology and evolution in the family Compositae, 267-281.
In: The Biology and Chemistry of the Compositae, eds. V.H. HEYWOOD,
J.B. HARBORNE, B.L. TURNER, Vol. I. London-New York: Academic Press
1977. - SOLBRIG, O.T., ROLLINS, R.C.: Evolution 31, 265-281 (1977). -
SØRENSEN, N.A.: Polyacetylenes and conservatism of chemical charactes
in the Compositae, 385-409. In: The Biology and Chemistry of the Com-
positae, eds. V.H. HEYWOOD, J.B. HARBORNE, B.L. TURNER, Vol. I. London-
New York: Academic Press 1977. - SORENSEN, P.D., TOTTEN, C.E., PIATAK,
D.M.: Biochem. Syst. Ecol. 6, 109-111 (1978). - SPEIRS, D.C.: Q. Bull.
Alp. Gard. Soc. 46(2), 167-168 (1978). - SPETA, F.: Candollea 32, 133-
163 (1977). - SPORNE, K.R.: Plant Syst. Evol. Suppl. 1, 33-51 (1977).
- STACE, C.A.: Breeding systems, variation patterns and species delimitation, 57-78. In: Essays in Plant Taxonomy, ed. H.E. STREET. London-
New York: Academic Press 1978. - STACE, H.M.: Aust. J. Bot. 26, 287-
307 (1978). - STACE, H.M., FRIPP, Y.J.: Aust. J. Bot. 25, 299-314
(1977a); - Aust. J. Bot. 25, 315-323 (1977b); - Aust. J. Bot. 25,
325-336 (1977c). - STEARN, W.T.: Ann. Mus. Goulandris 4, 83-198 (1978).
- STEBBINS, G.L.: Developmental and comparative anatomy of the Compositae, 91-109. In: The Biology and Chemistry of the Compositae, eds.
V.H. HEYWOOD, J.B. HARBORNE, B.L. TURNER, Vol. I. London-New York:
Academic Press 1977. - STELLEMAN, P.: The possible role of insect
visits in pollination of reputedly anemophilous plant, exemplified by
Plantago lanceolata, and syrphid flies, 41-46. In: The Pollination
of Flowers by Insects, ed. A.J. RICHARDS. Linn. Soc. Symp. Ser. 6
(1978). - STERLING, C.: Bot. J. Linn. Soc. 74, 345-354 (1977a); -
Bot. J. Linn. Soc. 74, 63-69 (1977b). - STERN, W.L.: Bot. J. Linn.
Soc. 76, 83-113 (1978). - STEYERMARK, J.A., KIRKBRIDE, J.H.: Brittonia
29, 191-198 (1977). - STONE, B.C.: Gard. Bull. Singapore 29, 137-144
(1977). - STRAKA, H., ALBERS, F.: Bot. Jahrb. 99, 363-369 (1978). -
STRALEY, G.B.: Ann. Mo. Bot. Gard. 64, 381-424 (1977). - STROHTER,
J.L.: Tageteae - systematic review, 769-783. In: The Biology and
Chemistry of the Compositae, eds. V.H. HEYWOOD, J.B. HARBORNE, B.L.
TURNER, Vol. II. London-New York: Academic Press 1977a; - Madroño 24,
129-139 (1977b); - Sida 7, 369-374 (1978). - STUCKY, J.: Am. J. Bot.
65, 125-133 (1978). - STUESSY, T.F.: Heliantheae - systematic review,
621-671. In: The Biology and Chemistry of the Compositae, eds. V.H.
HEYWOOD, J.B. HARBORNE, B.L. TURNER, Vol. II. London-New York: Academic
Press 1977a; - Fieldiana Bot. 38(6), 63-70 (1977b); - Rhodora 79,
1970-202 (1977c). - STUESSY, T.F., ESTABROOK, G.F.: Syst. Bot. 3,
145 (1978). - STYER, C.H.: J. Arnold Arbor. Harv. Univ. 58, 109-145
(1977). - SUSSMAN, R.W., RAVEN, P.H.: Science 200, 731-736 (1978). -
SWAIN, T., WILLIAMS, C.A.: Heliantheae - chemical review, 673-697. In:
The Biology and Chemistry of the Compositae, eds. V.H. HEYWOOD, J.B.
HARBORNE, B.L. TURNER, Vol. II. London-New York: Academic Press 1977.

TANAI, T.: J. Jpn. Bot. 53, 65-83 (1978). - TATEISHI, Y., OHASHI, H.:
Bull. Natl. Sci. Mus. Ser. B (Bot.) 3, 71-84 (1977). - TATEOKA, T.:
Bull. Natl. Sci. Mus. Ser. B (Bot.) 4, 1-3 (1978). - TATSUNO, A.,
SCOGIN, R.: Aliso 9, 185-188 (1978). - THOMPSON, M.F.: Bothalia 12,
429-435 (1978). - THORNE, R.F.: Plant Syst. Evol. Suppl. 1, 299-319
(1977a); - Gard. Bull. Singapore 29, 183-189 (1977b); - Notes R. Bot.
Gard. Edinburgh 36, 297-315 (1978). - TIFFNEY, B.H.: Nature (London)
265, 136-137 (1977). - TITZ, W.: Bot. Jahrb. 100, 110-139 (1978). -
TÖLKEN, H.R.: Contrib. Bolus Herbarium 8, 1-567 (1977). - TOMLINSON,
P.B.: Ann. Mo. Bot. Gard. 64, 685-693 (1977); - N. Z. J. Bot. 16,
299-309 (1978). - TOMLINSON, P.B., RAVEN, P.H.: Ann. Mo. Bot. Gard.
64, 657-658 (1977). - TOMB, A.S.: Lactuceae - systematic review,
1067-1079. In: The Biology and Chemistry of the Compositae, eds. V.H.
HEYWOOD, J.B. HARBORNE, B.L. TURNER, Vol. II. London-New York: Academic
Press 1977. - TOMB, A.S., CHAMBERS, K.L., KYHOS, D.W., POWELL, A.M.,
RAVEN, P.H.: Am. J. Bot. 65, 717-721 (1978). - TOWNER, H.F.: Ann. Mo.

Bot. Gard. 64, 48-120 (1977). - TSENG, C.C., SHOUP, J.R.: Am. J.
Bot. 65, 384-394 (1978). - TURALA-SZYBOWSKA, K.: Acta Biol. Cracov.
20, 1-9 (1977). - TURNER, B.L.: Fossil history and geography, 21-39.
In: The Biology and Chemistry of the Compositae, eds. V.H. HEYWOOD,
J.B. HARBORNE, B.L. TURNER, Vol. I. London-New York: Academic Press
1977a; - Am. J. Bot. 64, 78-80 (1977b); - Summary of the biology of
the Compositae, 1105-1118. In: The Biology and Chemistry of the Com-
positae, eds. V.H. HEYWOOD, J.B. HARBORNE, B.L. TURNER, Vol. II.
London-New York: Academic Press 1977c; - Ann. Mo. Bot. Gard. 64,
235-242 (1977d); - Madroño 25, 46-52 (1978). - TURNER, B.L., POWELL,
A.M.: Helenieae - systematic review, 699-737. In: The Biology and
Chemistry of the Compositae, eds. V.H. HEYWOOD, J.B. HARBORNE, B.L.
TURNER, Vol. II. London-New York: Academic Press 1977a; - Madroño 24,
1-5 (1977b).

UHL, C.H.: Rhodora 79, 95-114 (1977). - UHL, N.W.: Gentes Herbarum 11,
246-267 (1978a); - Gentes Herbarum 11, 268-283 (1978b); - Gentes Herbarum 11, 200-211 (1978c). - UHL, N.W., MOORE, H.E.: Biotropica 9,
170-190 (1977). - URBANSKA-WORYTKIEWICZ, K., LANDOLT, E.: Ber. Geobot.
Inst. Eidg. Tech. Hochsch. Stift. Rübel Zürich 45, 30-53 (1978).

VALADON, L.R.G.: Arctoteae and Calenduleae - chemical review, 989-998.
In: The Biology and Chemistry of the Compositae, eds. V.H. HEYWOOD,
J.B. HARBORNE, B.L. TURNER, Vol. II. London-New York: Academic Press
1977. - VALENTINE, D.H.: Ecological criteria in plant taxonomy, 1-18.
In: Essays in Plant Taxonomy, ed. H.E. STREET, London-New York: Academic Press 1978. - VAN CAMPO, M., NILSSON, S., LEEUWENBERG, A.J.M.:
Grana 18, 5-14 (1979). - VAN DER PIJL, L.: Reproductive integration
and sexual disharmony in floral functions, 79-88. In: The Pollination
of Flowers by Insects, ed. A.J. RICHARDS. Linn. Soc. Symp. Ser. 6
(1978). - VAN DER PLUYM, A., HIDEUX, M.J.: Plant Syst. Evol. 127,
55-85 (1977). - VAN STEENIS, C.G.G.J.: Gard. Bull. 29, 103-126 (1977);
- Notes R. Bot. Gard. Edinburgh 36, 317-323 (1978). - VAN VLIET,
G.J.C.M.: Acta Bot. Neerl. 27, 273-285 (1978). - VERDCOURT, B.,
HALLIDAY, P.: Kew Bull. 33, 191-227 (1978). - VERLAQUE, R.: Bull. Soc.
Bot. Fr. 124, 515-527 (1977a); - Bull. Soc. Bot. Fr. 124, 475-482
(1977c). - VIANO, J.: Candollea 33, 43-88 (1978a); - Candollea 33,
209-267 (1978b). - VIJAYARAGHAVAN, M.R., DHAR, U.: Bot. Not. 131, 127-
138 (1978). - VOGEL, S.: Apidologie 8, 321-335 (1977); - Evolutionary
shifts from reward to deception in pollen flowers, 89-96. In: The Pollination of Flowers by Insects, ed. A.J. RICHARDS, Linn. Soc. Symp.
Ser. 6 (1978a); - Flora 167, 329-366 (1978b); - Flora 167, 367-398
(1978c).

WAGENITZ, G.: Plant Syst. Evol. Suppl. 1, 375-395 (1977). - WAGNER,
H.: Cynareae - chemical review, 1017-1038. In: The Biology and Chemistry of the Compositae, eds. V.H. HEYWOOD, J.B. HARBORNE, B.L. TURNER,
Vol. II. London-New York: Academic Press 1977a; - Pharmaceutical and
economic uses of the Compositae, 411-433. In: The Biology and Chemistry
of the Compositae, eds. V.H. HEYWOOD, J.B. HARBORNE, B.L. TURNER,
Vol. I. London-New York: Academic Press 1977b. - WAGNER, P.: Bot. Not.
130, 383-402 (1977). - WATSON, T.J.: Brittonia 29, 199-216 (1977). -
WATSON, L., PATE, J.S., GUNNING, B.E.S.: Bot. J. Linn. Soc. 74, 123-130
(1977). - WCISLO, H.: Acta Biol. Cracav. 20, 154-165 (1977). - WEBER,
A.: Plant Syst. Evol. 127, 201-216 (1977); - Notes R. Bot. Gard.
Edinburgh 36, 355-368 (1978). - WEBERLING, F.: Beitr. Biol. Pflanz.
53, 61-89 (1977); - Bot. Jahrb. 99, 188-211 (1978). - WEBERLING, F.,
UHLARZ, H.: Plant Syst. Evol. 127, 217-242 (1977). - WEEDIN, J.F.,
POWELL, A.M.: Am. J. Bot. 65, 531-537 (1978). - WEIMARCK, G.: B. chro-
mosomes in Heracleum sphondylium s. lat. in Europe. In: Actes 2. Symp.
Int. Ombellifères, 293-303. eds. A.-M. CAUWET-MARC, J. CARBONNIER.
Perpignan 1978. - WELCH, H.J.: Taxon 27, 187-190 (1978). - WELLER, S.G.:

Syst. Bot. 3, 115-126 (1978). - WENDELBO, P.: Iran. J. Bot. 1, 113-
115 (1977). - WHALEN, M.: Madroño 24, 112-122 (1977). - WHEELER HAINES,
R., LYE, K.A.: Bot. Not. 130, 235-240 (1977). - WHITE, F.: Gard. Bull.
Singapore 29, 57-71 (1977). - WHITHAM, T.G.: Science 197, 593-595
(1977). - WIEHLER, H.: Selbyana 5, 11-60 (1978). - WILKEN, D.H.:
Am. J. Bot. 65, 896-901 (1978). - WILKINSON, H.P.: Kew Bull 32, 347-
360 (1978). - WILLIAMS, C.A., HARBORNE, J.B.: Biochem. Syst. Ecol.
5, 221-229 (1977a); - Biochem. Syst. Ecol. 5, 45-51 (1977b); - WILLIAMS,
M.C., BARNEBY, R.C.: Brittonia 29, 237-331 (1977a); - Brittonia 29,
310-326 (1977b). - WILLIAMS, N.H.: Selbyana 2, 310-322 (1978). -
WILLING, B., WILLING, E.: Willdenowia Beih. 11, 1-325 (1977). -
WILLSON, M.F., PRICE, P.W.: Evolution 31, 495-511 (1977). - WILSON,
A.C., CARLSON, S.S., WHITE, T.J.: Annu. Rev. Biochem. 46, 573-639
(1977). - WOODELL, S.R.J.: Directionality in bumblebees in relation
to environmental factors, 31-39. In: The Pollination of Flowers by
Insects, ed. A.J. RICHARDS. Linn. Soc. Symp. Ser. 6 (1978). - WOODELL,
S.R.J., MATTSON, O., PHILIPP, M.: Bot. Tidskr. 72, 15-29 (1977).

YEO, P.F.: Euphrasia: a taxonomically critical group with normal sexual
reproduction, 143-162. In: Essays in Plant Taxonomy, ed. H.E. STREET.
London-New York: Academic Press 1978.

ZABINSKA, D.: Acta Biol. Cracov. Bot. 20 (1-2), 133-146 (1977). -
ZANDONELLA, P., LECOCQ, M.: Pollen Spores 19, 119-141 (1977). - ZANONI,
T.A.: Phytologia 38, 433-454 (1978). - ZAVARIN, E., CRITCHFIELD, W.B.,
SNAJBERK, K.: Biochem. Syst. Ecol. 6, 267-278 (1978a). - ZAVARIN, E.,
SNAJBERK, K., LEE, C.-J.: Biochem. Syst. Ecol. 6, 177-184 (1978b).

Professor Dr. JÜRKE GRAU
Institut für Systematische Botanik
der Universität München
Menzinger Straße 67
D 8000 München 19

II. Paläobotanik

Bericht über die Jahre 1975–1978

Von Walter Jung und Friedemann Schaarschmidt

1. Allgemeines

Auf zwei größere Werke, die sich mit paläozoischen Pflanzen im Allgemeinen befassen, sei besonders hingewiesen. Es sind dies einmal "Die
Floren des Erdaltertums" von W. und R. REMY (1977), eine Weiterent-
wicklung der beiden früher erschienenen Bändchen über das paralische
und limnische Karbon. Es ist eine nützliche Einführung, die sich jedoch - wie aus dem Titel nicht hervorgeht - im wesentlichen auf Mitteleuropa beschränkt. Einem ganz anderen Zweck dient der von BOERSMA
und BROEKMEYER (1978) herausgegebene Index of figured plant megafos-
sils, der ein Verzeichnis aller im Zeitraum von 1971-1975 in Publikationen abgebildeten Karbonpflanzen enthält.

2. Schizophyta

Als Folge der Erkenntnis, daß Cyanophyceen bei der Sedimentbildung
eine bedeutende Rolle spielen, sind in den letzten Jahren zahlreiche
Arbeiten zu diesem Problem erschienen. Als Beispiel sollen die Untersuchungen von KEUPP (1977) am Solnhofener Plattenkalk (Jura) erwähnt
werden, dessen Sedimentation regelmäßig wechselnd durch Coccolitho-
phoriden und fädige Cyanophyceen erfolgt sein soll. Diesem Wechsel
liegen Zyklen zwischen normal marinen und übersalzenen Bedingungen zu-
grunde.

Einen Überblick über die Geschichte der Stromatolithen-Forschung und
den Wandel der Ansichten über ihre Entstehung bringt MONTY (in FLÜGEL,
1977). Während früher angenommen wurde, daß sie - sei es im Süßwasser,
der marinen Gezeitenzone oder im Flachmeer - wegen der Abhängigkeit
der Cyanophyceen von der Photosynthese nur in flachem Wasser entstanden sein könnten, haben Untersuchungen an rezenten Cyanophyceen ge-
zeigt, daß diese bis zu einer Meerestiefe von 1000 m lebensfähig sind.
Dabei ermöglicht bei geringem Lichtangebot das Phycoerythrin eine Absorption der blauen Reststrahlung; bei völliger Dunkelheit finden
chemosynthetische Vorgänge statt. Neues Licht in die Entstehung von
stromatolithischen Strukturen bringen Beobachtungen an rezenten Cyano-
bakterien durch KRUMBEIN und COHEN (in FLÜGEL, 1977).

Besonders faszinierend sind die Fortschritte in den Kenntnissen und
Vorstellungen über das frühe präkambrische Leben auf der Erde, das
über lange Zeit hinweg ausschließlich aus Schizophyten bestand. Die
meisten Arbeiten über dieses Gebiet wurden von BARGHOORN, SCHOPF und
ihren Mitarbeitern veröffentlicht. Das sich daraus ergebende Bild

---

*Abschnitt 1-6 von F. Schaarschmidt, Abschnitt 7 von W. Jung

stellt SCHOPF (1975a) in einer Übersicht zusammen. Danach sind die
ersten lebenden Systeme vor mehr als 3300 Mill. Jahren entstanden und
haben bereits vor mehr als 3000 Mill. Jahren das photo-autotrophe Stadium in Bakterien- und Cyanophyceen-ähnlichen Formen erreicht. Zeugen
hierfür sind kugelförmige *(Archaeosphaeroides)* und an Bakterien erinnern-
de Reste aus der Overwacht- und Fig-Tree-Gruppe in Transvaal (3200
Mill. Jahre).

Ähnlich früh sind vermutlich auch die ersten *Oszillatoria*-ähnlichen
Zellfäden entstanden, wie sie aus der Balawayan-Gruppe in Rhodesien
isoliert wurden. Nach der anfangs offensichtlich recht eintönigen
Cyanophyceen-Vegetation muß im mittleren Präkambrium (vor ca. 1000
Mill. Jahren) - mit dem Auftreten des ersten durch die Organismen
selbst produzierten freien Sauerstoffs - eine reiche Entfaltung dieser Schizophyten-Vegetation eingetreten sein. Gleichzeitig müssen neue
physiologische Mechanismen vor allem zum Schutz des sauerstoffempfind-
lichen Nitrogenase-Enzym-Komplexes und zur Einführung der O_2-Atmung
entwickelt worden sein. Für den gleichen Zeitraum (1300-900 Mill.
Jahre) wird die Entstehung der (anfangs haploiden) eukaryotischen Zellen angenommen. Als Hinweise hierauf werden etwas unklare Reste aus
dem Beck Spring Dolomite in Californien (1300 Mill. Jahre) und sichere
planktontische Algen aus der Bitter-Springs-Formation in Zentralaustra-
lien angeführt. Es ist jedoch kaum damit zu rechnen, daß der Zellinhalt
direkt sichtbar gemacht werden kann (GOLUBIC und BARGHOORN in FLÜGEL,
1977). In den folgenden 200 Mill. Jahren hätten sich dann Sexualität,
Meiose und Generationswechsel herausgebildet.

Zu diesen recht klaren und mit den geophysikalischen Erkenntnissen
übereinstimmenden Vorstellungen passen nur schwer die an Actinomyceten
(Witwateromyces) und Flechten *(Thuchomyces)* erinnernden Reste, die von
HALLBAUER (1975) und HALLBAUER et al. (1977) aus den 2500 Mill. Jahre
alten Witwatersrand-Konglomeraten Südafrikas beschrieben werden. Aus
den als Thucholit bezeichnenden Kohlebändern lassen sich immerhin rezenten Flechten täuschend ähnlich sehende säulenförmige Gebilde isolieren. Schwierig einzuordnen sind auch die von PFLUG (1976a) als
Ramsaysphaera beschriebenen kugelförmigen Gebilde, die sich aus feinsten,
0,3-3 μm großen Tröpfchen und bäumchenartig verzweigten Systemen auf-
bauen sollen. Sie werden als in flachem Wasser lebende Schleimkolonien
gedeutet.

3. Phycophyta

Im Rahmen dieses Vierjahresberichtes war es nicht möglich, die umfangreiche Literatur über fossile Algen vollständig zu berücksichtigen.
Es sei hingewiesen auf die Vorträge, die anläßlich des "First International Symposium on Fossil Algae" in Erlangen gehalten und von
FLÜGEL (1977) herausgegeben worden sind. Darin sind wichtige Arbeiten
über den derzeitigen Kenntnisstand der verschiedenen Algengruppen enthalten.

Hier können nur einige Probleme aufgegriffen werden. So untersuchte
FLAJS (1977) den mineralischen Bau der Skelette der verschiedenen Kalk-
algengruppen mit dem Rasterelektronenmikroskop und gibt eine Übersicht
über die Bautypen. Dabei kommt er zu dem Schluß, daß der Ordnungsgrad
der Mineralbildung von der Organisationshöhe der Alge abhängig ist.
Die Einordnung der vom Ordovizium bis ins Alttertiär vorkommende Sole-
noporaceen bei den Corallinaceen ist seit jeher unsicher, weil sie er-
heblich größere Zellen besaßen. Relikte der Primärstrukturen und die
Mg-Verteilung sprechen nach FLAJS für enge Beziehungen beider Gruppen.

Die Zuordnung des im Devon vorkommenden *Sycidium* zu den Charophyten
ist nach wie vor umstritten. Es ähnelt diesen zwar in der Form der
Oogonien, besitzt jedoch keine spiraligen Eindrücke, sondern gefel-
derte Längsstreifen. LANGER (1976) nimmt daher höchstens eine gemeinsame Wurzel beider an. Die von ihm beobachteten Poren werden von
MUSTAFA (1978) als Lösungserscheinungen gedeutet. MUSTAFA hält die
Sycidien für eindeutige Charophyten, die im Süß- bis Brackwasser leb-
ten. Die häufig zu beobachtende Massenentwicklung von Characeen in
der Erdgeschichte führen HILTERMANN und MÄDLER (1977) auf ökologische
Grenzsituationen zurück.

Von den für die Stratigraphie wichtigen sog. Hystrichosphären ist in
den letzten Jahrzehnten eine unübersehbare Zahl von Taxa beschrieben
worden. Nachdem ein großer Teil der postpaläozoischen Formen als Dinoflagellatenzysten erkannt worden ist, ist eine moderne Revision an
auserwählt gut erhaltenem Material dringend notwendig, wie es DAVEY
und VERDIER (1976) für die Kreide begonnen haben. Für das gleichalte
Palaeoperidinium pyrophorum wurde bisher angenommen, daß es sich um eine
Theke handeln könne. GOCHT und NETZEL (1976) weisen sehr anschaulich
nach, daß das Relief genau "umgekehrt" ist und daß somit sicher eine
proximate Zyste vorliegt. Diese Erscheinung wird in Anlehnung an ältere Autoren auf genetische Steuerung zurückgeführt. Mit Hilfe von
Dinoflagellatenzysten ist es HARLAN (1977) gelungen, verschieden warme
Wassermassen im Jungquartär des Schelfgebietes vor der englischen Küste und damit Veränderungen in den Strömungsverhältnissen nachzuweisen.

Die ebenfalls als Planktonorganismen gedeuteten Tasmaniten besitzen
dicke, von Poren oder Blasen durchsetzte Wände, in denen JUX (1975,
1977) ehemals mit Gas oder Öl gefüllte hydrostatische Apparate sieht,
die die Sinkgeschwindigkeit der kugeligen Organismen verändern konnten.

4. Mycophyta

Obgleich Pilzreste insbesondere von parasitisch lebenden Arten häufig
in fossilen Pflanzenresten vorkommen, werden sie bedauerlicherweise
nur selten näher untersucht. Daß auch botanisch interessante Ergebnisse zu erzielen sind, zeigen STIDD und COSENTINO (1975), die aus
dem amerikanischen Oberkarbon einen Pilz beschreiben, der dem an rezenten Cruciferen parasitierenden Weißrost *Albugo* (Oomycetes) ähnelt.
Er wächst in den Samenanlagen von *Nucellangium* - einer Cordaite - und
ruft hier die gleichen Krankheitserscheinungen hervor wie sein rezenter Verwandter. Dies weist darauf hin, daß auch die Paläobotanik ihren
Beitrag zum Problem Coevolution von Wirt und Parasit liefern könnte.
Mehr zweckgebunden ist der mycologische Beitrag von STRAUS (1977b),
dem es gelungen ist, außer tierischen auch pilzliche Gallen für die
Bestimmung jungtertiärer Pflanzenreste zu verwenden, indem er die entsprechenden Schadbilder an Blättern vergleicht. So hat er *Gymnosporangium*
(Uredinales) auf *Crataegus* und cf. *Taphrina* (Exoascaceae) auf *Parrotia* ge-
funden.

Schließlich hat JANSONIUS (1976) im kanadischen Paläogen in Deltasedi-
menten die so lange vernachlässigten und doch äußerst widerstandsfähi-
gen Pilzsporen für stratigraphische Zwecke verwendet.

5. Bryophyta

Moose sind nur unter ganz besonders günstigen Bedingungen erhaltungs-
fähig. Solche liegen offenbar in der Molteno-Formation (Obere Trias)
von Südafrika vor, aus der ANDERSON (1976) über den Fund von 49 Arten
berichtet. Drei von ihnen beschreibt sie näher: Zwei *Marchantia*-ähnli-
che Lebermoose und das Laubmoos *Muscites guescilini*.

6. Pteridophyta

a) Psilophytatae

Nachdem mehrere ehemals zu den Psilophyten gerechnete Pflanzen als
höher organisiert erkannt worden sind *(Drepanophycus* als Lycopodiale,
Pseudosporochnus als Cladoxylale und *Protopteridium* als Progymnosperme)
und uns anatomische Untersuchungen die verbleibenden Gattungen besser
kennen gelehrt haben, versucht BANKS (1975b) eine neue Gliederung und
gibt einen Überblick über den Kenntnisstand. Danach enthält die nun
Tracheophyta genannte Abteilung drei Unterabteilungen mit je einer
Familie.

Die ursprünglichste Familie sind die Rhyniaceen, die damit nach wie
vor die primitivsten Gefäß-Landpflanzen darstellen. Sie zeichnen sich
durch einfachen dichotomen Bau, endständige Sporangien und eine centrarche Stele aus. Zu den Rhyniaceen gehören als wichtigste Gattungen
Rhynia, Horneophyton und *Cooksonia*, vielleicht aber auch *Taeniocrada, Hick-
lingia* und *Yarravia*.

Bei den Zosterophyllaceen sind die Sporangien ährenartig seitenstän-
dig. Außer *Zosterophyllum* und *Gosslingia* gehört hierher auch *Sawdonia
ornata*, das frühere *Psilophyton princeps* var. *ornatum*. Die bei ihm gefundene exarche Stele weist auf Beziehungen zu den Lycophyten hin. Diese
Annahme wird auch durch Untersuchungen von MUSTAFA (1978) unterstützt,
der bei einer *Euthursophyton hambachense* genannten Pflanze im gesamten
Sproßsystem eine runde Stele beschreibt, während sie bei dem sehr ähn-
lichen *Thursophyton* (früher *Asteroxylon*) *elberfeldense* im unteren, beschupp-
ten Teil ähnlich wie bei *Asteroxylon mackiei* aus Rhynie zerklüftet ist.
Diese beiden Gattungen werden daher als primitive Lycophyten angesehen, während *Euthursophyton* noch zu den Zosterophyllaceen gestellt wird.

Die dritte Familie, die Trimerophytaceen, zeichnet sich durch reiche
dichotome und "trichotome" Verzweigung sowie eine centrarche Stele
aus. Die Sporangien stehen paarig in dichten Ständen. Die Familie ver-
mittelt zwischen den Rhyniaceen einerseits und den farnartigen Gruppen
der Progymnospermen, Cladoxylales und Coenopteriden sowie den Sphenopsiden andererseits. Zu ihr wird u.a. außer der namengebenden Gattung
vor allem das bereits seit 1859 bekannte *Psilophyton* gestellt.

Einige weitere Gruppen wie die Sciadophytaceen, die Barinophytaceen
und die Platyphyllales sind weniger gut untersucht und lassen sich
nicht in dieses Schema einordnen. Sie dürften nach Ansicht von BANKS
(1975b) auch keine Beziehungen zu den genannten drei Familien haben.
Von anderen Autoren wird diese Gliederung im wesentlichen übernommen,
so auch von W. und R. REMY (1977), die jedoch für die Trimerophytina
den Namen Psilophytina und als Oberbegriff für die drei Familien
Psilophyta weiterverwenden.

Wie schon von früheren Bearbeitern erwogen, hält es GENSEL (1977) -
trotz des Fehlens fossiler Bindeglieder - für möglich, daß die rezenten Psilotaceen Beziehungen zu einer dieser drei "Psilophyten"-Gruppen
haben. Insbesondere werden Vergleiche zwischen den noch nicht in Wedel
und Achsen differenzierten Sproßsystemen der frühen Farne und der Verzweigung von *Psilotum* und *Tmesipteris* angestellt.

Bemerkenswert erscheinen noch eine Reihe spezieller Arbeiten. So wird von
MUSTAFA (1978) *Sawdonia ornata* erstmals aus dem deutschen Devon angegeben.
BANKS et al. (1975) haben an pyritisiertem Material die Morphologie und Anatomie von *Psilophyton dawsonii* aus Gaspé (Kanada) eingehend untersucht. Die Verzweigung ist weitgehend dichotom, z.T. auch dreigabelig mit einem abortierten
Zweig. Die Protostele ist kaum höher entwickelt als bei *Rhynia*, die Rinde dagegen ist in einen inneren dünnwandigen und einen äußeren dickwandigen Bereich
differenziert. GENSEL (1976) beschreibt von Gaspé (Kanada) eine Pflanze als
Renalia hueberi, die *Cooksonia* ähnelt, sich jedoch pseudomonopodial verzweigt
und Sporangien mit einer an *Zosterophyllum* erinnernden Dehiszenz besitzt. Sie
dürfte zwischen beiden Gattungen vermitteln. Über einen Dehiszenz-Mechanismus
bei *Rhynia* berichtet REMY (1978): Die gesamte Sporangienwand besteht aus radial
verstärkten Zellwänden und konnte sich offenbar zusammenziehen.

Mit der alten Frage nach dem Alter der Landpflanzen beschäftigen sich
BANKS (1975a) und PRATT et al. (1978). Nach wie vor gilt als älteste
Land-Gefäßpflanze die von LANG (1937) beschriebene *Cooksonia* aus dem
Pridol (oberstes Silur) von Wales. In der folgenden Zeit, bis zum
Siegen (mittleres Unterdevon), fand eine rasche Entwicklung innerhalb
der Psilophyten (Rhyniophytina, Zosterophyllitina, Trimerophytina)
statt, und auch einfache Lycophyten bildeten sich bereits heraus. Als
Ausgangsbasis für die Evolution der frühen Gefäßpflanzen werden die
im Silur reich und hochorganisiert entwickelten Grünalgen (Dasyclad-
aceen, Characeen) angesehen. Als Stimulans können glacio-eustatische
Meeresspiegelschwankungen an der Wende Ordovizium/Silur angesehen werden bei gleichzeitigem Anstieg des O_2-Gehaltes der Luft und dem Auf-
bau eines UV-Schildes.

Das 40 Mill. Jahre während Silur war für die Grünalgen eine Zeit des
"Experimentierens", und verschiedene einzelne gefäßpflanzenartige
Merkmale sind offenbar unabhängig voneinander in verschiedenen Gruppen
entstanden. Wenn sie auch als Hinweise für eine terrestrische Lebensweise angesehen werden können, so haben doch die meisten dieser "Versuche" mit Sicherheit nicht zu echten Gefäßpflanzen geführt. Deren
Existenz kann daher nicht durch tracheidenartige Reste und Cutikular-
fetzen, wie sie PRATT et al. (1978) im amerikanischen Llandovery gefunden haben, und auch nicht durch das erste Auftreten trileter Sporen
belegt werden.

Von solchen landbewohnenden Thallophyten berichten verschiedene neuere
Arbeiten. So deuten CHALONER et al. (1974) *Spongiophyton* als dorsiventral wachsenden Landthallophyten mit extrem dicker Cuticula. NIKLAS
(1976a,b,c,d) führte an mehreren Resten solcher Grenzgruppen chemo-
taxonomische Untersuchungen durch. Dabei ergab sich, daß sich Hinweise
auf terrestrische Lebensweise sogar bei Phaeophyceen finden können.
So besitzt *Protosalvinia* Cutin und in den Vermehrungsorganen Sporopol-
lenin. Im Chemismus ähnlich *Protosalvinia* (wie auch der nicht als Phae-
ophycee gedeuteten *Taeniocrada*) ist der schon lange als Braunalge ange-
sehene *Prototaxites* (Besitz von Cutin und Suberin). Damit wird die bereits früher von LANG vertretene Ansicht, daß *Prototaxites* eine Land-
pflanze sei, wieder gestützt. Zu einer ausgestorbenen Linie der Grünalgen wird dagegen von NIKLAS (1976b) auf Grund gaschromatischer und
massenspektroskopischer Untersuchungen *Parka decipiens* aus dem Unteren
Old Red von Schottland gestellt.

b) Lycopodiatae

Daß die Lycophyten von den drei heute noch lebenden Pteriodophyten-
linien am weitesten - bis ins Unterdevon - zurückreichen, kann als
gesichert gelten. Daß aber im Unterdevon, zusammen mit *Drepanophycus*
und *Protolepidodendron*, auch bereits echte ligulate Lepidodendren vorkommen sollen, wie sie LEJAL-NICOL (1975) aus Libyen beschreibt,
klingt sehr unglaubhaft.

Im Gegensatz zu den anderen beiden Gruppen der Pteridophyten, bei denen es wie bei den Sphenopsiden noch Rätsel über ihren Ursprung gibt
oder die sich wie die Farne in letzter Zeit als eine ungeahnt vielge-
staltige Gruppe herausgestellt haben, erscheinen uns die Lycophyten
heute nach wie vor verhältnismäßig geschlossen und sowohl morphologisch wie anatomisch durch den Besitz einer exarchen Stele gut definiert. Auch ihre Ableitung von den Zosterophyllen, wie sie von BANKS
(1975b) erneut in Betracht gezogen wurde, ist einleuchtend. Wenn somit auch keine revolutionären Neuigkeiten zu berichten sind, so ist
doch manch Neues über die Bauverhältnisse der Lycophyten beigetragen
worden. So hat FAIRON-DEMARET (1978) festgestellt, daß bei *Drepanophycus*
neben isotomer Dichotomie auch Übergipfelung vorkommt, die sich auch
in der Stelenverzweigung ausdrückt. *Protolepidodendron scharianum* besitzt
nach MUSTAFA (1975) und SCHAARSCHMIDT (1976) ebenfalls eine exarche
Actinostele und nicht, wie früher von KRÄUSEL und WEYLAND beschrieben, eine mesarche Stele. Nach SCHAARSCHMIDT stehen die Blätter zu
acht in dekussierten Wirteln. Als Lycophyt wird von MUSTAFA auch
Brandenbergia angesehen, deren Blätter zweifach gegabelt sind.

Aus dem sibirischen Devon wird von JURINA und LEMOIGNE (1975) die Anatomie der verkieselten Zweige von *Lepidodendropsis kazachstanica*, einer
vermutlich baumförmigen, eligulaten Lepidophyte, beschrieben. Die
jüngeren Zweige besitzen exarche Protostelen, die älteren dagegen
Siphonostelen, ähnlich wie *Lepidodendron*. Die äußere Rinde besteht aus
dickwandigen Zellen und hatte offensichtlich statische Funktion, während die innere Rinde aus sternförmigen Zellen besteht, die ein typisches Aerenchym bilden. Daraus schließen die Autoren, daß die Pflanze
wie ihre karbonischen Verwandten sumpfige Standorte besiedelte.

> Eigenartig gestielte, dem Blattstiel aufsitzende Sporangien und gezähnte Sporo-
> phylle besitzt *Barsostrobus famennensis*, der von FAIRON-DEMARET (1977) aus dem
> belgischen Oberdevon bekannt gemacht wird. In den nordamerikanischen coal balls
> kommen zwei kleine Lycophyten vor: *Natalia sinuata* besteht nach BAXTER (1978)
> aus einem beblätterten Sproß, der nach unten in ein dünnes Rhizomorph ausläuft.
> Eigenartigerweise soll auch im Sproß eine endarche Siphonostele mit Sekundär-
> holz vorkommen, wie sie sonst bei Lycophyten nur aus Stigmarien bekannt ist.
> Die wenige Millimeter dicken Achsen von *Paralycopodites minutissimum*, die eine
> Siphonostele enthalten, werden von MOREY und MOREY (1977) als krautige ligulate
> Lycophyte gedeutet.

Eine bemerkenswerte anatomische Neuigkeit berichten CHALONER und COL-
LINSON von *Sigillaria*: Ihre Stammoberfläche ist so dicht mit Stomata
besetzt, wie die Blattunterseite von Dikotylen. Sie trug daher offensichtlich wesentlich zur Photosynthese bei und ermöglichte damit dem
massiven, unverzweigten Stamm die nötige Biomasse zu produzieren.

Ein recht seltsamer Lepidophyten-Zapfen ist der in den 20er Jahren
von BODE aus dem schlesischen Karbon beschriebene *Sporangiostrobus*. Nach
W. und R. REMY (1975a) sind die Sporophylle mit der Sporangienwand
verwachsen und fehlen daher scheinbar. Anhand von 2 Bruchstücken aus
dem Oberstephan Südspaniens rekonstruieren sie eine nicht oder wenig
verzweigte sukkulente Pflanze mit endständigen durchwachsenen Sporen-

zapfen. Unter reichem Material vom selben Fundort fanden WAGNER und
SPINNER (1976) in *Bodeodendron* die zugehörigen Stämme. Sie ähneln
Lepidodendron, besitzen jedoch kein Parichnossystem und vermutlich kei-
ne Ligula.

Viele botanische Termini werden in der Paläobotanik recht großzügig
verwendet, ohne daß man sich immer an die strenge Definition hält.
Dies hängt mit der oft unzulänglichen Erhaltung der Fossilien zusammen, die der Untersuchung Grenzen setzt. Es kann daher nur von Nutzen
sein, wenn sich ein Botaniker z.B. mit dem Problem der Dichotomie beschäftigt. So hatte SIEGERT festgestellt, daß es bei rezenten Selagi-
nellen keine echte Dichotomie gibt, sondern daß sie sich poly-hetero-
clad verzweigen: Das 1. Seitenzweigpaar wächst zur Laubtrieb-"Gabel"
aus, das zweite zum ersten gekreuzt stehende wird zum Wurzelträger.
Ganz ähnlich deutet er (1978) die Verzweigung bei *Stigmaria*, dem Basal-
organ der Lepidodendren. Seine Ansicht wird gestützt durch das gele-
gentliche Auftreten von "tap roots" in Stigmarien-Gabelungen. Während
die oberkarbonischen Stigmarien reich verzweigte Basalorgane waren,
besteht nach JENNINGS (1975b) die im Mississippian (Unterkarbon) von
Nordamerika vorkommende *Protostigmaria* lediglich aus 4 Auswüchsen und
erinnert damit an die Stammbasis von *Isoetes*, die von MÄGDEFRAU als re-
duzierte Stigmarie gedeutet worden ist.

Im Mesozoikum spielen die Lycophyten keine Rolle mehr. Lediglich *Pleuromeia* ist in der Trias, z.B. im deutschen Buntsandstein, weit verbreitet. KRASSILOV und ZAKHAROV (1975) halten eine ebenfalls in der unteren Trias Sibiriens vorkommende Art für fast identisch mit der mittel-
europäischen, deuten sie jedoch nicht wie bisher geschehen als Xero-
phyt, sondern als Mangrove-Pflanze. Wegen der Seltenheit der Gruppe
im Mesozoikum ist es bemerkenswert, daß KRASSILOV (1978b) noch 2 krau-
tige ligulate Lycopodien aus dem Mesozoikum Sibiriens beschreibt:
Lycopodites macrostomus und *Synlycostrobus tyrmensis*.

c) Equisetatae

Seitdem die früher als "Protoarticulaten" angesehenen mitteldevoni-
schen Gattungen *Hyenia* und *Calamophyton* als Farne erkannt worden sind,
gilt die bereits recht hoch organisierte *Pseudobornia ursina* aus dem
Oberdevon als früheste Sphenophyte. Die Suche nach älteren Vorläufern
ist daher verständlich. Nachdem bereits früher SCHWEITZER versucht
hatte, mit einem als *Equisetophyton praecox* bezeichneten nicht sehr über-
zeugenden Rest die Existenz prä-oberdevonischer Sphenophyten nachzuweisen, stellt nun MUSTAFA (1978) ebenfalls ein Einzelstück aus dem
Mitteldevon des Sauerlandes als *Honseleria verticillata* und als erste
"eindeutige" Sphenophyte vor. Das Stück besteht aus einem Stengel-
querbruch, der einen Markhohlraum enthalten soll und an dem ein Quirl
dreifach dichotom geteilter Blätter sitzt.

Zahlreiche Arbeiten beschäftigen sich mit karbonischen Calamiten. In
einem bemerkenswerten, wenn auch leider nur kurzgefaßten Beitrag entwickelt LEISTIKOW (1975) ein Konzept zur Phylogenie der Equisetales.
Darin wird der artikulierte Bauplan ausgehend vom rezenten *Equisetum*
funktionell begründet: Jeder Luftsproß wird vollständig als Knospe am
Rhizom angelegt und kann sich bei Beginn der Vegetationsperiode durch
Wasseraufnahme nach dem Teleskop-Prinzip sehr rasch strecken. Die er-
forderlichen großen Wassermengen können in den tracheenartigen Carinal-
kanälen befördert werden. Ganz ähnlich soll das Wachstum der Calamiten
vor sich gegangen sein. Ist diese Konstruktion für einen Geophyten wie
Equisetum sehr sinnvoll, so wird sie für die mit Sekundärholz ausgestat-
teten "rhizombürtigen Bäume" der Calamiten als materialverschwendende

widersinnige Lebensform angesehen, die sich nur stammesgeschichtlich
(die ursprünglich fehlende Wurzel wird durch das Rhizom ersetzt) erklären läßt und sich nur mangels Konkurrenz eine Zeit lang halten
konnte.

Als *Calamites* wurden bisher alle oberkarbonischen und permischen Calamiten bezeichnet. Daß sich darunter mehrere systematische Einheiten von Gattungsrang
verbergen, ist seit langem bekannt. Früh war daher auch der unterkarbonische
Archaeocalamites (al. *Asterocalamites*) abgetrennt worden. W. und R. REMY (1977,
1978a) haben nun *Calamites* in 3 Gattungen, *Mesocalamites*, *Calamitina* und *Cala-
mitopsis* (die sich nicht mit den bisher verwendeten Gruppen decken) aufgeteilt.
Den Namen *Calamites* übertragen sie dagegen auf die hiervon seit Jahrzehnten ab-
getrennte Gattung *Archaeocalamites*. Damit wird der Sinn der Nomenklaturregeln -
die Verständigung zu erleichtern - ins Gegenteil verkehrt und ein wirklicher
Fortschritt behindert.

Für *Calamites multiramis* bringen W. und R. REMY (1978a) eine neue Rekonstruktion: Während auf HIRMER zurückgehend die Crucicalamiten allgemein mit einer
riesigen, dicht verzweigten Krone dargestellt werden, nehmen die Autoren die
bereits von HALLE (1928) und GOOD (1976) vertretene Vorstellung auf, daß die
Blattquirle der zugehörigen *Annularia*-Beblätterung schräg zu den Hauptachsen
stehen und dadurch wedelartig abgeflachte laterale Sproßsysteme entstehen. Aus
Gründen des Lichtangebotes sollen jeweils gleichzeitig nur wenige dieser Wedel
apikal vorhanden gewesen sein und die Pflanze daher einen baumfarnartigen Habitus besessen haben.

GOOD (1976) hat erstmals strukturerhaltene *Annularia*-Blätter gefunden, die auf
der Unterseite in 2 Reihen etwas eingesenkt die schräg gestellten Spaltöffnun-
gen tragen. W. und R. REMY (1975c) fanden im Ruhrgebiet bei *Calamites astero-
pilosus* und einer auf derselben Platte vorkommenden *Annularia* kleine Sternhaare,
eine bei den Calamiten außergewöhnliche Erscheinung.

Während die meisten Arbeiten sich mit der Stammanatomie der Calamiten
befassen, mißt GOOD (1975) dieser nur geringen systematischen Wert
bei. Auch im übrigen Bau sieht er keine grundsätzlichen Unterschiede
zwischen den Equisetaceen und den Calamitaceen: Bei den letzteren
sind die Blätter wenigstens basal verwachsen und bei *Equisetum* treten
als Abnormität gelegentlich blattartige Sporophylle auf. Sekundäres
Dickenwachstum wird auch für die mesozoischen Equisetiden angenommen.
Noch wesentlicher ist es jedoch, daß ein weiterer Unterschied völlig
entfällt: Bisher wurde angenommen, daß die Sporen der Calamiten keine
Elateren besaßen, weil sie bisher niemals an fossilen Sporen gefunden
wurden. An besonders gut erhaltenem *Calamites*-Material konnte nun GOOD
feststellen, daß Elateren tatsächlich vorhanden waren. Er vermutet,
daß diese zum Ausschleudern der Sporen dienten und dabei abrissen. In
anderem Zusammenhang wird von GOOD (1977) darauf hingewiesen, daß die
für *Calamites* typische Sporenform *Calamospora* auch bei *Sphenophyllum* und
Discinites (Noeggeratiales) vorkommt.

In den Sporangienzapfen der Calamiten wechseln fertile und sterile
Blattquirle ab. Nach ihrer Stellung zueinander werden verschiedene
Gattungen unterschieden. R. und W. REMY (1975a) konnten zeigen, daß
sich *Palaestachya*, bei der die Sporangiophoren in der Achsel von Brak-
teen stehen, nicht wesentlich von *Calamostachys* unterscheidet, bei dem
die Sporangiophore in der Mitte zwischen zwei Blattquirlen stehen,
da diese Stellung auf Grund des Gefäßbündelverlaufes durch kongenitale
Verwachsung zustande kommt. In größeren Zapfenständen kommen sogar
beide Formen gemeinsam vor. Es gibt jedoch auch Fälle - wie *Schimperia* -
bei denen beide Wirtel echt miteinander abwechseln.

Alle diese Sporangiophore sind - wie es sich für Sphenophyten gehört -
peltat. Dies trifft jedoch nicht für die aus der Gondwanaflora von

Madagaskar von APPERT (1977) beschriebenen *Sakoarota palyangiata* zu. Bei
ihr stehen die Sporangien orthotrop an mehrfach dichotom geteilten
Seitenzweigen, wie sie ähnlich von der ebenfalls in der Gondwanaflora
vorkommenden *Phyllotheca* bekannt sind. Diese besitzt jedoch peltate
Sporangiophore.

Für die Lebensform von *Sphenophyllum* ist ein Befund von BATENBURG (1976,
1977) bedeutsam, der bei *Sph. emarginatum* aus dem Saarkarbon Blätter mit
Hakengrannen gefunden hat, die zum Klettern gedient haben können.

SCHABILION (1975) weist an coalball-Material von Kansas nach, daß sich
bei *Sphenophyllum* an jedem Knoten - ähnlich wie beim rezenten *Equisetum* -
ein interkalares Meristem befand, das dem Längenwachstum der Interno-
dien diente. Dies könnte ein Hinweis darauf sein, daß die oben von
LEISTIKOW geschilderten Vorstellungen auch für *Sphenophyllum* gelten.

Auf ein höheres als bisher angenommenes Alter für *Equisetum* -gleichsam
als Durchläufergattung - könnte die Beschreibung einer Art *(E. boureaui)*
aus der Trias von Kambodscha durch VOZENIN-SERRA und LAROCHE (1976)
hinweisen.

d) Filicatae

In den vergangenen 20 Jahren hat sich herausgestellt, daß das, was wir
heute unter Farnen verstehen, eine junge Gruppe ist, die nur bis ins
Oberkarbon zurückreicht. Die älteren farnartigen Pflanzen, die früher
Primofilices genannt wurden, haben sich inzwischen als eine so vielge-
staltige Gruppe erwiesen, daß es schwierig ist, sie in einem kurzen
Überblick zu behandeln, zumal die Zusammenhänge zwischen den einzelnen
Linien noch sehr unterschiedlich gesehen werden. Allgemein werden die
Progymnospermen wegen ihres hochorganisierten gymnospermenartigen vegetativen Baus von den Farnen getrennt. Man sollte sich jedoch darüber
im Klaren sein, daß sie es in der Vermehrung nicht über Heterosporie
hinausgebracht haben und somit als Pteridophyten zu betrachten sind.
Auch werden die Gymnospermen nicht dadurch monophyletisch, daß man sie
auf eine gemeinsame Ausgangsgruppe zurückzuführen versucht; denn sicher haben sich Cycadophyten- und Coniferophytenvorfahren auf einem
unteren Niveau innerhalb der Progymnospermen, etwa der Protopteridiales oder noch darunter, z.B. der Trimerophytinen, getrennt. Gerade in
diesem Bereich herrscht noch keine Einigkeit über die verwandtschaft-
lichen Beziehungen. So sehen ANDREWS und GENSEL (1975) zwei Linien
vom "*Psilophyton*-Komplex" ausgehen: Eine über *Trimerophyton* zu den Progymnospermen bis hin zu *Archaeopteris* und eine zweite über *Pertica* zu den
Coenopteridales und damit zu den Farnen. CORNET et al. (1976) dagegen
nehmen die Protopteridiales als Ausgangsgruppe sowohl für die Progymnospermen wie die Coenopteridales an, womit sie die Farne zu einem Sei-
tenzweig der Progymnospermen machen. W. und R. REMY (1977) trennen
beide Linien ebenfalls voneinander, stellen jedoch die Cladoxalales
zu den von ihnen so genannten Prospermatophyten und leiten von ihnen
die Cycadophyten ab. SCHECKLER (1976) dagegen möchte die Cycadophyten
auf die Aneurophytales (Protopteridiales) zurückführen.

α) Cladoxylales. SCHECKLER (1975a) beschreibt aus dem Oberdevon von New
York *Rhymokalon trichium*, das eine vielrippige Stele besitzt, von der
seitlich Xylemstränge zu den Zweigen abgehen. Er weist mit Recht darauf hin, daß natürlich auch die Polystele der Cladoxylales *eine* Stele
ist und wendet sich gegen eine Deutung der radiären Zellreihen im
Xylem als Sekundärholz, wie sie z.B. von REMY und REMY (1977) noch
vertreten wird. Damit werden Beziehungen zu den Medullosen unwahr-
scheinlicher, und SCHECKLER glaubt, daß ein Teil der Cladoxylales mit

den Coenopteriden verwandt ist. MUSTAFA (1978) führt scheinbare Arti-
kulierung bei Calamophyton auf Steinzellennester zurück und sieht Beziehungen zur Blattstellung. Auch für die bisher als zweifelhafte
Lycophyte gedeutete *Duisbergia* nimmt MUSTAFA polystelen Bau an und betrachtet sie als Cladoxylale.

β) *Coenopteridales*. Mehr und mehr gelingt es, den vor allem aus coal
balls bekanntgewordenen Achsen der Coenopteridales Blätter und Sporangien zuzuordnen und sich eine Vorstellung der Pflanzen zu machen.
Die vielleicht bemerkenswerteste Neuigkeit ist der von JENNINGS und
EGGERT (1977) erbrachte Beweis, daß *Senftenbergia* auf Grund des Achsen-
baues zu den Coenopteridales zu stellen ist. Die an *Pecopteris*-Blättern
gefundenen Sporangien besitzen eine apikale Kappe anulusartiger Zellen und wurden bisher als Schizaeaceen betrachtet. Dabei wurde übersehen, daß bei diesen die Sporangien stets einen einreihigen apikalen
Anulus haben. Nun bleibt mit *Oligocarpia* eine Gleicheniacee als einziger paläozoischer leptosporangiater Farn übrig. Der Sporangienbau von
Senftenbergia paßt zu anderen Coenopteriden-Sporangien, bei denen Gruppen von anulusähnlichen Zellen auftreten. Bei *Discopteris vuellersi* bestehen sie nach PFEFFERKORN (1978) sogar fast vollständig aus Anulus-
Zellen, die als Dehiszenz lediglich eine apikale Platte aussparen.

Zu den am besten bekannten Gattungen gehört der karbonische *Botryopteris*.
GALTIER und PHILLIPS (1977) setzen in einem 2. Teil ihre sehr gründ-
liche Revision fort. Dabei stellten sie fest, daß die ältesten Arten
im westeuropäischen Unterkarbon vorkommen, während jüngere Arten auch
in N-Amerika vertreten sind.

In speziellen Untersuchungen wurden neue komplizierte dreidimensionale Sproßsysteme gefunden. So besaß nach HOLMES und GALTIER (1976) *Botryopteris* aff.
hirsuta dekussiert stehende Seitenzweige, die dicht mit schwach entwickelten
Achsen besetzt sind. Bei *Psalixochlaena* treten dichotome und monopodiale Verzweigungen nebeneinander auf, letztere stets in Blattachseln. Nach HOLMES (1977)
war es ein am Boden kriechender Farn mit aufrechten Blättern. JENNINGS (1975a)
beschreibt erstmals den Achsenbau von *Alloiopteris sternbergi* mit schmetter-
lingsförmigen Xylem in den Hauptachsen und *Etapteris*-ähnlichem in den Blatt-
stielen. Zu den Coenopteridales wird schließlich von PFEFFERKORN (1976a) auch
Brittsia problematica gerechnet, ein bilaterales, doppelt gefiedertes Blatt,
bei dem die Achsen 1. Ordnung wie ein *Pecopteris*-Blatt fiederartig verbreitert
sind und die eigentlichen Fiedern 2. Ordnung schräg nach oben stehen. Es soll
sich um ein Schwimmblatt handeln, zu dem vermutlich *Biscalitheca*-Fruktifikatio-
nen gehören.

γ) *Eusporangiatae*. Die Stämme der baumförmigen jungpaläozoischen Marat-
tiacee *Psaronius* hatten kein sekundäres Dickenwachstum. Dieses wird
ersetzt durch ein mächtiges System von Luftwurzeln. EHRET und PHILLIPS
(1977) untersuchten an Hand einiger hundert coal balls den Bau und die
Ontogenie des Luftwurzelsystems. Die schräg nach unten abgehenden
Luftwurzeln verlaufen zunächst in der Stammrinde (innere Wurzelzone),
durchbrechen deren Sklerenchymring und bilden außerhalb des Stammes
die äußere Wurzelzone. Die Wurzeln bestehen aus einer Actinostele,
einer aerenchymatischen Innen- und einer parenchymatischen Außenrinde.
Es konnte sogar nachgewiesen werden, daß die Wurzeln Wurzelhauben
und - wie die rezenten Gattungen *Angiopteris* und *Marattia* - bis zu 4
Scheitelzellen hatten. Die Abdrücke der Stämme von *Psaronius* sind nach
PFEFFERKORN (1976b) als *Caulopteris* mit spiraliger Wedelstellung sowie
Megaphyton und *Artisophyton* mit zweiteiliger Wedelstellung bekannt. Für
die Unterscheidung ist außerdem der Stelenquerschnitt in den Blatt-
narben maßgebend. Sehr moderne Marattiaceensporangien beschreibt MIL-
LAY (1979) aus dem Pennsylvanian von Illinois. Daß Marattiaceen auch
in der Unteren Gondwana Indiens vorkommen, zeigen PANT und MISRA (1977)

mit *Trithecopteris*. Ein anderer eusporangiater Farn, *Damudosorus*, besitzt freie Sporangien mit einem unvollkommenen ringförmigen Anulus und wird daher als bedeutsam für die Evolution der leptosporangiaten Farne angesehen. Während die rezenten Marattiaceen monolete Sporen haben, sind durch VAN KONIJNENBURG-VAN CITTERT (1975) bei *Marattia anglica* aus dem Jura von Yorkshire trilete Sporen nachgewiesen worden.

Nachweise von Marattiaceen aus dem Tertiär sind selten. Es ist daher bemerkenswert, daß PALAMAREV et al. (1975) aus dem Miozän von Bulgarien einen an Hand der Epidermisstruktur bestimmten *Angiopteris ruziuciniana* angeben, der zusammen mit der leptosporangiaten *Lindsaea* vorkommt, die heute in SE-Asien ein ähnliches Areal wie die Marattiaceen besiedelt.

δ) *Protoleptosporangiatae*. Aus dem Rhäto-Jura des Iran bringt SCHWEITZER (1978) eine exzellente Rekonstruktion der Osmundacee *Todites princeps*. Es wird die Auffassung von WEBER bestätigt, daß es sich um einen kleinen, aufrechten Baumfarn handelte, der durch Ausläufer auf trockenfallendem Tonschlamm Tochterpflanzen erzeugen konnte. Dadurch war er in der Lage, ähnlich wie die rezente *Matteucia strutiopteris*, größere Flächen zu besiedeln. Eigenartigerweise ist der fossile *Todites* nur nordhemisphärisch verbreitet, während die rezenten nächsten Verwandten, *Todea* und *Leptopteris*, auf die Südhalbkugel beschränkt sind. KRASSILOV (1976b) nimmt an, daß im Mesozoikum Sibirien für die Gattung *Osmunda* ein Entwicklungszentrum war.

Der dem rezenten *Osmunda regalis* ähnliche *Osmundis döwkeri* wurde von VAN DER BURGH (1977) in der mizänen Braunkohle der Oberpfalz gefunden, wo er vermutlich ähnliche Biotype - saure Moorböden - bevorzugte.

ε) *Leptosporangiatae*. Aus dem iranischen Rhäto-Jura bringt SCHWEITZER (1978) sehr lebensechte Rekonstruktionen der Dipteridacee *Thaumatopteris brauniana* und der Matoniacee *Phlebopteris polypodioides*. Beide hatten offenbar recht ähnliche Lebensweisen mit kriechenden Rhizomen sowie schirmartigen Wedeln und bildeten größere reine Bestände. Der Wedel beider Familien ist nach SCHWEITZER primär dichotom geteilt, während HIRMER (1927) für die rezente *Motonia* zunächst einen monopodialen und nur für die Seitenteile dichotomen Bau angenommen hatte. *Phlebopteris brauneri* wird von REMY et al. (1975) aus dem Jura des nördlichen Südamerika angegeben.

Dicksoniaceen und Cyatheaceen waren bereits im Jura gut getrennte Familien, wie KRASSILOV (1978b) an sibirischem Material zeigt. Ein Farnrhizom aus der Kreide von Maryland wird von SKOG (1976) auf Grund seiner amphiphloischen Siphonostele zu den Loxsomaceae gestellt, einer kleinen leptosporangiaten Familie, die heute nur mit 2 Gattungen in Neuseeland und in Südamerika vorkommt.

ζ) *Progymnospermatae*. Die Progymnospermen werden von BECK (1976a) als Pflanzen mit Gymnospermenanatomie und Homo- bis Heterosporie definiert und gliedern sich in 3 Ordnungen, die Aneurophytales, die Protopityales und die Archaeopteridales. Er vermutet, daß auch im jüngeren Paläozoikum noch Pflanzen mit diesen Merkmalen vorkommen. Insbesondere komme hierfür *Noeggerathia* in Frage, wie dies von LEARY und PFEFFERKORN (1977) ausführlicher begründet wird. MUSTAFA (1975) stellt fest, daß *Protopteridium thomsoni* die gleiche dreieckige marklose Stele besitzt wie *Aneurophyton germanicum* und ordnet beide der Ordnung Protopteridiales zu.

CORNET et al. (1976) und SCHULTKA (1978) haben bei *Rhacophyton* nun eine knochenförmige mesarche Stele und Sekundärholz gefunden und stellen es zu den Protopteridiales. Die biseriate Verzweigung wird als Vor-

stufe zum flachen Wedel gedeutet. Nach SCHULTKA bildete *Rhacophyton*
reine Bestände an Süßgewässern, woe die mit Klimmhaken versehenen
dünnstämmigen Pflanzen sich gegenseitig stützten. Mit drei Gattungen,
die mit *Aneurophyton* verwandt sind *(Triloboylon, Tetraxylopteris* und *Proteo-
kalon)*, beschäftigt sich SCHECKLER in 2 Arbeiten (1975b, 1976), wobei
er der Ontogenie dieser dreidimensional wachsenden Sproßsysteme besondere Beachtung schenkt.

7. Spermatophyta

a) Gymnospermophytina

α) *Lyginopteridatae.* Seit der Entdeckung von *Callistophyton* (Fortschr. Bot.
17) war über die oberkarbonen Callistophytaceae immer wieder Interes-
santes zu berichten. Nun kann ROTHWELL (1975) den ersten Teil einer
Monographie vorlegen, die sich mit den vegetativen Strukturen beschäftigt. *Callistophyton* - die am besten bekannte Gattung - umfaßt verhältnismäßig kleine, am Boden dahinkriechende, reich verzweigte Sträucher,
mit maximal 3 cm dicken Achsen. An etwas angeschwollenen Knoten sitzen die sphenopteridischen, doppelt bis vierfach gefiederten Wedel
und entspringen auch die diarch gebauten sproßbürtigen Wurzeln. Dies
und kleine Dornen auf der Rinde legen den Gedanken nahe, daß *Callistophyton* teilweise auch als Spreizklimmer lebte. Der Stammbau ist lygino-
pteridisch-eustel mit sekundärem Leitgewebe und schmalem Periderm. Zumindest im Habitus dürfte *Microspermopteris* aus gleichalten Schichten
ähnlich gewesen sein (TAYLOR und STOCKEY, 1976): Mäßig dicke, verzweigte Achsen, besetzt mit zahlreichen Adventivwurzeln. In der Anatomie ähnelt diese Pteridosperme aber weniger *Lyginopteris* als vielmehr *Heterangium*: ist doch das Stadium der echten Eustele noch nicht erreicht. Es scheint eine gegenüber devonischen Verhältnissen fortge-
schrittene Protostele vorhanden, bei der die trennenden Markstrahlen
nur sehr dünne Stränge darstellen. Verschiedene Untersuchungen beschäftigen sich auch mit dem Verwandtschaftskreis *Heterangium - Rhodea -
Telangium* und *Crossotheca*. Nachdem früher (Fortschr. Bot. 37) *Sphenopteris*-
Laub als zu *Heterangium*-Achsen gehörig erkannt wurde, kann jetzt JEN-
NINGS (1976) auch *Rhodea*-Laub und *Telangium*-Synangien mit solchen Achsen
kombinieren. Daraus ergibt sich: Bei dieser Gattung erfolgte ebenfalls
wie bei Progymnospermen und Primofilices zwischen Unter- und Oberkarbon
eine Wedel-Planation und -Lamination. Über sterile und fertile Reste
aus dem gleichen Formenkreis berichten dann auch MILLAY und TAYLOR
(1977, 1978). Die neue Gattung *Feraxotheca* entspricht dem Abdruckgenus
Crossotheca: An pinnat verzweigten Achsen sitzen auf parenchymatischen
Polstern radiär- bis bilateral-symmetrische Synangien, an denen die
dicke periphere Wand einen vielleicht stammesgeschichtlich wichtigen
Unterschied zu *Telangium* darstellt. Der Verdacht liegt nahe, daß diese
fertilen und sterilen Teile nur Abschnitte eines überwiegend sterilen
Großwedels sind. Obwohl noch nicht abgeschlossen, sind die Untersuchungen amerikanischer Paläobotaniker an anderen männlichen Fruktifikatio-
nen oberkarboner Samenfarne sowohl phylogenetisch also auch taxonomisch
recht bedeutungsvoll. Die Untersuchungen von STIDD (1978b) z.B. an
strukturbietendem *Potoniea*-Material bestätigen, daß frühere Bearbeiter
den becherförmigen Bau aus den inkohlten Abdruckfossilien richtig gedeutet haben. Wichtiger ist, daß REM-Aufnahmen zeigen: Die zu *Linopteris/
Reticulopteris*-Laub und *Sutcliffia*-Achsen - nach STIDD et al. (1975) ähneln
letztere im Bau *Lyginopteris* und *Medullosa* - gehörenden Mikrosynangien
enthalten Sporen, die primitiven Angiospermenpollenkörnern im Exinen-
bau ähnlich sind. Nach STIDD (1978b) und STIDD et al. (1977) sind die
Synangien von *Dolerotheca* und der neu kreierten *Sullitheca* komplexe Ge-

bilde, bestehend aus je zwei miteinander verwachsenen, eingekrümmten
Reihen von Einzelsynangien, deren freie Enden so im Zentrum des Groß-
synangiums zu liegen kommen. Eine Synopsis solcher strukturbietender
Medullosaceen-Organe geben dann DENNIS und EGGERT (1978), wobei sie
allerdings die oben referierte Meinung über die Natur von *Sullitheca*
und *Dolerotheca* nicht teilen. Nach den Autoren lassen sich drei Gruppen
von Microsynangien bei dieser Familie unterscheiden, die durch fort-
schreitende Verwachsung und erhöhte Zahl der beteiligten Sporangien
gekennzeichnet sind. Aber auch sie sehen in *Dolerotheca* das komplexeste
Gebilde dieser Art. Im Rahmen einer vergleichenden Darstellung der
Evolution des Gymnospermenpollens gehen MILLAY und TAYLOR (1976) ausführlich auf Callistophytaceen und Medullosaceen ein, wobei aber nur
bereits früher Referiertes (Fortschr. Bot. 35, 37) zusammengestellt
ist. Mit dem ersten paläozoischen Nachweis eines syndetocheilen Spalt-
öffnungsbaus bei einer *Alethopteris*-Art im Oberkarbon der USA durch
STIDD und STIDD (1976) erfährt die heute allgemein verbreitete Ansicht
einer Pteridospermenabstammung der Angiospermen eine weitere Stütze.
Den berechtigten Anspruch der Paläobotanik, bei paläoklimatologischen
Fragen gehört zu werden, unterstreichen Untersuchungen von BUSCHE et
al. (1978) sowie von REMY und REMY (1978b). Die Autoren weisen darauf
hin, daß erstens für eine Klimaanalyse *alle* verfügbaren pflanzlichen
Reste, also Makro- *und* Mikroreste, herangezogen werden müssen, zum
anderen, wenn man dies befolgt, aus der *Callipteris*-Flora des Untersperm
z.B. nicht ohne weiteres auf arides oder semiarides Klima geschlossen
werden darf, weil ein flözferner Standort nicht gleichbedeutend sei
mit Wasserferne, sondern darin lediglich das Meiden eines sauren Moor-
bodens zum Ausdruck komme. Exzellent erhaltenes, weil echt versteiner-
tes Material von *Reticulopteris*- und *Neuropteris*-Laub - wiederum aus dem
nordamerikanischen Ober-Karbon - konnten REIHMAN und SCHABILION (1978)
studieren. Im Vergleich mit xeromorph ausgebildeten *Alethopteris*-Wedeln
beweisen diese hygromorphen Pteridophylle, daß nicht alle Samenfarne
trockene oder auch nur mesophile Standorte besiedelten. In Fortsetzung
ihrer revidierenden Arbeiten (Fortschr. Bot. 35, 37) beschreiben
DOUBINGER und GERMER (1975) sowie ALVAREZ RAMIS et al. (1978) die
Neuropterides bzw. die Sphenopterides des Saarkarbons. Weitere erwäh-
nenswerte Pteridophylla-Revisionen stammen aus der Feder von AMEROM
(1975) *(Eusphenopteris)*, GASTALDO und MATTEN (1978) *(Pecopteris)*, REMY
und REMY (1975b) *(Taeniopteris)* und von LAVEINE et al. (1977) *(Calli-
pterides)*, letztgenannte Arbeit mit einem vorwiegend phylogenetischen
Aspekt. Einen auch für Exkursionen auf dem europäischen Kontinent gut
geeigneten Pteridophyllen-Bestimmungschlüssel legen CHALONER und COL-
LINSON (1975b) vor.

Glossopteridales. Seit die Mitteilungen von PLUMSTEAD in der Paläobotanik
Furore machten (Fortschr. Bot. 17), ist die Diskussion über die Sexual-
organe dieser Gruppe nicht mehr verstummt (Fortschr. Bot. 19, 33,
35, 37). Dank der Bemühungen von indischen, südafrikanischen, nordamerikanischen und australischen Bearbeitern lichtet sich das Dunkel,
wenn auch nur etwas: Wie bereits früher mitgeteilt, lassen sich die
verschiedensten Typen unterscheiden (Prog. Bot. 37). Publikationen
von CHANDRA und SURANGE (1976, 1977a,b,c), WHITE (1978) und BENECKE
(1976) vermehren sie, zeigen aber zugleich die Mannigfaltigkeit selbst
bei *Glossopteris* sensu stricto. Gesichert ist, daß keine "Zwitterblüten"
auftreten. Ebenso herrscht weitgehend Übereinstimmung, daß bei den
typischen *Glossopteris*-Fruchtständen, z.B. *Dictyopteridium*, auch bei dem
klassischen *Scutum*, seien sie gestielt oder ungestielt, zweiteilige
Gebilde vorliegen. Gänzlich unumstritten ist dabei, daß diese Frucht-
gebilde der Mittelrippe von "normalen" Glossopteridales-Blättern entspringen. Wenn der Fruchtkomplex zweiteilig ist, dann ist die adaxiale
Hälfte ein steriles Blättchen, die abaxiale fertil, d.h. auf einem
"Receptaculum" sitzen spiralig angeordnet die Samen. Nach den schönen

Rekonstruktionen der indischen Bearbeiter stehen diese fruchttragen-
den *Glossopteris*-Blätter eingestreut zwischen sterilem Laub. Recht kon-
trovers wird die morphologische Terminologie gehandhabt: CHANDRA und
SURANGE (1976) verwenden bekannte Begriffe wie Braktee, Deckschuppe
bzw. Spatha und Rezeptakulum, während BENECKE (1976) neutral von fer-
tiler und steriler Hälfte spricht. Wenn auch KOVÁCS-ENDRÖDY (1974)
vielleicht etwas voreilig den Begriff "Megasporophyll" - für einen
Basiswulst, auf dem ein Same sitzt - in die Diskussion bringt, so
könnte seine Beschreibung eines einsamigen Fruchtkomplexes aus Pretoria ohne "sterile Hälfte" für das Verständnis der weiblichen Fortpflanzungsorgane von *Glossopteris* doch einmal wichtig werden. Etwas
einfach erscheint freilich seine Lösung des Problems, alle anderen beschriebenen Reste zu männlichen Organen zu erklären. Synoptisch faßt
schließlich SCHOPF (1976) unser derzeitiges Wissen über die männlichen
und weiblichen Organe der Glossopteridales zusammen, wobei er auch um
terminologische Klarstellung bemüht ist. Neu ist dabei sein phylogene-
tisches Konzept eines Anschlusses an die Cordaitidae über gewisse
platysperme Reste aus dem Gondwanabereich. Mit *Glossopteris*-Blättern
befassen sich KOVÁCS-ENRÖDY (1976) und DELEVORYAS und PERSON (1975)
(aus dem Mittel-Jura von Mexiko!) wie auch PANT und CHOUDHURY (1977).
Bei dem Material letzterer handelt es sich, obwohl morphologisch rezenten Farnen mit ungeteilten, pfeilförmigen Wedeln sehr ähnlich,
nach neueren vollständigen Funden, die anatomische Studien ermöglichten, wohl doch um eine glossopteridische Form.

Caytoniales. In Fortführung ihrer früher schon erwähnten Untersuchungen
(Fortschr. Bot. 35) beschäftigt sich REYMANOWNA (1973) mit fertilen
und vegetativen *Caytonia*-Resten aus dem polnischen Jura. Mikrotom-
Serienschnitte durch Früchte erlauben erstmalig Aussagen über den
Leitbündelverlauf in Frucht und Samen. Die Autorin konnte auch nachweisen, daß die Anatomie der Fruchtöffnung, des "Mundes" und die Be-
schaffenheit des Fruchtinneren die Bestäubungstropfen-Hypothes von
HARRIS (Fortschr. Bot. 14) sehr wahrscheinlich machen. Zugleich wird
die Frage zur Diskussion gestellt, ob nicht das fleischige Innere weniger ein Mittel der Endozoochorie als vielmehr eine Voraussetzung
für Autochorie nach Art der bekannten Explosionsmechanismen war. Dies
führt einmal mehr die Spezialisation dieser Samenfarngruppe vor Augen.
Prinzipiell ist dies auch die Ansicht von KRASSILOV (1977). Aber derzeit im Gegensatz zur überwiegenden Mehrzahl der Fachkollegen spielt
er mit dem Gedanken, ob die Caytoniales, die er nicht ohne eine gewisse logische Begründung von den Glossopteridales ableitet, nicht
doch direkte Angiospermen-Vorfahren sein könnten. Gegenüber diesen
allgemein interessanten Erwägungen berühren die Mitteilungen von
DOBRUSKINA (1975) über die Peltaspermaceen und von BARALE (1975) über
die Jura-Gattung *Cycadopteris* mehr speziell-taxonomische Probleme.

β) *Cycadatae.* Seine früheren Arbeiten über paläozoische Cycadophyten
(Fortschr. Bot. 33, 37) rundet MAMAY (1976) mit einer Gesamtbeschrei-
bung ab. So läßt sich nun die Klasse der Cycadeen über die Permgattun-
gen *Palaeocycas* und *Phasmatocycas* bis zu oberkarbonen Samenfarnen zurück-
verfolgen, und jene zählen somit neben den Koniferen bewiesenermaßen
zu den phylogenetisch ältesten lebenden Gymnospermen. Gleichzeitig er-
läutert der Autor Vorstellungen über die Entwicklung des modernen gefiederten Cycadeen-Wedel aus taeniopteridischen Vorfahren. Einen recht
anschaulichen Beitrag zur Frage nach dem Habitus fossiler Cycadophyten liefern KIMURA und SEKIDO (1975) mit ihrer Rekonstruktion der neu-
en Kreide-Gattung *Nilssoniocladus*, die an dünner Langtriebachse schopfig
beblätterte Kurztriebe mit Nilssonien-Wedeln trägt. Nur: ein neuer Fa-
milienname ist "Nilssoniaceae" wirklich nicht.

γ) *Bennettitae*. SHARMA (1976) setzt die Untersuchungen zur Ontogenie von *Williamsonia*-"Blüten" fort (Prog. Bot. 37). Aus der Tatsache, daß das Integument nicht verhärtet, wird der kühne Schluß gezogen, daß bei dieser Gattung offensichtlich weniger die Samenbildung, als vielmehr die vegetative Fortpflanzung die Regel war. In einer weiteren Publikation werden dann die Ergebnisse auch anderer Autoren, gewonnen an indischen Stücken aus dem Jura, zusammengefaßt (1977). Nach ASH (1976b) kommen in der Trias von USA Wedel vor, die wegen der fächerförmigen Anordnung der Segmente ursrünglich zu den Ginkgophyten gerechnet wurden, sich jetzt aber durch syndetocheilen Stomabau als Bennettiteen-Blätter mit stark verkürzter Rhachis zu erkennen gaben; ein Bau, der ja auch von gewissen Pterophyllen her bekannt ist. Schon früher haben DELEVORYAS und HOPE vor der verbreiteten Ansicht gewarnt, pachycaule Stämme seien für die mesozoischen Cycadophyten typisch (Fortschr. Bot. 35). Die gleichen Autoren (1976) untermauern nun mit neuen Beweisen diese frühere These. Die neue Williamsoniaceen-Gattung *Ischnophyton* aus der Obertrias von Nord-Carolina mit einem verhältnismäßig dünnen Stamm und Zamites-ähnlichen Wedeln stelle so direkt den Prototyp mesozoischer Cycadophyten dar.

δ) *Ginkgoatae*. Ist die neue dürftig bekannte Gattung *Thomasleslia* aus dem U.-Perm von Transvaal nicht nur dem seinerzeit von KRÄUSEL (Fortschr. Bot. 14) beschriebenen *Glossophyllum* ähnlich, sondern mit diesem auch verwandt und damit eine Ginkgophyte? LE ROUX (1975) ist geneigt, dies zu verneinen und denkt eher an Cycadophyten. Da auch die beigegebenen Photos wenig aussagekräftig sind, bleibt der Name ohne Inhalt. Anders als bei der Jura-Gattung *Stachyopitys* mit radiär gebauten Staubblättern sind nach BROWN (1975) bei der *Stenorhachis* genannten männlichen Ginkgophyten-Blüte aus der Unterkreide von Nordamerika wie bei dem rezenten *Ginkgo* die Mikrosporophylle bereits dorsiventral und nur mehr bisporangiat. Die Dichotomie der Aderung, die angedeutete Zweilappigkeit und bestimmte Kutikular-Charakteristika lassen in den derben Blättern der cenomanen *Podozamites obtusa* aus Böhmen somit eher die Blattreste einer Ginkgophyte erkennen. HLUŠTIK (1977a) schafft dafür das neue Genus *Nehvizdyas*. Die Ginkgoaceen waren also auch in der O-Kreide formenreicher als bisher angenommen.

ε) *Pinatae*. *Cordaitidae*. Schliffpräparate von *Nucellangium* aus dem Oberkarbon von Iowa, die STIDD und CONSENTINO (1976) untersuchten, zeigen eindrucksvoll fast alle Einzelheiten der Anatomie rezenter Coniferen-Samen: Megasporen-Membran, Prothallium und zwei Archegonien, in einem davon sogar Reste eines Proembryos. Dies und die Epidermis-Ähnlichkeit zwischen *Nucellangium* und gewissen *Cordaianthus*-Resten bestätigen überzeugend die von ANDREWS (Fortschr. Bot. 14) zuerst geäußerte Meinung der Cordaiten-Zugehörigkeit dieser Organgattung. Nomenklatorisch konsequent stellt GLUCHOVA (1978) für strukturbietende Cordaiten-Blätter in Ergänzung älterer Taxa die neuen Gattungen *Angophyllites* und *Europhyllites* auf, mit den Namen bereits auf die Verbreitung in der Angara-Gondwana-Provinz einerseits und in der euramerischen andererseits hinweisend.

Pinidae. Voltziales. Paläobotanische Befunde an paläozoischen und mesozoischen Koniferen verknüpft - stark spekulativ, wie der Autor selbst zugibt - HARRIS (1976a) miteinander, um einen gemeinsamen Ursprung der Taxidae und der Pinidae aus *Lebachia/Walchia*-Verwandten aufzuzeigen. Eine oftmalige Erfahrung wurde erneut durch GRAUVOGEL-STAMM und GRAUVOGEL (1975) bestätigt: Besser erhaltenes Fossilmaterial ermöglicht bessere Ergebnisse. In einer vorläufigen Notiz können die Genannten klarstellen, daß *Aethophyllum* aus der Unter-Trias der Vogesen nicht die Sporangienähre von *Schizoneura*, sondern der bis 22 cm lange Zapfen einer Konifere ist. Auch hier sind "alte" Merkmale - schmale 4-7 ner-

vige Blätter vom Typus *Poacordaites* - gekoppelt mit jüngeren. Die Zapfenschuppen sind fünfsamige Schuppenkomplexe, wie sie von *Swedenborgia*
beschrieben werden, zu der ja schmalblättrige Podozamiten gehören.
Man muß auf die ausführliche Darstellung gespannt sein. DELEVORYAS
und HOPE (1975) können zeigen, daß in der O.-Trias von Nord-Carolina
unter den Koniferen zwei Typen auftreten: Einmal Voltziaceen mit dem
typischen mehrzipfeligen Samen-Schuppenkomplex und daneben *Compsostrobus
neotericus* mit kompakten zweisamigen Samenschuppen und zweizeilig ge-
scheitelten Nadeln, vom Bau her bereits pinaceenhaft anmutend. Als
eine immer interessanter werdende Familie geben sich die Cheirolepidiaceae zu erkennen. Der hierher zu zählende *Classopollis* gehört in
den mesozoischen Floren (SWOBODA-PĚKNA, 1977, LIMA, 1976) zu den ver-
breitetsten Pollentypen. Diesem leichtkenntlichen Typ widmet neuerlich
SRIVASTAVA (1976) eine eingehende Studie. Neben der vorzüglichen Darstellung des Baus ist auch die paläogeographische und palökologische
Auswertung von allgemeinerem Wert. Danach sollen die Mutterpflanzen
trockene Standorte in Küstennähe bevorzugt haben. Das Aussterben im
Alttertiär wird mit der allgemeinen Klimaverschlechterung in Zusammenhang gebracht. Daß dieser hochentwickelte, häufig in Tetraden vorkommende (Fortschr. Bot. 27; SCHEURING, 1976) Pollentyp häufiger auftritt
als entsprechende Lieferanten gefunden wurden, ist bekannt. Ergebnisse
von WATSON (1977) und TSEYEN und CHENGYAO (1977) könnten dies erklären. Es kommen nämlich anscheinend dafür nicht nur die Gattungen
Hirmeriella und *Brachyphyllum*, sondern auch das neue Genus *Suturovagina*
sowie *Pseudofrenelopsis* und sogar *Frenelopsis* selbst in Frage. Aus den
Berichten der angeführten chinesischen Autoren geht auch hervor, daß
innerhalb der Cheirolepidiaceae eine allmähliche Hinwendung zur dekussierten Blattstellung zu bemerken ist, was die oft recht schwierige
Unterscheidung der Zweige von *Brachyphyllum, Cupressinocladus* bzw. *Palaeo-
cyparis* erklärt. Stammesgeschichtlich von Bedeutung könnte dabei, wie
den zusammenfassenden Arbeiten von DOLUDENKO (1978), REYMANOWNA und
WATSON (1976), von ALVIN (1977) und von HLUŠTIK und KONZALOVÁ (1976)
zu entnehmen ist, die Gattung *Frenelopsis* sein, die nunmehr nachgewie-
senermaßen zu den Cheirolepidiaceen gehört. Denn entweder könnte es
sich bei ihren dreizähligen Pseudowirteln um eine Parallelausbildung
einer ausgestorbenen Gruppe handeln, oder die Cheirolepidiaceen waren
eine Ahnengruppe für Cupressaceen und Taxodiaceen (auch Araukariaceen?)
zu denen ja vom Zapfenbau her Beziehung besteht.

Pinales. Araucariaceae. Fruchtzapfen von *Araucaria mirabilis* und von *Pararaucaria patagonica* aus der berühmten mittel- bis oberjurassischen Fos-
sillagerstätte Cerro Cuadrado in Patagonien hat STOCKEY (1977, 1978)
einer gründlichen anatomischen Revision unterzogen. Ihre Ergebnisse:
(1) WIELANDs männliche Zapfen von *A. mirabilis* sind unreife Fruchtzapfen oder bloße Zapfenachsen. (2) Die Fruchtzapfen dieser Art treten hauptsächlich in zwei Entwicklungsstadien auf: einjährige Zapfen
mit entwickeltem Prothallium und mehrjährige Zapfen, die reife Samen
mit vollentwickelten Embryonen enthalten. (3) Die Samen wurden durch
Zerfallen der Zapfen frei, frühestens im zweiten Jahr. Da die gefundenen Baueigentümlichkeiten nur noch bei der heute in Australien heimischen *A. bidwillii* auftreten, kann die fossile Art der gleichen Sektion *Bunya* zugewiesen werden, welche somit bis jetzt zwei Arten, eine
fossile und eine rezente, enthält. Bemerkenswert sind auch ihre Befunde an *Pararaucaria patagonica*. Deren ebenfalls in allen Einzelheiten
erhaltene Zapfen haben unter rezenten und fossilen Formen keine Paral-
lele. Der Gattungsname ist insofern irreführend, als mindestens zu
vier bis fünf Koniferenfamilien Beziehungen bestehen: Darunter auch
sehr deutliche zu den Cheirolepidiaceae, von deren besser bekannten
Gattungen *Hirmerella* und *Tomaxellia* aber leider keine intuskrustierten
Zapfen bekannt sind. Im Gegensatz zu den Embryonen von *A. mirabilis*
sind für *Pararaucaria* mehr als zwei - bis zu acht - Keimblätter nachge-

wiesen. Gleichfalls strukturbietendes Araukarien-Zapfenmaterial beschreiben SHARMA und BOHRA (1977) aus dem indischen Jura. Obwohl die
Autoren eine gewisse Ähnlichkeit mit den *A. mirabilis*-Zapfen erkennen,
geben sie selbst ihrem Bericht nur den Charakter einer vorläufigen
Mitteilung. Um wieviel schwieriger fossiles Abdruckmaterial zu bestimmen ist, zeigen die Ausführungen von BROWN (1977) über Araukarien-
Reste aus der südafrikanischen Unterkreide. So ist aus den meist nur
aus abgefallenen Zapfenschuppen bestehenden Resten nicht viel mehr
als nur die Gattungszugehörigkeit zu folgern. Von der angeblichen
Araukariaceen-Gattung *Aachenia*, seinerzeit von KNOBLOCH in Sammlungs-
beständen entdeckt (Fortschr. Bot. <u>35</u>), konnte nun WEBER (1975) auch
in der Kreide von Mexiko eine neue (?) Art an den kennzeichnenden
Zapfenschuppen und einem Zapfenbruchstück erkennen. Zu den mexikanischen Resten könnten extrem langnadelige Zweige gehören, die spiralig
gestellte Blattpolster tragen. Man kann der Meinung des Autors bei-
pflichten, daß *Aachenia* deshalb viel eher eine Pinacee als eine Arau-
kariacee ist; hat sie doch anscheinend freie Samen getragen.

Pinaceen. Wie MILLER (1976) feststellt, scheinen die Pinaceen zwar bereits in der Unteren Kreide vorhanden gewesen zu sein, aber nur mit
der Gattung *Pinus* selbst. Alle übrigen rezenten Gattungen einschließ-
lich des ausgestorbenen Genus *Pseudoaraucaria* müssen sich erst später
entwickelt haben. Diesen Schluß legt der Befund nahe, daß in der Kreide
bisher allein zwölf "Arten" des Organgenus *Pityostrobus* (MILLER 1975,
1977, 1978) gefunden wurden, die allesamt kaum echte generische Unterschiede im Zapfenbau zeigen, aber Nachweise von rezenten Typen aus dieser Zeitspanne nicht bekannt sind. Dies unterstreichen auch die Neu-
funde, die MILLER und ROBINSON (1975) sowie ROBINSON und MILLER (1977)
aus der Oberkreide bzw. Unterkreide von Nordamerika vermelden. Von der
heute monotypischen Gattung *Cathaya*, vor nicht allzulanger Zeit erst in
China (Fortschr. Bot. <u>21</u>) entdeckt und bekanntlich noch im Pliozän in
Mitteleuropa zuhause, häufen sich die Pollennachweise. Daß aber nach
Pollen zwölf Arten fossil nachweisbar sind, wie SIVAK (1976) meint,
erweckt Unglauben; selbst wenn man zugesteht, daß dieser Pollen groß-
teils als Pinus-Pollen verkannt wurde.

Taxidae. Sind schon derzeit die stammesgeschichtlichen Zusammenhänge
innerhalb der Pinidae nicht mehr recht erkennbar, so gilt dies besonders auch für die Taxidae, wie FERGUSON (1978) hervorhebt. Um hier Ab-
hilfe zu schaffen, haben jetzt FERGUSON et al. (1978) *Amentotaxus* aus
dem europäischen Tertiär monographisch untersucht.

Oft wird übersehen, mutmaßt HARRIS (1976b), daß auch nichtfruktifizie-
rende Koniferen-Reste nach der Art ihrer Blattstellung bzw. des Blatt-
wurfs eine taxonomische Zuordnung ermöglichen. So konnte dieser Autor
eine *Taxus*-Art aus dem Jura von Yorkshire als nahe Verwandte von *Amentotaxus* erkennen.

ζ) *Pinatae incertae sedis*. Die merkwürdige jungpaläozoische Gattung *Dicrano-
phyllum* hat BARTHEL (1977) einer Neu-Untersuchung unterzogen. Es liegt
ein typisches Merkmals-Mosaik vor aus altertümlichen (Gabelblätter
mit Spitzenwachstum) und mehr modernen (hypostomatische Längsstreifen,
Blattpolster) Kennzeichen, so daß man die Gattung sehr wohl an die
Wurzel einer ganzen Reihe von Koniferengruppen stellen kann. Eine zuvor nur recht ungenau bekannte merkwürdige Koniferengattung mit mehr-
samigen, geschlossenen Kapseln, etwas an *Leptostrobus* erinnernd, beschreibt MEYEN (1977) aus dem Perm Rußlands. Pollenfunde in der Mikro-
pyle beweisen, daß diese *Cardiolepis* eine echte Gymnosperme ist, von der
auch beblätterte Zweige und isolierte Samen vorliegen. Der taxonomische Rang und die systematische Stellung dieser "Cardiolepidaceae"
bleiben freilich unbekannt.

Über eine neue Konifere, *Searsolia*, mit breiten, opponiert stehenden Nadeln aus der älteren Gondwana-Formation Indiens berichten PANT und BHATNAGAR (1975). Wie bei anderen Koniferenresten aus dem gleichen Florengebiet lassen sich mit verschiedenen rezenten Taxa (z.B. auch mit den Podocarpaceen) hypothetische Zusammenhänge konstruieren. Aber die systematische Stellung muß schon wegen der ganz dürftig erhaltenen Fortpflanzungsorgane (keine Zapfen?) dahingestellt bleiben. Ursprünglich als Araukarien-Zapfen beschrieben, sind die *Dammarites*-Reste aus der Oberkreide der CSSR wohl überhaupt keine Fruktifikationsreste. Dies ergab eine Nachuntersuchung durch HLUŠTIK (1976). Zwar sieht der Autor gewisse äußerliche Ähnlichkeit mit Cordaiten (!); dessen ungeachtet erscheint einleuchtender aber die Deutung als pathologische Bildung, worauf der genannte Bearbeiter in einer weiteren Arbeit (1977b) selbst hinweist. So wären die band-förmigen Blätter an den angeschwollenen Zweigenden wohl eher von einem Fachmann für Pflanzengallen zu erklären.

b) Angiospermophytina

α) Allgemeines. Die Werke von BECK (1976b) und HUGHES (1976) geben einen ganz ausgezeichneten Überblick über unseren derzeitigen Wissens- und Forschungsstand in der Frage der Angiospermen-Entstehung. Auch in diesem Berichtszeitraum beschäftigen sich eine ganze Anzahl von Einzelarbeiten mit dem ewig jungen Thema "präkretazische Angiospermen". Hier ist in erster Linie die Abhandlung von SCHWEITZER (1977) über die rhätische "Zwitterblüte" *Irania hermaphroditica* zu nennen. Nun ist zwar Zwittrigkeit allein kein Angiospermen-Charakteristikum, aber in Zusammenhang mit den übrigen an *Irania* festzustellenden Merkmalen - schraubige Anordnung von wahrscheinlich bisporangiaten "Staubblättern" und darüber stehende, geschlossene, bei Reife aber aufplatzende Samen-kapseln (der Autor selbst gebraucht die neutralen Begriffe Mikro- und Megasporangiophore) - drängt sich angesichts des geologischen Alters die begründete Ansicht auf, bei *Irania* hätten wir es mit einer weiteren Vertreterin der sogenannten angiospermiden Pflanzengruppe zu tun. Dem Entdecker ist beizupflichten, wenn er vorläufig aber in dem Fossil eine Gymnosperme sieht und *Irania*, zu der anscheinend *Desmiophyllum*-Blätter gehören, mit den Caytoniales und Czekanowskiales vergleicht, die beide ja auch das "Proangiospermen"-Stadium erreicht haben. Ähnliches Interesse verdient eine Publikation von KRASSILOV (1975) über eine weitere angiospermide Familie aus dem Zeitraum O.-Jura bis U.-Kreide Sibiriens, die Dirhopalostachyaceae. Bei den gefundenen Resten handelt es sich um traubige Megasporangienstände, an denen paarweise einsamige, ventral dehiszierende Kapseln sitzen. Beachtenswert ist ferner der Hinweis, daß dazu vermutlich *Nilssonia*-Blätter gehören. Angesichts dieser überraschenden Entdeckungen verwundert auch das Ergebnis einer Nachuntersuchung der O.-Trias-Pflanze *Sanmiguelia* durch TIDWELL et al. (1977) nicht, welches besagt, daß nach wie vor (Fortschr. Bot. 21) am ehesten mit breitblättrigen Monokotylen Ähnlichkeit besteht. ASH (1976a) konnte sie übrigens auch für Texas nachweisen. In diesem Zusammenhang ist auch das Ergebnis einer Untersuchung von DOYLE et al. (1975) zu nennen, die feststellten, daß der mesozoische *Eucommiidites*-Pollen (Fortschr. Bot. 15) tatsächlich nach seiner Ultrastruktur zu keiner der lebenden Gymnospermengruppen und erst recht nicht zu den Angiospermen gehören kann. Auch hier wird an die Zugehörigkeit zu einem ausgestorbenen Taxon gedacht. Während in aller Regel Kreide-Angiospermen als Blattfossilien überliefert sind (FILIPPOVA, 1978; KNOBLOCH, 1978; RÜFFLE und KNAPPE, 1977; VACHRAMEEV und KOTOVA, 1977) machen Reste aus der Oberkreide von Kansas, über die DILCHER et al. (1976a) vorläufig referieren, eine seltene Ausnahme. Die Funde, kräftige, balgähnliche Früchte in spiraliger Anordnung an dicken Achsen, zu denen wahrscheinlich große Blätter (*Liriophyllum*) gehören, geben, weil mazerierfähig, die große Chance, einmal vollständige Kreidefrüchte

im Detail zu untersuchen. Jetzt schon ist die ja auch theoretisch zu
fordernde Primitivität der "Bälge", die wohl zu einer Angehörigen der
Magnoliidae oder Hamamelididae gehören, hervorzuheben. Dem Ziel, das
große Reservoir der zu Angiospermen gehörende Blattabdrücke auf ex-
akte Art und Weise nützen zu können, dienen in jüngster Zeit wieder
verschiedene Publikationen. MELVILLE (1976) unterbreitet dazu ein
Arsenal von recht brauchbaren Begriffen (Fortschr. Bot. 35, 37); DOLPH
(1976) stellt taximetrische Versuche an. Seine Ergebnisse sind aber
einstweilen eher negativ zu beurteilen. Sehr aufschlußreich verliefen
biochemische Parallel-Untersuchungen an der miozänen Succur-Creek-
Flora (Oregon, USA). Zwar sind die Unterschiede zwischen fossilen und
rezenten Arten gering, aber in einigen Fällen ließen sie NIKLAS und
GIANNASI (1977, 1978) andere Artverwandtschaften vermuten als bisher
auf Grund der Morphodiagnose angenommen.

Darüber hinaus liegen in größerer Zahl Spezialarbeiten über Angiosper-
men-Reste vor. Soweit sie systematisch-taxonomisch relevant erscheinen, sei im Folgenden ein kurzer Überblick gegeben.

β) *Magnoliatae*. BASINGER (1976) (Rosaceae); BIRADAR und MAHABALE (1976)
(Lythraceae); CREPET et al. (1975) (Juglandaceae); CHITALEY und KATE
(1977) (Lythraceae); CHITALEY und PATEL (1975) (Lythraceae); COLLINSON
und CRANE (1978) (Ericaceae); DILCHER et al. (1976b) (Juglandaceae);
DOLPH (1975) (Apocynaceae); FRIEDRICH und SIMONARSON (1976) (Aceraceae);
GREGOR (1977a,b, 1978a,b,c) (Myristicaceae, Theaceae, Rutaceae, Ole-
aceae, Mastixiaceae u.a.); HOLY (1975) (Mastixiaceae); JUNG (1978)
(Menispermaceae, Theaceae); JENTYS-SZAFEROWA (1975) (Betulaceae);
JÄHNICHEN und WALTHER (1974) (Oleaceae); KNAPPE und RÜFFLE (1975a,b)
(Platanaceae, Monimiaceae); LAŃCUCKA-ŚRODONIOWA (1975) (Saxifragaceae);
MAI (1975) (Theaceae); MAI und GIVULESCU (1976) (Annonaceae); MATSUO
(1975)(Hamamelidaceae); NIREI (1975) (Juglandaceae); PRAKASH (1975)
(Guttiferae); PRŮHAŽKA und BŮŽEK (1975) (Aceraceae); PUNT (1978)
(Loganiaceae); ROIRON und VERNET (1978) (Betulaceae); STRAUS (1977a)
(Myricaceae); TANAI (1977, 1978) (Fagaceae, Aceraceae, Nyssaceae);
TANAI und OZAKI (1977) (Aceraceae); TANAI und YOKOYAMA (1975) (Fag-
aceae); TIFFNEY und BARGHOORN (1976) (Vitaceae); VAUDOIS-MIÉJA (1978)
(Juglandaceae).

γ) *Liliatae*. BŮŽEK (1977) (PALMAE); CREPET (1978) (Araceae); DAGHLIAN
(1978) (Palmae); DILCHER und DAGHIAN (1977) (Araceae); HICKEY und
PETERSON (1978) (Zingiberaceae); KNOBLOCH (1977) (Potamogetonaceae);
KNOBLOCH und MAI (1975) (Sparganiaceae); MADISON und TIFFNEY (1976)
(Araceae); NAMBUDIRI und TIDWELL (1978) (Pandanaceae); PATIL und SINGH
(1978) (Pontederiaceae).

Literatur

ALVAREZ RAMIS, C., DOUBINGER, J., GERMER, R.: Palaeontographica B
165, 1-42 (1978). - ALVIN, K.L.: Palaeontology 20, 387-404 (1977). -
AMERON, H.W.J. van: Meded. Rijks Geol. Dienst Ser. C-III-1, 7, 5-208
(1975). - ANDERSON, H.M.: Palaeontogr. Afr. 19, 21-30 (1976). -
ANDREWS, H.N., GENSEL, P.G.: Can. J. Bot. 53, 1719-1728 (1975). -
APPERT, O.: Palaeontographica B 162, 1-50 (1977). - ASH, S.R.: J.
Paleontol. 50, 799-804 (1976a); - Am. J. Bot. 63, 1327-1331 (1976b);
- Palaeontology 20, 77-79 (1977).

BANKS, H.P.: Rev. Palaeobot. Palynol. 20, 13-25 (1975a); - Taxon 24,
401-413 (1975b). - BANKS, H.P., LECLERCQ, S., HUEBER, F.M.: Palae-

ontogr. Am. 8, 77-127 (1975). - BARALE, G.: Geobios 8, 181-184 (1975).
- BARTHEL, M.: Hallesches Jb. Geowiss. 2, 73-86 (1977). - BASINGER,
J.: Can. J. Bot. 54, 2293-2305 (1976). - BATENBURG, L.H.: Cour. Forsch.
Inst. Senckenberg 17, 73 (1976); - Rev. Palaeobot. Palynol. 24, 69-99
(1977). - BAXTER, R.W.: Palaeontographica, B 165, 79-84 (1978). -
BECK, CH.B.: Rev. Palaeobot. Palynol. 21, 5-23 (1976a); - (ed.): Origin
and early Evolution of Angiosperms. 341 S. New York, London: Columbia
Univ. Press 1976b. - BENECKE, A.K.: Palaeontogen. Afr. 19, 97-125
(1976). - BIRADAR, N.V., MAHABALE, T.S.: Palaeobotanist 23, 25-29
(1976). - BOERSMA, M., BROEKMEYER, L.M.: Spec. Publ. Labor. Palaebot.
Palynol. 1, 1-183 (1979). - BROWN, J.T.: J. Paleontol. 49, 724-730
(1975); - Palaeontogr. Afr. 20, 47-51 (1977). - BURGH, J. VAN DER:
Cour. Forsch. Inst. Senckenberg 24, 89-91 (1977). - BUSCHE, R., HASS,
R., REMY, W.: Argumenta Palaeobot. 5, 149-160 (1978). - BUTALA, J.R.,
CRIDLAND, A.A.: Taxon 27, 15-20 (1978). - BŮŽEK, Č.: Vestn. Ustred
Ustavu Geol. 52, 159-168 (1977).

CHALONER, W.G., COLLINSON, M.E.: Rev. Palaeobot. Palynol. 20, 85-101
(1975a). - CHALONER, W.G., MENSAH, M.K., CRANE, M.D.: Palaeontology
17, 925-947 (1974). - CHALONER, W.J., COLLINSON, M.E.: Proc. Geol.
Assoc. 86, 1-44 (1975b). - CHANDRA, S., SURANGE, K.R.: Palaeonto-
graphica B 156, 87-102 (1976); - Palaeobotanist 23, 161-175 (1977a);
- Palaeontographica B 164, 127-152 (1977b); - Palaeobotanist 24,
149-160 (1977c). - CHITALEY, S.D., KATE, U.R.: Rev. Palaeobot. Palynol.
23, 389-398 (1977). - CHITALEY, S.D., PATEL, M.Z.: Palaeontographica
B 153, 141-149 (1975). - COLLINSON, M.E., CRANE, P.R.: Bot. J. Linn.
Soc. 76, 195-205 (1978). - CORNET, B., PHILLIPS, T.L., ANDREWS, H.N.:
Palaentographica B 158, 105-129 (1976). - CREPET, E.L., DILCHER, D.L.,
POTTER, F.W.: Am. J. Bot. 62, 813-823 (1975). - CREPET, W.L.: Rev.
Palaeobot. Palynol. 25, 241-252 (1978). - CRIDLAND, A.A., BUTALA, J.R.:
Argumenta Palaeobot. 4, 79-82 (1975).

DAGHLIAN, Ch.: Palaeontographica B 166, 44-82 (1978). - DAVEY, R.J.,
VERDIER, J.P.: Rev. Palaeobot. Palynol. 22, 307-335 (1976). - DELE-
VORYAS, T., HOPE, R.C.: Rev. Palaeobot. Palynol. 20, 67-74 (1975);
- Rev. Palaeobot. Palynol. 21, 93-100 (1976). - DELEVORYAS, T.,
PERSON, C.P.: Palaeontographica B 154, 114-120 (1975). - DENNIS, R.L.,
EGGERT, D.A.: Bot. Gaz. 139, 117-139 (1978). - DILCHER, D.L., DAGHLIAN,
C.P.: Am. J. Bot. 64, 526-534 (1977). - DILCHER, D.L., CREPET, E.L.,
BEEKER, C.D.: Science 191, 854-856 (1976a). - DILCHER, D.L., POTTER,
F.W., CREPET, W.L.: Am. J. Bot. 63, 532-544 (1976b). - DOBRUSKINA,
I.A.: Paleontol. Zh. 4, 120-132 (1975). - DOLPH, G.: Palaeontographica
B 151, 1-51 (1975); - Palaeontographica B 156, 65-86 (1976). - DOLU-
DENKO, M.P.: Paleontol. Zh. 3, 107-121 (1978). - DOROFEEV, P.I.:
Paleontol. Zh. 1, 105-116 (1975). - DOUBINGER, J., GERMER, R.:
Palaeontographica B 153, 1-27 (1975). - DOYLE, J.A., VAN CAMPO, M.,
LUGARDON, B.: Pollen Spores 17, 429-486 (1975).

EHRET, D.L., PHILLIPS, T.L.: Palaeontographica B 161, 147-164 (1977).

FAIRON-DEMARET, M.: Palaeontographica B 162, 51-63 (1977); - Rev.
Palaeobot. Palynol. 26, 9-20 (1978). - FERGUSON, D.K.: Rev. Palaeo-
botan. Palynol. 26, 213-226 (1978). - FERGUSON, D.K., JÄHNICHEN, H.,
ALVIN, K.L.: Feddes Repert. 89, 379-410 (1978). - FILIPPOVA, G.:
Paleontol. Zh. 1, 138-144 (1978). - FLAJS, G.: Palaeontographica
B 160, 69-128 (1977). - FLÜGEL, E.: Verh. Geol. Bundesanst. 1974,
297-346 (1975); - (ed.): Fossil Algae. Recent results and developments.
375 S. Berlin-Heidelberg-New York: Springer 1977. - FRIEDRICH, W.L.,
SIMONARSON, L.A.: Palaeontographica B 155, 140-148 (1976). - FRIEDRICH,
W.L., STRAUCH, F.: Bot. Not. 128, 339-349 (1975).

GALTIER, J., PHILLIPS, T.L.: Palaeontographica B 164, 33-75 (1977).
- GASTALDO, R., MATTEN, L.C.: Palaeontographica B 165, 43-52 (1978).
- GENSEL, P.G.: Rev. Palaeobot. Palynol. 22, 19-37 (1976); - Brittonia 29 (1), 14-29 (1977). - GLUCHOVA, L.V.: Paleontol. Zh. 4, 115-
121 (1978). - GOCHT, H., NETZEL, H.: Neues Jahrb. Geol. Palaeontol.
Abh. 3, 380-413 (1976). - GOOD, CH.W.: Palaeontographica B 153,
28-99 (1975); - Am. J. Bot. 63, 719-725 (1976); - Geobotany 1977,
43-64 (1977). - GRAUVOGEL-STAMM, L., GRAUVOGEL, L.: Geobios 8, 143-
146 (1975). - GREGOR, H.-J.: Palaeontol. Z. 51, 199-226 (1977a);
- Mitt. Bayer. Staatssamml. Palaeontol. Hist. Geol. 17, 249-265
(1977b); - Palaeontol. Z. 52, 198-204 (1978a); - Mitt. Bayer. Staatssamml. Palaeontol. Hist. Geol. 18, 143-166 (1978b); - Feddes Repert.
88, 645-653 (1978c).

HALLBAUER, D.K.: Miner. Sci. Eng. 7, 111-131 (1975). - HALLBAUER, D.K.,
JAHNS, H.M., BELTMANN, H.A.: Geol. Rundsch. 66, 477-491 (1977). -
HALLE, T.G.: Svensk bot. Tidskr. 22 (1-2), 230-255 (1928). - HARLAN,
R.: Palaeontographica B 164, 87-126 (1977). - HARRIS, T.M.: Rev.
Palaeobot. Palynol. 21, 119-134 (1976a); - Am. J. Bot. 63, 902-910
(1976b). - HICKEY, L.J., PETERSON, R.K.: Can. J. Bot. 56, 1136-1152
(1978). - HILTERMANN, H., MÄDLER, K.: Palaeontol. 51, 135-144 (1977).
- HIRMER, M.: Handbuch der Paläobotanik. 1. Thallophyten, Bryophyten,
Pteridophyten. 780 S. München, Berlin: Oldenbourg 1927. - HLŮSTIK, A.:
Sb. Nar. Muz. Praze Rada B 30, 49-70 (1976); - Sb. Nar. Muz. Praze
B 30, 173-186 (1977a); - Vestn. Ustred. Ustavu Geol. 52, 359-366
(1977b). - HLŮSTIK, A., KONZALOVÁ, M.: Vestn. Ustred. Ustavu Geol.
51, 37-45 (1976). - HOLMES, J.C.: Palaeontographica B 164, 33-75
(1977). - HOLMES, J., GALTIER, J.: Rev. Palaeobot. Palynol. 22, 207-
224 (1976). - HOLÝ, F.: Sb. Nar. Muz. Praze Rada B 31, 109-238 (1975).
- HUGHES, N.F.: Palaeobiology of Angiosperms Origins. 242 S. Cambridge,
London, New York, Melbourne: Cambridge Univ. Press 1976.

JÄHNICHEN, H., WALTHER, H.: Feddes Repert. 85, 17-41 (1974). -
JANSONIUS, J.: Geosci. Man 15, 129-132 (1976). - JENNINGS, J.R.:
J. Paleontol. 49, 52-57 (1975a); - Palaeontology 18, 19-24 (1975b);
- Am. J. Bot. 63, 1119-1133 (1976). - JENNINGS, J.R., EGGERT, D.A.:
Rev. Palaeobot. Palynol. 24, 221-225 (1977). - JENTYS-SZAFEROWA, J.:
Acta Palaeobot. 16, 3-70 (1975). - JUNG, W.W., SCHLEICH, H., KÄSTLE,
B.: Mitt. Bayer. Staatssamml. Palaeontol. Hist. Geol. 18, 131-142
(1978). - JURINA, A., LEMOIGNE, Y.: Palaeontographica B 150, 162-168
(1975). - JUX, U.: Palaeontographica B 149, 113-138 (1975); - Palaeonto-
graphica B 160, 1-16 (1977).

KEUPP, H.: Palaeontol. Z. 51, 102-116 (1977). - KIMURA, T., SEKIDO,
S.: Palaeontographica B 153, 111-118 (1975). - KLAUS, W.: Beitr.
Paläontol. Österr. 3, 105-127 (1977). - KNAPPE, H., RÜFFLE, L.:
Wiss. Z. Humboldt Univ. Berlin, Math. Naturwiss. Reihe 24, 487-492
(1975a); - Wiss. Z. Humboldt Univ. Berlin, Math. Naturwiss. Reihe
24, 493-499 (1975b). - KNOBLOCH, E.: Cas. Miner. Geol. 22, 29-42
(1977); - Palaeontographica B 166, 93-98 (1978). - KNOBLOCH, E., MAI,
D.H.: Cas. Miner. Geol. 20, 141-147 (1975). - KONIJNENBURG-VAN CITTERT,
J.H.A. VAN: Rev. Palaeobot. Palynol. 20, 205-214 (1975); - Rev.
Palaeobot. Palynol. 26, 125-141 (1978). - KOVÁCS-ENDRÖDY, E.: Palae-
ontogr. Afr. 17, 11-14 (1974); - Palaentogr. Afr. 19, 67-95 (1976). -
KRÄUSEL, R., WEYLAND, H.: Senckenbergiana 14, 391-403 (1932). -
KRASSILOV, V.: Palaeontographica B 153, 1-27 (1975); - Rev. Palae-
obot. Palynol. 24, 155-178 (1977); - Rev. Palaeobot. Palynol. 26,
113-124 (1978a); - Palaeontographica B 166, 16-29 (1978b). - KRASSILOV,
V.A., ZAKHAROV, YU.D.: Rev. Palaeobot. Palynol. 19, 221-232 (1975).

ŁAŃCUCKA-ŚRODONIOWA, M.: Acta Palaeobot. 16, 103-112 (1975). - LANGER,
W.: Palaeontol. Z. 50, 209-221 (1976). - LAVEINE, J.P., COQUEL, R.,
LOBOZIAK, S.: Geobios 6, 757-848 (1977). - LEARY, R.L., PFEFFERKORN,
H.W.: Ill. State Geol. Surv. Circ. 500, 1-77 (1977). - LEISTIKOW, K.U.:
Cour. Forsch. Inst. Senckenberg 13, 140-143 (1975). - LEJAL-NICOL, A.:
Palaeontographica B 151, 1-96 (1975). - LE ROUX, S.F.: Palaeontogr.
Afr. 18, 31-34 (1975). - LIMA, M.R. de: Ameghiniana 13, 226-234 (1976).

MADISON, M., TIFFNEY, B.H.: J. Arnold Arbor. 57, 185-201 (1976). -
MAI, D.H.: Wiss. Z. Friedrich Schiller Univ. Jena, Math. Naturwiss.
Reihe 24, 463-476 (1975). - MAI, D.H., GIVULESCU, R.: Rev. Roum.
Ser. Geol. 20, 277-281 (1976). - MAMAY, S.H.: Prof. Paper U.S. Geol.
Surv. 934, 1-76 (1976). - MUTSUO, H.: Jpn. J. Geol. Geogr. 45, 1-8
(1975). - MELVILLE, R.: Taxon 25, 549-561 (1976). - MEYEN, S.V.:
Paleontol. Zh. 3, 128-138 (1977). - MILLAY, M.A.: Am. J. Bot. 64,
223-229 (1979). - MILLAY, M.A., TAYLOR, T.N.: Rev. Palaebot. Palynol.
21, 65-91 (1976); - Am. J. Bot. 64, 177-185 (1977); - Rev. Palaeobot.
Palynol. 25, 151-161 (1978). - MILLER, C.N.: Am. J. Bot. 62, 706-713
(1975); - Rev. Palaeobot. Palynol. 21, 101-117 (1976); - Bull Torrey
Bot. Club 104, 5-8 (1977); - Bot. Gaz. 139, 284-287 (1978). - MILLER,
Ch.N., ROBINSON, C.R.: J. Paleontol. 49, 138-150 (1975). - MOREY, E.D.,
MOREY, P.R.: Palaeontographica B 162, 64-69 (1977). - MUSTAFA, H.:
Argumenta Palaeobot 4, 101-133 (1975); - Argumenta Palaeobot. 5, 31-
56 (1978).

NAMBUDIRI, E.M., TIDWELL, W.D.: Palaeontographica B 166, 30-43 (1978).
- NAUTIYAL, A.CH.: Curr. Sci. 45, 609-611 (1976). - NIKLAS, K.J.:
Rev. Palaeobot. Palynol. 21, 187-203 (1976a); - Rev. Palaeobot.
Palynol. 21, 205-217 (1976b); - Rev. Palaeobot. Palynol. 22, 1-17
(1976c); - Rev. Palaeobot. Palynol. 22, 265-279 (1976d). - NIKLAS,
K.J., GIANNASI, D.E.: Science 196, 877-878 (1977); - Am. J. Bot. 65,
943-952 (1978). - NIKLAS, K.J., PHILLIPS, T.L., CAROZZI, A.V.: Palao-
ontographica B 155, 1-30 (1976). - NIREI, H.: Geoscienc. Osaka City
Univ. 19, 31-61 (1975).

PALAMAREV, E., PETROVA, A., USMOVA, K.: Phytology 2, 25-33 (1975). -
PANT, D.D., BHATNAGAR, S.: Palaeontographica B 152, 191-199 (1975).
- PANT, D.D., CHODHURY, A.: Paleontographica B 164, 153-166 (1977).
- PANT, D.D., MISRA, L.: Palaeontographica B 164, 76-86 (1977). -
PATIL, G.V., SINGH, R.R.: Palaeontographica B 167, 1-7 (1978). -
PFEFFERKORN, H.W.: Fieldiana Geol. 33, 315-322 (1976a); - Ill. State
Geol. Surv. Circ. 492, 1-31 (1976b); - Argumenta Palaeobot. 5, 167-
193 (1978). - PFLUG, H.D.: Palaeontographica B 158, 130-168 (1976a);
- Palaeont. Z. 50, 15-26 (1976b). - PRAKASH, U.: Palaeobotanist 22,
63-75 (1975). - PRATT, L.M., PHILLIPS, T.L., DENNISON, J.M.: Rev.
Palaeobot. Palynol. 25, 121-149 (1978). - PROHAŹKA, M., BUŽEK, Č.:
Rozpr. Ustred. Ustavu Geol. 41, 5-86 (1975). - PUNT, Rev. Palaeobot.
Palynol. 26, 312-335 (1978).

REIHMAN, M.A., SCHABILION, J.T.: Am. J. Bot. 65, 834-844 (1978). -
REMY, R., REMY, W.: Argumenta Palaeobot. 4, 83-92 (1975). - REMY, W.:
Argumenta Palaeobot. 5, 23-30 (1978). - REMY, W., REMY, R.: Argumenta
Palaeobot. 4, 13-29 (1975a); - Argumenta Palaeobot. 4, 135-138 (1975c);
- Argumenta Palaeobot. 4, 31-37 (1975b); - Die Floren des Erdaltertums.
468 S. Essen: Glückauf 1977; - Argumenta Palaeobot. 5, 1-10 (1978a);
- Argumenta Palaeobot. 5, 133-147 (1978b). - REMY, W., REMY, R.,
PFEFFERKORN, H.W., VOLKHEIMER, W., RABE, E.: Argumenta Palaeobot. 4,
55-77 (1975). - REYMANÓWNA, M.: Acta Palaeobot. 14, 45-87 (1973). -
REYMANÓWNA, M., WATSON, J.: Acta Palaeobot. 17, 17-26 (1976). -
ROBINSON, C.R., MILLER, C.N.: Am. J. Bot. 64, 770-779 (1977). -
ROIRON, F., VERNET, J.L.: Geobios 11, 799-816 (1978). - ROTHWELL, G.W.:

Palaeontographica B 151, 171-196 (1975). - RÜFFLE, L., KNAPPE, H.:
Z. Geol. Wiss. Berlin 5, 269-302 (1977).

SCHAARSCHMIDT, F.: Cour. Forsch. Inst. Senckenberg 17, 90 (1976). -
SCHABILION, J.T.: Rev. Palaeobot. Palynol. 20, 103-108 (1975). -
SCHECKLER, S.E.: Can. J. Bot. 53, 25-38 (1975a); - Am. J. Bot. 62,
923-934 (1975b); - Can. J. Bot. 54, 202-219 (1976). - SCHEURING, B.W.:
Pollen Spores 18, 611-639 (1976). - SCHOPF, J., OEHLER, D.Z.: Science
193, 47-49 (1976). - SCHOPF, J.M.: Rev. Palaeobot. Palynol. 21, 25-64
(1976). - SCHOPF, J.W.: Annu. Rev. Earth Planet Sci. 3, 213-249 (1975a);
- Endeavour 34 (122), 51-58 (1975b). - SCHULTKA, H.: Argumenta Palae-
obot. 5, 11-22 (1978). - SCHWEITZER, H.-J.: Palaeontographica B 161,
98-145 (1977); - Palaeontographica B 168, 17-60 (1978). - SHAILA, C.,
SURANGE, K.R.: Palaeobotanist 24, 195-201 (1977). - SHARMA, B.D.:
Geobios 4, 503-506 (1976); - Acta Palaeobot. 18, 19-29 (1977). -
SHARMA, B.D., BOHRA, D.R.: Acta Palaeobot. 18, 31-36 (1977). - SIEGERT,
H.: Cour. Forsch. Inst. Senckenberg 34, 107-108 (1978). - SIVAK, J.:
Pollen Spores 18, 243-288 (1976). - SKOG, J.E.: Am. Fern J. 66, 8-14
(1976). - SRIVASTAVA, S.K.: Lethaia 9, 437-457 (1976). - STIDD, B.M.:
Am. J. Bot. 65, 677-683 (1978a); - Am. J. Bot. 65, 243-245 (1978b).
- STIDD, B.M., COSENTINO, K.: Science 190, 1092-1093 (1975); - Bot.
Gaz. 137, 242-249 (1976). - STIDD, B.M., OESTRY, L.L., PHILLIPS, T.L.:
Rev. Palaeobot. Palynol. 20, 55-56 (1975). - STIDD, B.M., LEISMAN,
G.A., PHILLIPS, T.L.: Am. J. Bot. 64, 994-1002 (1977). - STIDD, L.L.,
STIDD, B.M.: Science 193, 156-157 (1976). - STOCKEY, R.: Am. J. Bot.
64, 733-744 (1977); - Palaeontographica B 166, 1-15 (1978). - STRAUS,
A.: Cour. Forsch. Inst. Senckenberg 24, 87 (1977a); - Verh. Bot. Ver.
Provinz Brandenburg 113, 43-80 (1977b). - SWOBODA-PĚKNÁ, M.: Vestn.
Ustred. Ustavu Geol. 52, 352-358 (1977).

TANAI, T.: J. Fac. Sci. Hokkaido Univ. Ser. IV 17, 505-516 (1977); -
J. Fac. Sci. Hokkaido Univ. Ser. IV 18, 243-282 (1978). - TANAI, T.,
OZAKI, K.: J. Fac. Sci. Hokkaido Univ. Ser. IV 17, 575-606 (1977).
TANAI, T., YOKOYAMA, A.: J. Fac. Sci. Hokkaido Univ. Ser. IV 17,
129-141 (1975). - TAYLOR, T.N., STOCKEY, R.: Am. J. Bot. 63, 1302-
1310 (1976). - TIDWELL, W.D., SIMPER, A.D., THAYN, G.: Palaeonto-
graphica B 163, 143-151 (1977). - TIFFNEY, B.H., BARGHOORN, E.S.:
Rev. Palaeobot. Palynol. 22, 169-191 (1976). - TSEYEN, C., CHENGYAO,
T.: Acta Palaeontol. Sin. 16, 180-189 (1977).

VACHRAMEEV, V.A., KOTOVA, I.Z.: Paleontol. Zh. 4, 101-109 (1977). -
VAUDOIS-MIÉJA, N.: Rev. Palaeobot. Palynol. 25, 269-294 (1978). -
VOZENIN-SERRA, C., LAROCHE, J.: Palaeontographica B 159, 158-166
(1976).

WAGNER, R.H., SPINNER, E.: C. R. Acad. Sci. Ser. D 282, 353-356
(1976). - WATSON, J.: Palaeontology 20, 715-750 (1977). - WEBER, R.:
Palaeontographica B 152, 76-83 (1975). - WHITE, M.E.: Aust. Mus. 31,
473-504 (1978).

Professor Dr. WALTER JUNG
Institut für Paläontologie und historische
Geologie der Universität
Richard-Wagner-Straße 10/II
D 8000 München

Dr. FRIEDEMANN SCHAARSCHMIDT
Forschungsinstitut Senckenberg
Senckenberganlage 25
D 6000 Frankfurt am Main

E. Geobotany

I. Areal- und Florenkunde (Floristische Geobotanik)

Von Eckehart J. Jäger

Im diesjährigen Bericht wird über die floristische Erkundung Afrikas, Australiens und Ozeaniens berichtet, dazu über neue Arbeiten zur Florenstruktur und Florengenese Afrikas und des westlichen Nordamerika. Über Floren aus der Holarktis und Südamerika vgl. Prog. Bot. 40, 413 und wieder in den nächsten Berichten, ebenso über ökologische Areal-Interpretation.

1. Forschungstrends

Die vergleichende Florenkunde hat jetzt besonders reiche Arbeitsmöglichkeiten, weil die letzten Gebiete der Erde zugänglich werden, ihre Floren aber noch den natürlichen Bestand erkennen lassen. Die Erforschung der *floristischen Grundlagen* hält daher unvermindert an. Auch in den Tropen sind die Länder Ausnahmen, in denen während des letzten Jahrzehnts nicht an einer neuen Standardflora gearbeitet wurde.

Damit ist Material für exaktere *Vergleiche der Floren*, ihres Arten- und Endemitenreichtums und für Untersuchungen ihrer geographischen Beziehungen gegeben.

Rasch schreitet die *Kartierung der Pflanzenareale* voran. Von den 70 Mill. km² des Holarktischen Florenreiches sind schon ziemlich genau 10% in mehr oder weniger vollständigen Kartenfloren bearbeitet. Auch der Kartierung der Pilze, Flechten, Algen und Moose wird immer mehr Aufmerksamkeit gewidmet. Von tropischen und südhemisphärischen Sippen erscheinen in taxonomischen Monographien viele gute Punktkarten.

Die *ökologische Interpretation* der Areale wird durch Untersuchungen des Konkurrenzverhaltens, der Populationsökologie, der Produktionsbiologie, der Klimaresistenz, der endogenen Ruheperioden und der Ausbreitungsbiologie bereichert.

Die *genetische Phytochorologie* profitiert von der besseren Kenntnis der Areal-Differenzierungsmuster und der geographischen Beziehungen der Floren, aber auch von den engen Wechselbeziehungen zur Evolutionsforschung, Cytogeographie und Chemotaxonomie. Die Entwicklung karpologischer, palynologischer und cuticularanalytischer Methoden verbessert die Bestimmungsmöglichkeiten von Fossilien, so daß Spekulationen über Ursprungszentren und über Migration allmählich durch gesicherte Rekonstruktionen der Chorogenese der Taxa, der Genese der Floren und der Entwicklung der Florengebiete abgelöst werden.

Die *Erfassung und Prognose anthropogener Florenveränderungen* bleibt weiterhin ein Hauptgegenstand der chorologischen Forschung. Für Maßnahmen zum Schutz aussterbender Arten werden (vorläufig allerdings fast ausschließlich für Gefäßpflanzen) anhand von Verbreitungskarten und Populationsstudien differenzierte Hinweise gegeben. In einigen Ländern werden die Vorkommen der gefährdeten Arten bereits regelmäßig kontrolliert.

2. Taxonomische Grundlagen

Sichere taxonomische Grundlagen sind das A und O der chorologischen
Arbeit. Erst bei weiter Fassung der eventuell in Subspecies unter-
gliederten Arten und bei sorgfältiger Abgrenzung auch von Sektionen
und Series werden viele Zusammenhänge erkennbar.

Wenn der am Oranje-River gefundene *Rumex garipensis* mit dem russischen
R. marschallianus identisch ist (MERXMÜLLER, 1976), erhebt sich sofort
die Frage nach dem Zustandekommen der weiten Disjunktion (eventuell
synanthrop?). Dagegen führt die Aufdeckung der Verwandtschaft von
Pelargonium endlicherianum (sect. *Jenkinsonia*) mit den Sektionen *Cortusina*
und *Eumorpha* zur wesentlichen Verringerung des Abstandes der disjunk-
ten Arealteile des Verwandtschaftskreises (Orient + Erythräa statt
Orient + Südafrika, MARKGRAF, 1976). Wenn alle Vorkommen von *Dryopteris
austriaca*, *D. dilatata*, *D. extremiorientalis* und *D. alexeenkoana* aus dem Kauka-
sus, Sibirien und dem Fernen Osten von PIĘKOS-MIRKOWA (1977) zu
Dryopteris assimilis gestellt werden, entsteht ein ganz neues Bild der
geographischen Gliederung dieses Komplexes. Nachdem *Himantothallus grandi-
folius* zu den Desmarestiales gestellt wurde, gibt es in allen antark-
tischen Meeren keine Laminariales mehr (MOE und SILVA, 1977).

> Andererseits haben chorologische Untersuchungen heuristischen Wert für die
> Taxonomie. Dafür gibt es auch aus der letzten Zeit Beispiele von den Moosen
> (Vikariismus, Differenzierungsmuster und ökologische Charakteristik decken
> Verwandtschaftsbeziehungen und Sippengrenzen auf, ABRAMOVA und ABRAMOV, 1978)
> bis hinauf zu den Asteraceae (natürlichere Gliederung der Lactuceae aufgrund
> der Chorologie; ökogeographische Charakterisierung der Anthemideae durch Kar-
> tierung der Sippenhäufung, HEYWOOD et al., 1977).

3. Floristische Grundlagen

a) Afrika

Afrika und Madagaskar beherbergen fast ein Viertel aller Gefäßpflan-
zen-Arten. Noch erfolgt dort die Zerstörung der tropischen Vegetation
rascher als die Erforschung der Flora; deshalb ist in den weniger bekannten Teilen intensivere Sammelarbeit nötig (PRANCE, 1978). Da die
Schaffung nationaler Forschungsstätten Zeit braucht und da das Herbar-
material meist in Europa oder Amerika liegt, wird - bis auf Ägypten,
Libyen und Südafrika - die Arbeit an den Floren noch von den ehemali-
gen Kolonialmächten getragen. Übersichten über die taxonomische und
floristische Arbeit geben außer dem jährlich erscheinenden AETFAT-
Index (Brüssel) PRANCE (1978), GOLDBLATT (1979), BRENAN (1979),
QUEZEL (1979).

Die afrikanischen Tropen sind unter den drei großen Tropengebieten
der Erde am besten erforscht. Charakteristisch für den - auch in Süd-
und Nordafrika - raschen Fortschritt ist es, daß fast alle in der
Floren-Bibliographie von BLAKE und ATWOOD (1942) genannten afrikanischen Standardfloren inzwischen durch neue ersetzt sind oder werden.
Das Gebiet des Kontinents wird von Norden nach Süden durch die unten
aufgeführten Länder- und Regionalfloren fast vollständig abgedeckt.
Diese Floren wurden meist auch in den letzten Jahren fortgesetzt. Etwa
35% der bekannten afrikanischen Gefäßpflanzenarten sind darin bereits
bearbeitet. Stand und Publikationstempo gehen aus der folgenden Übersicht hervor (nach GOLDBLATT, 1979; BRENAN, 1979; QUEZEL, 1979; PRANCE,

1978, und den Floren selbst; geschätzte Gesamtartenzahl schwankend, z.B. in Südafrika nach KILLICK, 1976: 17.000, nach GOLDBLATT, 1979: 18.500).

Titel und Herausgeber	publiz. Teile	bearb.	Arten von insgesamt
Flore de l'Afrique du Nord (MAIRE)	Bd. 1-14; 1952-1977	40%	6000
Students Flora of Egypt (TÄCKHOLM)	vollst. 1974	100%	2000
Flore du Sahara (OZENDA)	vollst. 1977	100%	1100
Flora of West Tropical Africa (HUTCHINSON und DALZIEL)	vollst. 1954-1971	100%	7400
Flora of East Tropical Africa (TURRILL und MILNE-REDHEAD)	1952-1978	40%	10-11000
Flore du Cameroun (AUBRÉVILLE)	Lf. 1-20; 1963-1978	16%	6500
Flore du Gabon (AUBRÉVILLE)	Lf. 1-24; 1961-1977	40%	3500
Flore du Congo, du Rwanda et du Burundi (ROBYNS)	Bd. 1-10; 1948-1963	40%	10000
Fortsetzung: Flore d'Afrique Centrale (LAWAIRÉE)	41 Lfn. 1967-1972		
Conspectus Florae Angolensis (EXELL et al.)	Bd. 1-4; 1937-1970 Farne 1977	40%	4600
Flora Zambesiaca (EXELL und WILD)	Bd. 1, 2, 3.1, 4, 10.1; 1960-1978	35%	6000
Flora of Southern Africa (CODD et al.)	Bd. 1, 9, 13, 16.1+2, 22, 26; 1963-1979	10%	18500

Von den in diesen Floren nicht berücksichtigten Gebieten liegt für Äthiopien und Somalia (zusammen 6300 spec.) der vollständige Conspectus von CUFODONTIS (1953-1972) vor; 29 kleine Familien wurden ausführlich in der "Adumbratio Florae Aethiopicae" bearbeitet (neu: DEFILIPP, 1976: Oxalidaceae; ROTI MICHELOZZI, 1978: Olacaceae). Für Sudan gibt es außer der älteren Flora von ANDREWS (1950-1956) die neue Lokalflora des Djebel Marra von WICKENS (1976b). Von Congo-Brazzaville und der Zentralafrikanischen Republik fehlen neue Floren.

Die Aufgabe der *Regionalfloren* muß vor allem die Klärung der Sippengrenzen sein; ihr Fortschritt ist daher oft mit der Einziehung von Arten verbunden. In Afrika ist die Zahl der jährlich eingezogenen Arten auf über die Hälfte der jährlich neu beschriebenen angestiegen (PRANCE, 1978).

Die Berechtigung von *Länder- und Lokalfloren* innerhalb der großen Regionalfloren ergibt sich aus dem rascheren Abschluß, der besseren Benutzbarkeit beim Bestimmen und der Möglichkeit genauerer chorologischer und ökologischer Angaben. Zum Gebiet der "Flore d'Afrique du Nord" gehört die libysche Flora, von der 37 Familien in Lieferungen bearbeitet sind (ALI et al., 1976-1977). Von BOULOS (1977) wurde für Libyen der erste Teil einer Florenliste zusammengestellt. Aus dem Gebiet der "Flora of West Tropical Africa" gibt es einige Länderfloren, z.B. bereits 4 von 10 Bänden der "Flore Illustrée du Sénégal" (BERHAUT, 1971-1975) oder den Katalog der Gefäßpflanzen von Niger (FABREGUES und LEBRUN, 1976) mit Verbreitungsangaben und einigen Punktkarten).

In Ostafrika wurden im ersten Band einer neuen, illustrierten "Flore du Rwanda" (TROUPIN, 1978) über 50% der behandelten 489 Dicotylen (bis Caesalpiniaceae) erstmals für Rwanda angegeben. Die "Flora de Moçambique" (FERNANDES, 1969-1973, im Gebiet der "Flora Zambesiaca", 36 Lieferungen) wurde wohl nach 1973 nicht fortgesetzt. Da die große Südafrika-Flora nicht bald vollendet werden kann, ist es besonders erfreulich, daß nach dem Abschluß der SW-Afrika-Flora (MERXMÜLLER, 1966-1974), sowie der Floren von Lesotho (GUILLARMOD, 1971) und Natal (ROSS, 1972) nun eine neue Flora von Swaziland (COMPTON, 1976, 2118 Arten) und eine neue Bearbeitung der 600 Asteraceae von Natal erschienen ist (HILLIARD, 1977; 12% der Flora, 124 *Senecio*, 123 *Helichrysum*). Auch für die Gehölze Südafrikas gibt es ein neues Florenwerk (PALGRAVE, 1977), das reich illustriert ist und für jede der 1155 heimischen Arten eine recht detaillierte Verbreitungskarte enthält. Die zweibändige Übersicht über alle Spermatophyten-Gattungen dieses Gebietes (DYER, 1975-1976) ist eine wichtige Orientierungs-hilfe, allerdings sind die angegebenen Artenzahlen nach einer Liste von GOLD-BLATT (1979) z.T. zu korrigieren. In der Flora von Madagaskar (HUMBERT, 1963-1975) sind noch eine ganze Reihe großer Familien zu bearbeiten (600 Leguminosen, 300 Melastomataceae, 600 Euphorbiaceae, 100 Araliaceae, 600 Rubiaceae u.a.). Mit zwei Lieferungen (Goodeniaceae und Campanulaceae, BOSSER et al., 1976) wurde 100 Jahre nach dem Erscheinen der letzten Flora der Maskarenen eine neue Flora dieser Inseln begonnen. Für Mauritius gibt es bereits eine neue Übersicht über die Farne (mit florengeographischer Analyse, LORENCE, 1978).

b) Malesien und Ozeanien

Eine ausgezeichnete Informationsquelle über alle floristischen Forschungsvorhaben und Veröffentlichungen in diesem Raum ist das "Flora Malesiana Bulletin" (Leiden, Bd. 32, 1979). Über den Stand der Forschungen berichtet auch PRANCE (1978). Die Zahl der Aufsammlungen weist innerhalb Malesiens Süd-Sumatra, Celebes, Zentralborneo und die Andamanen als die am wenigsten erforschten Gebiete aus. In Ozeanien ist auf den Cocos-Inseln und den Wallis- und Horne-Inseln nur sporadisch gesammelt worden. Zu den gut bekannten Floren gehören dagegen die von Java, der Malayischen Halbinsel und der Fidschi-Inseln (Vorarbeiten zu einer bereits fertiggestellten Flora: SMITH, 1978; Farnflora: BROWNLIE, 1977). In den bisher erschienenen 8 Spermatophytenbänden der "Flora Malesiana" (STEENIS, 1948-1978) sind von insgesamt schätzungsweise 25000 Arten 3800, also 15%, bearbeitet worden, von den 2500 Farnen etwa 18%. Zuletzt erschienen die Farne der *Lomariopsis*-Gruppe, die Ulmaceae (K aller Gattungen, diese deutlich zonal differenziert: *Ulmus* extratropisch, eine Art bis Flores und Celebes, *Celtis* austral-submeridional, *Trema* tropisch-subtropisch), Symplocaceae, Crypteroniaceae, Bignoniaceae, Onagraceae, Cornaceae, Iridaceae, Lamiaceae und Anacardiaceae. Schneller wird wahrscheinlich die malesische Gehölzflora abgeschlossen werden, von der der 3. Band vorliegt (WHITMORE, 1978), vielleicht auch das neu begonnene Handbuch der Flora von Papua-Neuguinea (1. Band mit 164 Arten aus 11 Familien, WOMERSLEY, 1978).

Die Revision der eigenartigen Flora Neukaledoniens wurde mit einem reich illustrierten Orchideenband fortgesetzt (HALLE, 1977, 172 Verbreitungskarten). Von Mikronesien gibt es 3 Lieferungen einer neuen Flora (zuletzt Convolvulaceae, FOSBERG und SAChET, 1977).

c) Australien

Eine neue Übersichtsflora nach Art des alten "BENTHAM" ist nach
CAROLIN (1978) für Australien nicht vorgesehen. So ist die taxonomische Abgrenzung und Gruppierung Monographien vorbehalten, die meistens
auch gute Punktkarten enthalten, wie die der Chloanthaceae (MUNIR,
1978, PK aller Arten; Mannigfaltigkeitszentrum: Südwestaustralien)
oder Combretaceae (BYRNES, 1977). Die südlichen und östlichen Teile
des Kontinents sind schon länger ziemlich gut erforscht. Von der
"Flora of New South Wales" sind nach dem Erscheinen der Plantaginaceae
(BRIGGS et al., 1977) und Tremandraceae (THOMPSON, 1978) 46 von 218
Familien bearbeitet. Queensland hat ein neues Handbuch der Farnpflan-
zen (ANDREWS, 1979) und den ersten Band einer Flora des Südostens des
Bundesstaates (Dicotylen: Casuarinales - Sapindales - STANLEY et al.,
1979). Zu der neuen Flora von Victoria gibt es bereits wieder Änderungen und Ergänzungen (BEAUGLEHOLE, 1978). Der erste Band der Flora
Südaustraliens (Pteridophyta und Monocotylen) liegt in einer völlig
revidierten und illustrierten Neuauflage vor (JESSOP, 1978). Erstmalig
gibt es eine vollständige illustrierte Flora mit Verbreitungsangaben
auch für Westaustralien (BLACKALL und GRIEVE, 1954-1975; Bericht über
die Forschungsarbeit: GRIEVE, 1975). Schlecht bekannt ist noch die
Westküste Nordaustraliens.

> Für das Studium synanthroper Arealveränderungen ist die Liste der synanthropen
> Arten in Neuseeland wichtig (SYKES, 1978: Lamiales; WEBB, 1978: Umbelliferae),
> da sie in dem neuen Florenwerk fehlen. Auch unter den 203 Farnpflanzen Neusee-
> lands sind 13 Synanthrope (und 83 Endemiten; Übersicht über die Florenbeziehun-
> gen: PARRIS, 1976). Neue Florenlisten gibt es aus diesem Raum für die kleine-
> ren Inseln Kermadec (SYKES, 1977, auch Synanthrope und Verwandtschaft der
> Flora), Lord-Howe-I. (RODD in RECHER, 1974), Campbell (MEURK, 1975), Auckland
> (JOHNSON und CAMPBELL, 1975, 187 von 257 Taxa heimisch, davon 30% endemisch,
> 14 Arten neu) und für die antarktischen Inseln (70 Arten, davon 10 synanthrop,
> auf South Georgia, Prince Edward-I., Crozet, Kerguelen, Heard- und Mc Donald-I.,
> Macquarie, GREENE und WALTON, 1975).

Das wachsende Interesse an den Kryptogamen spiegelt sich im Erscheinen
mehrerer Mooskonspekte. Die Hauptarbeit bereitet hier die taxonomische
Prüfung der oft unkontrolliert beschriebenen Arten. Nach kritischer Bearbeitung der Moose von Samoa erhielt SCHULTZE-MOTEL (1975a) eine Reduktion der Artenzahl gegenüber der nach der Literatur zusammengestell-
ten Liste um 20%. Solche kritischen Floren, wie sie auch für die Moose
Südaustraliens (SCOTT und STONE, 1976), Polynesiens (MILLER et al.,
1978) und der Gesellschafts-Inseln (WHITTIER, 1976) neu erschienen
sind, gestatten schon Vergleiche der (im einzelnen sicher noch etwas
zu korrigierenden) Artenzahlen. So gibt es an Laubmoos-Arten auf Neu-
guinea 930, auf Neukaledonien 539, auf den Philippinen 512, auf den
Fidschi-Inseln 307, auf Hawaii 213 und auf den niedrigeren mikronesi-
schen Inseln nur 142 (SCHULTZE-MOTEL, 1975a). In ganz Westafrika wach-
sen 964 Arten (SCHULTZE-MOTEL, 1975b, 130 *Fissidens*!), also auf wesentlich größerer Fläche nur etwa genauso viele wie in Zentraleuropa. Eine
vorläufige Liste aller Moose Borneos verzeichnet 607 Arten (TOUW, 1978).
Von den 78 bisher bekannten Lebermoosarten der Seychellen sind 35
außerdem nur im afrikanischen Raum, nur 10 sonst nur in Asien verbreitet. Das unterstreicht die Zugehörigkeit der Inselgruppe zu Afrika.
Die im Mediterrangebiet dominierenden Marchantiales fehlen hier völlig,
der Anteil der Lejeuneaceae ist - wie in Tropenfloren gewöhnlich -
sehr hoch (60%, GROLLE, 1978).

4. Kartenatlanten, Pilzkartierung (vgl. Prog. Bot. 38, 318; 40, 416)

Gegenwärtig ist es wohl der wichtigste Fortschritt der Chorologie, daß Verbreitungskarten, die wichtigsten Grundlagen der chorologischen Arbeit, in wachsendem Tempo und immer besserer Qualität erarbeitet werden. Neue Bibliographien verzeichnen z.B. für Kuba fast 1000 Karten (MANITZ, 1978, mit Angabe des Ausschnittes der Karte), für das herzynische Gebiet jetzt schon etwa 15000 Karten (HILBIG, 1978).

Der 2. Band der "Vergleichenden Chorologie der zentraleuropäischen Flora" (MEUSEL et al., 1978) enthält Gesamtarealkarten von etwa 2500 Taxa, die gegenüber dem ersten Band viel detaillierter und mit ausführlicherem Begleittext versehen sind. Auf dieser Grundlage wird die ökogeographische Stellung und die Genese der zentraleuropäischen Flora analysiert. Mit der vergleichenden Methode werden typische Gemeinsamkeiten und Unterschiede in der ökologischen Konstitution der Arten herausgearbeitet. Die Kartierung ganzer Verwandtschaftskreise (viele Karten von Gattungen und Sektionen) ermöglicht eine tiefergehende ökologische und chorogenetische Ausdeutung der Areale.

> Vollständige Kartenatlanten aller Gefäßpflanzenarten gibt es schon für etwa 30 Länder und Gebiete der Holarktis, außerdem liegen einige vollständige Gehölzkarten-Atlanten vor. Ein neuer Atlas mit 2217 Karten wurde für die Great Plains im mittleren Westen der USA von einer Gruppe von Botanikern erarbeitet (MC GREGOR und BARKLEY, 1977), also für ein Gebiet, das größtenteils nicht in modernen Floren erfaßt war. Tropische Gefäßpflanzen werden häufig in der "Flora Malesiana" (70 Karten in Bd. 8, STEENIS, 1978) und für Afrika in den "Distributiones Plantarum Africanarum" kartiert (jetzt 487 Punktkarten, zuletzt Rubiaceae, Zingiberaceae, *Diospyros*), außerdem erschien eine Kartenreihe von 45 taxonomisch klaren afrikanischen Arten von SCHULTKA (1971) und ein Verbreitungskarten-Atlas von LEBRUN (1977). Der erste Band eines neuen Gehölz-Arealatlas der Sowjetunion (Gymnospermen, Monocotylen, Salicales-Aristolochiales, SOKOLOV et al., 1977, 91 Karten) wurde ausschließlich von Spezialisten der taxonomischen Gruppen erarbeitet, was bei der nomenklatorischen Verwirrung z.B. in den Gattungen *Salix* und *Betula* eine Voraussetzung zur Klärung der Verbreitungsbilder war.

Schließlich sei die aktive Arbeit an der Kartierung der Pilze in Mittel- und Nordeuropa erwähnt. Für die ersten 150 kartierten Arten sind in der Bundesrepublik Deutschland bereits 26530 Fundmeldungen eingegangen (KRIEGLSTEINER, 1979, 31 Musterkarten). Eine Kartenserie von 21 Geastrales in der DDR (DÖRFELT et al., 1979) enthält ausführliche Kommentare zu den oft xerothermen Arealen und Gesamtareal-Diagnosen nach dem Verfahren von MEUSEL et al. (1978).

5. Areal- und Florengeschichte

Florengeschichtliche Aussagen werden entweder von den immer besser bestimmten und in immer größerer Funddichte vorliegenden Fossilien oder von den immer besser bekannten und sicherer deutbaren geographischen Beziehungen der rezenten Floren abgeleitet. Den Grad der Kenntnis fossiler Areale zeigt z.B. *Engelhardia*. Von dieser Regen- und Lorbeerwald-Juglandacee sind aus dem europäischen Tertiär schon über 90 Fundorte bekannt (JÄHNICHEN et al., 1977).

a) Neunachweise ostasiatisch + ostamerikanischer Laubwald-Elemente im europäischen Tertiär

Neu nachgewiesen wurden im zentraleuropäischen Miozän/Pliozän zwei Oleaceen-Gattungen: die heute nur ostamerikanische *Forestiera* (JÄHNICHEN und WALTHER, 1974) und *Chionanthus*, dessen Areal heute die bekannte A. GRAY-Disjunktion (Ostamerika + Ostasien) aufweist (GREGOR, 1978). Die karpologische Neubearbeitung der berühmten eozänen Geiseltalflora ergab unter 28 Arten 10 neue, darunter den Erstnachweis der heute nur chinesischen Staphyleacee *Tapiscia* in Europa, und als korrigiertes Vegetationsbild einen Kiefern-Lorbeerwald in subtropisch semihumidem, wintertrockenem Klima (MAI, 1976). Reich an arktotertiären Elementen waren die zentraleuropäischen Wälder erstmals im Mitteloligozän (Haselbacher Serie, 132 Arten, darunter neu die heute nur mittelamerikanische Hamamelidacee *Matudaea*; genaue Rekonstruktion von Klima und Vegetation: MAI und WALTHER, 1978).

b) Herkunft der Angiospermen

Ob die ältesten Fossilien von Angiospermen (nicht vor Barremien, vor 120 Mill. Jahren) schon ausreichen, um das Ursprungsgebiet dieser Gruppe zu erschließen, wie das immer wieder versucht wird (RAVEN und AXELROD, 1978: Westgondwana; AUBRÉVILLE, 1976: polytop in den Tropen) ist zweifelhaft. Für eine Rekonstruktion aufgrund rezenter Areale ist seither sicher viel zuviel geschehen. Wenn auch die "pflanzengeographisch unmöglichen" (STEENIS, 1979) Angaben von *Pinus* sect. *Cembrae, Picea, Abies* und *Cathaya* aus dem Spätpliozän von Timor (!) vielleicht durch Kontamination des Bohrmaterials zustandegekommen sind, so wird doch auch aus anderen Angaben immer deutlicher, welch gewaltige Entfernungen schon in dieser kurzen Zeit (etwa 1 Mill. Jahre) von Pflanzenarten überwunden wurden und in welch riesigen Arealen große Verwandtschaftskreise ganz ausstarben.

c) Entwicklung der Florengebietsgliederung

Dagegen reicht das paläofloristische Material aus, um die Ausbildung der Florenregionen in der Erdgeschichte zu rekonstruieren. VAKHRAMEEV et al. (1978) bringen dazu eine umfangreiche Synthese der vorliegenden Literatur für das Paläozoikum und Mesozoikum Eurasiens. 30 Eurasienkarten illustrieren die Lage der untersuchten Floren und die Grenzen der Florengebiete. Für *Pleuromeia* sowie charakteristische Bennettitales und Cycadaceae werden auch Arealkarten beigegeben. Im Paläo-, Meso und Neophytikum brachten jeweils zunächst Perioden starker GebirgsEinebnung ein humides Klima und ein zonal wie regional wenig gegliedertes Vegetationsbild, es folgten jedesmal Perioden starker Gebirgsbildung mit rascher Aridisierung, z.T. Vereisung, ausgeprägter Zonalität und rascher Evolution von Pflanzensippen. Indien gehörte bis zur Grenze Trias/Jura zum Gondwana-Gebiet, von da an zum indisch-europäischen Gebiet.

d) Disjunktionen Alte + Neue Welt: Kontinentaldrift oder Fernverbreitung?

Zur Erklärung vieler Angiospermen-Areale wird die Kontinentaldrift herangezogen (z.B. AUBRÉVILLE, 1976; SCHNELL, 1970-1977). Wahrscheinlich ist z.B. die atlantische Mangrove wegen des geringen Alters des Atlantiks so viel artenärmer als die pazifische. Durch die ehemals südlichere Lage Afrikas konnte dieser Kontinent auch im Süden nicht umwandert werden (HADAČ, 1976; SCHNELL l.c.).

Für die allermeisten südamerikanisch + afrikanischen Disjunktionen
(z.B. 12 Familien) läßt sich aber aus dem Grad der taxonomischen Dif-
ferenzierung und aus der Verbreitungsbiologie Fernverbreitung als
Ursache wahrscheinlich machen (GOLDBLATT, 1979; BRENAN, 1979; THORNE,
1978). Zum Teil erfolgte diese Ausbreitung wohl noch über den schmalen,
inselreichen eozänen Uratlantik. Die Zahl dieser Disjunktionen ist
übrigens nicht größer als die der Verbindungen der afrikanischen Tropen mit dem östlichen Tropengebiet. Ebenso lassen sich die 25 südafri-
kanisch + australischen Gattungsdisjunktionen alle durch Fernverbreitung erklären (GOLDBLATT, 1979). Auch die Flora Madagaskars, in der
viele in Afrika durch die Aridisierung ausgestorbene Sippen erhalten
geblieben sind, ist wahrscheinlich durch Ferntransport wesentlich
bereichert worden (neue Analyse der geographischen Beziehungen der
Familien und Gattungen, große Ähnlichkeit mit Ost- und Südostafrika:
LEROY, 1979). Nur wenige Disjunktionen von südafrikanischen Gymnospermen (*Podocarpus*, *Widdringtonia*-Verwandtschaft) gehen nach GOLDBLATT (1979)
vielleicht auf das Auseinanderdriften der Kontinente zurück.

Von den kalifornisch + südeurasischen Disjunktionen sind nach RAVEN
und AXELROD (1978) einige wohl nicht durch nördlichere Verbindungen
zustandegekommen (z.B. *Filago*- und *Lotus*-Verwandtschaft), sondern Schol-
len von vor der Drift geschlossenen Arealen. Die Zahl dieser Verbindungen ist aber gering, weil nach der Ausbildung des mediterranoiden
Klimas in Kalifornien (dort sehr spät: Pliozän) nie mehr ein direkter
Kontakt zu anderen Mediterranklima-Gebieten bestand.

e) Nord + südamerikanisch semiaride Disjunktionen

Unter allen Mediterranklima-Gebieten hat mit der kalifornischen die
chilenische Provinz die engsten Florenbeziehungen, da die Vogelzug-
richtung Ferntransport ermöglicht (RAVEN und AXELROD, 1978). Für
extrem xerische Sippen wie *Larrea* (aus Südamerika nach N) oder die
Boraginacee *Tiquilia* (aus Nordamerika nach S) ist Ferntransport nach
WELLS und HUNZIKER (1976) bzw. RICHARDSON (1977) die einzig vorstell-
bare Erklärung der nord + südamerikanischen Wüstendisjunktion (vgl.
auch SOLBRIG, 1976). Die mexikanischen Grasländer haben sich wahrscheinlich schon seit dem Frühtertiär autochthon entwickelt (RZEDOWSKI,
1975). Im Pliozän hat dann die Migration grasfressender Säuger manche
nord + südamerikanische Disjunktion zustandegebracht (WILLIAMS, 1975,
Fossilfunde der *Piptochaetium*-Verwandtschaft).

Semiaride Fallaubwald-Disjunktionen können aber auch auf klimabedingte
Reduktion ehemals zusammenhängender Areale zurückgehen (WEBSTER und
POVEDA, 1978), für *Jatropha*. So traten z.B. im Quartär Trockenperioden
auf, die im Amazonasgebiet von Regenwaldelementen in isolierten Refu-
gien überdauert wurden (PRANCE, 1973, für Chrysobalanaceae u.a. Familien).

f) Geschichte der Floren des westlichen Nordamerika

Nach einer umfassenden Listenübersicht der Herkunft der kalifornischen
Gattungen bei RAVEN und AXELROD (1978) gehören über die Hälfte der
Gattungen der kalifornischen Florenprovinz zu arktotertiären Verwandt-
schaftskreisen, etwa 25% sind von der Madrotertiär-Flora abzuleiten,
13,5% von warmtemperierten und subtropischen Wüsten.

Die Areale der arktotertiären Sippen differenzierten sich schon im
Tertiär deutlich zonal und peripher-zentral, z.B. *Sequoia*: nördlich
humid, *Sequoiadendron*: südlich semiarid. Bei der miozän/pliozänen

Gebirgsbildung, Abkühlung und Aridisierung zogen sich die Wälder an die Küste zurück. Die höhere Feuchtigkeit der Küstengebirge erklärt den Reichtum an Relikten. Nach der extremen Xerothermperiode vor 8000-4000 Jahren konnten sich manche reliktischen Baumarten - nun z.T. in Reinbeständen - in die waldfreie Gebiete wieder ausbreiten. Einige Relikte, z.B. *Lyonothamus*, erhielten sich nur auf den vorgelagerten Inseln, deren Rolle also der der Kanaren im Mediterranraum ähnelt.

Die Madrotertiärflora geht auf präadaptierte Besiedler von trockenen Sonderstandorten im Komplex trockener kretazischer *Araucaria*-Wälder des Chihuahua-Sonora-Gebietes zurück (AXELROD, 1979). Die Ausbildung einer ausgedehnten ariden Vegetation begann aber erst im Mittelpliozän und setzte sich in den Interglazialen fort. Heute sind Klima und Vegetation des Sonora-Gebietes extremer xerisch als je zuvor. Entsprechend der langen Entwicklung enthält die Sonora heute sowohl alte, isolierte Gattungen (*Fouquieria, Simmondsia, Yucca, Nolina*) als auch jüngere Xerophyten, die ihre Anpassung erst im Mitteleozän erreichten (*Bursera, Acacia*) und viele Kräuter pliozänen und pleistozänen Alters. Aus dem schon längere Zeit semiariden mediterran-orientalischen Raum konnten sich über 100 Arten synanthrop in der Sonora ausbreiten (AXELROD, 1979).

Aus dem zunächst sommertrockenen Entfaltungszentrum drang die Madrotertiärflora mit Gattungen wie *Arctostaphylos, Ceanothus, Rhus, Quercus* und *Cupressus* nach Kalifornien ein, als sich dort im Pliozän das semiaride Winterregenklima ausbildete. Einwanderungen aus verschiedenen anderen Quellen trugen schließlich dazu bei, daß die Flora des kalifornischen Florengebietes heute mit 4452 Gefäßpflanzenarten (davon 47,7% Endemiten) und 50 endemischen Gattungen die eigenartigste und reichste des extratropischen Nordamerika ist. Die Einstufung des Florengebietes als Provinz (RAVEN und AXELROD, 1978) erscheint daher als zu niedrig.

Die Florengeschichte von Ost-Washington und Idaho wurde von DAUBENMIRE (1975) aufgrund der publizierten Fossilfloren und der Interpretation rezenter Areale analysiert. Hier haben sich sehr viele arktotertiäre Gattungen erhalten können, allerdings keine immergrünen Angiospermen. Durch die nördlichere Lage und die vom Oligozän bis Pleistozän erfolgende Hebung der Rocky Mountains kam hier ein starker Einfluß von borealen Elementen hinzu, seit dem Pliozän außerdem ein xerisches Element (*Artemisia, Agropyron* u.a.). Die Verarmung der Laubwaldflora führt DAUBENMIRE (1975) sicher zu Recht auf die in ausreichend humiden Gebieten zu geringe Sommerwärme zurück. Das Verhalten synanthroper Arten aus Laubwaldgebieten zeigt, daß nur ein schmaler Höhengürtel in den sommerwärmeren Rocky Mountains heute die für solche Arten geeignete Klimakombination aufweist, und erklärt die beschränkte Verbreitung der Relikte.

g) Afrikanisch-südeurasische Florenbeziehungen und spättertiär-quartäre Aridisierung

α) *Tropisches Afrika*. Die Sonderstellung der afrikanischen Tropenflora, besonders ihre relative Artenarmut, ist ebenso wie viele nord + südliche Disjunktionen semiarid afrikanischer Sippen auf starke spättertiär-quartäre Feuchteschwankungen zurückzuführen, die auch paläobotanisch nachgewiesen wurden (BRENAN, 1979; GOLDBLATT, 1979; SCHNELL, 1970-1977). Auch einige west + ostafrikanisch humide Disjunktionen (z.B. *Mansonia* und *Coleotrype*, BRENAN, 1979) und das Fehlen südamerikanisch + madagassischer Sippen in Afrika lassen sich so erklären, ebenso innerhalb Westafrikas die durch den semi-humiden Korridor von Togo-Dahome getrennten "sassandrischen" Disjunktionen (SCHNELL, 1970-1977). Auffällig ist in Afrika das Fehlen oder die Seltenheit gerade von

hygrophilen Familien wie Theaceae, Lauraceae, Myrtaceae, Melastomataceae, Araliaceae, aber auch der Palmen (117 Arten im tropischen Afrika
gegenüber 1385 in den östlichen Tropen).

Im tropischen Afrika gibt es nur 9, in Südamerika 34 endemische Familien. Nach einer Karte des Endemitenreichtums bei BRENAN (1979) sind
die Regenwaldgebiete Afrikas noch relativ endemitenreich (in Kamerun
und Gabun über 20 Gattungen, viele auch in Zaire, sowie durch die
lange Isolation seit dem Pleistozän in Usambara, Uluguru und Nguru).
Die meisten Savannenelemente sind dagegen weit verbreitet.

β) *Sahara*. Auch im ganzen Gebiet der Sahara gibt es nur 12% endemische
Arten (QUEZEL, 1979), das wird durch die extreme und junge Aridisierung erklärt. Noch im Paleozän trug die Sahara tropischen Regenwald.
Im Miozän setzte die Aridisierung und die Ausbildung der Hartlaub-
vegetation ein, vor 4-5 Mill. Jahren schließlich war die Aridisierung
so weit fortgeschritten, daß *Tamarix*- und *Retama*-Gehölze vorkamen. Zu
Beginn des Pliozän hatte die Öffnung der Straße von Gibraltar das
Eindringen des Meeres in das vorher trockene Mediterranbecken ermöglicht; daher trug das mediterrane Nordafrika im Pliozän noch Wälder,
die den heutigen hyrkanischen ähnelten (mit *Juglans* und *Pterocarya*);
Olea, Quercus ilex und *Qu. suber* waren mehr lokal verbreitet.

Seit dem Miozän entstanden viele Florenbeziehungen zwischen Orient
und Nordafrika (Listen bei QUEZEL, 1979; HEDGE und WENDELBO, 1978
in einer Analyse der iranischen Florenelemente) und über den Gebirgs-
rand Ostafrikas bis nach Südafrika bzw. umgekehrt. Meist handelte es
sich dabei um xerische Sippen (Vahliaceae, Resedaceae, Moringaceae,
Neuradaceae, Balanitaceae, Salvadoraceae, Pedaliaceae, Ixioideae,
Ericoideae: GOLDBLATT, 1979) oder Gebirgspflanzen wie die orienta-
lisch + ostafrikanisch verbreitete *Primula* sect. *Sphondylium* (HEDGE und
WENDELBO, 1978). Nach LAVRANOS (1978) ist diese Palästina-Arabien-
Route durch einen Spezialmonsum mit Winterregen auch heute für mediterrane Elemente gangbar.

Der Einfluß der afrikanischen Tropenflora im Norden blieb dagegen
wegen der Frostempfindlichkeit gering (BOBROW, 1978: keine *Welwitschia*-
Flora in Mittelasien!, QUEZEL, 1979; STOCKER, 1976), wenn sich auch in
trockenen Quartärperioden die an Trockenheit besonders gut angepaßten
Tropenxerophyten bis nach Marokko und an der Küste des Roten Meeres
nach Norden ausbreiten konnten.

Erst diese quartären Trockenperioden eliminierten in Nordafrika die
hygrophilen Sippen, die im Hyrkan und auf den Kanaren z.T. erhalten
blieben. Gleichzeitig breiteten sich asiatische *Artemisia* -Steppen in
der Nordsahara aus, mit abgeleiteten Arten von Sippen aus dem mittel-
asiatischen Entfaltungszentrum der asiatisch ariden Flora (HEDGE und
WENDELBO, 1978; BOBROW, 1978). Um die saharischen Gebirge kam noch
Hartlaub-Waldland vor. Die wenigen borealen Arten sind dorthin wohl
durch Ferntransport gelangt (WICKENS, 1976a,b). Im Postglazial (vor
9000 Jahren) erlangte semiaride Dornbusch- und Hartlaubwald-Vegetation
noch einmal wesentlich weitere Verbreitung als heute (WICKENS, 1976b).

So hat sich in der Sahara ein deutlicher Nord-Süd-Gradient der Florenelemente herausgebildet. Den Anteil dieser Elemente berechnet FRANKEN-
BERG (1978) für Hunderte von Quadraten eines 80 km-Rasters und versucht
so, die Grenzen und die Gliederung der Sahara zu präzisieren. (Die
zugrundegelegte "homogenisierte Florenliste" aus allen saharischen
Lokalfloren wurde leider nicht veröffentlicht.) Früher wurde dieses
größte Wüstengebiet ganz zur Holarktis gestellt, jetzt rechnet man
meist die Nordhälfte zur Holarktis, die Südhälfte zur Paläotropis

(vgl. SCHNELL, 1970-1977). FRANKENBERG (1978) stellt sie dagegen als
Saharo-arabische Übergangsregion wegen zahlreicher eigener Elemente
(z.B. *Nucularia, Randonia, Cornulaca monacantha*, K bei BRENAN und QUEZEL
l.c.) zwischen die Florenreiche. Von QUEZEL (1979) wird sie sogar
dreigeteilt: er stellt den Norden als Saharo-arabische Region zur
Holarktis, den Süden als Saharo-afrikanische Unterregion zur Paläotropis, das Zentrum als Saharo-sudanische Komplexzone zwischen die
Florenreiche. Die Gliederung der Erde in Florenreiche wird durch diese
Lösungen aber unklar. Solche Übergangsgebiete könnten an den meisten
Florenreichsgrenzen und anderen Florengebietsgrenzen ausgegliedert
werden.

γ) *Südafrika*. Das Gebiet der "Flora of South Africa" ist ungewöhnlich
artenreich (18500 Arten nach der neuen Zählung von GOLDBLATT, 1979).
Dafür sind die Standortmannigfaltigkeit, der unterschiedliche Nieder-
schlagsrhythmus und die explosive Artbildung durch Klima-Fluktuationen
noch keine ausreichende Erklärung. Auch die Selbständigkeit ist mit
29% Endemiten von 1930 Gattungen, 80% endemischen Arten und 10 endemischen Familien sehr groß. Besonders betrifft das die Kapregion. Der
Endemitenanteil unterscheidet sich übrigens nicht nur von Familie zu
Familie (hoch bei den Asteraceae: 1801 von 2072, Aizoaceae 1896/2020,
Ericaceae: 797/799, Iridaceae: 816/840, Liliaceae: 803/907 und Orchidaceae: 371/461, über letztere auch SCHELPE, 1978; niedrig bei den
Poaceae: 330 von 743), sondern auch bei den verschiedenen Vegetations-
formen und Lebensformen (niedrig z.B. bei Wasserpflanzen).

Außerordentlich hoch ist in Südafrika die mittlere Artenzahl je Gattung:

Kapregion	8,9	Kalifornien	5,7
Europa	7,8	Nordostamerika	5,2
Südafrika	7,7	Texas	3,9
Hawaii	7,5	Sonora	3,3

Das ist als Ausdruck explosiver Artbildungsvorgänge in isolierten Gattungen anzusehen. Dagegen liegt der Anteil der 10 größten Gattungen
an der Flora, der auf Hawaii 42,1% ausmacht, in Südafrika mit 15,1%
nicht besonders hoch (GOLDBLATT, 1979). Obwohl für Südafrika nur wenige
Fossilfloren ausgewertet werden konnten, bedeutet auch diese Analyse,
die durch vollständige, kritisch geprüfte Listen aller endemischen
Gattungen und aller Familien (mit Artenzahlen) gut fundiert ist, einen
wesentlichen Fortschritt unserer Kenntnisse über Struktur und Genese
der afrikanischen Floren.

Literatur

ABRAMOVA, A.L., ABRAMOV, I.I.: Novosti sist. nizš. rast. 15, 189-194
(1978). - ALI, S.I., JAFRI, S.M.H., ALAVI, S.A., GHAFOOR, A., SIDDIQI,
M.A.: Flora of Libya, Fasc. 1-37. Koenigstein: Koeltz 1976-1977. -
ANDREWS, F.W.: The Flowering Plants of the Sudan. Vol. 1-3. Arbroath:
Buncle 1950-1956. - ANDREWS, S.B.: Handbook to the Ferns and Fern
Allies of Queensland. Brisbane: Herbarium 1979. - AUBRÉVILLE, A.:
Flore du Gabon. Fasc. 1-24. Paris: Mus. Nat. Hist. Natur. 1961-1977;
- Flore du Cameroun. Fasc. 1-20. Paris: Mus. Nat. Hist. Natur. 1963-
1978; - Adansonia sér. 2, 16, 297-354 (1976). - AXELROD, D.I.: Age and
Origin of Sonoran Desert Vegetation. Occas. Pap. Calif. Acad. Sci.
(San Francisco) 132, 1-74 (1979).

BEAUGLEHOLE, A.C.: Victorian Nat. 95, 67-74 (1978). - BERHAUT, J.:
La Flore Illustrée du Sénégal. Vol. 1-4.Dakar:Gvt.du Sénégal 1971-1975.
BLACKALL, W.E., GRIEVE, B.J.: How to Know Western Australian Wild-
flowers. Parts 1-4. Nedlands: Univ. of Western Australia 1954-1975.
BLAKE, S.F., ATWOOD, A.C.: Geographical Guide to Floras of the World.
Vol. 1. New York-London: Hafner 1942, Repr. 1963. - BOBROV, E.G.:
Bot. Zh. 63, 1393-1402 (1978). - BOSSER, J., CADET, T., JULIEN, H.R.,
MARAIS, W.: Flore des Mascareignes. Fasc. 110, 111. Paris: Office
Rech. Sci. Techn. Outremer, 1976. - BOULOS, L.: Publ. Cairo Univ.
Herb. 7-8, 115-141 (1977). - BRENAN, J.P.M.: Ann. Missouri Bot. Gard.
65, 437-478 ("1978", 1979). - BRIGGS, B.G., CAROLIN, R.C., PULLEY,
J.M.: Flora of New South Wales, 181 Plantaginaceae. New South Wales
Dept. Agric. (o. Ort) 1977. - BROWNLIE, G.: The Pteridophyte Flora
of Fiji. Vaduz: Cramer 1977. - BYRNES, N.B.: Contr. Queensl. Herb. 20,
1-88 (1977).

CAROLIN, R.C.: ASBS Newsl. 16, 2-3 (1978). - CODD, L.E.W., WINTER,
B. DE, KILLICK, D.J.B., RYCROFT, H.B.: Flora of Southern Africa.
Vol. 1, 9, 13, 16.1, 16.2, 22, 26. Pretoria: Government Printer 1963-
1979. - COMPTON, R.H.: The Flora of Swaziland. J. S. Afr. Bot. Suppl.
11 (1976). - CUFODONTIS, G.: Enumeratio plantarum Aethiopiae Spermato-
phyta. Bull. Jard. Bot. Natl. Belg. Suppl. 23-42 (1953-1972).

DAUBENMIRE, R.: J. Biogeogr. 2, 1-18 (1975). - DEFILIPP, R.A.: Webbia
30, 177-190 (1976). - Distributiones Plantarum Africanarum. Fasc. 13,
14. Meise: Jardin Bot. Natl. Belg. 1978. - DÖRFELT, H., KREISEL, H.,
BENKERT, D.: Hercynia (Leipzig) 16, 1-56 (1979). - DYER, R.A.: The
Genera of Southern African Plants. Vol. 1-2. Pretoria: Government
Printer 1975, 1976.

EXELL, A.W., WILD, H.: Flora Zambesiaca. Vol. 1, 2, 3.1, 4, 10.1.
London: Crown Agents Oversea Gvts. Administr. 1960-1978. - EXELL,
A.W., CARISSO, W., MENDONÇA, F.A.: Conspectus Florae Angolensis.
Vol. 1-4 und Pteridophyta. Lisboa: Junta Invest. Cient. Ultramar
1937-1977.

FABREGUES, B.P., LEBRUN, J.P.: Catalogue des plantes vasculaires du
Niger. Alfort: Inst. Elév. Méd. Vét. Pays Tropicaux 1975 (1976). -
FERNANDES, A.: Flora de Moçambique. Fasc. 1-36. Lisboa: Junta Invest.
Cient. Ultramar 1969-1973. - FOSBERG, F.R., SACHET, M.-H.: Flora of
Micronesia. Fasc. 3: Convolvulaceae. Smithson. Contrib. Bot. 36, 1-34
(1977). - FRANKENBERG, P.: Bonner Geogr. Abh. 58, 1-136 (1978).

GOLDBLATT, P.: Ann. Missouri Bot. Gard. 65, 369-436 ("1978", 1979).
- GREENE, S.W., WALTON, D.W.H.: Pol. Rec. 17, 473-484 (1975). -
GREGOR, H.-J.: Feddes Repert. 88, 645-653 (1978). - GRIEVE, B.J.:
J. R. Soc. West. Aust. 58, 33-53 (1975). - GROLLE, R.: Wiss. Z.
Friedrich-Schiller-Univ. Jena, Math.-Nat. 27, 7-17 (1978). - GUILLARMOD,
A.J.: Flora of Lesotho (Basutoland). Lehre: Cramer 1971.

HADAČ, E.: Folia Geobot. Phytotaxon. (Praha) 11, 213-216 (1976). -
HALLÉ, N.: Orchidacées. (Flore de la Nouvelle Calédonie et dépendances
8) Paris: Mus. Nat. Hist. Natur. 1977. - HEDGE, I.C., WENDELBO, P.:
Notes R. Bot. Gard. Edinburgh 36, 441-464 (1978). - HEYWOOD, V.H.,
HARBORNE, J.B., TURNER, B.L.: The Biology and Chemistry of the Com-
positae. Vol. 1-2. London-New York-San Francisco: Academic Press 1977.
- HILBIG, W.: Wiss. Z. Martin-Luther-Univ. Halle, Math.-Nat. 28,
41-81 (1979). - HILLIARD, O.M.: Compositae in Natal. Pietermaritz-
burg: Univ. of Natal Press 1977. - HUMBERT, H.: Flore de Madagascar
et des Comores. Paris: Mus. Nat. Hist. Natur. 1936-1975. - HUTCHINSON,
J., DALZIEL, J.M.: Flora of West Tropical Africa. Vol. 1-3 und
Pteridophyta. London: Crown Agents Oversea Gvts. Administr. 1954-1971.

JÄHNICHEN, H., WALTHER, H.: Feddes Repert. 85, 17-41 (1974). -
JÄHNICHEN, H., MAI, D.H., WALTHER, H.: Feddes Repert. 88, 323-363
(1977). - JESSOP, J.P.: Flora of South Australia. Part 1. Adelaide:
Gvt. Printer 1978. - JOHNSON, P.N., CAMPBELL, D.J.: N. Z. J. Bot. 13,
665-720 (1975).

KILLICK, D.J.B.: Boissiera 24, 633-634 (1976). - KRIEGLSTEINER, G.J.:
Z. Mykologie 45, 73-128 (1979).

LAVRANOS, J.J.: Bot. Jahrb. 99, 152-167 (1978). - LAWALRÉE, A.:
Flore d'Afrique Centrale: Zaïre-Rwanda-Burundi. Bruxelles: Jard. Bot.
Belgique 1967-1972 (51 Lieferungen). - LEBRUN, J.P.: Eléments pour un
atlas des plantes vasculaires de l'Afrique sèche, 1. Maisons Alfort:
Inst. Elév. Méd. Vét. Pays Trop. 1977. - LEROY, J.-F.: Ann. Missouri
Bot. Gard. 65, 535-589 ("1978", 1979). - LORENCE, D.H.: Bot. J. Linn.
Soc. 76, 207-257 (1978).

MAI, D.H.: Abh. Zentr. Geolog. Inst. 26, 93-149 (1976). - MAI, D.H.,
WALTHER, H.: Abh. Staatl. Mus. Mineral. Geol. Dresden 28, 1-200 (1978).
- MAIRE, R.: Flore de l'Afrique du Nord. Vol. 1-14. Paris: Lechevalier
1952-1977. - MANITZ, H.: Wiss. Beitr. Friedrich-Schiller-Univ. Jena
(Beitr. Phytotaxonomie, 6. F.), 171-221 (1978). - MARKGRAF, F.: Bot.
Jahrb. 95, 401-405 (1976). - MC GREGOR, R.L., BARKLEY, T.M.: Atlas of
the Flora of Great Plains. Ames: Iowa State Univ. Press 1977. -
MERXMÜLLER, H.: Prodromus einer Flora von Südwestafrika. Lehre: Cramer
1966-1972; - Mitt. Bot. München 12, 351-356 (1976). - MEURK, C.D.:
N. Z. J. Bot. 13, 721-742 (1975). - MEUSEL, H., JÄGER, E., RAUSCHERT,
S., WEINERT, E.: Vergleichende Chorologie der Zentraleuropäischen
Flora. Band 2 (Text und Karten). Jena: Fischer 1978. - MILLER, H.A.,
WHITTIER, H.O., WHITTIER, B.A.: Prodromus Florae Muscorum Polynesiae.
Lehre: Cramer 1978. - MOE, R.L., SILVA, P.C.: Science 196, 1206-1208
(1977). - MUNIR, A.A.: Brunonia 1, 407-692 (1978).

OZENDA, P.: Flore du Sahara, 2. ed. Paris: Centre Nat. Rech. Sci. 1977.

PALGRAVE, K.C.: Trees of Southern Africa. Cape Town-Johannesburg:
Struik 1977. - PARRIS, B.S.: Fern Gaz. 11, 231-245 (1976). - PIĘKOS-
MIRKOWA, H.: Acta Soc. Bot. Pol. 46, 577-585 (1977). - PRANCE, G.T.:
Acta Amazonica 3, 5-28 (1973); - Ann. Missouri Bot. Gard. 64, 659-
684 ("1977", 1978).

QUEZEL, P.: Ann. Missouri Bot. Gard. 65, 479-534 ("1978", 1979).

RAVEN, P.H., AXELROD, D.I.: Origin and Relationships of the Californian
Flora. Berkeley-Los Angeles-London: Univ. of Calif. Press 1978. -
RECHER, H.F.: Environmental Survey of Lord Howe Island. Sydney: Austral
Mus. 1974. - RICHARDSON, A.T.: Rhodora 79, 467-572 (1977). - ROBYNS,
W.: Flore du Congo Belge, du Rwanda et du Burundi. Vol. 1-10.
Bruxelles: Inst. Nat. Agr. Congo Belge 1948-1963. - ROSS, J.H.: The
Flora of Natal. Mem. Bot. Surv. S. Afr. 39, 1-418 (1972). - ROTI
MICHELOZZI, G.: Webbia 32, 417-453 (1978). - RZEDOWSKI, J.: Taxon
24, 67-80 (1975).

SCHELPE, E.A.C.L.E.: Bot. Jahrb. 99, 146-151 (1978). - SCHNELL, R.:
Introduction à la phytogeographie des pays tropicaux. Vol. 1-4.
Paris: Gauthier-Villars 1970-1977. - SCHULTKA, W.: Oberhessische
Naturwiss. Z. 38, 123-126 (1971). - SCHULTZE-MOTEL, W.: Lindbergia 3,
57-59 (1975a); - Willdenowia 7, 473-535 (1975b). - SCOTT, G.A.M.,
STONE, I.G.: The Mosses of Southern Australia. London: Academic Press
1976. - SMITH, A.C.: Allertonia 1, 331-414 (1978). - SOKOLOV, S.JA.,
SVJAŽEVA, O.A., KUBLI, V.A.: Areály derev'jev i kustarnikov SSSR. Tom 1

Leningrad: Nauka 1977. - SOLBRIG, O.T.: The origin and floristic affinities of the South American temperate desert and semidesert regions, 7-50. In: Evolution of Desert Biota, ed. D.W. GOODALL. Austin-London: Univ. of Texas Press 1976. - STANLEY, T.D., ROSS, E.M., REYNOLDS, S.T., PEDLEY, L.: Handbook to the Flora of South-Eastern Queensland. Vol. 1. Brisbane: Herbarium 1979. - STEENIS, C.G.G.J. VAN: Flora Malesiana. Ser. I Vol. 8. Leyden: Noordhoff 1974-1978; - Ser. II Vol. 1 part 4. Leyden: Noordhoff 1978; - Flora Malesiana Bull. 32, 3357 (1979). - STOCKER, O.: The Water-Photosynthesis Syndrome and the Geographical Plant Distribution in the Saharan Deserts, 506-521. In: Water and Plant Life, eds. O.L. LANGE, L. KAPPEN, E.-D. SCHULZE. Berlin-Heidelberg-New York: Springer 1976. - SYKES, W.R.: N. Z. DSIR Bull. 219, 1-216 (1977); - N. Z. J. Bot. 16, 391-395 (1978).

TÄCKHOLM, V.: Students Flora of Egypt. Beirut. Coop. Printing Co. 1974. - THOMPSON, J.: Flora of New South Wales, 111. Tremandraceae. National Herbarium of New South Wales 1978. - THORNE, R.F.: Notes R. Bot. Gard. Edinburgh 36, 297-315 (1978). - TOUW, A.: J. Hattori Bot. Lab. 44, 147-176 (1978). - TROUPIN, G.: Flore du Rwanda. Vol. 1. Tervuren: Mus. R. Afr. Centr. 1978. - TURRILL, W.B., MILNE-REDHEAD, E.: Flora of Tropical East Africa. London: Crown Agents Overs. Gvts. Administr. 1952-1978.

VAKHRAMEEV, V.A., DOBRUSKINA, I.A., MEYEN, S.V., ZAKLINSKAJA, E.D.: Paläozoische und mesozoische Floren Eurasiens und die Phytogeographie dieser Zeit. Jena: Fischer 1978.

WEBB, C.J.: N. Z. J. Bot. 16, 387-390 (1978). - WEBSTER, G.L., POVEDA, L.J.: Brittonia 30, 265-270 (1978). - WELLS, P.V., HUNZIKER, J.H.: Ann. Missouri Bot. Gard. 63, 843-861 (1976). - WHITMORE, T.C.: Tree Flora of Malaya. Vol. 3. Malaysia: Longman 1978. - WHITTIER, H.O.: Mosses of the Society Islands. Grainsville: Univ. Press 1976. - WICKENS, G.E.: Kew Bull. 31, 105-150 (1976a); - The Flora of Jebel Marra (Sudan Republic) and its Geographical Affinities. Kew: Bot. Gardens 1976b. - WILLIAMS, D.E.: J. Biogeogr. 2, 75-85 (1975). - WOMERSLEY, J.S.: Handbook of the Flora of Papua New Guinea. Vol. 1. Melbourne: Univ. Press 1978.

Dr. ECKEHART J. JÄGER
Sektion Biowissenschaften der
Martin-Luther-Universität
Neuwerk 21
DDR 402 Halle/Saale

II. The History of Flora
and Vegetation During the Quaternary

By Burkhard Frenzel

1. General Problems

During the last years general problems of pleistocene vegetation history have been discussed here repeatedly. The same holds true for this
report, because the previous painstaking analysis and interpretation
of past changes in flora and vegetation of various parts of the world
lead to the consequence that more background information of more general biological or geological significance is needed, this in its turn
stimulating criticism of previously held views.

a) The Stratigraphical Division of the Pleistocene

As in the last report (see Prog. Bot. 40, 429, 1978), the question
as to the division of the pleistocene remains of some interest: It
must be remembered that the hitherto held view of a fourfold division
of the pleistocene in the surroundings of the Alps originated in geomorphological observations on river activity. Phases of widespread aggradation seemed to have been caused by cold climates, i.e., by glacial periods, whereas phases of linear fluvial erosion were thought
to have been the consequence of interglacial, warm and moist climates.
Following these suggestions the hypothesis of only four glacial and
three interglacial phases in southern Central Europe seemed to have
been warranted. Yet meanwhile even geomorphological observations show
that many more glacial and interglacial phases must have existed (see
LÖSCHER, 1976). KUKLA (1977, 1978) has pointed out that probably some
misinterpretations exist of geomorphological and stratigraphical observations, so that it might be wise to abandon as soon as possible
the classical names of the Günz, Mindel, Riß and Würm glaciations:
Some of the classical river terraces might comprise more than only
one glacial to interglacial cycle and according to a reappraisal of
older observations and according to modern absolute datings at least
the so-called last interglacial, the Eemian, should be divided into
three different interglacials of various ages.

It is suggested that during the last $1.7 \cdot 10^6$ years at least 17 glacial
to interglacial cycles had occurred (8 within the last 800,000 years).
It is recommended to accept the $^{18}O/^{16}O$-curves of deep sea sediments
as a firm base for the chronostratigraphical division of the pleistocene, though the correlation between these isotope curves and the sequence of glacial and interglacial phases on the continents is extremely difficult to establish. Though MANGERUD et al. (1979a,b) doubt
that KUKLA's (1977, 1978) speculations as to the real age of the Eemian
interglacial were well founded (see also BOWEN, 1979) the problem of a
presumably much larger number of glacial to interglacial cycles than
has been accepted up till now remains unsolved. Moreover ZAGWIJN (1974)
has shown that some of the interglacial beds within the Netherlands,
which have been dated as belonging to the Holsteinian interglacial,

are in reality of different ages. This holds true for the so-called
zone of Rosmalen and for the lowermost interglacial in the Noordbergum
boring. They are now held as belonging to the Cromer complex, which is
said to have consisted of four interglacials, this number being en-
larged by the still older Leerdam and Bavel interglacials (ZAGWIJN
and DOPPERT, 1978), the stratigraphical position of which is situated
between the Waalian interglacials and the Cromer complex. The curve
of the mean July temperatures, given by ZAGWIJN and DOPPERT for the
whole of the pleistocene in the Netherlands thus favors the view of
approximately 10 to 16 interglacials and of about 10 more significant
interstadials.

If these suggestions should prove to be correct, the question must be
answered, why up till now such an astonishingly great number of interglacial phases could have been overlooked. The answer seems to be that
during the pleistocene, at least within the Netherlands, several phases
of longer hiatuses have existed, most of all at the very boundary between the pliocene and the pleistocene and twice within the Cromer
complex (ZAGWIJN and DOPPERT, 1978). Another explanation may be found
in the observation given by ZAGWIJN (1974) that the vegetation history
of several interglacials of the Cromer complex and of the Holsteinian
proper seem to have resembled each other, thus making botanical discrimination extremely difficult. It is possible, though not proven,
that this very fact has complicated the stratigraphical division of
the pleistocene in southern Central Europe so much (SERET and ROUCOUX-
WOILLARD, 1978; WELTEN, 1978; URBAN, 1978a; FRENZEL, 1978a,b,c). In
the Netherlands and in the adjacent regions of Northwestern Germany
the situation is still more complicated by the big rivers repeatedly
changing their beds (ZAGWIJN and DOPPERT, 1978; BOENIGK, 1978b) and
by tectonic subsidence and uplift (ZAGWIJN and DOPPERT, 1978; BOENIGK,
1978a,b; cf. SEMMEL, 1978; BIBUS and SEMMEL, 1977; BRUNNACKER et al.,
1978; ŠIBRAVA, 1978b). These interrelations between tectonic movement
and river activity may have caused the formation of river sediments,
the fabrics of which from a mere sedimentological point of view may
resemble each other strongly, though their fossil plant remains point
to divergent ages (URBAN, 1978b). Still another source of error is
the fact that very often only shorter parts of a warm-period vegetation history are found, so that the question whether they might belong
to true interglacials or only to more significant interstadials remains unanswered (e.g., "Fortuna Oscillation" during the Pretiglian
in northwestern Germany: URBAN, 1978b).

Last but not least, some hitherto well-known interglacials seem to
disintegrate now into several distinct interglacials of various ages.
This had already been stated for the Cromer complex, but the same seems
to hold true for the Tiglian: Frechen interglacial I to III, and van
Eyck interglacial (URBAN, 1978b), and ŠIBRAVA (1978b) stresses the
same situation as having happened during the Holstein interglacial.
To me the situation today seems to resemble a boat sailing between
Scylla and Charybdis: The paleobotanist has to find his way cautiously between the tempting stimulus of rapidly increasing the number of
interglacials and the retarding factor of sticking too much to old
axioms. In view of these difficulties the need for absolute dating
methods increases. The methods available are discussed by ŠIBRAVA
(1978a), and DREIMANIS et al. (1978) report on successes and handi-
caps of the thermoluminescence method, whereas GROOTES (1978) sum-
marizes the advantages and results of an improved ^{14}C-dating method.
This very method is handicapped by an obviously changing radiocarbon
content of the atmosphere, these changes consisting of long and short-
term oscillations, respectively. Both these types of changing [14]C-
content might render exact datings difficult. PEARSON et al. (1977)

stressed that according to their dendrochronological and radiocarbon investigations of Irish oaks the short-term ^{14}C-oscillations had never existed, whereas SUESS (1978, 1979), using dendrochronologically dated tree-rings from Central Europe and from the US, pointed to the correctness of his previous research work. Moreover it could be shown that an interrelation seems to exist between changes in sunspot activity, changes in the radiocarbon content, and climate. If so, deviations may exist between real and radiocarbon ages of unknown length, yet in phase with changes of climate. This might hinder exact datings enormously; e.g. SCHOVE (1978) suggests that the postglacial did not begin at about 10300 b.p., as is accepted by most scientists, but at about 11250 b.p. The consequences for an exact determination of immigration rates, rates in plant succession etc. are evident, provided that these interrelations between solar activity, radiocarbon production, and content and climate had really existed.

Besides paleobotany the classical method for relative dating of the past is paleontology. LOŽEK (1978) discusses the previous research work on mollusk stratigraphy in pleistocene sediments of north-central Europe, showing convincingly the advantages of this method not only for dating purposes but for paleoecological investigations as well, e.g., for the Holsteinian interglacial. TOBIEN (1975) stresses the good agreement between paleobotanical (FRENZEL, 1968) and paleontological reconstructions of the late tertiary plant geographical formations, and points to the various steps in the evolution of the European cold-climate faunas and of the distribution patterns of this type of fauna during the younger part of the pleistocene: During the Mindelian glaciation a mammalian fauna, characteristic of cold steppe or tundra communities, can be followed from the north and east only up to the 45° n.l.; during the Rissian glaciation the southern part of France, northwestern Yugoslavia and the Crimean peninsula were reached, and only during the Würmian glaciation had nearly the whole of the mediterranean region been conquered by this fauna type, the typical mediterranean fauna only then immigrating into northern Africa.

2. Dendrochronology, Paleoclimatology

Summaries of dendrochronological and dendroclimatological methods are given by MUNAUT (1978) and by FRITTS (1978). Dendrochronology, up till now in general only relatively seldom used in Europe, merits increased interest in view of its broad and varied applicability. In Ireland (PILCHER et al., 1977; BAILLIE, 1977a,b), in Scotland (BAILLIE, 1977c), and in Skåne, Southern Sweden (BARTHOLIN and BERGLUND, 1975) regional chronologies of some millenia's duration each could be constructed. They repeatedly show regional differences in the annual increment of synchroneously grown trees, facilitating a reconstruction of past growth regions in various parts of Europe. The same holds true for Tasmania (OGDEN, 1978). Moreover, a multi-factorial analysis of climatic parameters and tree growth, done on *Pinus halepensis* (SERRE, 1977), *Larix decidua* (SERRE, 1978), elm (BRETT, 1978), and oak (PILCHER, 1976; HUGHES et al., 1978) makes a cautious reconstruction of accurately dated past climates in various parts of Europe possible, stimulating speculations on changing circulation patterns of the atmosphere and in its turn possibly enabling not only a reconstruction but forecast of climate as well, if a periodicity in changes of climate did exist. Using this method, the authors cited rely on changes in the width of tree-rings, whereas SCHWEINGRUBER et al. (1978) stress that changes in the X-ray density of tree-rings are much more reliable. Yet of course nonclimatic influences on tree-growth, like epidemics, must be duely taken into consideration (BRUBAKER, 1978).

These paleoclimatological aspects of dendrochronology may prove of great importance even for planning and other practical purposes since on the one hand impact of man on climate is continuously increasing, and on the other paleobotanical observations do exist pointing to pos- sibly extremely rapidly acting changes from interglacial to glacial climates (at a rate of some 100 to 300 years only). The implications of both these factors are discussed by FLOHN (1977), though to me the probability of such dramatically acting changes of climate appears to be extremely small. The present-day state of knowledge of recent changes of climate is summarized by KUKLA et al. (1977), whereas PFISTER (1977) has tried to reconstruct climate in south-central Europe for the period 1683 to 1738.

Problems of paleoclimatology are discussed for Central Europe, using paleobotany and mollusk analyses as indicators, by LOŽEK (1979) and by JÄGER and LOŽEK (1978a). It is suggested that during subboreal times the climate of the central-European lowlands had experienced at about 1250 to 650 B.C. a long-lasting drought, causing the formation of widespread rendzina soils, whereas this drought could not be de- tected in the mountainous regions of Europe. Whether this drought really was of climatical origin or whether it was simulated by strong clearances only, may be questioned. These interrelations between cli- mate, as deduced from paleoecological investigations, and the activi- ty of man, are investigated by BARRY et al. (1977) for the Eastern Canadian arctic. Here, too, strange coincidences occur, but the authors doubt whether they all really mean strong dependencies of man on cli- mate.

NICHOLS et al. (1978) have tried to reconstruct palynologically the paleowindsystems in Baffin Island during the younger part of the holo- cene. These investigations are based on pollen-influx calculations to certain lake basins. It is suggested that by this means not only the prevailing southerly wind directions can be established, but the peri- odicity of about 210 to 250 years for the displacements of the stable Rossby waves as well. These deductions are paralleled by observations on the productivity of lake and terrestrial ecosystems in Subarctic Labrador-Ungava, aiming at an evaluation of the climatic causes for the productivity. It is suggested that the astonishingly slow immigra- tion rate of forest trees into the area might have been caused by re- peatedly occurring cold spells, though the overall climatic conditions may have been relatively good. If these conclusions should prove to be correct, this might serve as an aid to understanding the discrepancies sometimes suggested between character of climate and climax communities during some of our interstadials.

The repeatedly discussed problem of a long-lasting cold spell during preboreal times, the equivalent of the so-called Piottino-oscillation, is discussed again by BEHRE (1978b), pointing to a period between about 10,000 and 9600 b.p., yet BIRKS and MATHEWES (1978) could not find any indication for this oscillation in Central Scotland. LAUER and FRANKENBERG (1979) have made the interesting attempt to recon- struct the annual precipitation rates, temperatures, and the poten- tial evaporation at about 18,000 and 5500 b.p. for the Western Sahara, and to calculate from this by means of statistical methods the number of species involved during both these dates, per 6400 km^2, and of the florogeographical relationships. These calculations are based on pre- sent-day conditions. It may be asked whether they can really serve as a good starting point for theoretical reconstructions of the past.

3. History of Distribution Areas

The surprising fact that within the western part of the mediterranean
area pollen grains of *Cedrus* are repeatedly met with in late-glacial
sediments may be explained according to de BEAULIEU and REILLE (1973)
by long-distance transport rather than by growth of this conifer in
the area under consideration. It is held that the same holds true for
e.g., *Platanus*. The spontaneous occurrence and the history of the distribution area of *Xanthium* in southeastern and Central Europe is investigated by BRANDE (1976b). *Quercus ilex* seems to have occurred in
southern Europe since the oligocene (PONS and VERNET, 1971). Beginning
with miocene times this oak is thought to have occupied its present-
day distribution area, and it may be suggested that *Quercus ilex* developed cold resistance from the Villafranchium onward. It is interesting to note, that this oak during interglacial phases has never
been of general importance in the mediterranean area, only being
triggered by the impact of man during the last millenia.

Beech seems to have spread on the French coast of the British Channel
at about 400 or even 950 A.D., thus pointing to remarkable differences
in age between the beginning and the end of comparable phases in the
vegetation history of different areas (Ardenne Mts. at about O B.C.;
MUNAUT and GILOT, 1976). According to GUILLET et al. (1972) the present-day border between the oak-beech forests and the beech-fir forests
on the west flank of the Vosges Mts. seems to have been established at
about the end of subboreal times. The ensuing history of fir is said
to have been governed most of all by the impact of man. The history of
the distribution of the European spruce in south-central Europe is investigated by KRAL (1977), using the classical ages for the zone boundaries in pollen diagrams as means for dating the migration only. In
the peatbog of Pechschnait, near Traunstein, Bavaria, *Betula nana* ssp.
flabiliforma WARNUNG was found in late-glacial or early postglacial sediments, i.e., in the same area where the dwarf birch is growing now
(SCHMEIDL, 1977).

RYBNIČEK and RYBNIČKOVÁ (1976) stress that several of the plant species, so characteristic for the present-day xerothermic vegetation
of Cschechoslovakia, had only immigrated there thanks to human activity since medieval times, and in general this floristic element is fed
by different sources, some dating from the late-glacial steppe vegetation, others from neolithic and bronze age crops, coming from the Near
East, and still others having immigrated from the mediterranean area.
The flora of montane spruce forests and of their fen vegetation in the
surrounding mountainous regions seems also to have been strongly favored by man. For the distribution area of such interesting plants as
*Adonis aestivalis, Fumaria vaillantii, Glaucium corniculatum, Vaccaria pyramidata,
Caucalis lappula, Galium tricorne*, and *Papaver rhoeas* during the eigth to tenth
century A.D., see OPRAVIL (1978).

4. Paleobotanical Methods

The important point of reliably reconstructing the past vegetation by
means of comparison between pollen analysis and present-day vegetation
has been investigated again several times: Eastern Washington and Northern Idaho: MACK et al. (1978); Belgium: MULLENDERS et al. (1978);
LEJOLY-GABRIEL (1978); DAMBLON (1979); Marseille: TRIAT (1967);
Białowieża Forest, Eastern Pland: DĄBROWSKI (1975); *Tamarix*, TRIAT

(1971); *Pinus contorta*, PERRY (1978). The relationship between modern
pollen rain and the halophytic vegetation near the coasts of the sea
are investigated by MUNAUT (1975, English Channel) and by HEIM (1975,
surroundings of Venice). KOIVO and RITCHIE (1978) have found that the
diatom assemblages from various lakes in the boreal-arctic transition
near the Mackenzie Delta, N.W.T., Canada, differ remarkably from one
another only where strong chemical differences in the mineral content of lake waters occur. Difficulties arising from redeposition of
older pollen is discussed by OHNGEMACH (1976); BRANDE (1976b) stresses
the errors in pollen analytical work, caused by preparation in heavy
liquids, if during the diagenesis accumulation of pyrite had occurred.
Coprolithes from prehistoric settlements as tools for the investigation of plants, having been digested by man and animals, were thoroughly analysed by PAAP (1976).

The volcanic ashes and tuff of Laacher See, eastern Eifel, Western
Germany, which serve as a widely applicable means of dating in Central
European late-glacial sediments, seem to be not as unique as was pre-
viously thought, since JUVIGNE (1978) has found tuff layers of the
same sedimentological composition in sediments of the much older Arcy-
Kesselt-interstadial of the Grande Pile peat bog, southeastern France.

> The value of scanning electron microscopy for paleobotanical investigations is
> shown by BLAHA and BORTENSCHLAGER (1972), by GUGGENHEIM (1975) for *Tilia*, and
> by MATHEWES (1978) for some western Canadian *Myriophyllum* species. A new coring
> device: KLAUS (1975).

5. Paleoethnobotany

Paleoethnobotany is proving to become a very useful tool in not only
evaluating the evolution and importance of prehistoric agriculture,
but in establishing genetic processes of the past as well. The present
state of knowledge is discussed by WILLERDING (1978a,b), HOPF (1975),
KÖRBER-GROHNE (1973, 1974, Southern Germany), SCHWEINGRUBER (1975,
1976a,b, charcoal analysis). In these papers the advantages and pitfalls of the methods used are discussed very competently.

WRIGHT (1976) and WHYTE (1977) stress the strange parallelism between
the evolution of domesticated plants and changes of climate. It is argued that climate played a decisive role in the formation of cultivated
plants, whereas KNÖRZER (1978) points to the problem of coevolution of
cultivated plants and weeds, using *Camelina sativa* as an example. Very
interesting observations are reported by HANSEN (1978) on the history
and evolution of cultivated plants nowadays in Greece, the previous
earliest remains of which date from full glacial times. These plants
like *Lithospermum arvense*, *Alkanna* cf. *orientalis*, *Anchusa* sp., *Lens*, *Pistacia*
aff. *lentiscus*, *Prunus amygdalus* etc. may have been directly or indirectly
gathered by man, and even *Avena* sp. or *Hordeum* cf. *spontaneum* occur there
at already 12500 b.p. Of course, it was a long way from these earliest
findings to modern cultivated plants or weeds. Important observations
are reported by HOPF (1974) on the cultivated plants of younger neolithic times in Yugoslavia, of neolithic to medieval times in Bulgaria
(1973), and of neolithic times in the surroundings of Ludwigsburg,
Germany (1977). It could be further shown that even during neolithic
times in southern Europe (HOPF, 1978) as well as in the southwestern
part of the USSR (JANUSHEVICH, 1978) regional differences in the agricultural activities of man existed. For the history of cultivated
plants in Caucasia during neolithic to iron age times, see LISITSINA

(1978). ZÓLYOMI (1971) has tried to evaluate cautiously the surface
of the cultivated area in the surroundings of lake Balaton, beginning
with early neolithic times. Some observations as to the history of
synanthropic plants in late-bronze age and iron-age times, see
HAJNALOVA (1978) Bratislava; WILLERDING (1974b) Pipinsburg, Osterode;
KÖRBER-GROHNE and WILMANNS (1977) and LANGE (1978) Central Europe.

Several authors reported on p. pte. extremely rich macrofossil floras
of weeds and cultivated plants, dating from medieval times: Haithabu:
ECKSTEIN (1977), ECKSTEIN and SCHIETZEL (1977), KÖRBER-GROHNE (1977a),
SEEHANN (1977, fungi), BEHRE (1978a, *Prunus domestica*); Lübeck: KROLL
(1978, *Prunus avium, Pr. cerasus*); northern Germany: WILLERDING (1974a,
1978b,c), BEHRE (1975, 1976); southwestern Germany: KÖRBER-GROHNE
(1977, 1978); southern Poland: WASYLIKOWA (1978), KOSINA (1978). In
these papers, as well as in OPRAVIL's (1978) article on synanthropic
plants of medieval times in southern Czechoslovakia very interesting
attempts are made, aiming at the reconstruction of past plant communities and at their identification with modern communities. Doing so,
only the thanatocoenosis can be used, the actual equivalent of which
may sometimes be questioned in view of the deviating agricultural
techniques and other competitors.

6. Paleoecology and the Activity of Man

The evolution of soils, mollusk faunas and geomorphological processes
clearly indicates the early disturbances of the previous equilibrium
between climate, pedogenesis and evolution of the vegetation, caused
by neolithic man in Central Europe (SMOLIKOVA and LOŽEK, 1978; JÄGER
and LOŽEK, 1978). This is in strong contrast to the only very faint
influence of man on the biosphere in some parts of Northern America
(BRUGAM, 1978). According to GROENMAN-van WAATERINGE (1978a), analyz-
ing impact of neolithic man on the landscape in the Netherlands, the
clearly observable regional differences in the response of the vegetation to the activity of man seem to have been governed most of all
by the natural conditions of the areas under consideration rather than
by different activities. These interrelations between the nutrient
status of soils and spontaneous vegetation on the one hand and the activity of man on the other is discussed intensively by KRAMM (1978)
for some regions of northwestern Germany (Ems-Hase-area). Here changes
in the distribution area of the then existing settlements, the changes
in the prevailing agricultural activities, and the strange parallelism
of ash influx and peat formation or the impact of man and the strong
increase of the amount of *Fagus* pollen grains is discussed. According
to REILLE (1977), who gives a summary of the observations pertaining
to the consequences of the impact of man on the natural vegetation of
Corsica, the same interrelations or interdependencies between man and
the history of beech forests can be observed also on this island.

The papers of BASTIN (1977), HEIM (1976), GROENMAN-van WAATERINGE
(1978b), JANSMA (1978) for Belgium and for the Netherlands, of WILLER-
DING (1975) for Kreis Osterode, Harz Mts., of KÖRBER-GROHNE (1977,
1978), and HEIM (1978) for the southwestern part of Central Europe,
of BOREL (1977) for southeastern France and of GRÜGER (1976) for
Macedonia strongly contribute to a better understanding of the zonation of agriculture in Europe during neolithic to Roman times. Besides
of this, BASTIN (1977) points to the astonishingly strong contribution
of linden tree to the neolithic vegetation in Belgium, and WILLERDING
(1975) stresses that according to the weed vegetation during iron age

times no indications pointing to an impoverishment of the soils by
agriculture can be found.

During the last years several teams have tried to reconstruct the
plant associations in the vicinity of which the Swiss lake-dwellings
were constructed (SCHWEINGRUBER, 1975, 1976a,c; PAWLIK and SCHWEIN-
GRUBER, 1976; BAUDAIS-LUNDSTROM, 1978; ARNOLD and SCHWEINGRUBER,
1975; WEGMÜLLER, 1976; RUOFF, 1976; HEITZ-WENIGER, 1977, 1978). Here,
too, the strong share of *Tilia* in the existing forests is stressed
(HEITZ-WENIGER, 1977, 1978), but sometimes pollen analysis and den-
droloy seem to deviate a little from one another in their results
(HEITZ-WENIGER, 1977; PAWLIK and SCHWEINGRUBER, 1976). The duration
of one of these settlements, amounting to 150 to 200 years, is interesting (RUOFF, 1976). This, together with the evaluation of the surface area of the cleared land, seems to be of some importance for
paleoecological conclusions, because clearances seem to have strengthened runoff, sediment load of the rivers and extent and periodicity
of river floods considerably, thus causing repeatedly a new biotope,
the riverine forests of recent days with their dynamics of soil formation, aggradation and development of vegetation (Moravia: OPRAVIL,
1977; the river Danube: BECKER, 1978, FRENZEL, 1978d, KOHL, 1978;
river Main: BECKER and SCHIRMER, 1977; Belgium: HAESAERTS, 1977; perhaps lower part of the Vistula river: DROZDOWSKI and BERGLUND, 1978).

References

ARNOLD, B., SCHWEINGRUBER, F.H.: Bull. Soc. Neuchatel. Sci. Nat. 98,
175-193 (1975).

BAILLIE, M.G.L.: Tree-Ring Bull. 37, 1-12 (1977a); - Tree-Ring Bull.
37, 13-20 (1977b); - Tree-Ring Bull. 37, 33-44 (1977c). - BARRY, R.G.,
ARUNDALE, W.H., ANDREWS, J.T., BRADLEY, R.S., NICHOLS, H.: Arct. Alp.
Res. 9, 193-210 (1977). - BARTHOLIN, TH.S., BERGLUND, B.E.: Fornvännen
70, 201-208 (1975). - BASTIN, B.: Diss. Archaeol. Gandenses 17, 31-43
(1977). - BAUDAIS-LUNDSTRÖM, K.: Ber. Dtsch. Bot. Ges. 91, 67-83
(1978). - DE BEAULIEU, J.-L., REILLE, M.: Le Quaternaire; géodynamique,
stratigraphie et environnement. Trav. franç. récents. 9e Congr. Int.
de L'INQUA, Christchurch, Déc. 1973, 198-199 (1973). - BECKER, B.:
Beiträge zur postglazialen Landschaftsentwicklung des Donautales, 23-35.
In: Beiträge zur Quartär- und Landschaftsforschung. Festschrift zum
60. Geburtstag von J. FINK. Wien: Hirt 1978. - BECKER, B., SCHIRMER,
W.: Boreas 6, 303-321 (1977). - BEHRE, K.-E.: Folia Quat. 46, 49-62
(1975); - Neue Ausgrabungen und Forschungen in Niedersachsen 10,
197-224 (1976); - Ber. Dtsch. Bot. Ges. 91, 161-179 (1978a); -
Petermanns Geogr. Mitt. 1978, Nr. 2, 97-102 (1978b). - BIBUS, E.,
SEMMEL, A.: Catena 4, 385-408 (1977). - BIRKS, H.H., MATHEWES, R.W.:
New Phytol. 80, 455-484 (1978). - BLAHA, J., BORTENSCHLAGER, S.:
Beitr. elektronenmikroskop. Direktabb. Oberfl. 5, 845-854 (1972). -
BOENIGK, W.: Kölner Geogr. Arbeiten 36, 59-68 (1978a); - Eiszeitalter
Ggw. 28, 1-9 (1978b). - BOREL, J.L.: Bull. Mus. R. Art Hist. 46, 237-
243 (1977). - BOWEN, D.Q.: Nature 277, 171-172 (1979). - BRANDE, A.:
Bot. Jahrb. Syst. 95, 406-410 (1976a); - Flora 165, 95-101 (1976b).
- BRETT, D.W.: Tree-Ring Bull. 38, 35-44 (1978). - BRUBAKER, L.B.:
Tree-Ring Bull. 38, 49-60 (1978). - BRUGAM, R.B.: Ecology 59, 19-36
(1978). - BRUNNACKER, K., SCHIRMER, W., SPOERER, H., TILLMANNS, W.:
Schriftenr. Bayer. Landesamt Wasserwirtsch. 7, 26-31 (1978).

DĄBROWSKI, M.J.: Biul. Geologiczny 19, 157-172 (1975). - DAMBLON, F.: Lejeunia, N.S. 95, 1-65 (1979). - DREIMANIS, A., HÜTT, G., RAUKAS, A., WHIPPEY, P.W.: Geosci. Canada 5, 55-60 (1978). - DROZDOWSKI, E., BERGLUND, B.E.: Boreas 5, 95-107 (1976).

ECKSTEIN, D.: Ber. Ausgrabungen Haithabu 11, 112-119 (1977). - ECKSTEIN, D., SCHIETZEL, K.: Ber. Ausgrabungen Haithabu 11, 141-164 (1977).

FLOHN, H.: Umschau 77, 561-569 (1977). - FRENZEL, B.: Grundzüge der pleistozänen Vegetationsgeschichte Nord-Eurasiens. 326 S. Wiesbaden: Steiner 1968; - Das Problem der Riß/Würm-Warmzeit im deutschen Alpenvorland, 103-114. In: Führer zur Exkursionstagung des IGCP-Projektes 73/1/24 "Quaternary Glaciations in the Northern Hemisphere", 5.-13. Sept. 1976, ed. B. FRENZEL. Bonn-Bad Godesberg, DFG 1978a; - Über das geologische Alter einiger Interglazialvorkommen im südlichen Mitteleuropa, 172-180. ibid. 1978b; - Das Interglazial vom Pfefferbichl bei Buching, Landkreis Füssen, 181-184. ibid. 1978c; - Zur postglazialen Palökologie der Donau und ihrer südlichen Zuflüsse im deutschen Alpenvorland, 124-126. ibid. 1978d. - FRITTS, H.C.: Naturwissenschaften 65, 48-56 (1978).

GROENMAN-VAN WAATERINGE, W.: CBA Res. Rep., Nr. 21, 124-146 (1978a); - Palynologische Untersuchungen dreier Grabenprofile aus der römerzeitlichen Siedlung in Rijswijk (Z.H.), 452-456. In: De Bult, eine Siedlung der Cananefaten (Nederlandse Oudheden 8), Amersfoort 1978b. - GROOTES, P.M.: Science 200, 11-15 (1978). - GRÜGER, E.: Pollen Analysis, 294-299. In: Monumenta Archaeologica. ed. M. GIMBUTAS, Vol. I, Los Angeles 1976. - GUGGENHEIM, R.: Flora 164, 287-338 (1975). - GUILLET, B., HASSKO, B., JAEGY, R.: C.R. Acad. Sci. Paris 274, sér. D, 2966-2968 (1972).

HAESAERTS, P.: Diss. Archaeol. Gandenses 17, 14-28 (1977). - HAJNALOVA, E.: Ber. Dtsch. Bot. Ges. 91, 85-96 (1978). - HANSEN, J.M.: Ber. Dtsch. Bot. Ges. 91, 39-46 (1978). - HEIM, J.: Colloq. Phytosociol. 4, 463-469 (1975); - Glain et Salm, Haute Ardenne, Nr. 5, 67-74 (1976); - Rev. Archéol. de l'Est et du Centre-est 29, Nr. 1,2, 56-63 (1978). - HEITZ-WENIGER, A.: Bauhinia 6, Nr. 1, 61-81 (1977); - Bot. Jahrb. Syst. 99, 48-107 (1978). - HOPF, M.: Jahrb. Röm.-Germ. Zentralmus. Mainz 20, 1-56 (1973); - Jahrb. Röm.-Germ. Zentralmus. Mainz 21, 1-11 (1974); - Monogr. Röm.-Germ. Zentralmus. Mainz 1, Teil III, 166-173 (1975); - Forsch. Ber. Vor-. und Frühgesch. Baden-Württ. 8, 3-10 (1977); - Ber. Dtsch. Bot. Ges. 91, 31-38 (1978). - HUGHES, M.K., LEGGETT, P., MILSOM, S.J., HIBBERT, F.A.: Tree-Ring Bull. 38, 15-23 (1978).

JÄGER, K.-D., LOŽEK, V.: Mitteleuropäische Bronzezeit, 211-229, Berlin 1978a; - Petermanns Geogr. Mitt. 1978, 145-148 (1978b). - JANSMA, M.J.: Diatomeenanalysen einiger Grabenprofile in der römischen Siedlung von Rijswijk (Z.H.), 447-451. In: De Bult, eine Siedlung der Cananefaten (Nederlandse Oudheden 8), Amersfoort 1978. - JANUSHEVICH, Z.V.: Ber. Dtsch. Bot. Ges. 91, 59-66 (1978). - JUVIGNÉ, E.: Bull. Soc. Géogr. Liège 14, 205-210 (1978).

KLAUS, W.: Jahrb. Oberösterr. Musealver. 120, I, 345-350 (1975). - KNÖRZER, K.-H.: Ber. Dtsch. Bot. Ges. 91, 187-195 (1978). - KÖRBER-GROHNE, U.: Archäol. Korrespondenzblatt 3, 381-386 (1973); - Archäol. Korrespondenzblatt 3, 271-276 (1974); - Ber. Ausgrabungen Haithabu 11, 64-111 (1977); - Fundber. Baden-Württ. 3, 579-584 (1977); - Forsch. Ber. Archäol. Mittelalters Baden-Württ. 3, 184-201 (1978). - KÖRBER-GROHNE, U., WILMANNS, O.: Magdalenenberg 5, 51-68, Villingen:

Neckar-Verlag 1977. - KOHL, H.: Zur Jungpleistozän- und Holozänstrati-
graphie in den oberösterreichischen Donauebenen, 269-290. In: Beiträge
zur Quartär- und Landschaftsforschung. Festschrift zum 60. Geburtstag
von J. FINK. Wien: Hirt 1978. - KOIVO, L.K., RITCHIE, J.C.: Can. J.
Bot. 56, 1010-1020 (1978). - KOSINA, R.: Ber. Dtsch. Bot. Ges. 91,
121-127 (1978). - KRAL, F.: Verbreitungsgeschichte der Fichte im Alpen-
raum, 180-186. ed. SCHMIDT-VOGT, H. In: Die Fichte, Vol. I. Hamburg-
Berlin: Parey 1977. - KRAMM, E.: Abh. Landesmus. Naturkunde Münster
Westf. 40, H. 4, 49 S. (1978). - KROLL, H.: Ber. Dtsch. Bot. Ges. 91,
181-185 (1978). - KUKLA, G.J.: Earth Sci. Rev. 13, 307-374 (1977); -
Transact. Nebraska Acad. Sci. 6, 57-93 (1978). - KUKLA, G.J., ANGELL,
J.K., KORSHOVER, J., DRONIA, H., HOSHIAI, M., NAMIAS, J., RODEWALD,
M., YAMAMOTO, R., IWASHIMA, T.: Nature 270, 573-580 (1977).

LANGE, E.: Ber. Dtsch. Bot. Ges. 91, 197-204 (1978). - LAUER, W.,
FRANKENBERG, P.: Abh. Math.-Nat. Kl. Akad. Wiss. Lit., Mainz 1979,
Nr. 1, 61 pp. (1979). - LEJOLY-GABRIEL, M.: Acta Geogr. Lovaniensia
13, 278 pp. (1978). - LISITSINA, G.N.: Ber. Dtsch. Bot. Ges. 91,
47-57 (1978). - LÖSCHER, M.: Heidelberger Geogr. Arb. 45, 1-157 (1976).
- LOŽEK, V.: Československý Kras 29, 7-25 (1979); - Schriftenr. Geol.
Wiss. 9, 121-136 (1978).

MACK, R.N., BRYANT, V.M., PELL, W.: Bot. Gazette 139, 249-255 (1978).
- MANGERUD, J., SØNSTEGAARD, E., SEJRUP, H.-P.: Nature 277, 289-192
(1979a); - Nature 277, 108 (1979b). - MATHEWES, R.W.: Can. J. Bot.
56, 1372-1380 (1978). - MULLENDERS, W., PLASMANNE, B., DIRICKX, M.:
Lab. Palynol. Phytosociol., Louvain-La-Neuve, 14 pp. (1978). - MUNAUT,
A.V.: Coll. Phytosociol. 4, 471-477 (1975); - Lejeunia, N.S. 91, 47 pp.
(1978). - MUNAUT, A.V., GILOT, E.: Bull. Soc. R. Bot. Belg. 109,
231-237 (1976).

NICHOLS, H., KELLY, P.M., ANDREWS, J.T.: Nature 273, 140-142 (1978).

OGDEN, J.: Tree-Ring Bull. 38, 1-13 (1978). - OHNGEMACH, D.: Jahrb.
Ges. Naturkunde Württ. 131, 125-139 (1976). - OPRAVIL, E.: Antro-
pozoikum 11, 171-182 (1977); - Ber. Dtsch. Bot. Ges. 91, 97-106
(1978).

PAAP, N.A.: Ber. Rijksdienst Oudheidkundig Bodemonderzoek 26, 127-132
(1976). - PAWLIK, B., SCHWEINGRUBER, F.H.: Jahrb. Schweizer. Ges.
Ur- und Frühgesch. 59, 77-91 (1976). - PEARSON, G.W., PILCHER, J.R.,
BAILLIE, M.G.L., HILLAM, J.: Nature 270, 25-58 (1977). - PERRY, D.A.:
Ann. Bot. 42, 1001-1002 (1978). - PFISTER, Chr.: Vierteljahresschr.
Naturforsch. Ges. Zürich 122, 447-471 (1977). - PILCHER, J.R.: Tree-
Ring Bull. 36, 21-27 (1976). - PILCHER, J.R., HILLAM, J., BAILLIE,
M.G.L., PEARSON, G.W.: New Phytol. 79, 713-729 (1977). - PONS, A.,
VERNET, J.L.: Bull. Soc. Bot. France 118, 841-850 (1971).

REILLE, M.: Bull. Assoc. Franc. Etude Quatern., Suppl. Nr. 47, 329-
342 (1977). - RUOFF, U.: Jahrb. Schweizer Ges. Ur- und Frühgesch.
59, 67-75 (1976). - RYBNÍČEK, K., RYBNÍČKOVÁ, E.: Studie ČSAV 13,
61-65 (1976).

SCHMEIDL, H.: Telma 7, 267-270 (1977). - SCHOVE, D.J.: Palaeogeogr.,
Palaeoclimat., Palaeoecol. 25, 209-233 (1978). - SCHWEINGRUBER, F.H.:
Helv. Archaeol. 6, Nr. 21, 2-15 (1975); - Acad. Helv. 2, 106 pp.
(1976a); - Cour. Forsch.-Inst. Senckenberg 17, 29-58 (1976b); - Bota-
nische Untersuchungen der Holzreste aus Egolzwil 5, 151-162. In:
Das jungsteinzeitliche Jäger-Bauerndorf von Egolzwil 5 am Wauwiler-
moos. Zürich: Schweiz. Landesmus. 1976c. - SCHWEINGRUBER, F.H., FRITTS,
H.C., BRÄKER, O.U., DREW, L.G., SCHÄR, E.: Tree-Ring Bull. 38, 61-91

(1978). - SEEHANN, G.: Ber. Ausgrabungen Haithabu 11, 120-140 (1977). - SEMMEL, A.: Geol. Jahrb. Hessen 106, 291-302 (1978). - SERET, G., ROUCOUX-WOILLARD, G.: The glaciations in the Vosges Lorraines, 1-30. In: Führer zur Exkursionstagung des IGCP-Projektes 73/1/24 "Quaternary Glaciations in the Northern Hemisphere", 5. bis 13.9.1976, ed. B. FRENZEL, Bonn-Bad-Godesberg: DFG 1978. - SERRE, F.: Tree-Ring Bull. 37, 21-31 (1977); - Tree-Ring Bull. 38, 25-34 (1978). - SHORT, S.K., NICHOLS, H.: Arct. Alp. Res. 9, 265-290 (1977). - ŠIBRAVA, V.: Isotopic Methods in Quaternary Geology, 165-169. In: Contributions to the Geologic Time Scale; Studies in Geology Nr. 6; Amer. Assoc. Petrol. Geologists, 1978a; - Schriftenr. Geol. Wiss. 9, 81-90 (1978b). - SMOLIKOVA, L., LOŽEK, V.: Die nacheiszeitlichen Bodenabfolgen von Poplze und Štětí als Beleg der Boden- und Landschaftsentwicklung im böhmischen Tschernosemgebiet, 531-549. In: Beiträge zur Quartär- und Landschaftsforschung. Festschrift zum 60. Geburtstag von J. FINK. Wien: Hirt 1978. - SUESS, H.E.: Radiocarbon 20, Nr. 1 (1978); - Umschau 79, 312-317 (1979).

TOBIEN, H.: Quartärpaläontologie 1, 221-233 (1975). - TRIAT, H.: Ann. Fac. Sci. Marseille 39, 183-190 (1967); - Ann. Univ. Provence Sci. 46, 151-154 (1971).

URBAN, B.: Pollenanalytische Untersuchungen am Interglazial von Seibranz-Fischweiher bei Bad-Wurzach (Schwäbisches Alpenvorland), 94-102. In: Führer zur Exkursionstagung des IGCP-Projektes 73/1/24 "Quaternary Glaciations in the Northern Hemisphere", 5. bis 13.9.1976. ed. B. FRENZEL. Bonn-Bad-Godesberg: DFG 1978a; - Sonderveröff. Geol. Inst. Univ. Köln 34, 165 pp. (1978).

WASYLIKOWÁ, K.: Ber. Dtsch. Bot. Ges. 91, 107-120 (1978). - WEGMÜLLER, S.: Pollenanalytische Untersuchungen über die Siedlungsverhältnisse der frühneolithischen Station Egolzwil, 141-150. In: Das jungsteinzeitliche Jäger-Bauerndorf von Egolzwil 5 im Wauwilermoos. ed. R. WYSS. Zürich: Schweiz. Landesmus. 1976. - WELTEN, M.: Das Jüngere Quartär im nördlichen Alpenvorland der Schweiz aufgrund pollenanalytischer Untersuchungen, 54-75. In: Führer zur Exkursionstagung des IGCP-Projektes 73/1/24 "Quaternary Glaciations in the Northern Hemisphere", 5. bis 13.9.1976, ed. B. FRENZEL, Bonn-Bad-Godesberg: DFG 1978. - WHYTE, R.O.: Hum. Ecol. 5, 209-222 (1977). - WILLERDING, U.: Kasseler Beitr. Ur. und Frühgesch. 4, 191-214 (1974a); - Nachr. Niedersachsens Urgesch. 43, 134-137 (1974b); - Nachr. Niedersachsens Urgesch. 44, 107-112 (1975); - Ber. Dtsch. Bot. Ges. 91, 3-30 (1978a); - Ber. Dtsch. Bot. Ges. 91, 129-160 (1978b); - Mittelalterliche Pflanzenreste aus der Wüstung Oldendorp bei Einbek, Kreis Northeim, 228-248. In: Die Wüstung Oldendorp bei Einbek. Archäologisch-historische Untersuchungen zur Siedlungsgeschichte des mittleren Leinetales. ed. E. PLÜMER, Einbek 1978c. - WRIGHT, H.E.: Science 194, 385-389 (1976).

ZAGWIJN, W.H.: Bull. Assoc. Franc. Etude Quatern. 11, 105-107 (1974). - ZAGWIJN, W.H., DOPPERT, J.W.Chr.: Geol. Mijnb. 57, 577-588 (1978). - ZÓLYOMI, B.: IIIeme Congr. Internat. des Musées d'Agriculture. Budapest 1971; Résumées des Comm. présentees, 194-195 (1971).

Professor Dr. BURKHARD FRENZEL
Botanisches Institut der Universität
D 7000 Stuttgart 70 (Hohenheim)

III. Vegetation Science (Sociological Geobotany)

By Rüdiger Knapp

1. Reviews, Textbooks, Bibliographies

Beside the articles in scientific journals and the monographs within
publication series covering most of the original progress in vegetation science, a growing number of reviews and outlines on particular
topics are collected in symposium reports and similar volumes; they
recently treat mosaics of plant communities (TÜXEN, 1978a), vegetation
and climate (DIERSCHKE, 1977), vegetation on wet sites of inland areas
(DYKYJOVÁ and KVĚT, 1978; GÉHU, 1978a; GOOD et al., 1978) and of coastal regions (CHAPMAN, 1977; CLARK, 1977), portions of Africa (WERGER,
1978) and some other fields mentioned later. Systematic and complete
texts on ample fields can be attained in volumes by single experienced
authors. One example is the volume on the vegetation of central Europe
including the Alps by ELLENBERG (1978), emphasizing in the new revised
edition physiologic-ecologic processes within vegetation. The amplified
integration of vegetation science in university curricula is reflected
by the highly augmented parts on vegetation (EHRENDORFER, 1978) in the
new edition of the well-known botany textbook of *Strasburger*. A new Rus-
sian textbook (RABOTNOV, 1978a) accentuates coenologic aspects within
vegetation (also RABOTNOV, 1978b), whereas the work of DAUBENMIRE
(1978) refers mainly to North America. Numerical considerations continue to be frequently published in books of mostly limited size (e.g.,
FRENKIEL and GOODALL, 1978; INNIS, 1978; ORLOCI, 1978; DE WIT and
GOUDRIAAN, 1978). The central position of vegetation science in nature
conservation is documented in the texts of BOHN et al. (1978), GODWIN
(1978), HILLESHEIM-KIMMEL et al. (1978), LONDO (1977), and TRAUTMANN
et al. (1978). The increasing amount of publications and information
makes bibliographies on particular fields of vegetation science more
and more indispensable, representing recent root studies in vegetation (TÜXEN and WILMANNS, 1978), paleosociology (TÜXEN and WOJTERSKA,
1979), syntaxonomy (TÜXEN et al., 1979), hydrology in plant communities (TÜXEN and GROOTJANS, 1978; also TÜXEN, 1978e), symphytosociology
(TÜXEN, 1978f), vegetational productivity (KNAPP and TÜXEN, 1979),
dynamics and reproduction in vegetation (KNAPP, 1978a, 1979a), plant
communities of bryophytes (VON HÜBSCHMANN and TÜXEN, 1978) and in bogs
(TÜXEN, 1978d). A series of bibliographies collects the literature on
several countries (France: GÉHU et al., 1978; India: MEHER-HOMJI and
GUPTA, 1978; outline on Yemen: HEPPER, 1977; Senegal, Mauretania,
Mali, Upper Volta, Chad, East Africa: KNAPP, 1979a; Chile: RAMIREZ,
1979).

2. General Results and Methods

a) Numerical and Classic Methods in Vegetation Classification

Numerical methods and computer applications imply new impulses in
vegetational ordination and classification on the basis of the species

composition of stands (relevés) or plots in several countries. Such
methods are being considered also in Russia (e.g. ALEXANDROVA, 1978;
MIRKIN, 1977; MIRKIN and ROSENBERG, 1978). Numerical studies concern
often only the vegetation of relatively small areas and of few plant
communities. An example of an ample scale numerical ordination is a
treatment of the peatland vegetation of Finland with factor analysis
and reciprocal averaging (PAKARINEN and RUUHIJÄRVI, 1978; the multi-
dimensional trends in mire classification attempt an integration on the
basis of a factor loading matrix. There are new efforts to combine
numerical methods and Sigmatist's classic tabulation techniques (FEOLI,
1977; HOLZNER et al., 1978; VAN DER MAAREL, 1978; VAN DER MEULEN et
al., 1978; NEUHÄUSL, 1977; STRENG and SCHÖNFELDER, 1978; SCHÖNFELDER,
1978, cf. also CAMPBELL, 1978, for comparison WHITTAKER, 1978). In so
far as adequate comparisons have been evaluated (e.g., VAN DER MAAREL
et al., 1978; STRENG and SCHÖNFELDER, 1978), they continue to confirm
the reliability of the classic methods. The programs and techniques
are now so advanced that computer print tables would be suitable for
publication without retyping. Most authors of clustering-type programs,
however, concede that a final manual rearrangement of the table is
recommended (e.g., VAN DER MEULEN et al., 1978). Ample systems of plant
communities on the basis of characteristic and differential species
recently reported (e.g., ELLENBERG, 1978; HORVÁT, 1977; KÁRPÁTI, 1978;
OBERDORFER, 1978; PASSARGE, 1978; TÜXEN and WOJTERSKA, 1979) have been
developed with classic syntaxonomic methods.

b) Vegetation Dynamics

Long-term permanent plot and similar studies yield more and more inductive results on vegetational succession (BECKING and OLSON, 1978;
BOCK et al., 1978; BRÅKENHIELM, 1977; DAWKINS and FIELD, 1977; HILL
and JONES, 1978; HUNDT, 1978; ISÉPY, 1978; JOENJE, 1978; JOHNSON,
1977; KRAUSE, 1978; LONDO, 1978; VAN DER MAAREL, 1978; POISSONET et
al., 1978; SCHMIDT, 1979; SCHWAAR, 1977; in Sweden also with detailed
use of ancient documents: DRAKENBERG, 1978; ZACKRISSON, 1978). Many of
these results reveal decisive action of biotic or abiotic influences,
till now neglected or not perceived in successional considerations.
For example, species important for the progress of successions can be
highly inhibited by substances from certain dominants in earlier
stages (HORSLEY, 1977; KNAPP, 1979b; PAUL, 1978; STEPHENSON, 1977).
All these various influences add to the difficulty of adequate numerical modeling of successional processes (AUSTIN, 1977; DUMAS, 1978;
KAUPPI et al., 1978; POISSONET et al., 1978; READER and THOMAS, 1978).
The biomathematic modeling of successions and other vegetation changes
has to consider the dynamics and interactions of the species populations within the communities, but also the various delay kernels of
their theoretical extension rates caused by several parameters and
interferences, based on amplified Lotka-Volterra and equivalent functions or including time lag models becoming Stieltjes integrals
(CUSHING, 1977). Since many biological or additional environmental
parameters fluctuate seasonally or for other reasons, the models have
to be mostly nonautonomous. Shifts from changing conditions to stabili-
zation in vegetation can be deduced theoretically by such modeling
(YODZIS, 1978). The interrelations of diversity and successional state
are explainable by quantitative inclusion of competitional changes in
connection with differences in stability (HUSTON, 1979).

c) Distributional Patterns and Mosaics of Plant Communities

Intensified structure analysis, typification, and classification concern assemblages of plant communities coexisting in certain patterns

and mosaics. More detailed definitions of plant communities in connection with methodical progress and the corresponding separation of
special associations at edges, on small openings, and in highly shaded
parts within forest ecosystems and similar differentiations in other
vegetational formations necessitate such studies, as also the increased application of vegetation mapping. Often designated as vegetational
complexes or mosaics, these phenomena are now defined under the terms
of geosyntaxa, sigmassociations, or sigmeta (TÜXEN, 1978a; DIERSCHKE,
1978; PIGNATTI, 1978). The field of studies concerned has been named
by TÜXEN (1978b) synsociology or symphytosociology. A broad spectrum
of European examples in such studies has recently been presented for
forests (GÉHU and GÉHU, 1978), for aquatic, littoral, and bog vegetation (MERIAUX and GÉHU, 1978; J. TÜXEN, 1978; TÜXEN, 1978c; WEBER-
OLDECOP, 1978), for coastal plant communities (BABCZYNSKA and CELINSKI,
1978; GÉHU, 1978b; WOJTERSKI, 1978), for xerothermic and submediter-
ranean formations (GILS and HUITS, 1978; WILMANNS and TÜXEN, 1978;
ZOLLER et al., 1978), for high mountains (BALCERKIEWICZ and WOJTERSKA,
1978; RIVAS MARTÍNEZ, 1978), for agricultural regions (GÉHU and GÉHU,
1978; HARMS and DAMEN, 1978; THALEN, 1978; WIEGERS, 1978, after their
abandonment: MARINČEK and ŽUPANČIČ, 1978), and for settlement areas
(HÜLBUSCH, 1978; KIENAST, 1978). Results of such studies have been
presented also for Japan (MIYAWAKI, 1978) and Argentina (WERNER, 1978).

d) Vegetation Mapping

Augmented apparative procedures are now used in vegetation mapping for
special purposes, e.g., multispectral scanning forming the survey basis
electronically by measuring the reflected solar radiation and the thermal emission of the vegetation surface via satellites (VAN LEEUWEN,
1977; additional possibilities of vegetational mapping with aerial
methods: FERRARI et al., 1978; WALLEN and JACKSON, 1978). Also more
statistical and numerical evaluations, partly with computer methods,
are applied in vegetation mapping (GRACIA et al., 1978; PABST, 1978),
promoted by surveys in certain grids (HEGG, 1978). Connections between
soil properties or ecotopes and vegetation in maps are emphasized by
KÜCHLER (1975), STUMPEL and KALKHOVEN (1978). Many vegetation mapping
projects concern environmental and conservational protection or land
use and settlement planning (e.g., HÜLBUSCH, 1978; KIENAST, 1978;
KRAUSE et al., 1977; MEISEL, 1977b; TRAUTMANN et al., 1977); a few of
these maps have been published, e.g., some informative examples on
coastal vegetation in Germany (DAHL and HECKENROTH, 1978a,b) and in
Florida (RICHARDSON, 1977). Some new maps represent vegetation only
scantily surveyed before in this respect: parts of Iran and Afghanistan
(FREY and PROBST, 1978; HUSS, 1978), of the Colorado mountains (KOMAR-
KOVÁ and WEBBER, 1978), Australia (CARNAHAN, 1976), and particular
areas of the Alps (MEISEL, 1977a) and of the Carpathians (BALCERKIEWICZ
and WOJTERSKA, 1978).

3. Cold-Resistant Vegetation of North America, Asia and Europe

a) Aquatic and Littoral Vegetation in North America

The aquatic and littoral vegetation is now being studied intensively
in North America (GOOD et al., 1978; HASLAM, 1978), partly with emphasis on its often alarming destruction by environmental deteriora-
tion (BRUGAM, 1978; BUMBY, 1977; TERRELL et al., 1978), its indication of detrimental factors (BRISTOW et al., 1978), and its potenti-
ality to remove substances causing undesired water eutrophication

(MUZTAR et al., 1978). This vegetation is also being studied as habitat and nutrition basis of water birds in numerous North American wildlife research projects (ANDERSON, 1977; SWANSON, 1978; TAYLOR, 1978). Thus, results on the wetland vegetation have been presented for several regions, e.g., for New York State (BLOOMFIELD, 1978; FOREST, 1977), Ontario (CROWDER et al., 1978), New Jersey (FERREN and GOOD, 1977) and North Dakota (LARSON, 1977).

b) Weed and Ruderal Vegetation

In central Europe and adjacent areas, the diversity of weed vegetation is highly reduced by influences of herbicides and by copious mineral fertilization. But detailed surveys augment the differentiation of weed communities (e.g., BACHTHALER, 1978; DUVIGNEAUD, 1978; ELLENBERG, 1976, 1978; HADAČ, 1978; HOLZNER, 1978; KIENAST, 1978; KOPECKÝ and HEJNÝ, 1978; MALMGREN, 1978; OLSSON, 1978; WOJCIK, 1977) and can be applied to land use planning and agriculture. The development of new weed genotypes by hybridization, mutations, and selection, acting relatively rapidly in certain anthropogenic vegetations (RAUBER, 1977; WARWICK and BRIGGS, 1978) implies the formation and characterization of new plant communities (KNAPP, 1978b). This development enhances the applications just mentioned, but necessitates more elaborate and time-consuming identification in connection with the field surveys. In northern Eurasia and North America, covered before by natural vegetation, sites appropriate for weed and ruderal communities have originated from new anthropogenic influences during the last decades. The actual immigration problems of nitrophilous species and the development of weed and ruderal vegetation along with certain indigenous plants (apophytes) are now being studied in such regions in northern Asia (eastern Siberia: NECHAEVA et al., 1978; ULJANOVA, 1978a; Sakhalin: ULJANOVA, 1978b, Chukotka: DOROGOSTAISKAYA, 1978).

c) Vegetation of Central Asia and Adjacent Areas

Besides regional monographs (Mongolia: LAVRENKO and RACHOVSKAYA, 1976; Sajan, Tuva and adjacent mountain areas: KUMINOVA, 1976; VLASENKO, 1978; Manchuria: KITAGAWA, 1979; parts of Afghanistan: FREY and PROBST, 1978; GILLI, 1977; HUSS, 1978; and of north-eastern Iran: RECHINGER, 1977), the typology and dynamics of steppes (FORSHKOVA and BARBASHOVA, 1977; MALYSHEV, 1977; YURTSEV, 1978) the dominant patterns (MAKHMETOV, 1978; MOORE and BHADRESA, 1978) and floro-genetic impacts (BELOVA, 1978; BOBROV, 1978; KISELEVA, 1978) are discussed extensively in connection with central Asian vegetation. The plant communities of the river valleys and around springs have been proved to be partly centers of diversification and preservation beside their productivity importance in the vegetation of the climatically arid basins (in Mongolia: KASHAPOV et al., 1977; in Kazakhstan: DEMINA, 1979; at the Irtiysh: PROKOPIEV, 1978; at the Amudarya: GLADYSHEV and RODIN, 1977; in central Iran: CARLE and FREY, 1977). The problems of the status and the dynamics of certain dry woodlands covering much of southern Central Asia are considered in connection with studies on one of their most important dominants: *Pistacia vera* (LOVELIUS et al., 1977; POPOV, 1976). Special regional and local climatic conditions and dispersal adaptations contribute to the enormous rise of the upper vegetation limits in the high mountain regions of the Pamirs (HUSS, 1978; NOSOVA, 1977).

4. Meridional, Subtropical and Related Vegetation

a) Mediterranean Vegetation

The vegetation of the areas around the Mediterranean Sea continues
to supply interesting new results, due to the extreme diversity in
sites, in species numbers, and in anthropogenic influences since earliest periods. These results are meaningful also for many other countries, since the Mediterranean vegetation has been proved to be an
expansion reservoir of useful plants, but also of weeds, in earlier
periods and still at the present. Beside already thoroughly studied
areas (e.g., southern France, review by WRIGHT and WANSTALL, 1977,
also POISSONET et al., 1978), many parts of the Mediterranean and adjacent vegetation are still insufficiently known in detail. Recent
vegetation studies on areas not well known before concern parts of
Turkey (AKMAN et al., 1978; ÖNER and OFLAS, 1977; USLU, 1977), of
Greece (DAFIS and LANDOLT, 1976; ADAMADIADOU et al., 1978; ZOLLER et
al., 1977), of Spain (DE BOLÔS, 1977; RIVAS-MARTÎNEZ, 1977) and of
the Crimea (GOLUBEV and KOBECHINSKAYA, 1978; SHELJAG-SOSONKO and
DIDUKH, 1978). The majority of the Mediterranean species endemic in
small areas occur mostly in plant communities of rock fissures or
other open mountain vegetation (BRULLO and GRILLO, 1978; FEOLI-
CHIAPELLA and FEOLI, 1977; GAMISANS, 1977; HRUSKA DELL'UOMO, 1976;
RIVAS-MARTÎNEZ, 1978; UBALDI, 1978). On the basis of recent special
results, comparisons of the Mediterranean vegetation with climatically similar areas in other continents afford new perspectives (GIGON,
1978).

b) Vegetation of Central America and Mexico

Continental Central America is outstanding in interpenetrations of
plant communities dominated by species of neotropic or temperate
(holarctic) affinities, respectively. Many genera of mainly tropical
distribution exist as far north as in parts of north-eastern Mexico.
On the other hand, species of *Pinus* are dominant in certain mountain
forests of Guatemala (VEBLEN, 1978a), Honduras and Nicaragua; they
prevail even in vast lowland areas of Nicaragua. *Quercus* species domi-
nate beyond the southern *Pinus* limits in the mountain forests of great
parts of Costa Rica and Panama (*Clidemio-Quercetalia* etc.). The lowland
rain forests of these countries (*Swietinio-Brosimetalia*, et al.) are, however, purely tropical in structure (e.g., in Costa Rica: ALLEN, 1977).
The mountain ranges and the interspersed basins have highly diversified altitudinal zonations, due not only to differences in elevation,
but also to extreme contrasts in the humidity or aridity (in Mexico:
ERN, 1976; LAUER, 1978; PUIG and STRESSER-PEAN, 1977; ROBERT, 1977;
SEGOVIA and HELGUERAS, 1976; in Guatemala: VEBLEN, 1978b; in Costa
Rica: MACEY, 1975). The vegetation of arid northern Mexico and adjacent regions is dominated often by few species. Their function and
dynamics are decisive for the regional vegetation character, as studied for *Prosopis* (Mimosaceae) (SIMPSON, 1977) and for *Larrea* (Zygophyllaceae) (MABRY et al., 1977), considering also the sociology of these
shrubs in South American semi-deserts. A cyclic dominance change in
the Chihuahuan desert is induced by burrowing rodents, damaging the
roots and causing mortality of *Opuntia leptocaulis* (YEATON, 1978). The
open places, formerly settled by *Opuntia*, are colonized by *Larrea tridentata*. The *Larrea* aggregations serve again as site for *Opuntia* establish-
ment. Finally, the spreading *Opuntia* can eventually suppress and replace *Larrea*. The vegetation of the Antillean Islands is decidedly neotropical in character, with relatively few holarctic irradiations. The

structure (TANNER, 1977), dynamics (BYER and WEAVER, 1977), and epi-
phytes (RUSSELL and MILLER, 1977) of montanic Antillean forests have
been studied in some well-preserved stands. The Bermuda Islands are
the northern-most area within the range of the Atlantic Ocean with
vegetation of mostly tropical affinities, due to the influence of the
Gulf Stream (KNAPP, 1979c).

5. Tropical Vegetation of Africa and South America

a) African Miombo Woodland and Termitaria Vegetation

The miombo woodlands, prevalent in large areas from Mozambique to
Angola, Zaire and southern Kenya, concentrated in the Zambezian region (also WERGER, 1978; WERGER et al., 1978), are dominated by trees
up to about 20 m tall (mostly *Caesalpiniaceae, Piliostigmo-Brachystegietea*).
The trees usually shed their leaves during cold spells in dry seasons
and remain leafless until the temperature rises. In cases of persist-
ing mild weather conditions, most of the trees retain their leaves
until the flush period, which can start already 10 weeks before the
beginning of the rains. Burning experiments, started in Zambia in
1933 and maintained until 1970, have been evaluated numerically with
principal components analysis and with similarity coefficients (LAWTON,
1978). Of the four species groups differentiated, the miombo dominants
are resistant to weak fires; but they need some fire protection to survive the sapling stage. This protection can be provided by the second
group, with *Uapaca* species (Euphorbiaceae), small trees with relatively
large undivided leaves, highly reducing the inflammable herb layer by
their shade. Fire-sensitive tree species of the third group, attain-
ing in remnant forest patches heights until about 30 m, have been
proved in these experiments to establish very slowly in the miombo
woodland after total fire protection. Twenty seven years after the
start of this protection one individual tree of this group only had
reached the heights of the miombo canopy. Beside micro-organisms and
fire, termites are important decomposers of the miombo tree litter,
falling partly at the end of the dry season (leaves, about 3 t/ha/yr),
partly more evenly over the year (dead twigs, about 1 t/ha/yr) (MALAISSE
et al., 1975). Termites are also decisive for certain dense thicket communities growing on and around termite mounds (termitaria, MALAISSE,
1976, 1978; VAN DER MEULEN, 1979).

b) South American Savanna and Dry Woodland Vegetation

Certain parts of the South American savanna vegetation are considered
for an ancient natural plant formation (SARMIENTO and MONASTERIO,
1975). Pollen spectra, indicating savanna, up to more than 30,000
years old have been found in the Amazon basin. On the Venezuelan
plains, savanna vegetation is assumed at early periods of the same
age on the basis of geomorphologic and pedologic data. Decisive anthropogenic influences can be excluded at such early periods in South
America. Most present-day savannas are modified by human action; they
are partly totally anthropogenic on former and potentially natural
forest sites. The primary or secondary status of the savannas can be
recognized by their species composition. In addition to other favorable factors (fires, ungulate herbivores etc.), certain edaphic conditions promote savanna vegetation, e.g., ill-drained heavy soils
with strongly alternating saturation and deficiency of water with the
change of dry and rainy seasons, or very poor soils with low water-

retaining capacity (SARMIENTO and MONASTERIO, 1975). Before spreading
of anthropogenic effects, extinct large herbivorous mammals contrib-
uted highly to conditions favorable for savannas. For example, giant
ground sloths (Mylodon) have been found to become exterminated soon
after the first evidence of human immigration (MOORE, 1978). More
studies are necessary in several South American savanna areas similar
to those done in the campina Amazonica (LISBOA, 1976; MACEDO, 1977;
PRANCE et al., 1975) and in the campo cerrado (HEISEKE, 1977; SILBER-
BAUER-GOTTSBERGER, 1977).

Other fields of vegetation science will be considered in subsequent
volumes (explanation in Prog. Bot. <u>34</u>, 419). Publications quoted in
our reviews in former volumes of Progress in Botany could not be cited
here again.

References

ADAMADIADOU, S., SIAFACA, L., MARGARIS, N.: Flora <u>167</u>, 561-573 (1978).
- AKMAN, Y., BARBÉRO, M., QUÉZEL, P.: Phytocoenologia <u>5</u>, 1-79 (1978).
- ALEXANDROVA, V.D.: Bot. Zh. <u>63</u>, 1694-1702 (1978). - ALLEN, P.H.:
The Rain Forests of Golfo Dulce. 417 pp. Stanford: Univ. Press 1977.
- ANDERSON, C.M.: Great Basin Nat. <u>37</u>, 24-34 (1977). - AUSTIN, M.P.:
Vegetatio <u>35</u>, 165-176 (1977).

BABCZYNSKA, B., CELINSKI, F.: Ber. Symp. Int. Ver. Vegetationskd.
<u>1977</u>, 51-65 (1978). - BACHTHALER, G.: Angew. Bot. <u>52</u>, 43-55 (1978). -
BALCERKIEWICZ, S., WOJTERSKA, M.: Ber. Symp. Int. Ver. Vegetationskd.
<u>1977</u>, 161-178 (1978). - BECKING, R.W., OLSON, J.S.: O.R.N.L. Environ.
Sci. Div. Publ. <u>111</u>, 1-94 (1978). - BELOVA, V.A.: Bot. Zh. <u>63</u>, 1341-
1344 (1978). - BLOOMFIELD, J.A. (ed.): Lakes of New York State. 499+
473 pp. New York: Academic Press 1978. - BOBROV, E.G.: Bot. Zh. <u>63</u>,
1393-1402 (1978). - BOCK, J.H., RAPHAEL, M., BOCK, C.E.: J. Appl. Ecol.
<u>15</u>, 597-602 (1978). - BOHN, U., KRAUSE, A., LOHMEYER, W., TRAUTMANN,
W., WOLF, G.: Schriftenr. Landesanst. Ökol. Landsch. Forstpl. Nord-
rhein-Westfalen <u>3</u>, 1-103 (1978). - BOLÒS, O. DE: Stud. Phytol. (Pécs)
<u>1977</u>, 17-21 (1977). - BRÅKENHIELM, S.: Acta Phytogeogr. Suec. <u>63</u>,
1-116 (1977). - BRISTOW, J.M., CROWDER, A.A., KING, M.R., VANDERKLOET,
S.: Nat. Can. <u>104</u>, 465-473 (1978). - BRUGAM, R.B.: Ecology <u>59</u>, 19-36
(1978). - BRULLO, S., GRILLO, M.: Not. Soc. Ital. Fitosociol. <u>13</u>,
23-61 (1978). - BUMBY, M.J.: Trans. Wis. Acad. <u>65</u>, 120-151 (1977). -
BYER, M.D., WEAVER, P.L.: Biotropica <u>9</u>, 35-47 (1977).

CAMPBELL, M.: Vegetatio <u>37</u>, 101-109 (1978). - CARLE, R., FREY, W.:
Arb. Sonderforschungsber. Tübingen <u>19</u>, 1-58 (1977). - CARNAHAN, J.A.:
In: Atlas of Australian Resources. Canberra: Dep. Nat. Resources 1976.
- CHAPMAN, V.J. (ed.): Ecosyst. World (Amsterdam) <u>1</u>, 1-428 (1977). -
CLARK, J.R. (ed.): Coastal Ecosystem Management. 928 pp. New York:
Wiley 1977. - CROWDER, A.A., BRISTOW, J.M., KING, M.R., VANDERKLOET,
S.: Nat. Can. <u>104</u>, 441-464 (1978). - CUSHING, J.M.: Lect. Notes Bio-
math. <u>20</u>, 1-196 (1977).

DAFIS, S., LANDOLT, E.: Veröff. Geobot. Inst. Eidg. Tech. Hochsch.
Stift. Rübel Zürich <u>56</u>, 1-242 (1976). - DAHL, H.-J., HECKENROTH, H.:
Naturschutz Landschaftspfl. Niedersachsen 6, 1-214 (1978a); - Natur-
schutz Landschaftspfl. Niedersachsen <u>7</u>, 1-176 (1978b). - DAUBENMIRE,
R.: Plant Geography. 352 pp. London-New York: Academic Press 1978.
- DAWKINS, H.C., FIELD, D.R.B.: A Long-Term Surveillance System.
124 pp. Oxford: Dep. For. Univ. 1977. - DEMINA, O.M.: Bot. Zh. <u>64</u>,

58-63 (1979). - DIERSCHKE, H. (ed.): Ber. Symp. Int. Ver. Vegetationskd. 1975, 1-613 (1977); - Ber. Symp. Int. Ver. Veget. 1977, 264-
265 (1978). - DOROGOSTAISKAYA, E.V.: Bot. Zh. 62, 246-253 (1978). -
DRAKENBERG, B.: Sven. Bot. Tidskr. 72, 103-123 (1978). - DUMAS, J.-C.:
Nat. Can. 104, 395-400 (1978). - DUVIGNEAUD, J.: Bull. Soc. R. Bot.
Belg. 111, 27-35 (1978). - DYKYJOVÁ, D., KVĚT, J. (eds.): Ecol. Stud.
28, 1-464 (1978).

EHRENDORFER, F.: Geobotanik, 856-987, 1006-1012. In: Lehrbuch der
Botanik für Hochschulen. Stuttgart, New York: Fischer 1978. - ELLEN-
BERG, H.: Bayer. Landwirtsch. Jahrb. 53, 51-59 (1976); - Vegetation
Mitteleuropas mit den Alpen. 2. Aufl. 981 pp. Stuttgart: Ulmer 1978.
- ERN, H.: Willdenowia Beih. 10, 1-128 (1976).

FEOLI, E.: Vegetatio 33, 147-152 (1977). - FEOLI-CHIAPELLA, L.,
FEOLI, E.: Vegetatio 34, 21-39 (1977). - FERRARI, C., MANDRIOLI, P.,
RINALDI, A.: Not. Soc. Ital. Fitosociol. 13, 1-11 (1978). - FERREN,
W.R., GOOD, R.E.: Bull. Torrey Bot. Club 104, 392-395 (1977). -
FOREST, H.S.: Folia Geobot. Phytotaxon. 12, 329-341 (1977). - FRENKIEL,
F.N., GOODALL, D.W.: Simulation Modelling of Environmental Problems,
112 pp. New York: Wiley 1978. - FREY, W., PROBST, W.: Tübinger Atlas
Vord. Orient Beih. Naturwiss. 3, 1-109 (1978).

GAMISANS, J.: Phytocoenologia 4, 35-41, 133-179, 317-376 (1977). -
GÉHU, J.-M. (ed.): Colloq. Phytosociol. 5, 1-370 (1978a); - Ber. Symp.
Int. Ver. Vegetationskd. 1977, 77-82, 267-272 (1978b). - GÉHU, J.-M.,
GÉHU, J.: Ber. Symp. Int. Ver. Vegetationskd. 1977, 179-187, 303-307
(1978). - GÉHU, J.-M., FOUCAULT, B. DE, GÉHU-FRANCK, J.: Excerpta Bot.
B 17, 118-248 (1978). - GIGON, A.: Ber. Geobot. Inst. Eidg. Tech.
Hochsch. Stift. Rübel Zürich 45, 74-133 (1978). - GILLI, A.: Feddes
Repert. 88, 375-387 (1977). - GILS, H. v., HUITS, P.: Ber. Symp. Int.
Ver. Vegetationskd. 1977, 13-26 (1978). - GLADYSHEV, A.I., RODIN, L.E.:
Bot. Zh. 62, 1572-1584 (1977). - GODWIN, H.: Fenland, 200 pp. Cambridge:
Univ. Press 1978. - GOLUBEV, V.N., KOBECHINSKAYA, V.G.: Bot. Zh. 63,
1788-1794 (1978). - GOOD, R.E., SIMPSON, R.L., JACKSON, C.G. (eds.):
Freshwater Wetlands. 378 pp. New York: Academic Press 1978. - GORSHKOVA,
A.A., BARBASHOVA, L.G.: Zj. Obshch. Biol. Akad. (Moskva) 38, 365-371
(1977). - GRACIA, C.A., GRANGER, D., GUILLERM, J.L., ROMANE, F.: Ber.
Symp. Int. Ver. Vegetationskd. 1977, 415-434 (1978).

HADAČ, E.: Folia Geobot. Phytotaxon. 13, 129-163 (1978). - HARMS, W.B.,
DAMEN, M.: Ber. Symp. Int. Ver. Vegetationskd. 1977, 435-450 (1978). -
HASLAM, S.M.: River Plants, Macrophytic Vegetation. 296 pp. New York:
Cambridge Univ. Press 1978. - HEGG, O.: Geogr. Helv. 33, 45-50 (1978).
- HEISEKE, D.R.: Forstarchiv 48, 4-9 (1977). - HEPPER, F.N.: Publ.
Cairo Univ. Herb. 718, 307-322 (1977). - HILL, M.O., JONES, E.W.:
J. Ecol. 66, 433-456 (1978). - HILLESHEIM-KIMMEL, U., KARAFIAT, H.,
LEWEJOHANN, K., LOBIN, W.: Schriftenr. Inst. Naturschutz Darmstadt
11(3), 1-395 (1978). - HOLZNER, W.: Vegetatio 38, 13-20 (1978). -
HOLZNER, W., WERGER, M.J.A., ELLENBROEK, G.A.: Vegetatio 38, 157-164
(1978). - HORSLEY, S.B.: Can. J. For. Res. 7, 205-216 (1977). - HORVÁT,
A.O.: Nat. Can. 104, 61-73 (1977). - HRUSKA DELL'UOMO, K.: Not. Soc.
Ital. Fitosociol. 12, 19-30 (1976). - HÜBSCHMANN, A. v., TÜXEN, R.:
Excerpta Bot. B 17, 276-308 (1978). - HÜLBUSCH, K.-H.: Ber. Symp. Int.
Ver. Vegetationskd. 1977, 321-327 (1978). - HUNDT, R.: Vegetatio 38,
1-12 (1978). - HUSS, H.: Über Flora und Vegetation, 167-192. In:
Grosser Pamir. Graz: Akad. Verlagsanst. 1978. - HUSTON, M.: Am. Nat.
113, 81-101 (1979).

INNIS, G.S. (ed.): Ecol. Stud. 26, 1-298 (1978). - ISÉPY, I.: Vegetatio
37, 187-190 (1978).

JOENJE, W.: Vegetatio 38, 95-102 (1978). - JOHNSON, W.C.: Bull. Torrey
Bot. Club 104, 334-346 (1977).

KÁRPÁTI, I.: Keszthelyi Agric. Egyetem Növ. Tansz. 20(5), 1-62 (1978).
- KASHAPOV, R.S., MANIBAZAR, N., MIRKIN, B.M., ONISHCHENKO, L.I.:
Bot. Zh. 62, 699-703 (1977). - KAUPPI, P., HARI, P., KELLOMÄKI, S.:
Oikos 30, 100-105 (1978). - KIENAST, D.: Ber. Symp. Int. Ver. Vegeta-
tionskd. 1977, 329-345 (1978). - KISELEVA, A.A.: Bot. Zh. 63, 1647-
1655 (1978). - KITAGAWA, M.: Flora Veget. Mundi 4, 1-716 (1979). -
KNAPP, R.: Excerpta Bot. B 18, 57-108 (1978a); - Oberhess. Naturwiss.
Z. 44, 71-129 (1978b); - Excerpta Bot. B 18, 301-320, 19, 11-60, 123
148 (1979a); - Mitt. Florist. Soziol. Arbeitsgem. 21, 97-104 (1979b);
- Documents Phytosociol. N. Sér. 3, 20 pp. (1979c). - KNAPP, R., TÜXEN,
R.: Excerpta Bot. B 18, 267-300 (1979). - KOMÁRKOVÁ, V., WEBBER, P.J.:
Arct. Alp. Res. 10, 1-29 (1978). - KOPECKÝ, K., HEJNÝ, S.: Vegetatio
36, 43-52 (1978). - KRAUSE, A.: Vegetatio 36, 119-122 (1978). - KRAUSE,
A., MEISEL, K., TRAUTMANN, W.: Schriftenr. Naturschutz Landschaftspfl.
14, 60-69 (1977). - KÜCHLER, A.W.: Ber. Symp. Int. Ver. Vegetationskd.
1969, 467-488 (1975). - KUMINOVA, A.V. (ed.): Vegetation Cover of
Khakasia, 424 pp. Novosibirsk: Izd. Nauka 1976.

LARSON, G.E.: Prairie Nat. 8, 1-18 (1977). - LAUER, W.: Erdkunde 32,
101-109 (1978). - LAVRENKO, E.M., RACHOVSKAYA, E.I. (eds.): Main Ecosystems in Mongolian P.R., 199 pp. Leningrad: Izd. Nauka 1976. -
LAWTON, R.M., J. Ecol. 66, 175-198 (1978). - LEEUWEN, C.G. VAN (ed.):
Rijksinst. Natuurbeheer Jaarverslag 1977, 27-34, 72-73 (1977). -
LISBÔA, P.L.B.: Acta Amazonica 6, 193-211 (1976). - LONDO, G.: Natuur-
tuinen en -parken. 134 pp. Zutphen: Thieme 1977; - Vegetatio 38, 185-
190 (1978). - LOVELIUS, N.V., GORELOVA, T.G., KUZNETSOV, V.I.: Bot.
Zh. 62, 1491-1497 (1977).

MAAREL, E. VAN DER: Vegetatio 38, 21-28 (1978). - MAAREL, E. VAN DER,
JANSSEN, J.G.M., LOUPPEN, J.M.W.: Vegetatio 38, 143-156 (1978). -
MABRY, T.J., HUNZIKER, J.H., DIFEO, D.R. (eds.): U.S./I.B.P. Synth.
Ser. (Stroudburg) 6, 1-304 (1977). - MACEDO, M.: Acta Amazonica 7
(Suppl. 1), 1-69 (1977). - MACEY, A.: Rev. Biol. 23, 239-255 (1975).
- MAKHMETOV, B.B.: Bot. Zh. 63, 697-699 (1978). - MALAISSE, F.:
Boissiera 24, 505-513 (1976); - Monograph. Biol. 31, 1279-1300 (1978).
- MALAISSE, F., FRESON, R., GOFFINET, G., MALAISSE-MOUSSET, M.: Ecol.
Stud. 11, 137-152 (1975). - MALYSHEV, L.I.: Ecology and Pasture Degres-
sion of Zabaicalie Steppe Communities. 192 pp. Novosibirsk: Izd. Nauka
1977. - MALMGREN, U.: Sven. Bot. Tidskr. 72, 137-142 (1978). -
MARINČEK, L., ZUPANČIČ, M.: Ber. Symp. Int. Ver. Vegetationskd. 1977,
213-221 (1978). - MEHER-HOMJI, V.M., GUPTA, R.K.: Excerpta Bot. B 17,
309-320, 18, 1-56 (1978). - MEISEL, K.: Bundesforschungsanst. Natur-
schutz Landschaftsoekol. Jahresber. 1977, 13-14 (1977a); - Schriftenr.
Naturschutz Landschaftspfl. 14, 187-191 (1977b). - MERIAUX, J.-L.,
GEHU, J.-M.: Ber. Symp. Int. Ver. Vegetationskd. 1977, 97-116 (1978).
- MEULEN, F. VAN DER: Diss. Bot. 49, 1-191 (1979). - MEULEN, F. VAN
DER, MORRIS, J.W., WESTFALL, R.: Vegetatio 38, 129-134 (1978). - MIRKIN,
B.M.: Bot. Zh. 62, 1697-1708 (1977). - MIRKIN, B.M., ROSENBERG, G.S.:
Zh. Obshch. Biol. Akad. Nauk (Moskva) 39(2)(1978). - MIYAWAKI, A.:
Ber. Symp. Int. Ver. Vegetationskd. 1977, 241-265 (1978).-MOORE, D.M.:
Bot. J. Linn. Soc. (London) 77, 177-202 (1978). - MOORE, P.D., BHADRESA,
W.: J. Appl. Ecol. 15, 837-846 (1978). - MUZTAR, A.J., SLINGER, S.J.,
BURTON, J.H.: Can. J. Plant Sci. 58, 829-862 (1978).

NECHAEVA, T.I., VERHOLAT, V.P., POLICHUK, Y.S.: Bot. Zh. 62, 387-388
(1978). - NEUHÄUSL, R.: Vegetatio 35, 115-122 (1977). - NOSOVA, L.I.:
Bot. Zh. 62, 568-574 (1977).

OBERDORFER, E. (ed.): Süddeutsche Pflanzengesellschaften, Bd. II, 2. Aufl., 355 pp. Stuttgart, New York: Fischer 1978. - ÖNER, M., OFLAS, S.: Vegetatio 34, 55-62 (1977). - OLSSON, H.: Vegetatio 36, 65-82 (1978). - ORLOCĪ, L.: Multivariate Analysis in Vegetation Research. 2nd ed., 450 pp. The Hague: Junk 1978.

PABST, H.: Forstwiss. Zentralbl. 97, 251-256 (1978). - PAKARINEN, P., RUUHIJÄRVI, R.: Ann. Bot. Fenn. 15, 147-157 (1978). - PASSARGE, H.: Feddes Repert. 89, 134-189 (1978). - PAUL, P.: Bull. Soc. Bot. Fr. 125, 493-505 (1978). - PIGNATTI, S.: Ber. Symp. Int. Ver. Vegetationskd. 1977, 27-41 (1978). - POISSONET, P., ROMANE, F., THIAULT, M., TRABAUD, L.: Vegetatio 38, 135-142 (1978). - POPOV, K.P.: Bot. Zh. 61, 177-186 (1976). - PRANCE, G.T., LISBÔA, P.L.B., ANDERSON, A.B., ALBUQUERQUE, B.W.P. DE, BRAGA, M.M.N., BRAGA, P.I.S.: Acta Amazonica 5, 207-260 (1975). - PROKOPIEV, E.P.: Bot. Zh. 63, 752-758 (1978). - PUIG, H., STRESSER-PEAN, G.: Veget. Huasteca (México) 5, 1-544 (1977).

RABOTNOV, R.A.: Fitocenologija, 384 pp. Moskva: Izd. Univ. 1978a; - Verh. K. Ned. Akad. Wet. Afd. Natuurkd. 2e R. 70, 1-26 (1978b). - RAMĪREZ, C.: Excerpta Bot. B 19, 63-92 (1979). - RAUBER, R.: Z. Pflanzenkr. Sonderh. 8, 37-55 (1977). - READER, R.J., THOMAS, A.G.: Can. J. Bot. 55, 3075-3079 (1978). - RECHINGER, K.H.: Iran. J. Bot. 1, 155-180 (1977). - RICHARDSON, D.R.: Fla. Sci. 40, 281-330 (1977). - RIVAS-MARTĪNEZ, S.: Phytocoenologica 4, 14-34 (1977); - Ber. Symp. Int. Ver. Vegetationskd. 1977, 189-212 (1978). - ROBERT, M.-T.: Bull. Soc. Bot. Fr. 124, 197-216 (1977). - RUSSELL, K.W., MILLER, H.A.: J. Arnold Arbor. 58, 1-24 (1977).

SARMIENTO, G., MONASTERIO, M.: Ecol. Stud. 11, 223-250 (1975). - SCHMIDT, W.: Scripta Geobot. (1979, forthcoming). - SCHÖNFELDER, P.: Naturschutz Landschaftspfl. Niedersachsen 8, 1-108 (1978). - SCHWAAR, J.: Geol. Jahrb. F 4, 125-139 (1977). - SEGOVIA, A.O., HELGUERAS, A.L.: Publ. Inst. Invest. Rec. Biot. 1, 1-44 (1976). - SHELJAG-SOSONKO, J.R., DIDUKH, J.P.: Bot. Zh. 63, 1285-1301, 1430-1439 (1978). - SILBER-BAUER-GOTTSBERGER, L., MORAWETZ, W., GOTTSBERGER, G.: Biotropica 9, 253-261 (1977). - SIMPSON, B.B. (ed.): U.S./I.B.P. Synth. Ser. (Stroudburg) 4, 1-272 (1977). - STEPHENSON, A.G.: Mich. Bot. 16, 15-18 (1977). - STRENG, R., SCHÖNFELDER, P.: Hoppea (Regensburg) 37, 407-433 (1978). - STUMPEL, A.H.P., KALKHOVEN, J.T.R.: Vegetatio 37, 163-174 (1978). - SWANSON, G.A.: J. Wildl. Manage. 42, 426-428 (1978).

TANNER, E.V.J.: J. Ecol. 65, 883-918 (1977). - TAYLOR, T.S.: J. Wildl. Manage. 42, 900-903 (1978). - TERRELL, E.E., EMERY, W.H.P., BEATY, H.E.: Bull. Torrey Bot. Club 105, 50-57 (1978). - THALEN, D.C.P.: Ber. Symp. Int. Ver. Vegetationskd. 1977, 491-514 (1978). - TRAUTMANN, W., KORNECK, W., KRAUSE, A., VOGGENREITER, V.: Bundesforschungsanst. Naturschutz Landschaftsoekol. Jahresber. 1977, 8-10 (1977). - TRAUTMANN, W., SUKOPP, H., MEISEL, K., LOHMEYER, W., KRAUSE, A.: Flora und Vegetation, 245-293. In: Natur- und Umweltschutz, ed. G. OLSCHOWY. Hamburg, Berlin: Parey 1978. - TÜXEN, J.: Ber. Symp. Int. Ver. Vegetationskd. 1977, 67-76 (1978). - TÜXEN, R. (ed.): Ber. Symp. Int. Ver. Vegetationskd. 1977, 1-535 (1978a); - Ber. Symp. Int. Ver. Vegetationskd. 1977, 3-11 (1978b); - Ber. Symp. Int. Ver. Vegetationskd. 1977, 273-286 (1978c); - Bibliogr. Phytosociol. Syntaxon. 32, 1-224 (1978d); - Doc. Phytosociol. N. Sér. 2, 453-467 (1978e); - Excerpta Bot. B 17, 45-49 (1978f). - TÜXEN, R., GROOTJANS, A.P.: Excerpta Bot. B 17, 50-80 (1978). - TÜXEN, R., WILMANNS, O.: Excerpta Bot. B 17, 33-44 (1978). - TÜXEN, R., WOJTERSKA, M.: Excerpta Bot. B 18, 145-196 (1979). - TÜXEN, R., KOZLOWSKA, A., WOJTERSKA, M.: Excerpta Bot. B 18, 197-266 (1979).

UBALDI, D.: Not. Soc. Ital. Fitosociol. 13, 13-22 (1978). - ULJANOVA, T.N.: Bot. Zh. 62, 214-222 (1978a); - Bot. Zh. 63, 1004-1015 (1978b). - USLU, T.: Commun. Fac. Sci. Univ. Ankara 21 (1), 1-60 (1977).

VEBLEN, T.T.: Unasylva 29, 25-32 (1978a); - Geogr. Rev. 68, 417-434 (1978b). - VLASENKO, V.I.: Bot. Zh. 63, 1474-1484 (1978).

WALLEN, V.R., JACKSON, H.R.: Agron. J. 70, 922-923 (1978). - WARWICK, S.I., BRIGGS, D.: New Phytol. 81, 711-738 (1978). - WEBER-OLDECOP, D.W.: Ber. Symp. Int. Ver. Vegetationskd. 1977, 83-95 (1978). - WERGER, M.J.A. (ed.): Monograph. Biol. 31, 1-1439 (1978). - WERGER, M.J.A., WILD, H., DRUMMOND, B.R.: Vegetatio 37, 79-90 , 151-162 (1978). - WERNER, D.J.: Ber. Symp. Int. Ver. Vegetationskd. 1977, 223-239 (1978). - WHITTAKER, R.H. (ed.): Classification of plant communities. 408 pp. The Hague: Junk 1978. - WIEGERS, J.: Ber. Symp. Int. Ver. Vegetationskd. 1977, 455-480 (1978). - WILMANNS, O., TÜXEN, R.: Ber. Symp. Int. Ver. Vegetationskd. 1977, 287-302 (1978). - WIT, C.T. DE, GOUDRIAAN, J.: Simulation of Ecological Processes, 175 pp. New York: Wiley 1978. - WOJCIK, Z.: Prace Inst. Geogr. Pol. Akad. Nauk 121, 1-111 (1977). - WOJTERSKI, T.: Ber. Symp. Int. Ver. Vegetationskd. 1977, 43-50 (1978). - WRIGHT, L.W., WANSTALL, P.J.: Dep. Geogr. Univ. London Occas. Papers 9, 1-44 (1977).

YEATON, R.I.: J. Ecol. 66, 651-658 (1978). - YODZIS, P.: Lect. Notes Biomath. 25, 1-191 (1978). - YURTSEV, B.A.: Bot. Zh. 63, 1566-1578 (1978).

ZACKRISSON, O.: Sven. Bot. Tidskr. 72, 205-226 (1978). - ZOLLER, H., GEISSLER, P., ATHANASIADIS, N.: Bauhinia (Basel) 6, 215-255 (1977). - ZOLLER, H., BÉGUIN, C., HEGG, O.: Ber. Symp. Int. Ver. Vegetationskd. 1977, 117-150 (1978).

                              Professor Dr. RÜDIGER KNAPP
                              Botanisches Institut I
                              der Universität
                              Senckenbergstr. 17-25
                              D 6300 Giessen

Subject Index

Encyclopedia of Plant Physiology

New Series

Editors: A. Pirson, M. H. Zimmermann

Distribution rights for India:
UBS Publishers' Distributors Pvt., Ltd.,
New Delhi

Volume 5
Photosynthesis I
Photosynthetic Electron Transport and
Photophosphorylation

Editors: A. Trebst, M. Avron
With contributions by numerous experts

1977. 128 figures. XXIV, 730 pages
ISBN 3-540-07962-9

Contents: History. – Electron Transport. –
Energy Conservation. – Structure and
Function. – Algal and Bacterial Photo-
synthesis.

Volume 7
Physiology of Movements
Editors: W. Haupt, M. E. Feinleib
With contributions by numerous experts

1979. 185 figures, 19 tables. XVII, 731 pages
ISBN 3-540-08776-1

Contents: General Aspects of Plant Movement. – Intracellular Movements. – Loco-
motion in Microbial Plants. – Movements
Using Turgor Mechanisms. – Growth Movements.

Volume 8
Secondary Plant Products
Editors: E. A. Bell, B. V. Charlwood
With contributions by numerous experts

1979. Approx. 300 figures, schematas and
equations, 88 tables. Approx. 630 pages
ISBN 3-540-09461-X

Contents: Historical Introduction. – The
Possible Significance of Secondary Com-
pounds in Plants. – Expression and Control
of Secondary Metabolism. – The Alkaloids. –
The Isoprenoids. – Plant Phenolics. Non-
Protein Amino Acids in Plants. – Plant
Amines. – Cyanogenic Glycosides. – Gluco-
sionolates. – Betalains. – Plant Lipids of
Taxonomic Significance. – Plant Carbo-
hydrates.

Volume 6
Photosynthesis II
Photosynthetic Carbon Metabolism and
Related Processes

Editors: M. Gibbs, E. Latzko
With contributions by numerous experts

1979. 75 figures. XX, 587 pages
ISBN 3-540-09288-9

Contents: Introduction. – CO_2 Assimilation:
The Reductive Pentose Phosphate Cycle. The
C_4 and Crassulacean Acid Metabolism Pathways. Factors Influencing CO_2 Assimilation.
Regulation and Properties of Enzymes of
Photosynthetic Carbon Metabolism. Metabolism of Primary Products of Photosynthesis. Glycolic Acid and Photorespiration. –
Ferredoxin-Linked Reactions. Author Index. – Subject Index.

Springer-Verlag
Berlin
Heidelberg
New York

Monographs on Theoretical and Applied Genetics

Editors: R. Frankel (Coordinating Editor)
G. A. E. Gall, M. Grossman, H. F. Linskens,
D. de Zeeuw

Springer-Verlag
Berlin
Heidelberg
New York